Essentials of Food Chemistry

Jianquan Kan · Kewei Chen
Editors

Essentials of Food Chemistry

 Springer

Editors
Jianquan Kan
College of Food Science
Southwest University
Chongqing, China

Kewei Chen
College of Food Science
Southwest University
Chongqing, China

ISBN 978-981-16-0612-0   ISBN 978-981-16-0610-6   (eBook)
https://doi.org/10.1007/978-981-16-0610-6

This Springer imprint is published by the registered company Springer Nature Singapore Pte Ltd.
The registered company address is: 152 Beach Road, #21-01/04 Gateway East, Singapore 189721, Singapore

Preface

Food chemistry is one of the basic professional courses for students majoring in food science and engineering. Food chemistry is the study related to the chemical composition, structure, physical and chemical properties, nutritional and safety properties of food from the viewpoint of chemical and molecular level, as well as the changes that occur during production, processing, storage, and transportation of food and the impact of these changes on food quality and safety, which is considered as one of the basic applied sciences. Therefore, for an undergraduate or graduate student majoring in food science and engineering, it is necessary to master the basic knowledge and research methods of food chemistry to deal well with the complicated problems in the field of food processing and preservation.

Food chemistry is an emerging subject that is interpenetrated by multiple disciplines. Food, chemistry, biology, agriculture, medicine, and materials science are constantly providing new research fruits into food chemistry, and they are also using the research results from food chemistry. Food chemistry shows rapid development in the field of food science and various engineering disciplines. New research methods and results continuously turn up, which enable many new research results in this area to appear in this book to fully reflect the current situations in the field of food chemistry. Therefore, in the process of editing this book, some influential textbooks and documents related to food chemistry are referred to, the most important of which is *Fennema's Food Chemistry* edited by Damodaran, Parkin, and Fennema, and *Food Chemistry* edited by Belitz, Grosch, and Schieberle. These two books have had a great influence on the education of students majoring in food science.

This book focuses on the basic theories of food chemistry and related practical knowledge, which is divided into 11 chapters. The main content includes the 6 major nutritional components in food and other food components such as pigments, flavor components and harmful constituents, as well as changes of these food components in food processing and storage and their influences on food quality and safety. This book also contains the introduction of enzymes in food and their application in the food industry, and overall it pays attention to reflect the latest research results in food chemistry. Each chapter gives abstracts and keywords, as well as necessary linking questions and references, to help students better understand and master the key and difficult points of the chapter. Therefore, the content of this book is novel, with the basic theory combined with their practical use in the food industry, which gives the readers a comprehensive picture of the food chemistry network.

The book is divided into 11 chapters. Among them, Jianquan Kan and Kewei Chen from Southwest University jointly wrote Chap. 1 (Introduction); Lichao Zhao and Mingyue Song from South China Agricultural University, Chap. 2 (Water); Xinhuai Zhao from Northeast Agricultural University, Yu Fu from Southwest University, and Chunli Song from Qiqihar University, Chap. 3 (Proteins); Jie Pang from Fujian Agriculture and Forestry University and Fusheng Zhang from Southwest University, Chap. 4 (Carbohydrates); Hui He and Tao Hou from Huazhong Agricultural University, Chap. 5 (Lipids); Yali Yang from Shaanxi Normal University, Chap. 6 (Vitamins); Aidong Sun from Beijing Forestry University and Hui Li from Institute of Quality Standard and Testing Technology for Agro-products of CAAS, Chap. 7 (Minerals); Hui Shi from Southwest University, Chap. 8 (Enzymes); Kewei Chen from Southwest University, Chap. 9 (Pigments); Liyan Ma and Jingming Li from China Agricultural University, Chap. 10 (Food Flavor Substances); and Kewei Chen and Jianquan Kan from Southwest University, Chap. 11 (Harmful Food Constituents). The whole book was edited by Jianquan Kan and Kewei Chen.

Professor Zongdao Chen from Southwest University, Prof. Bolin Zhang from Beijing Forestry University, Prof. Mouming Zhao from South China University of Technology, and Prof. Dongfeng Wang from Ocean University of China provided valuable opinions on the compilation of this textbook, and we also extend our sincere thanks to Springer publishers for the great support to the smooth publication of this book.

Chongqing, China   Jianquan Kan
August 2020   Kewei Chen

Brief Introduction

This book focuses on the basic theories of food chemistry and related practical knowledge. The book is divided into 11 chapters. The main content includes the 6 major nutrients in food, the structure and properties of food pigments and flavor components, and harmful constituents, and their changes in food processing and storage as well as their influences on food quality and safety. Besides, the application of enzymes in the food industry is also introduced, with the latest research results related to food chemistry. Each chapter gives abstracts and keywords, as well as necessary relevant questions and references, to help students better understand and master the key and difficult points of the chapter.

This book can be used not only as a textbook for undergraduates majoring in food science and engineering or food quality and safety in universities, but also for teachers and students with majors like food science and engineering, as well as scientific and technical personnel or managerial staff engaged in the production and processing of agricultural products.

Contents

Editors and Contributors

About the Editors

Dr. Jianquan Kan is a professor and doctoral supervisor at College of Food Science, Southwest University, China. He is currently the Director of the Quality and Safety Risk Assessment Laboratory for Agricultural Products Storage and Preservation of the Ministry of Agriculture and Rural Affairs (Chongqing), and the academic leader in the discipline of "Food Science and Engineering" belonging to Chongqing's key disciplines. Dr. Jianquan Kan is also the leader of the "Excellent Teaching Team" in the field of Food Science and Safety (Chongqing), and he was awarded "Person of the Year (2015)" as Scientific Chinese. He is a special expert candidate for high-level talents in Sichuan Province of the "Thousand Talents Program" and currently is the chief scientist in the Innovation Team of Chongqing Modern Special High-efficient Agricultural Flavoring Industrial Technology System. Dr. Jianquan Kan has been the Editor-in-Chief for 6 textbooks related to food chemistry, for instance, Practical Chemistry of Oil and Fat (1997, Southwest Normal University Press: Chongqing), Food Chemistry (2002 1st edition, 2008 2nd edition, 2016 3rd edition, China Agricultural University Press: Beijing; 2006 1st edition, 2017, 3rd edition, New Wenjing Development and Publishing Co., Ltd.: Taiwan; 2009, 1st edition, China Metrology Press: Beijing; 2012, NOVA Science Publishers, Inc.: New York), Advanced Food Chemistry (2009, Chemical Industry Press: Beijing), etc. He has published more than 150 SCI and EI papers, and obtained or applied in total 32 national invention patents, and presided over 56 provincial or ministerial research projects, besides 71 research projects commissioned by enterprises. He won the first prize for technological invention from the Ministry of Education of the People's Republic of China, the first prize for technological invention from Chinese Institute of Food Science and Technology, two second prizes and four third prizes for Chongqing Science and Technology Progress.

Dr. Kewei Chen is an associate professor and master supervisor at College of Food Science, Southwest University, China. He obtained his doctorate from Universidad de Sevilla, Spain, in 2016, and from 2012 to 2016, he also did his research work at Instituto de la Grasa (CSIC, Spain). He has participated in several international book chapters about the food industry and technology. His major research interests and areas are related to food chemistry and nutrition, including micronutrient bioavailability, function evaluation, and food safety, and he has presided over four provisional and administerial projects related to food nutritional changes and food safety evaluation. Dr. Kewei Chen has published more than 10 research papers on Food Chemistry, Journal of Functional Foods, Molecular Nutrition & Food Research, etc.

Contributors

Kewei Chen College of Food Science, Southwest University, Chongqing, China

Yu Fu College of Food Science, Southwest University, Chongqing, China

Hui He College of Food Science and Technology, Huazhong Agricultural University, Wuhan, China

Tao Hou College of Food Science and Technology, Huazhong Agricultural University, Wuhan, China

Jianquan Kan College of Food Science, Southwest University, Chongqing, China

Hui Li Institute of Quality Standards and Testing Technology for Agro-Products, Chinese Academy of Agricultural Sciences, Beijing, China

Jingming Li College of Food Science & Nutritional Engineering, China Agricultural University, Beijing, China

Liyan Ma College of Food Science & Nutritional Engineering, China Agricultural University, Beijing, China

Jie Pang College of Food Science, Fujian Agriculture and Forestry University, Fuzhou, China

Hui Shi College of Food Science, Southwest University, Chongqing, China

Chunli Song College of Food and Bioengineering, Qiqihar University, Qiqihar, China

Mingyue Song College of Food Science, South China Agricultural University, Guangzhou, China

Aidong Sun College of Biological Science and Technology, Beijing Forestry University, Beijing, China

Yali Yang College of Food Engineering and Nutritional Science, Shaanxi Normal University, Xian, China

Fusheng Zhang College of Food Science, Southwest University, Chongqing, China

Lichao Zhao College of Food Science, South China Agricultural University, Guangzhou, China

Xinhuai Zhao School of Biology and Food Engineering, Guangdong University of Petrochemical Technology, Maoming, China

Chapter 1
Introduction

Jianquan Kan and Kewei Chen

Abstract In this chapter, the introduction of food chemistry is given including the main content and research field, history of this independent discipline, and the associated applications in the food industry. After these, information about the main changes that occurred in the food industry is provided and the future trend for the development of the food industry is also discussed, to help readers obtain an overall comprehension of what food chemistry is, how it researches, and why it is important for food industries.

Keywords Food chemistry · Introduction · Application · Food industry

Foodstuff refers to an edible material containing various nutrients, and nutrients are substances necessary for our bodies to maintain normal growth and metabolism which can be divided into six categories, namely as proteins, lipids, carbohydrates, minerals, vitamins, and water. Due to the importance of dietary fiber in the human diet, it can be included as an additional category although it shows carbohydrate nature. Most of the foodstuff is processed before human consumption and we usually call processed foodstuff as food, while the concept of food always covers any edible foodstuff as well.

Food chemistry is a science that uses the theory and method of chemistry to study the nature of food, that is, to study the chemical composition, structure, physicochemical properties, nutritional values, and safety parameters of foods from the chemical and molecular levels; meanwhile, it also contains investigations about the effect of processing, storage, transportation, and distribution on food quality and safety. Food chemistry belongs to food science, and is also a branch of applied chemistry. It forms a discipline that lays a theoretical foundation for improving food quality, developing new food resources, scientifically adjusting dietary structure, and innovating food technologies related to processing, storage and transportation, packaging, food quality control, etc.

J. Kan (✉) · K. Chen
College of Food Science, Southwest University, Chongqing 400715, China
e-mail: ganjq1965@163.com

© The Author(s), under exclusive license to Springer Nature Singapore Pte Ltd. 2021
J. Kan and K. Chen (eds.), *Essentials of Food Chemistry*,
https://doi.org/10.1007/978-981-16-0610-6_1

1.1   A Brief History of the Development of Food Chemistry

Food chemistry is an independent discipline formed in the early twentieth century with the development of chemistry, biochemistry, and the rise of the food industry. It is closely related to human life and food production practices. Although in some sense the origin of food chemistry can be traced back to ancient times, food chemistry evolved as a discipline in the late nineteenth century with some important scientific achievements revealed.

Swedish famous chemist Carl Wilhelm Scheele (1742–1786) isolated and studied the properties of lactic acid (1780), separated citric acid from lemon juice (1784) and gooseberry (1785), and obtained malic acid (1784) from apples. He also determined citric acid and tartaric acid (1785) in 20 kinds of common fruits, which is considered as the beginning of quantitative analysis for food chemistry. French chemist Antoine Laurent Lavoisier (1743–1794) first carried out the elemental investigations of ethanol (1784). Meanwhile, Swiss chemist Nicolas Theodore de Sanssure (1767–1845) used the ashing method to determine the mineral content of plants and completed the elemental composition analysis of ethanol (1807).

British chemist Humphrey Davy (1778–1829) published the book named as "Elements of Agricultural Chemistry, in a Course of Lectures for the Board of Agriculture" in 1813, which discussed some relevant aspects of food chemistry. The classic study by French chemist Michel Eugene Chevreul (1786–1889) on animal fats led to the discovery and naming of stearic acid and oleic acid. French chemist Jean Baptiste Duman (1800–1884) suggested in 1871 that a diet consisting only of protein, carbohydrates, and fat is not sufficient to sustain human life. German chemist Justus Von Liebig (1803–1873) classified food ingredients in 1842 as nitrogenous parts (plant proteins, casein, etc.) and nitrogen-free compounds (fats, carbohydrates, etc.) and published the first book focused on food chemistry in 1847 as "Researches on the Chemistry of Food", while at that time food chemistry discipline has yet been established.

Until the early twentieth century, the food industry has become an important industry in developed countries and some developing countries. Most of the food materials have been analyzed for their compositions by chemists, biologists, and nutritionists, which set the foundation for the formation of the food chemistry discipline. Meanwhile, different branches in the food industry have created their own chemical foundations, such as grain and oil chemistry, fruit and vegetable chemistry, dairy chemistry, sugar chemistry, meat and poultry egg chemistry, aquatic chemistry, additive chemistry, and flavor chemistry, which provided sufficient research materials for the systematic establishment of food chemistry. In the middle of the twentieth century, journals including "Journal of Food Science", "Journal of Agricultural and Food Chemistry", and "Food Chemistry", which had world-wide influence, were successively issued, marking the formal establishment of food chemistry as a discipline.

In the past 20 years, many books focused on the field of food chemistry have been published, such as "Food Science (in B. A. Fox eds.)", "Food Chemistry (in O. R.

Fennema eds; in H. D. Belitz, W. Grosch & P. Schieberle eds.)", " Food Processing Handbook (in J. G. Brennan eds.)", "Carbohydrates (in R. V. Stick & S. J. Williams eds.)", "Food Protein Chemistry (in N. K. Murthy eds.)", and "Functional Properties of Proteins in Food (in Z. E. Sikorski eds.)", which show the contemporary knowledge about food chemistry, and they are considered as the important reference book for students.

1.2   Content and Scope of Food Chemistry

The research field of food chemistry includes firstly, the chemical composition, properties, structure and function of nutrients, color, aroma, taste, harmful components and bioactive substances in foods, and related analytical techniques; then changes among food ingredients during processing, storage, and distribution, that is, elucidation of the course of chemical reactions, the structure of intermediates and final products, and their impact on food quality, hygiene, and safety; what's more, research on new technologies for food storage and processing, development of new products, new food resources, and new food additives, etc.

According to different research scope, food chemistry mainly includes food nutrient chemistry, color, aroma and flavoring chemistry, chemistry in food processing, food physical chemistry and hazardous ingredients in food, and food analysis technology. According to the classification of substances in the research, food chemistry mainly includes carbohydrate chemistry, food oleo chemistry, food protein chemistry, food enzymology, food additives, vitamin chemistry, mineral chemistry, condiment chemistry, food flavor chemistry, pigment chemistry, food poison chemistry, and bioactive compound chemistry. In addition, the research field of food chemistry also covers the treatment of edible water, environmental protection during food production, separation and extraction of useful ingredients, deep processing and comprehensive utilization of agricultural resources, and development of green and organic foods.

Food chemistry is closely related to chemistry, biochemistry, physiology, botany, zoology, nutrition, medicine, technology, hygiene, and molecular biology, and it relies on the knowledge of the above disciplines to effectively investigate human food. Different from bioscientists, food chemists are primarily concerned with dead or dying biological material (e.g., post-harvest plants and post-mortem muscles) and their exposure to changes in a variety of environmental conditions, which include not only life compatible conditions, e.g., the proper conditions for maintaining the freshness of fruits and vegetables during storage and transportation such as low temperature and modified atmosphere packaging, but also the undesirable environment for life maintenance, e.g., the employment of heat, freezing, concentrating, dehydrating, irradiating, and preservative addition, to investigate changes under these conditions and evaluate their consequent impact on food quality and safety. In addition, food chemists also care about destroyed food tissues (flour, fruit, and vegetable juices, etc.), single-cell materials (eggs, algae, etc.), and some important biological fluids (milk, etc.). These research materials show different characteristics as other research topics, which are crucial for food processing and preservation.

1.3 Main Chemical Changes in Food

From raw material production, storage, transportation, processing to product sales, each process involves a series of chemical changes. Table 1.1 lists the possible changes in foods during processing and storage, while Table 1.2 shows some of the chemical and biochemical reactions that cause changes in food quality and safety.

Table 1.1 Possible changes during food processing or storage

Attribute	Changes
Texture	Loss of solubility or water holding capacity, texture hardening or softening, etc.
Flavor	Rancid odor, burnt odor, aroma formation, off-flavor, etc.
Color	Browning (dark), bleaching (fading), formation of unusual colors and attractive colors, etc.
Nutrition	Degradation or loss of proteins, lipids, vitamins, and minerals and related bioavailability changes
Safety	Formation or inactivation of poisons, formation of substances interfering physiological regulation

Table 1.2 Some chemical reactions and biochemical reactions influencing food quality or safety

Reactions	Examples
Maillard reaction	Formation of color, aroma, and taste of baked food
Enzymatic browning	Color changes in peeled fruits
Oxidation	Odor formation in lipid storage, vitamin degradation, pigment fading, and decrease of protein nutritional value
Hydrolysis	Hydrolysis of lipids, proteins, vitamins, carbohydrates, pigments, etc.
Metal Reaction	Color change of anthocyanin, magnesium removal in chlorophyll, catalytic auto-oxidation
lipids Isomerization	Formation of trans-unsaturated fatty acids and conjugated fatty acids
Lipid cyclization	Formation of monocyclic fatty acid
Lipid polymerization	Foam generation and viscosity increase in fried oil
Protein denaturation	Ovalbumin coagulation and enzyme inactivation
Protein cross-linking	Processing of proteins under alkaline conditions reduces their nutritional value
Glycolysis	Anaerobic respiration of post-mortem animal tissues and post-harvest plant tissues

The chemical changes that occur during processing and storage of foods generally include enzymatic changes during physiological ripening and aging; changes due to variations in water activity; enzymatic changes and chemical reactions due to mixing of raw materials or tissues; decomposition, polymerization, and denaturation caused by processing conditions (e.g., heat); oxidation reactions caused by oxygen or other oxidants; photochemical changes during light exposure; and changes caused by the component migration from the packaging material to food. The most important of these changes are enzymatic browning, non-enzymatic browning, lipid hydrolysis, lipid oxidation, protein denaturation, protein cross-linking, proteolysis, hydrolysis of oligosaccharides and polysaccharides, glycolysis and natural pigment degradation, etc., which influence greatly food quality and safety. For instance, the oxidation of lipids will lead to the odor of fat-containing foods; enzymatic browning is the reason for the color change of damaged fruit.

The interaction among the main components in food during food processing and preservation also has an important impact on the quality of the food (Fig. 1.1). As shown in Fig. 1.1, active carbonyl compounds and peroxides are extremely important reaction intermediates, which are derived from chemical reactions of lipids, carbohydrates, and proteins, and in turn cause changes in pigments, vitamins, and flavors, resulting in a variety of undesirable changes in food.

The factors affecting these reactions come either from the food itself, such as the composition, water activity, pH values, and glass transition temperature (Tg) or from the surrounding environment such as temperature, special treatment, atmospheric composition, and light, where temperature, special treatment, pH values, water activity, and food composition turn out to be the most important. In general,

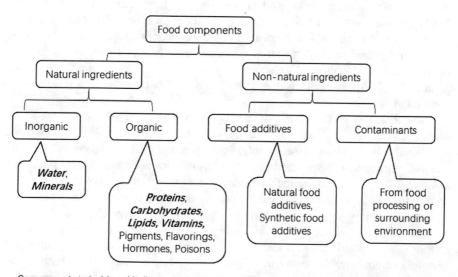

Components in bold and italic mean the basic nutrients

Fig. 1.1 Illustration of food chemical composition

within the medium temperature range, the rate constant of the chemical reaction varies with temperature, in accordance with the Arrhenius equation in classical chemistry:

$$k = A \exp(-Ea/RT),$$

where k is the velocity constant at temperature T; A and R are constants; T is the absolute temperature; Ea is the activation energy of the reaction.

While, the equation will deviate at high or low temperatures where the reaction enzymes will be inactivated, leading to changes of reaction pathway or other side reactions. Therefore, understanding the chemical and biochemical reactions and studying these influential factors can help to control the changes in food quality and safety during food processing and preservation, which establishes the basic principle for the investigations of food chemistry.

1.4 Research Methods Used in Food Chemistry

The research method used in food chemistry is to analyze and comprehensively understand the changes of food substances at the molecular level through lab investigations and theoretical deduction.

Different from pure chemistry research, food chemistry always relates the chemical composition, physical and chemical properties, and changes of food with food quality and safety. Therefore, food chemistry research always aims at revealing food quality or safety changes and considers both concrete food materials and processing conditions as an important basis for experimental design. Due to the complex constitution in food, it is very difficult to define all of the concomitant reactions. Frequently, a simplified, simulated food material system is usually applied for investigations. However, the results obtained are sometimes not reasonable to explain the real situation in the food system. Thus, the deficiencies of the research method should be clarified in advance.

Food chemistry related to a concrete food product can be roughly divided into four aspects as follows: (1) property determination including chemical composition, nutritional value, functional ingredients, and safety and quality concerns; (2) chemical and biochemical changes and their reaction kinetics during food processing and storage; (3) key factors affecting food quality and safety in the above changes; (4) application of the research results to the processing and storage of food. Therefore, food chemistry investigations should include physical and chemical tests and sensory evaluation. The physicochemical test is mainly to carry out the component analysis and structural identification of the food, that is, to analyze changes of nutrients, harmful components, pigments, and flavors; the sensory evaluation is a comprehensive and visual review to analyze changes in texture, flavor, and color of the tested food.

According to the experimental results and theoretical calculation, a chemical reaction equation can be established and a reasonable hypothesis can be obtained to predict the possible consequences on food quality and safety. After that, practical processing assays are carried out to verify the supposed hypothesis. Further investigations are reaction kinetics, either to understand the reaction mechanism or to explore the factors affecting the reaction rate, so as to seek possible control measurements. Chemical reaction kinetics is to investigate the effect of material concentration, collision probability, space barrier, activation energy barrier, reaction temperature and pressure, and reaction time on reaction rate and reaction equilibrium, through which the reaction intermediates, catalytic factors, and reaction pathways affected by various conditions will be clarified. According to these theoretical foundations, food chemists will be able to select the appropriate conditions in food processing and storage and control food quality and safety.

Eventually, food chemistry investigations will be translated into: reasonable ingredient combination, effective accesses or barrier for reactant contact, proper protection or catalytic measures, optimal conditions for reaction time, temperature, light exposure, oxygen content, water activity, and pH values, etc., resulting in an ideal method for food processing and storage.

1.5 Food Chemistry for the Development of Food Industry Technology

Traditional foods are no longer able to satisfy consumers for high-level foods, and the modern food industry shows the trend of nutrition strengthening, safety guarantee, and health care. The basic theory and applied research results of food chemistry form the main principles in the rapid growth of the food industry where many important reactions were elucidated including the Maillard reaction, caramelization reaction, lipid oxidation, enzymatic browning, starch gelatinization and aging, hydrolysis of polysaccharides, protein hydrolysis, protein denaturation, pigment degradation, vitamin variation, metal catalytic reaction, enzyme catalytic reaction, lipid hydrolysis and transesterification, lipid thermal oxidation and polymerization, and flavor change reaction and post-harvest reactions, all of which play a vital role for the modern food processing and storage technologies (Table 1.3).

The agricultural and food industry is one of the most widely applied areas of bioengineering. The development of bioengineering has broadened the path for the quality transformation of edible agricultural products, the exploration of new foods and food additives, and enzyme preparations, but the success of bioengineering in food applications relies heavily on food chemistry. First of all, it is necessary to use food chemistry research to indicate the physical properties of the original biological raw materials to obtain the key points for transformation and to get a picture that what kind of food additives and enzyme preparations are urgently needed; secondly,

Table 1.3 Influence of food chemistry on technological progress in food industries

Branches	Technologies and achievements
Fruit and vegetable	Chemical peeling, color protection, texture control, vitamin retention, deodorization, waxing, chemical preservation, modified atmosphere storage, active packaging, enzymatic juice extraction, filtration and clarification, chemical antisepsis, etc.
Meat	Post-mortem treatment, preservation and tenderization, color protection and fixation, improvement of the emulsifying power, gelation and viscoelasticity of meat paste, fresh meat packaging in supermarkets, production and application of fumigant, production of artificial meat, comprehensive utilization of internal organs, etc.
Beverage industry	Instant drink, stable protein drink, water treatment, stabilization of meat juice, juice color protection, clarification control, flavor modification, liquor control in spirit, beer clarification, removal of beer foam and bitterness, beer odor elimination, juice debittering, odor removal in soy drinks, etc.
Dairy	Stabilization of yogurt and mixed milk with juice, rennet substitutes and reconstituted cheese, whey utilization, nutritional enhancement of dairy products, etc.
Baking	high-efficiency leavening agents, crispiness increase, improvement of bread color and texture, and prevents starch retrogradation and mildew
Edible oil	Refining, winterization, and temperature regulation of oil, fat modification, isolation and utilization of DHA, EPA, and MCT, production of edible emulsifiers, antioxidants addition, reducing oil content of fried foods, etc.
Condiment	Production of meat soup, nucleotide flavoring agent, iodized salt and organic selenium salt, etc.
Fermented food	Post-treatment of fermented products, flavor control during post-fermentation, comprehensive utilization of by-products and residues, etc.
Basic food	Flour improvement, nutrient fortification of refined grain products, hydrolysis of cellulose and hemicellulose, production of high fructose syrup, starch modification, hydrogenated vegetable oil, production of new sweeteners, new oligosaccharides and new food additives, oil modification, separation of plant proteins and functional peptides, development of microbial polysaccharides and single-cell proteins, development and utilization of edible resources from wild and marine environment, or for drug and food use, etc.
Food analysis	Development of test standards, rapid analysis, and biosensors, etc.

products from bioengineering process sometimes do not exactly satisfy the requirements for their application in food, and they need further separation, purification, combination, or chemical modification, which need the guidance of food chemistry results; finally, it is needed in food chemistry to evaluate the potential hazard and effective impact of bioengineering materials that have not been used in traditional food processing.

In this word, food chemistry also boosts the high-tech progress in food science and engineering. In the past 20 years, many new technologies or materials have been pushed to the food industry. For example, biodegradable packaging materials, microwave processing technology, irradiation preservation technology, supercritical extraction and molecular distillation technology, membrane separation technology, active packaging technology, microcapsule technology, etc., have to accord the principles established by food chemistry to exert their influence in the food industry.

1.6 Future Trend for Food Chemistry

Due to the new modern analytical methods, as well as the advancement of biological theory and applied chemistry theory, we have a further understanding of the structure and reaction mechanism of food ingredients. The use of biotechnology and modern industrial technology to change the composition, structure, and nutrition of food and the in-depth study of the physiological activities of functional factors in functional foods at the molecular level will lead to new breakthroughs in the theory and application of food chemistry in the future. Currently, investigations have been focused on the analysis of the chemical composition, properties, and changes of different raw materials in food processing and storage and their impact on food quality and safety; as well, some of the technical problems that handicap the food industry, such as discoloration, odor, textural roughness, and quality deterioration have been solved. All of these investigations need to be carried on as new resources or materials are emerging in our daily life and more industrial barriers or technology defects are encountered in the food industry. Therefore, the future research lines of food chemistry will have the following aspects:

(1)   Exploration of new food resources: removal of potential harmful components while maintaining beneficial ingredients;
(2)   Research on functional factors in food including the composition, structure, properties, physiological activity, and their analysis and isolation methods for comprehensive development of functional food;
(3)   Auxiliary chemical treatment, membranes, and active packaging materials in modern storage and preservation technology;
(4)   Food flavor chemistry and processing technology combined with modern analytical and new technologies;
(5)   New food additives including the development, production, and application of enzyme preparations;
(6)   Development of rapid quantitative and qualitative analysis methods or new detection techniques;
(7)   Intensive processing and comprehensive utilization of food resources;
(8)   Modification technology of food raw materials;
(9)   Study on the interactions between chemical components in foods and their impact on food quality and safety;

(10)   Technical barriers during the industrialization and modernization of traditional foods;
(11)   Study on mechanisms of typical chemical reaction in foods and elucidation of functional or hazard properties of the yielded intermediates and products.

As more modern technologies are involved in food processing and new discoveries are revealed about the food ingredients or components, new branches of food chemistry will be put forward including carbohydrate chemistry and food colloid chemistry, food microwave chemistry, food fermentation chemistry, food ultra-high-pressure chemistry, etc.

Questions

1.   What is food chemistry? What are included in the research of food chemistry?
2.   What are the main chemical changes in food? How about their impact on food quality and safety?
3.   What are the characteristics of research methods in food chemistry?
4.   What do you think will be the research hotspots of food chemistry in the future?

Bibliography

1. Belitz, H.D., Grosch, W., Schieberle, P.: Food Chemistry. Springer-Verlag Berlin, Heidelberg (2009)
2. Carocho, M., Morales, P., Ferreira, I.C.F.R.: Antioxidants: reviewing the chemistry, food applications, legislation and role as preservatives. Trends Food Sci. Technol. **71**, 107–120 (2018)
3. Cheung, P.C.K., Mehta, B.M.: Handbook of Food Chemistry. Springer-Verlag Berlin, Heidelberg (2015)
4. Crawford, E., Crone, C., Horner, J., Musselman, B.: Food packaging: strategies for rapid phthalate screening in real time by ambient ionization tandem mass spectrometry. In: Benvenuto, M.A., Ahuja, S., Duncan, T.V., Noonan, G.O., Roberts-Kirchhoff, E.S. (eds.) Chemistry of Food, Food Supplements, and Food Contact Materials: From Production to Plate, pp. 71–85. Amer Chemical Soc, Washington (2014)
5. Damodaran, S., Parkin, K.L., Fennema, O.R.: Fennema's Food Chemistry. CRC Press/Taylor & Francis, Pieter Walstra (2008)
6. Diez-Simon, C., Mumm, R., Hall, R.D.: Mass spectrometry-based metabolomics of volatiles as a new tool for understanding aroma and flavour chemistry in processed food products. Metabolomics **15**, 41 (2019)
7. Filice, M., Aragon, C.C., Mateo, C., Palomo, J.M.: Enzymatic transformations in food chemistry. Curr. Org. Chem. **21**, 139–148 (2017)
8. Hoehn, E., Baumgartner, D.: Fruits and vegetables. In: Aliani, M., Eskin, M.N.A. (eds.) Bitterness: Perception, Chemistry and Food Processing, pp. 53–82. John Wiley & Sons Inc., Hoboken (2017)
9. Kan, J.: Food Chemistry. China Agricultural University Press, Beijing (2016)
10. Lagana, P., Avventuroso, E., Romano, G., Gioffré, M.E., Patanè, P., Parisi, S., Moscato, U., Delia, S.: Classification and technological purposes of food additives: the European point of view. In: Lagana, P., Avventuroso, E., Romano, G., Gioffré, M.E., Patanè, P., Parisi, S., Moscato, U., Delia, S. (eds.) Chemistry and Hygiene of Food Additives, pp. 1–21. Springer-Verlag, Berlin, Berlin (2017)

11. Lima Pallone, J.A., dos Santos Carames, E.T., Alamar, P.D.: Green analytical chemistry applied in food analysis: alternative techniques. Current Opinion in Food Science **22**, 115–121 (2018)
12. Ritzoulis, C.: Introduction to the Physical Chemistry of Foods. CRC Press/Taylor & Francis Group, Boca Raton (2013)
13. Rodriguez-Amaya, D.B.: Effects of processing and storage. In: Rodriguez-Amaya, D.B. (ed.) Food Carotenoids: Chemistry, Biology and Technology, pp. 132–173. West Sussex, John Wiley & Sons Ltd (2016)
14. Rodriguez-Amaya, D.B.: Qualitative and quantitative analyses. In: Rodriguez-Amaya, D.B. (ed.) Food Carotenoids: Chemistry, Biology and Technology, pp. 47–81. West Sussex, John Wiley & Sons Ltd (2016)
15. Rychlik, M.: Challenges in food chemistry. Front. Nutr. **2**, 11 (2015)
16. Schreier, P.: Food chemistry links chemistry with biology and medicine. Mol. Nutr. Food Res. **50**, 339 (2006)
17. Ustunol, Z.: Applied Food Protein Chemistry. West Sussex, John Wiley & Sons Ltd (2015)

Dr. Jianquan Kan is a professor and doctoral supervisor at College of Food Science, Southwest University, China. He is currently the Director of the Quality and Safety Risk Assessment Laboratory for Agricultural Products Storage and Preservation of the Ministry of Agriculture and Rural Affairs (Chongqing), and the academic leader in the discipline of "Food Science and Engineering" belonging to Chongqing's key disciplines. Dr. Jianquan Kan also leads the "Excellent Teaching Team" in the field of Food Science and Safety, and he was awarded "Person of the Year (2015)" as Scientific Chinese. He is a special expert candidate for high-level talents in Sichuan Province of the "Thousand Talents Program" and currently is the chief scientist in the Innovation Team of Chongqing Modern Special High-efficient Agricultural Flavoring Industrial Technology System. Dr. Jianquan Kan has been the Editor-in-Chief for 6 textbooks related to food chemistry, for instance, Practical Chemistry of Oil and Fat (1997, Southwest Normal University Press: Chongqing), Food Chemistry (2002 1st edition, 2008 2nd edition, 2016 3rd edition, China Agricultural University Press: Beijing; 2006 1st edition, 2017, 3rd edition, New Wenjing Development and Publishing Co., Ltd.: Taiwan; 2009, 1st edition, China Metrology Press: Beijing; 2012, NOVA Science Publishers, Inc.: New York), Advanced Food Chemistry (2009, Chemical Industry Press: Beijing), etc. He has published more than 150 SCI and EI papers, and obtained or applied in total 32 national invention patents, and presided over 56 provincial or ministerial research projects, besides 71 research projects commissioned by enterprises. He won the first prize for technological invention from the Ministry of Education of the People's Republic of China, the first prize for technological invention from Chinese Institute of Food Science and Technology, two second prizes and four third prizes for Chongqing Science and Technology Progress.

Dr. Kewei Chen is an associate professor and master supervisor at College of Food Science, Southwest University, China. He obtained his doctorate from Universidad de Sevilla, Spain, in 2016, and from 2012 to 2016, he also did his research work at Instituto de la Grasa (CSIC, Spain). He has participated in several international book chapters about the food industry and technology. His major research interests and areas are related to food chemistry and nutrition, including micronutrient bioavailability, function evaluation, and food safety, and he has presided over four provisional and administerial projects related to food nutritional changes and food safety evaluation. Dr. Kewei Chen has published more than 10 research papers on Food Chemistry, Journal of Functional Foods, Molecular Nutrition & Food Research, etc.

Chapter 2
Water

Lichao Zhao and Mingyue Song

Abstract Water is found to be the most abundant component in all living things and, consequently, is in almost all foods. The presence of water has a significant impact on the properties of foods and is the cause of food spoilage or deterioration. By controlling the water in foods, we can predict and control the rate of many chemical and biochemical reactions, which contribute to food stability and quality retention. In this chapter, you will learn about the important role of water in foods, the structure and nature of water and ice, categories of water in foods, the concept and significance of water activity and moisture sorption isotherms, the relationship between water activity and food stability, the definition of molecular mobility and its role in food stability, as well as the moisture transfer law in foods.

Keywords Water · Categories of water · Water activity · Molecular mobility · Glass transition temperature · Food stability

2.1 Overview

Water is the most abundant and widely distributed substance on earth. It is not only concentrated in the lakes, rivers, and oceans, but also in most organisms. The water content in the human body is generally 70%, which is an indispensable component for maintaining life activities and regulating metabolic processes. Under normal circumstances, each person needs to absorb 2 ~ 2.7 L of water from foods every day. The excess water is discharged in the form of sweat and urine to maintain the balance of water content in the human body.

Water is a substantial component in foods (Table 2.1). The content, distribution, and state of water in foods significantly impact the structure, appearance, texture, flavor, color, fluidity, freshness, and spoilage of foods. Therefore, water content is a major quality indicator in many statutory food quality standards. Since most fresh

L. Zhao (✉) · M. Song
College of Food Science, South China Agricultural University, Guangzhou 510642, China
e-mail: ZLC@scau.edu.cn

© The Author(s), under exclusive license to Springer Nature Singapore Pte Ltd. 2021
J. Kan and K. Chen (eds.), *Essentials of Food Chemistry*,
https://doi.org/10.1007/978-981-16-0610-6_2

Table 2.1 Water content of some typical foods (%)

Product	Water content	Product	Water content	Product	Water content
Tomato	95	Milk	87	Jam	28
Lettuce	95	Potato	78	Honey	20
Cabbage	92	Banana	75	Cream	16
Beer	90	Chicken	70	Rice or flour	12
Citrus	87	Meat	65	Milk powder	4
Apple juice	87	Bread	35	Shortening	0

foods usually have high water content, it is necessary to take effective preservation methods to control it for long-term storage. But the inherent characteristics of foods will significantly change with the process of normal dehydration or low-temperature freeze-dry dehydration. Theoretically and technically, there are many problems remaining to be further resolved in the aspects of thawing, rehydration, controlling water migration of water within foods, controlling moisture content or water activity for adapting many physical and chemical changes, and utilizing the appropriate interactions between water and non-water components (especially proteins and polysaccharides) to generate health-promoting functions. Therefore, the relationship between water and food is one of the important contents of food science and has great significance to food stability and preservation.

2.2 Water and Ice

2.2.1 The Physical Properties of Water and Ice

Water is a substance that can be converted into solid, liquid, and gas states. By comparing the property of water with that of similar molecular weight and atomic composition, such as HF, CH_4, H_2F, H_2Se, and NH_3, water has unusually high melting and boiling point temperatures, exhibiting unusually large values of permittivity, heat capacity, and heat for phase transformation (fusion, vaporization, and sublimation).

These special thermal properties of water have a huge impact on the freezing and drying processes in food processing. Water has a low density, when it freezes, the volume of transformed ice increases, exhibiting an abnormal swelling characteristic, which usually causes damage to the structure of the food. The thermal conductivity of water is large compared with most of the other liquids, and the thermal conductivity of ice is slightly larger than nonmetallic solid. It is noteworthy that the thermal conductivity of ice at 0 °C is approximately quadruple that of liquid water at the same temperature, indicating that ice will conduct thermal energy at a much greater rate than immobilized (e.g., tissue) water. Since thermal diffusivity is indicative of the rate at which a material will undergo a change in temperature, we would expect that

Table 2.2 The physical constants of water and ice

Item	Physical constants			
Molecular weight	18.0153			
Melting and boiling point at 101.3 kPa	0.00 °C and 100.00 °C			
Critical temperature and pressure	373.99 °C and 22.06 MPa (218.60 atm)			
Triple point (temperature and pressure)	0.01 °C and 611.73 Pa (4.59 mmHg)			
ΔH_{fusion} (0 °C)	6.012 kJ (1.436 kcal)/mol			
$\Delta H_{vaporization}$ (100 °C)	40.657 kJ (9.711 kcal)/mol			
$\Delta H_{sublimation}$ (0 °C)	50.91 kJ (12.06 kcal)/mol			
	20 °C (H_2O)	0 °C (H_2O)	0 °C (Ice)	−20 °C (Ice)
Density (g/cm^3)	0.99821	0.99984	0.9168	0.9193
Viscosity (Pa·s)	1.002×10^{-3}	1.793×10^{-3}	–	–
Interfacial tension (N/m, relative to air)	72.75×10^{-3}	75.64×10^{-3}	–	–
Vapor pressure (kPa)	2.3388	0.6113	0.6113	0.103
Heat capasity [J/(g·K)]	4.1818	4.2176	2.1009	1.9544
Thermal conductivity [W/(m·K)]	0.5984	0.561	2.24	2.433
Thermal diffusivity (m^2/s)	1.4×10^{-7}	1.3×10^{-7}	11.7×10^{-7}	11.8×10^{-7}
Permittivity	80.2	87.9	~90	~98

ice, in a given thermal environment, will undergo temperature change at a rate of nine times greater than that for liquid water. These differences in thermal conductivity and diffusivity values for water and ice provide a good basis for understanding why tissues freeze more rapidly than they thaw under symmetrically applied temperature differentials. The physical constants of water and ice are shown in Table 2.2.

2.2.2   The Water Molecule and Their Association

To explain the physical and chemical properties of water, it is better first to consider the nature of a single water molecule, and then to consider the cluster characteristics of water molecules with increasing size, finally considering the nature of the bulk system. The electronic structure of the hydrogen atom in the water molecule is $1s^1$, and the electronic structure of the oxygen atom in the water molecule is $1s^2 2s^2 2p2p2p$ with two unpaired 2p electrons. When a chemical bond is forming between the oxygen atom and hydrogen atom, the oxygen is sp^3 hybridized to generate four sp^3 hybrid orbitals. Two of them are occupied by the lone pair of electrons of the oxygen atom itself (Φ, Φ), and the other two sp^3 hybrid orbitals overlap with the 1s orbital of two hydrogen atoms to form two covalent sigma (σ) bonds of 40% ionic character (Φ + H, Φ + H). A pyramidal structure is then formed, with the oxygen atom at the center of the tetrahedron. Two of the four vertices in the tetrahedron are occupied by

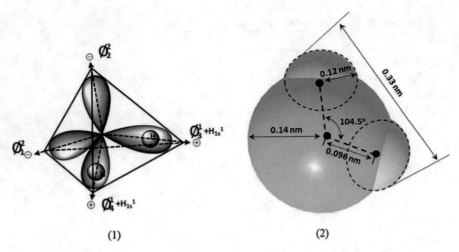

(1)   (2)

Fig. 2.1 Schematic model of a single HOH molecule. (1) Possible sp³ configuration; (2) Van der Waals radii for a molecule in the vapor state

hydrogen atoms, and the remaining two are occupied by the two pairs of lone-pair electrons from oxygen atoms which squash the pair of bonding electrons. So, the angle between the two O–H bonds is 104.5°, which is somewhat different from the angle of 109° 28′ in the typical tetrahedron. The dissociation energy of each O–H bond is 4.614×10^2 kJ/mol (110.2 kcal/mol). The internuclear distance between oxygen and hydrogen atom is 0.096 nm, and the Van der Waals radii of oxygen and hydrogen are 0.14 nm and 0.12 nm, respectively (Fig. 2.1).

The highly electronegative oxygen of the water molecule can be visualized as it partially draws away the single electrons from the two covalently bonded hydrogen atoms, thereby leaving each hydrogen atom with a partial positive charge and a minimal electron shield; that is, each hydrogen atom assumes some characteristics of a bare proton, resulting in a polarized state of water molecule and a dipole moment in the vapor state of 1.84 D for pure water. At the same time, its hydrogen atom highly tends to form a hydrogen bond with the lone pair of electrons on the outer layer of the oxygen atom of another water molecule, which results in considerable intermolecular attractive forces, and hence water molecules are associated with considerable tenacity.

Within a water molecule, the hydrogen–oxygen bonding orbitals are located on two of the axes of an imaginary tetrahedron (Fig. 2.1(1)), and these two axes can be considered as representing lines of positive force (hydrogen bond donor sites). Oxygen's two lone-pair orbitals can be considered as residing along the remaining two axes of the tetrahedron, representing lines of negative force (hydrogen bond acceptor sites). By these four lines of force in a tetrahedral orientation, each water molecule has the potential to form a hydrogen bond with a maximum of four others (Fig. 2.2). As each water molecule has an equal number of hydrogen bond donor and acceptor site, it is found that the attractive forces among water molecules are

Fig. 2.2 Hydrogen bonding of water molecules in a tetrahedral configuration. Blue circles are oxygen atoms, and gray circles are hydrogen atoms. Hydrogen bonds are represented by dashed lines

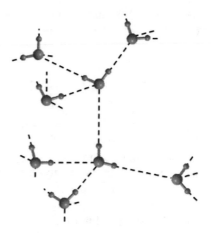

unusually large, even when compared with those other small molecules that also engage in hydrogen-bonding associations.

This ability of water to engage in extensive three-dimensional hydrogen bonding provides a logical explanation for many of its unusual properties. All of these can be related to the additional energy necessary to break large numbers of intermolecular hydrogen bonds. The permittivity (dielectric constant) of water is also influenced by hydrogen bonding. It appears that hydrogen-bonded molecular clusters give rise to multimolecular dipoles, effectively increasing the permittivity. This explains that water shows a lower viscosity characteristic as the hydrogen-bonded arrangements of water molecules are highly dynamic, allowing the individual molecule to alter their hydrogen-bonding relationships with neighboring molecules within a time frame of nano- to picoseconds, thereby facilitating mobility and fluidity.

The hydrogen bond association of water molecules can also explain many of their unusual physical properties. To explain the characteristics of density change in water, it is also necessary to understand the structural change of water during temperature change. Ice at 0 °C has a coordination number (number of nearest neighbors) of 4.0, with the nearest neighbor distance being 0.276 nm. As the temperature increases, melting occurs to broke many hydrogen bonds and destroy the integrated crystal structure, which splits into multiple tetrahedrons and a small amount of freely moving water molecules and fills the gaps in the original ice structure. Therefore, the density of ice increases and its volume decreases when melting. For example, as the temperature is raised, the coordination number increases from 4.4 in water at 1.5 °C to 4.9 at 83 °C. However, the thermal motion increases when the temperature continues to rise, and the collision frequency of water molecules also increases, resulting in volume expansion. For instance, the distance between nearest neighbors increases from 0.276 nm in ice at 0 °C to 0.29 nm in water at 1.5 °C then to 0.305 nm at 83 °C. It is evident, therefore, that the influence of the coordination number is predominant at temperatures between 0 and 4 °C, and the Brownian motion is predominant when the temperature continues to rise, resulting in a lower density of water. The net result of both factors is that the water density is maximal at 3.98 °C.

2.2.3 The Structure of Water (Liquid)

In liquid water, the relative positions between molecules are constantly changing because of the continuous thermal motion, hence it is impossible to form a single, defined rigid structure like a crystal, but far more organized than that of the vapor state. X-ray diffraction analysis revealed that water is a microscopic crystal that is "ordered" at short distances and has a structure like ice in short periods. When several water molecules are hydrogen bonded to form a cluster of water molecules $(H_2O)n$, the orientation and movement of water molecules will be significantly affected by other surrounding ones. In addition to the irregular distribution and ice fragments, there are many dynamically balanced, incomplete polyhedron connections among water molecules. Therefore, the structure of pure water cannot be described in a single way and must rely on a certain theoretical model.

Three general types of model for liquid water have been proposed: mixture models, interstitial models, and continuum models (also termed homogeneous or uniformist models). All these models agree with the concept that each water molecule frequently changes its binding arrangement, that is, a hydrogen bond is rapidly terminated and replaced with a new hydrogen bond. The system will maintain hydrogen bonding and network structure to a certain degree under constant temperature. The main structural feature is that liquid water molecules form hydrogen bond associations in a short, twisted tetrahedron.

2.2.4 The Structure of Ice

Water can crystallize into various structural forms of ice. Ice is a crystal formed by the ordered arrangement of water molecules, which are hydrogen bonded to form a very loose (low density) rigid structure (Fig. 2.3). The analysis of this rigid structure of ice can help us better understand the structure of water.

The O–O internuclear nearest neighbor distance, in ice, is 0.276 nm, and the O–O–O bond angle is about 109°, very close to the perfect tetrahedral angle of 109.28°. According to this manner, each water molecule can associate with four others to form a tetrahedral structure. When several unit cells are combined and viewed from the top (down the c-axis), the hexagonal symmetry of ice is apparent. As shown in Fig. 2.4a, the tetrahedral substructure is evident from molecule W and its four nearest neighbors, with 1, 2, and 3 being visible and the fourth lying below the plane of the paper, directly under molecule W. When Fig. 2.4a is viewed in three dimensions, as in Fig. 2.4b, it is evident that two planes of molecules are involved (open and filled circles). These two planes are parallel, very close together, and they move as

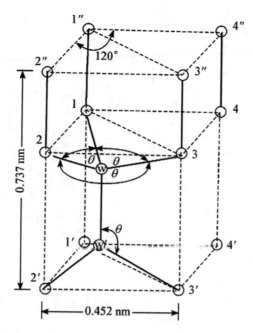

Fig. 2.3 Unit cell of ordinary ice at 0 °C. Cycles represent oxygen atoms of water molecules

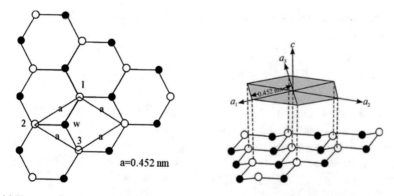

(a) Hexagonal structure viewed down the c-axis   (b) Three-dimensional view of the basal plane

Fig. 2.4 The basic plane of ice is a combination of two planes with slightly different heights. Each cycle represents the oxygen atom of a water molecule. O and ● represent, respectively, oxygen atoms in the upper and lower layers of the basal plane

a unit during the "slip" or flow of ice under pressure, as in a glacier. Pairs of planes with this type comprise the basal planes of ice. By stacking several basal planes, an extended structure of ice is obtained. Three basal planes have been combined to form the structure shown in Fig. 2.5. Viewed down the c-axis, the appearance is the same as that shown in Fig. 2.4a indicating that the basal planes are perfectly aligned. Ice

Fig. 2.5 The extended structure of ordinary ice. Each cycle represents the oxygen atom of a water molecule. O and ● represent, respectively, oxygen atoms in the upper and lower layers of the basal plane

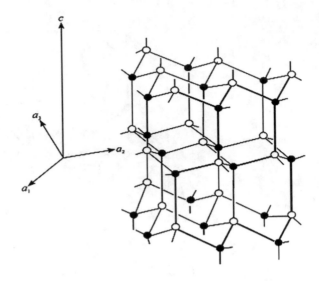

is monorefringent in this direction, whereas it is birefringent in all other directions. This c-axis is therefore the optical axis of ice.

2.3 Interaction Between Water and Non-Aqueous Component

2.3.1 Interaction of Water with Ions and Ionic Groups

The depolarization of water molecules is the basis for the formation of three-dimensional network structures, and enables water molecules to undergo electrostatic interaction (dipole–ion) with ions or ionic groups, such as Na^+, Cl^-, $-COO^-$, and $-NH_3^+$, entering the network structure. The example in Fig. 2.6 illustrates the possible interactions of water molecules adjacent to NaCl. The interaction energy of Na^+ with

Fig. 2.6 Likely arrangement of water molecules adjacent to sodium chloride ion pair. Only water molecules in the plane of the paper are represented

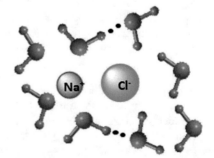

water molecules (83.68 kJ/mol) is approximately four times that of hydrogen bonds between water molecules (20.9 kJ/mol). Water combined with ionic or ionic groups is the most closely integrated part of the food.

The addition of ion or ionic groups has important effects on the stability of the water structure. Based on the normal structure of water, the addition of solute that can dissociate will break the normal tetrahedral arrangement of pure water, which can prevent icing at 0 °C.

In a dilute salt solution, the effect may be positive or reverse. Some negative ions with a weak electric field and positive ions with large ionic radius, such as K^+, Rb^+, Cs^+, NH_4^+, Cl^-, Br^-, I^-, NO_3^-, BrO_3^-, IO_3^-, and ClO_4^-, can disrupt the normal structure of water and fail to impose a compensating amount of new structure. Therefore, such salt solutions are more fluid than pure water. An ion or multivalent ion with a strong electric field and small ionic radius, such as Li^+, Na^+, H_3O^+, Ca^{2+}, Ba^{2+}, Mg^{2+}, Al^{3+}, F^-, and OH^-, can strongly interact with four to six first-layer water molecules, causing them to be less mobile, so that they can stabilize the water network structure. Therefore, the solution of such ions is less fluid than pure water.

In addition to affecting the structure of water, ions can interact with water resulting in changing the dielectric constant of water, modifying the thickness of the double electron layer around the colloid, and significantly affecting the association between water and other non-aqueous solutes or suspended substances. Therefore, the conformation of proteins and the stability of colloids (salting in and salting out) will be affected by the type and number of coexisting ions, such as salting in and salting out processes.

2.3.2 Interaction of Water with Neutral Groups Capable of Hydrogen Bonding

Water can also covalently bond to proteins, starches, pectin, cellulose, etc., in foods. These interactions are weaker than dipole–ion interactions and almost show the same strength as those of water–water hydrogen bonds. Therefore, solutes capable of hydrogen bonding can maintain the structure of pure water, and at least do not disrupt the network structure of water. In some instances, it is found that the distribution and orientation of the hydrogen-bonded sites of the solute are geometrically incompatible with those existing in pure water. Such solutes frequently have a disruptive influence on the normal structure of pure water. Urea is a good example of a small hydrogen-bonding solute that, for geometric reasons, may have a marked disruptive effect on the normal structure of water.

Although such interactions are weak and limited in food, they are important to maintain the characteristics of the food. For example, a "water bridge" composed of several water molecules can be formed between two positions of a biomacromolecule or between two macromolecules to maintain its specific configuration. Figures 2.7

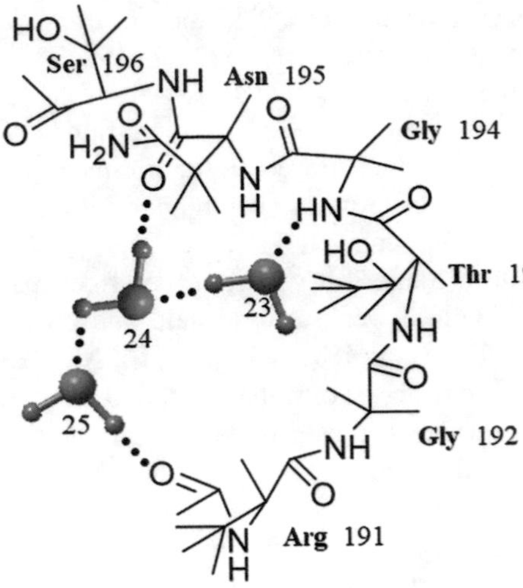

Fig. 2.7 Example of a
three-molecule water bridge
in papain

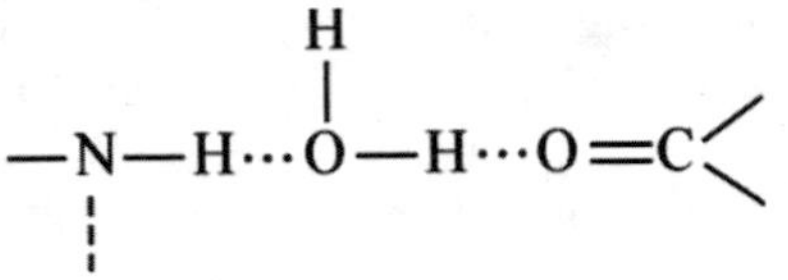

Fig. 2.8 Hydrogen bonding
(dotted lines) of water
molecules to two kinds of
functional groups commonly
occurring in proteins

and 2.8 show a three HOH bridge between backbone peptide units and the hydrogen bond formed between two functional groups in the protein molecule.

It has been found that the distance between the hydrophilic groups in many crystalline macromolecules is equal to that between the nearest O–O in pure water. If this spacing is dominant in hydrated macromolecules, it will promote the formation of hydrogen bonds between the first and second layers of water.

2.3.3 Interaction of Water with Nonpolar Substances

By adding hydrophobic substances such as hydrocarbons, rare gases, fatty acids, amino acids, and the apolar groups of proteins to water, their hydrophobic parts will repel with surrounding water molecules, resulting in enhanced water–water hydrogen bonds around the hydrophobic groups. The difference in polarity leads to a decrease in entropy of the system, which is thermodynamically unfavorable ($\Delta G > 0$). This process has been termed as "hydrophobic hydration" (Fig. 2.9(1)). It should be outlined as two special structural changes, that is, the formation of clathrate hydrates and hydrophobic interactions in proteins.

Clathrate hydrates, which is an ice-like inclusion compound wherein water, the host substance, form a hydrogen-bonded cage-like structure that physically entraps

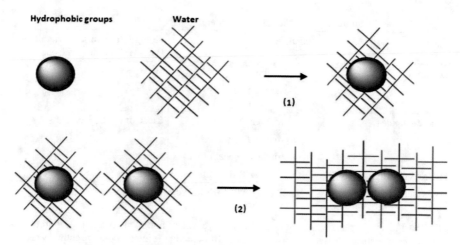

Fig. 2.9 1 Hydrophobic hydration 2 hydrophobic association. Open circles are hydrophobic groups. Hatched areas are water

a small apolar molecule known as the guest molecule. The "host" of the clathrate hydrate is generally composed of 20 to 74 water molecules, and the "guest" is a low molecular weight compound, such as low molecular weight hydrocarbons, rare gases, short-chain primary, secondary and tertiary amines, alkylammonium salts, halogenated hydrocarbons, carbon dioxide, sulfur dioxide, ethylene oxide, ethanol, sulfonium, and phosphorus salts. The cage-like structures should show suitable shape and size so that the gust molecule can be trapped. The interaction between the host water molecule and the guest molecule is generally a weak Van der Waals force, but in some cases, it is an electrostatic interaction. In addition, macromolecular guests such as proteins, sugars, lipids, and other substances in biological cells can also form clathrate hydrates with water, reducing the freezing point of the hydrates.

The microcrystals of clathrate hydrates are very similar to the crystals of ice, but when large crystals are formed, the original tetrahedral structure gradually becomes a polyhedral structure, which differs greatly from the structure of ice in appearance. The clathrate hydrate crystals can maintain a stable crystal structure above 0 °C and under appropriate pressure. It has been evidenced that there are naturally similar clathrate hydrate structures in biological materials, and they are likely to affect the conformation, reactivity, and stability of biomacromolecules such as proteins. The clathrate hydrate crystals have not yet been developed and utilized and may have good application prospects in seawater desalination, solution concentration, and prevention for oxidation.

Hydrophobic interaction means that hydrophobic groups gather together as much as possible to reduce their contact with water molecules. This is a thermodynamically favorable ($\Delta G < 0$) process, which is a partial reversal of hydrophobic hydration (Fig. 2.9b). In most proteins, 40% of amino acids have nonpolar side chains, such as alanine methyl, phenylalanine phenyl, valine isopropyl, cysteine thiomethyl,

Table 2.3 Classification of types of water–solute interactions

Type	Example	Strength (in comparison with H_2O–H_2O Hydrogen Bond[1])
Dipole-ion	H_2O-free ion	Strong[2]
	H_2O-charged substituent on organic molecule	
Dipole–Dipole	H_2O-protein NH	Close or equal
	H_2O-protein CO	
	H_2O-protein side chain OH	
Hydrophobic hydration	$H_2O + R^{[3]} \rightarrow R$ (hydration)	Much less ($\triangle G > 0$)
Hydrophobic interaction	R (hyd) + R (hyd) $\rightarrow R_2$ (hyd) + H_2O	Incomparable[4] ($\triangle G < 0$)

[1]12–25 kJ/mol; [2]Far below the strength of a single covalent bond; [3]R is alkyl; [4]Hydrophobic interactions are entropy driven, while dipole–ion and dipole–dipole interactions are enthalpy driven.

the second butyl in isoleucine, and leucine isobutyl. Other compounds including alcohol, fatty acids, and nonpolar groups of free amino acids can also participate in the hydrophobic interaction, while the effect of hydrophobic interaction in these compounds is not as important as that in protein.

In summary, the water–solute interaction is very important, and the bonding forces existing between water and various kinds of solutes are summarized in Table 2.3.

2.3.4 Types of Water in Foods

According to the nature and degree of the interaction of water and non-aqueous substances, the water in the food can be categorized into bound water and bulk water.

Bound water, also named as fixed water, generally refers to the water present near a solute or other non-aqueous component that is chemically bonded to a solute molecule. It has special properties that are significantly different from bulk water in the same system, such as low fluidity, not freezing at -40 °C, not capable of dissolving any added solute, and broadening the width of the hydrogen spectrum in hydrogen nuclear magnetic resonance (HNMR) analysis. According to the strength of bonding, the bound water can be further classified into:

(1)   Chemically compound water, or constituent water, refers to the water that binds most strongly to the non-aqueous substance and forms as the whole with the non-aqueous substance, such as those located in the pores inside the protein molecule or as a chemical hydrate. They cannot be frozen at -40 °C, neither dissolve any added solute nor utilized by microorganisms. Chemically compound water only accounts for a small part of the water in food.

(2) Vicinal water refers to those in the first layer around the most hydrophilic group of non-water components, and the main binding force is the association between water–ion and water–dipole. The water associated with the ion or ionic group is the most tightly binding vicinal water, including monolayer water and those in microcapillaries (<0.1 μm diameter). It cannot be frozen at −40 °C, or dissolve any added solute.

(3) Multilayer water refers to the water located at the remaining position of the first layer mentioned above and some layers of water formed on the outer layer of monolayer water, mainly depending on the effect of hydrogen bonding between water and water or solute. Although the multilayer water is not as firmly bound as the vicinal water, it is still very tightly bound to the non-aqueous components, and its properties are not the same as those of pure water. That is, most of the multilayer water still does not freeze at −40 °C, even if it freezes, the freezing point is greatly reduced. It has solvent capacity but partially reduced.

Bulk water, or free water, refers to the water other than bound water in foods. It can be divided into the following three forms:

(1) Entrapped water is the water that is retained by the microscopic or submicroscopic structures or membranes in tissues. Entrapped water is unable to flow freely, so sometimes it is called immobilized water or intercepting water.

(2) Capillary water is trapped by capillary force in the intercellular space of biological tissues and food structure. It is also called intercellular water in biological tissues. Its physical and chemical properties are identical to entrapped water.

(3) Free flow water refers to the free moving water in the plasma, lymph, and urine of animals, the conduits of plants, and in the vacuoles of cells.

It is difficult to quantify the boundaries between bound water and bulk water, and they can only be qualitatively differentiated according to their physical and chemical properties (Table 2.4).

Therefore, the amount of bound water has a relatively fixed proportional relationship with the number of polar groups of organic macromolecules in foods. For example, on average 100 g of protein can hold as high as 50 g of water, and 100 g of starch, from 30 to 40 g of water. Bound water is regarded as an inherent part of food and plays an important role in the flavor of food, and the flavor and quality of the food changes when the bound water is forcibly separated. However, the separation is not easy as the combination is relatively tight. Since the vapor pressure of the bound water is much lower than bulk water, bound water is unable to be separated from food at a certain temperature (100 °C).

Bound water is also hard to freeze (freezing point below −40 °C), and nearly does not dissolve any solute, or incapable of being utilized by microorganisms, which differs from bulk water. Due to these facts, spores of plants and other microorganisms can maintain their vitality at very low temperatures, and the cellular structure of succulent tissues (fresh fruits, vegetables, meat, etc.) are often destroyed by ice crystals after freezing. As a result, the tissue collapses to varying degrees after thawing.

Table 2.4 Properties of water in foods

	Bound water	Bulk water
General description	The water exists near the solute or other non-aqueous component, including chemically compound water, vicinal water, and almost all multilayer water	The water that is away from non-aqueous components in position exist by water–water hydrogen bonding
Freezing point(compared with pure water)	Greatly decreased, does not freeze at −40 °C	Slightly decreased, while capable of freezing
Solvent power	None	None
Microbial utilization	Partially utilizable	Utilizable
Average molecular motion	Greatly decreased to none	Slightly
Enthalpy of vaporization(compared with pure water)	Increased	Basically unchanged
Total amount in high water content foods (%)	<0.03–3	Approximate 96

Note The quantitative determination of water content is based on the reduction of the weight of a sample under 105 °C heat treatment until the weight is constant

2.4 Water Activity

2.4.1 Definition of Water Activity

It has long been recognized that a close relationship exists between the water content of food and its perishability. However, the stability and safety of food are not directly related to the water content, but to the "state" of the water, or to the "availability" of the water in foods. It is evidenced that different types of foods have significant differences in their degree of spoilage and deterioration even if the water content is the same. The intensity between water and non-aqueous constituents is different, and the water that is engaged in strong associations is less likely to be able to support degradative activities such as the growth of microorganisms and hydrolytic chemical reactions. Therefore, the term "water activity" (A_w or a_w) was developed.

Water activity refers to the ratio between the vapor pressure of water in food and the saturated vapor pressure of pure water at the same temperature. It can be expressed by the following Eq. (2.1):

$$A_w = p/p_0 \tag{2.1}$$

where A_w is the water activity, p is the partial pressure of water vapor when a certain food reaches equilibrium in a closed container, and p_0 is the saturated vapor pressure of pure water at the same temperature. p/p_0 can also be referred to as relative vapor pressure. This is an easy formula for measuring. The following formula (2.2) can help to fully understand the meaning of water activity:

$$A_w = f/f_0 \tag{2.2}$$

This equation is a conceptual expression of water activity that is derived from the law of equilibrium thermodynamics, where f is the fugacity of water in the food (the fugacity is the tendency of the solvent to escape from the solution) and f_0 is the fugacity of the pure water under the same conditions. At low temperatures (e.g., room temperature), the difference between f/f_0 and p/p_0 is less than 1%, so the definition of A_w in terms of p/p_0 is clearly justifiable.

If we consider pure water as a food, the water vapor pressures p and p_0 are equal, so $A_w = 1$. However, foods contain not only water but also non-aqueous components. The vapor pressure of foods is lower than that of pure water, that is, p is always smaller than p_0, so $A_w < 1$.

Relative vapor pressure is related to percent equilibrium relative humidity (% ERH) of the product environment as follows (2.3):

$$p/p_0 = \mathrm{ERH}/100 = N = n_1/(n_1 + n_2) \tag{2.3}$$

It shows that the water activity of the food is numerically equal to the equilibrium relative humidity divided by 100, where N is the mole fraction of solvent (water) and n_1 and n_2 is the number of moles of solvent and solute, respectively. n_2 can be calculated by measuring the freezing point of the product according to the following formula (2.4):

$$n_2 = G\Delta T_f/(1000 \times K_f) \tag{2.4}$$

where G is the number of grams of solvent in sample; ΔT_f is the temperature decrease (°C) that drops from the freezing point; K_f is the constant of molar freezing point drop of water (1.86).

It must be emphasized that water activity is an intrinsic property of the sample. The equilibrium relative humidity is the atmospheric property that is in equilibrium with the sample, and they are only numerically equal. It takes a long time to balance the small sample (less than 1 g) with the environment, while for many samples at temperatures below 50 °C, it is almost impossible to balance the environment.

There are many methods for measuring the water activity of foods, such as cryoscopy method, diffusion method, equilibrium relative humidity method, and water activity meter measurement.

Last but not the least, it should be noted that water activity correlates sufficiently better with rates of food spoilage than other water content index, and it is still not a totally reliable predictor since other factors such as oxygen concentration, pH, water fluidity, and types of solute also have a strong influence on the rate of food deterioration.

2.4.2 Relationship Between Water Activity and Temperature

The temperature must be indicated when determining the water activity, as both p and p_0 are influenced by temperature in the expression of water activity, and thus the water activity also changes with temperature. The modified Clausius–Clapeyron equation accurately represents the relationship between water activity and temperature.

$$d(\ln A_w)\big/d(1/T) = -vH/R \text{ or } \ln A_w = -k\Delta H\big/[R(1/T)] \qquad (2.5)$$

where R is the gas constant; T is the thermodynamic temperature; ΔH is the net absorption heat of the water in the sample (the latent heat of vaporization of pure water); k is a function of the nature and concentration of the non-aqueous component in the sample, and is also a function of temperature. However, in some case where the sample is constant and the temperature variation range is narrow, k can be regarded as a constant, which can be expressed by the following Eq. (2.6):

$$k = \frac{\text{Asolute temperature of sample} - \text{Absolute temperature when the vapor pressure of pure water is p}}{\text{Absolute temperature when the vapor pressure of pure water is p}}$$

$$(2.6)$$

From Eq. 2.5, linear plots of $\ln A_w$ versus $1/T$ for native starch at various moisture contents are shown in Fig. 2.11. It is apparent that they showed a good linear relationship in a certain temperature range, and the dependence of A_w on temperature is a function of water content.

The $\ln A_w$—$1/T$ plot is not always a straight line if the temperature range is extended. When ice begins to form, a sharp break occurs in the plot at the freezing point of the sample, and the change rate of $\ln A_w$ with $1/T$ becomes significantly larger below the freezing point and is no longer affected by the non-water component in food (Fig. 2.12). As at subfreezing temperatures, the latent heat vaporization of

Fig. 2.11 Clausius–Clapeyron relationship of water activity of native starch and temperature. Water content values displayed after each line are expressed as g H$_2$O/g dry starch

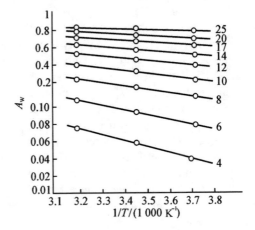

Fig. 2.12 Relationship of water activity and temperature for aqueous systems above and below freezing

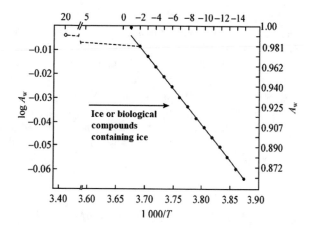

water should be replaced by the sublimation heat of ice, that is, the ΔH value in the Eq. (2.5) is greatly increased. For this reason, the water activity of the sample after frozen should be calculated according to the following equation:

$$A_w = p_{(ff)}/p_{0(scw)} = p_{0(ice)}/p_{0(scw)} \tag{2.7}$$

where $p_{(ff)}$ is the partial pressure of water in partially frozen food, $p_{0(scw)}$ is the vapor pressure of pure undercooled water, and $p_{0(ice)}$ is the vapor pressure of pure ice.

In Table 2.5, it has been illustrated below 0 °C, the vapor pressures of pure ice and supercooled water, and the calculated A_w values of the frozen foods at different temperatures.

Table 2.5 Vapor pressure and water activity of water, ice, and food at various temperatures below freezing point

Temperature/°C	Vapor pressure of liquid water[1]/kPa	Vapor pressure of ice[2] and ice-containing food/kPa	A_w
0	0.6104[2]	0.6104	1.00[4]
−5	0.4216[2]	0.4016	0.953
−10	0.2865[2]	0.2599	0.907
−15	0.1914[2]	0.1654	0.864
−20	0.1254[2]	0.1034	0.82
−25	0.0806[3]	0.0635	0.79
−30	0.0509[3]	0.0381	0.75
−40	0.0189[3]	0.0129	0.68
−50	0.0064[3]	0.0039	0.62

[1]supercooled water; [2]observed data; [3]calculated data; [4]only suitable for pure water.

At subfreezing temperatures, A_w becomes independent of sample composition, and depends solely on temperature as in the presence of the ice phase, A_w values are not influenced by the kind or ratio of solutes present. Consequently, in the subfreezing event, physical, chemical, and biochemical changes that is influenced by solute presence cannot be accurately forecasted based on A_w. Meanwhile, the significance of A_w is different in terms of food stability at the temperature above and below the freezing point. For example, in a food with an A_w of 0.86 at $-15\ ^\circ C$, the microorganisms cannot grow and multiply, and the chemical reaction proceeds slowly. But at 20 $^\circ C$, some microorganisms start to grow and the chemical reaction can be carried out relatively quickly in the food under the same A_w value.

2.4.3 Relationship Between Water Activity and Moisture Content

2.4.3.1 Moisture Sorption Isotherms of Water

At a constant temperature, a plot of water content (expressed as a mass of water per unit mass of dry material) of food versus its water activity is known as a moisture sorption isotherm (MSI). Shown in Fig. 2.13 is a schematic MSI for a high moisture food plotted to include the full range of moisture content from normal to dry. However, the data of those in the low moisture region are not detailed depicted. The omission of the high moisture region and expansion of the low moisture region, as is normal practice, yields an MSI that is much more useful (Fig. 2.14). Moisture sorption isotherms exhibit a variety of shapes as shown in Fig. 2.15, which is determined by many factors, including ingredients, physical structure, pretreatment of food, temperature, and method of plotting isotherms.

Fig. 2.13 Schematic MSI encompassing a broad range of moisture contents

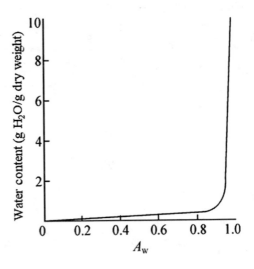

Water activity is temperature dependent, and thus MSIs must also exhibit temperature dependence (Fig. 2.16). At a certain moisture content, the water activity increases with increasing temperature, frequently in conformity with the Clausius–Clapeyron equation and the laws of various changes in food.

To understand the meaning and practical application of MSI, it can be divided into three zones for discussion (Fig. 2.14 and Table 2.6).

Zone I: Water in this zone is strongly associated with the non-aqueous components in foods by the water–ion or water–dipole interaction so the A_w is the lowest, generally between 0 and 0.25. This part of water cannot be used as a solvent, most of which remains unfrozen at $-40\ ^{\circ}C$, has no significant plasticizing effect on food solid, and can be simply regarded as part of the food solid. The water at the high moisture end of Zone I (junction of zones I and II) corresponds approximately to the monolayer

Fig. 2.14 Generalized MSI for the low moisture segment of food (20 °C)

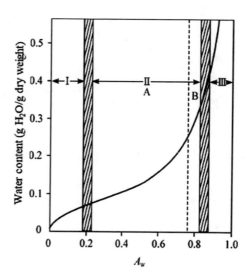

Table 2.6 Characteristics of water in different zones on MSI

	Zone I	Zone II	Zone III
A_W	0–0.25	0.25–0.85	>0.85
Moisture content%	0–7	7–27.5	>27.5
Freezability	Unfreezable	Unfreezable	Normal
Solvent power	None	Slight ~ Moderate	Normal
Water state	Monomolecular water layer adsorption	Multimolecular water layer condensation	Capillary water
			Free mobile water
	Chemical association water	Physical adsorption	
Microbial availability	Unavailable	Partially available	Available

Fig. 2.15 Resorption isotherms for various foods and biological substances (Temperature 20 °C except for number 1, which is 40 °C) 1. confection (major component powdered sucrose); 2. spray-dried chicory extract; 3. roasted coffee; 4. pig pancreas extract powder; and 5. native rice starch

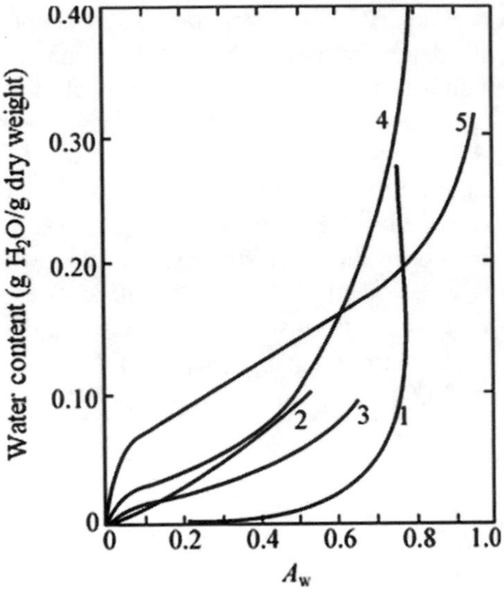

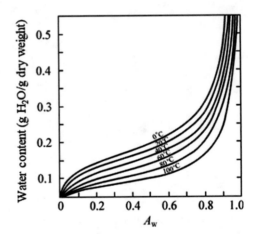

Fig. 2.16 Moisture sorption isotherms for potatoes at various temperatures

water content in food, which can be regarded as an approximate amount of water required to form a monolayer around accessible strong polarity groups of dry matter.

Zone II: Water in zone II occupies the remaining positions of the first layer of the surface of the non-aqueous component and the other layers around the hydrophilic group (such as amino group and hydroxyl group), including multilayer water, entrapped water, and water in a capillary with diameter <1 μm. Water activity in this zone is between 0.25 and 0.85. As the water activity increases, the dissolution is initiated and the reactants are free to mobile, and the rate of most chemical reactions also increases. Meanwhile, it exerts a significant plasticizing action and causes incipient swelling of the solid matrix.

Zone III: This part of the water is the bulk water, which shows the weakest association and best fluidity. Water activity in this zone is between 0.8 to 0.99. Bulk water is also physically trapped by food so that macroscopic flow is impeded. However, in all other aspects, this water has properties like those of water in a dilute salt solution. The enthalpy of vaporization of the water in this zone is basically the same as that of pure water. This water can be frozen, used as a solvent, and facilitates the progress of chemical reactions and the growth of microorganisms.

Although the MSI is divided into three intervals, the position of the boundary line of each zone cannot be accurately determined, and the water can be exchanged within a zone or between zones except chemically compound water. In addition, when water is added to the dry food, the property of the original water in it can be slightly changed, such as the swelling and dissolution process. However, when adding water in zone II, the water in zone I remains almost unchanged. Similarly, the addition of water in zone III also leaves the properties of zone II water almost unchanged (Fig. 2.14). Thus, the water that is most weakly bound in food plays an important role in the stability of food.

Therefore, the study of MSI is very meaningful for understanding: (1) the relationship between sample dehydration and relative vapor pressure during concentration and drying; (2) how to combine food to prevent moisture transfer between ingredients; (3) determining the moisture resistance; (4) predicting the water activity at which the growth of microorganisms can be inhibited; (5) predicting the relationship between chemical and physical stability and water content of the product; (6) determining the association strength between water and non-aqueous components in different foods.

2.4.3.2 Hysteresis of Moisture Sorption Isotherm (MSI)

An MSI prepared by the addition of water (resorption) to a dry sample will not necessarily be superimposable on an isotherm prepared by desorption. This lack of superimposability is referred to as "hysteresis" and a schematic example is shown in Fig. 2.17. The MSI of many foods shows hysteresis. The magnitude of hysteresis, the shape of the curves, and the inception and termination points of the hysteresis loop can vary considerably depending on factors such as the nature of the food, the physical changes it undergoes when water is removed or added, temperature, the rate of desorption, and the degree of water removal during desorption.

Generally, when the A_w value is constant, the moisture content of the food during desorption is greater than that in resorption. Therefore, the food obtained by desorption must maintain a lower A_w to be stable compared with the same A_w level of food obtained by resorption, and the food obtained by resorption is relatively high in cost. In practice, the MSI can be applied to the observation and study of moisture absorption products, and the moisture desorption isotherm can be used to research the drying process.

The cause of sorption hysteresis can be analyzed from many aspects. For instance, some water-absorbent parts interact with non-aqueous components and thus cannot

Fig. 2.17 Hysteresis of MSI

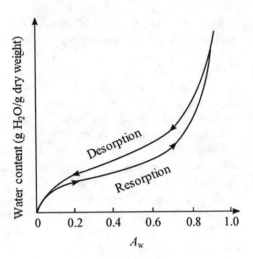

release water during the process of food desorption. A series of capillary phenomena caused by the irregular shape of food needs different vapor pressure to fill or evacuate water. That is, $p_{inside} > p_{outside}$ permits evacuation, and $p_{outside} > p_{inside}$ allows filling. The food tissue will be changed during desorption, and the water cannot be tightly bound when reabsorbed, resulting in high A_w values. While a comprehensive and definitive explanation has yet to be formulated.

2.4.4 Water Activity and Food Stability

It has often been demonstrated that food stability and water activity are closely related in many situations. Data in Fig. 2.18 are typical relationships between reaction rate and A_w in the temperature range of 25–45 °C, which show that the reaction rates and the positions and shapes of the curves can be altered by sample composition, physical state and structure (capillarity) of the sample, the composition of the atmosphere (especially oxygen), temperature, and by hysteresis effects.

2.4.4.1 Relationship Between A_w and Growth of Microorganisms

The growth and reproduction of microorganisms in food are related to the moisture content that can be utilized, that is, the A_w of the food determines the reproduction, metabolism (including toxin production), resistance, and survival of the microorganism in the food. Different microorganisms require different A_w when they are propagated in food. Generally, bacteria are most sensitive to low A_w, yeast is the second, and mold is the least sensitive. As shown in Table 2.7, microorganisms

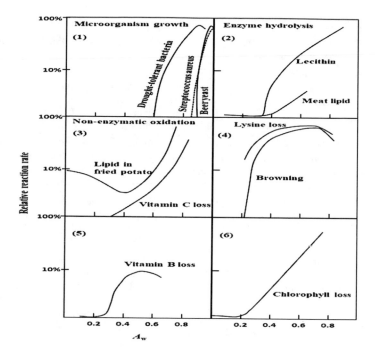

Fig. 2.18 Relationships between reaction rate and A_w

cannot grow when A_w is lower than the minimum A_w threshold required for growth. When A_w is below 0.6, the growth of most microorganisms is inhibited.

Decreasing the A_w value can inhibit the growth of microorganisms, thereby reducing the rate of food spoilage, food toxins, and microbial metabolic activity. However, the A_w value required to suspend different metabolic processes is different. For example, the A_w required for bacterial spore formation is higher than that of their growth. The A_w threshold for Bacillus welchii reproduction is 0.96, and the optimum A_w for sporulation is 0.993. There is almost no spore formation when the Aw is slightly lower than 0.97. The production of toxins is the most important microbial metabolic activity related to human health. Generally, the Aw required for the growth of toxin-producing molds is lower than that for the toxin formation. Thus, in some foods where the microbial growth is controlled by adjusting A_w, possibly microorganism grows but does not produce toxins. For example, the A_w threshold required for the growth of Aspergillus flavus is 0.78–0.8, and that for toxin production is required to be 0.83.

The requirement for water by microorganisms is also affected by other coexistence factors such as pH, nutrients, and oxygen. By selecting suitable conditions (A_w, pH, humidity, preservative, etc.), microorganisms can be inhibited or killed, thereby improving food stability and safety. A_w is surely not only related to the harmful microorganisms which cause food spoilage, but also to some beneficial microorganisms required for fermented foods. In the processing of fermented foods, it is

Table 2.7 Relationship between A_w and the growth of microorganisms in foods

Range of A_w	Microorganisms generally inhibited by lowest A_w of the range	Foods Generally within this Range of A_w
1.00 ~ 0.95	*Pseudomonas, Escherichia Proteus, Shigella, Klebsiella, Bacillus, Clostridium perfringens, some yeasts*	Highly perishable (fresh) foods, canned fruits, vegetables, meat, fish, and milk; cooked sausages and breads; foods containing up to 7% (w/w) of sodium chloride or 40% of sucrose
0.95 ~ 0.91	*Salmonella, Vibrio parahaemolyticus, C. Botulinum, Serratia, Lactobacillus, some molds, yeasts (Rhodotorula, Pichia)*	Some cheeses, cured meats, some fruit juice concentrates, foods containing up to 15% (w/w) of sodium chloride or saturated (65%) sucrose
0.91 ~ 0.87	*Many yeasts (Candida, Torulopsis, Hansenula, Micrococcus)*	Fermented sausages (salami), sponge cakes, dry cheeses, margarine, foods containing up to 15% (w/w) of sodium chloride or saturated (65%) sucrose
0.87 ~ 0.80	*Most molds (mycotoxigenic penicillia), Staphylococcus aureus, most Saccharomyces (bailii) spp., Debaryomyces*	Most fruit juice concentrates, sweetened condensed milk, chocolate syrup, maple and fruit syrups; flour, rice, pulses with 15–17% of moisture content; fruit cake; country style ham, fondants
0.80 ~ 0.75	*Most halophilic bacteria, mycotoxigenic aspergilli*	Jam, marmalade, marzipan, glace fruits, some marshmallows
0.75 ~ 0.65	*Xerophilic molds, Saccharomyces bisporus*	Rolled oats of 10% moisture content; grained nougats, fudge, marshmallows, jelly, molasses, raw cane sugar, some dried fruits, nuts
0.65 ~ 0.60	*Osmophilic yeasts, few molds*	Dried fruits with 15–20% of moisture content, toffees and caramels, honey
0.50	No microbial proliferation	Pasta with 12% of moisture content, spices with 10% of moisture content
0.40	No microbial proliferation	Whole egg powder with 5% of moisture content
0.30	No microbial proliferation	Cookie, crackers, bread crusts, and so forth with 3–5% of moisture content
0.20	No microbial proliferation	Whole milk powder with 2–3% of moisture content; dried vegetables with 5% of moisture content; corn flakes with 5% of moisture content, country style cookies, crackers

necessary to increase A_w above the value required for beneficial microbial growth, reproduction, and metabolism.

2.4.4.2 Relationship Between A_w and Chemical Reaction that Caused Food Deterioration

Like the inhibition of microorganisms, low A_w can inhibit chemical changes of foods and stabilize their quality. Most chemical reactions must be carried out in an aqueous solution. Many chemical reactions in the food are inhibited with decreased A_w. For example, many chemical reactions are ionic reactions that occur under the condition where the reactants must first be ionized or hydrated, which require enough bulk water to proceed.

Water is not only a medium for chemical reactions, but also is the reactant of many others (such as hydrolysis). If A_w is decreased, the amount of bulk water participating in the reaction is reduced, and the rate of the chemical reaction is slowed down. In many enzymatic reactions, in addition to being a reactant, water acts as a transport medium for the diffusion of substrates to enzymes and promotes the activation of enzymes and substrates by hydration. The activity of most enzymes is inhibited when the A_w is lower than 0.8. If the A_w value falls within the range of 0.25–0.30, the amylase, polyphenol oxidase, and peroxidase in the food are strongly inhibited or lose their activity, while lipase remains active at A_w 0.05–0.1.

In general, low A_w inhibits chemical changes in foods, but there are exceptions. The relationship between lipid oxidation and A_w is shown in Fig. 2.18(3). Obviously, extremely high or low A_w will accelerate lipid oxidation. Only when the A_w is close to the boundary of the isotherm zones I and II, the oxidation rate of the lipid is the lowest. The reasonable explanation regarding this behavior is that the first water added to a very dry sample (zone I) is believed to bind hydroperoxides, interfering with their decomposition, and thereby hindering the progress of oxidation. In addition, this water hydrates metal ions that catalyze oxidation, apparently reducing their effectiveness. The addition of water beyond the boundary of zones I and II results in increased rates of oxidation by increasing the solubility of oxygen and allowing macromolecules to swell, thereby exposing more catalytic sites. At still greater A_w (>0.8), the added water may retard rates of oxidation, which is due to the dilution of catalysts that reduces their effectiveness, and the decreased concentration of the reactant.

For other examples, the relationships between the Maillard reaction, vitamin B_1 degradation, and A_w are shown in Fig. 2.18(4) and (5), respectively. This type of water-mediated reaction generally does not initiate when the A_w is lower than the isotherm boundary of zones I and II. At medium to high A_w, it shows the maximum conversion rates. While in the food with medium to high moisture content, the reaction rate decreases sometimes when A_w continues to increase. The possible reasons are: (1) For those reactions in which water is a product, an increase in water content can result in product inhibition; (2) When the water content of the sample is such that the solubility, accessibility (surfaces of macromolecules), and mobility of rate-enhancing constituents are no longer the limiting factor, further addition of water serves only to dilute rate-enhancing constituents and thus decrease the reaction rate.

Generally, reducing the A_w of food retards the progress of both enzymatic and non-enzymatic browning, reducing nutrient destruction and preventing the decomposition of water-soluble pigments. However, the extremely low A_w can accelerate lipid oxidation and rancidity. To maintain the moisture content necessary for the highest stability of the food, it is preferred to keep A_w in the range of bound water. In this way, chemical changes can be difficult to occur without losing the water absorption and recovery properties of the food.

It is suggested from the above examples that the maximum reaction rate of food chemical reactions generally occurs with moderate water content (A_w is from 0.7 to 0.9), and the minimum reaction rate generally first appears between the zones I and II of the isotherm. The reaction rate is kept at a minimum when the A_w is further decreased except for the lipid oxidation. The water content corresponds to monolayer water in this situation. Therefore, the monolayer value of a food frequently provides a good first estimate of the water content providing maximum stability of a dry product, and knowledge of this value is of considerable practical importance. Based on the MSI data, it could be used the Eq. (2.9) developed by Brunauer et al. to compute the monolayer value:

$$\frac{A_W}{M(1 - A_W)} = \frac{1}{m_1 C} + \frac{C - 1}{m_1 C} \times A_W \tag{2.9}$$

where A_w is water activity, M is moisture content (g H_2O/g dry weight), m_1 is the monolayer value (g H_2O/g dry weight), and C is a constant.

From this equation, it is apparent that a plot of $A_w/[M(1 - A_w)]$, known as a BET plot, should yield a straight line. An example, A BET plot for native potato starch is shown in Fig. 2.19. The linear relationship is well fit within 0–0.35 of A_w value.

The BET monolayer value can be calculated as follows:

$$\text{Monolayer value } (m_1) = 1/((y \text{ intercept}) + (\text{slope}))$$

From Fig. 2.19, the y intercept is 0.6, the slope is 10.7, thus

Fig. 2.19 BET plot for native potato starch (resorption temperature, 20 °C)

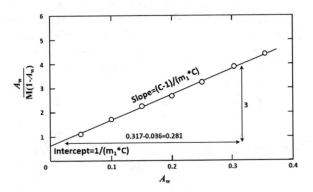

$$m_1 = 1/(0.6 + 10.7) = 0.88 \text{ g H}_2\text{O/g dry matter.}$$

In this particular example, the BET monolayer value corresponds to a A_w of 0.2.

In addition to chemical reactions and microbial growth, A_w also influences the texture of dry and semidry foods. For example, suitably low A_w is necessary to maintain crispness of crackers, popped corn, and potato chips, to avoid caking of granulated sugar, dry milk, and instant coffee, and to prevent stickiness of hard candy, etc. The reasonable A_w for dry materials without incurring a loss of desirable properties ranges from 0.35 to 0.5, depending on the product. Furthermore, suitably high A_w values of soft-textured foods are needed to avoid undesirable hardness.

2.5 Molecular Mobility

2.5.1 Definition of Molecular Mobility

In addition to water activity, molecular mobility (*Mm*) is also an important indicator for predicting food stability. Mm is the total measurement of the rotational and translational motion excluding the vibration of a molecule. For metastable or non-equilibrium states, dynamic methods are more suitable than thermodynamic methods for understanding, predicting, and controlling their properties. Mm is causatively related to the diffusion-limited properties of foods, therefore is considered as a suitable kinetic method for this purpose.

Complex food systems, like other biomacromolecules (polymers), often contain amorphous regions in their structure, such as gelatin, elastin, amylopectin, amylose, and sucrose, which can exist in an amorphous state. Amorphous refers to a non-equilibrium, non-crystalline state in which a substance is present. When the saturation condition prevails and the solute remains non-crystalline, the solid formed is amorphous. The amorphous regain exists in a metastable equilibrium or non-equilibrium state. Although its stability is not high, it exhibits excellent food quality properties. Therefore, the major issue of food processing is to find a balance between food quality and stability, that is, to maintain food quality simultaneously keeping the food in a relatively stable non-equilibrium state.

The analysis of the food state clearly shows that the *Mm* is zero when the substance is in a complete crystalline state, and the *Mm* value is almost zero when the substance is in a completely glassy state (amorphous). Under other circumstances, the *Mm* value is greater than zero. The main factors that determine the *Mm* value are water and the non-aqueous component that is dominant in food. It should be emphasized that there are two important states in the amorphous state; glassy state and rubbery state. The former refers to that it has a certain shape and volume like a solid, while the arrangement of molecules like a liquid is only approximately ordered, and the latter, the state when a macromolecular polymer is transformed into a soft and elastic solid. The transformation temperature in food from a glassy state to a rubbery state is called glass transition temperature (Tg). Water molecules are small in volume, liquid at room temperature, and low in viscosity. Therefore, when the temperature of the food system is at Tg, the water molecules can still rotate and move; while as the main components of food, macromolecular polymers, such as proteins and carbohydrates, are not only the determinants of food quality, but also affect the viscosity and diffusion properties of food, so they also determine the molecular mobility of food. Usually, the Mm value of most foods is not equal to zero.

2.5.2   Molecular Mobility and Food Stability

It is appropriate to use molecular mobility to predict the chemical reaction rate of food systems which includes enzyme-catalyzed reactions, protein folding changes, proton transfer changes, and free radical binding reactions. According to the theory of chemical reaction, the rate of a chemical reaction is controlled by three factors: the diffusion coefficient (factor) D (for a reaction to occur, first the reactants must be able to contact each other), the collision frequency factor A (the number of collisions per unit time), and the activation energy Ea of the reaction (the effective energy must exceed the activation energy to cause the reaction when two properly oriented reactants collide). If D is more restrictive to the reaction than A and Ea, then the reaction is a diffusion-limited reaction; in addition, under general conditions, the reaction that natively is not a diffusion-limited reaction may also transform into a diffusion-limited one when the water activity or system temperature is lowered. This is due to that a decrease in moisture leads to an increase in the viscosity of the food system or a decrease in temperature reduces the mobility of molecules. Therefore, it is useful to use molecular mobility to predict the rate of diffusion-limited reactions in food. Such foods include starchy foods (such as dough, candies, and snacks), protein-based foods, medium moisture foods, and dried or freeze-dried foods. It also can be used to predict the quality of the food during processing such as crystallization, baking, extrusion, drying, and freezing. The examples of food properties and changes controlled by Mm are shown in Table 2.8. However, for those reactions and changes that are not diffusion limited, it is not appropriate to apply molecular mobility, such as the growth of microorganisms.

When in the frozen state, the physical, chemical, and biochemical reactions of food cannot be explained by A_w due to the complex changes. During the freezing process, as the ice crystals continue to precipitate, the concentration of the solute in the unfrozen phase increases continuously, and the freezing point gradually decreases until the non-aqueous component of the food begins to crystallize (the temperature at this time may be referred to as the eutectic temperature T_E). After the formation of the eutectic, the frozen concentration is terminated. Since the composition of most foods is quite complicated, its T_E is lower than the initial ice crystallization temperature, so its unfrozen phase can maintain a supersaturated state of viscous liquid for a long time with the decrease of temperature where the viscosity does not increase significantly, referring as a rubbery state. At this moment, physical, chemical, and biochemical

Table 2.8 Food properties and characteristics associated with Mm

Dry or semi-dried food	Frozen food
Flow property and viscosity	Moisture migration (crystallization of ice)
Crystallization and recrystallization	Crystallization of lactose (sand crystals in frozen confections)
Chocolate frosting	Enzyme activity per
Food cracking during drying	sists during freezing, sometimes with appar ent improvement
Texture of dry and medium moisture foods	The collapse of the amorphous structure occurs during the first stage of freeze-drying
Food structure collapse during freeze-drying	Volume shrinkage of food (partial collapse of the foamy structure in the frozen dessert)
Escaping of volatile substances by encapsulations	
Activity of enzyme	
Maillard reaction	
Gelatinization of starch	
The change of baked foods caused by retrogradation	
Fragmentation of baked foods during cooling	
Heat killing of microbial spores	

reactions persist and cause food spoilage. The viscosity of the high-concentration solute in the unfrozen phase begins to increase significantly with the temperature further decreasing and limits the molecular movement of the solute nucleus as well as the diffusion of the moisture.

As mentioned before, the temperature where the food system transforms from a glassy state to a rubber state is corresponding to Tg, which is greatly influenced by the moisture content and solute type in food. For every 1% increase in water, Tg decreases by 5 to 10 °C. For example, when the moisture content of freeze-dried strawberries is 0%, the Tg is 60 °C; while the moisture content increases to 3, 10, and 30%, Tg has dropped to 0 °C, −25 °C, and −65 °C, respectively. Tg is proportionally increased with the increase of molecular weight of the solute when its molecular weight is not more than 3000 D and Tg no longer depends on molecular weight beyond 3000 D. For different types of starch, the shorter the side chain of the amylopectin, and the greater the number, the lower the Tg. For example, when wheat amylopectin is compared with rice amylopectin, the side chain of wheat amylopectin is more and shorter. Therefore, in similar moisture content, its Tg is also smaller than that of rice starch.

The unfrozen water in a glass state is not bonded by hydrogen bond, and the molecules are trapped in a glass state having a high viscosity caused by a high solute viscosity, and this water is not reactive. The entire food system exists as a

Table 2.9 Tg determination method

Determination method	Nature
Thermal difference method (DTA); Differential scanning calorimetry (DSC)	Change of thermodynamic properties
Thermal expansion method; Refractive index method	Volume change
Dynamic mechanical analysis (DMA); Dynamic mechanical thermal analysis (DMTA)	Change of mechanical properties
Nuclear magnetic resonance (NMR); Relaxation map analysis (MA)	Electromagnetic effect

non-reactive, amorphous solid. Therefore, at Tg, all changes due to diffusion limitations (including many metamorphic reactions) are severely limited, and food is highly stable. The preservation temperatures of many foods are higher than Tg, and Mm is relatively high, which causes instability. Therefore, the stability of the low-temperature frozen food can be determined by the difference between the Tg of the food and the storage temperature $T(T - Tg)$, and the greater the difference, the worse the stability of the food.

When discussing the relationship between Mm and food properties, the following exceptions should be noted: (1) the conversion rate is not a chemical reaction that is significantly affected by diffusion; (2) the desirable properties of foods can be achieved by specific chemical actions (such as adjusting pH or partial pressure of oxygen); (3) the Mm of sample estimated here is based on the polymer component (Tg of the polymer), while the key factors determining the important properties of the product are the small molecules infiltrating into the polymers; (4) the cell growth of nutritional microorganisms as A_w is more reliable to estimate compared with Mm at this situation.

Determination of Tg is a key point in controlling food quality and stability. There are many methods for determining Tg of foods, as shown in Table 2.9. The DSC and the NMR method are more frequently used. DSC and NMR are analytical techniques and detection instruments based on the principles of calorimetry and molecular motion, respectively. Both have advantages and disadvantages when studying the glass transition of foods. Since the Tg value is highly dependent on the conditions and the method used, so it is generally possible to carry out the research using different methods. The Tg of complex systems is difficult to measure, and only the Tg of a simple system can be easily determined.

2.5.3 Comparison of A_w and Mm in Predicting Food Stability

The A_w and Mm methods are two complementary methods for studying food stability. The A_w method mainly focuses on the effectiveness (availability) of water in foods, such as the ability of water as a solvent; the Mm method is mainly to study the microviscosity of foods and the diffusion ability of chemical components and is related to many food properties that are limited by diffusion. The Mm method is significantly more effective in estimating the properties limited by diffusion, such as the physical properties of frozen foods, the optimal conditions for freeze-drying, and physical changes including crystallization, gelation, and starch aging. The A_w value is noneffective in predicting the physical and chemical properties of frozen foods. The Mm method and A_w method have substantially the same effects when estimating physical changes such as aggregation, adhesion, and brittleness caused by food preservation near room temperature. The Mm method is significantly less practical and unreliable in estimating the microbial growth and non-diffusion-limited chemical reaction rates (such as high activation energy reactions and reactions in lower viscosity media), while in these cases, the A_w method is more effective.

The Mm method cannot achieve or exceed the level of the A_w method in practicality until the quick, correct, and economical technique of determining the Mm and Tg of the food is well established. However, the Tg of the food system is a new idea and a new method for predicting the stability of foods. The research in the future will focus on how to combine the important critical parameters such as Tg, moisture content, and water activity with the existing technical method, and apply it to the optimization of processing and storage processes of various foods.

2.6 Water Transfer and Food Stability

The water transfer in food can be divided into two stations: (1) the position transfer of water in different parts within the same food or between different foods, resulting in the change of the original water distribution; (2) the phase transfer of water in food, in particular, the transfer of gas phase and liquid phase water leads to a change in the moisture content of the food, which has a great influence on the storage, processability, and commercial value of the food.

2.6.1 Position Transfer of Water in Food

According to the thermodynamics laws, the chemical potential (μ) of water in food can be expressed by the formula (2.12):

$$\mu = \mu^0(T, p) + RT \ln A_W \tag{2.12}$$

where $\mu^0 (T, p)$ is the chemical potential of pure water at a certain temperature (T) and pressure (p), and R is a gas constant. It can be seen from the above formula that if the temperature (T) or water activity (A_w) of different foods or different parts within the same food is different, the chemical potential of the water will be different, and the water will move along the chemical potential gradient, resulting in water transfer in food. In theory, the position transfer of water in food will not be terminated until the chemical potential of the water in all parts of the food is completely equal, that is, the thermodynamic equilibrium is reached.

The water transfer caused by the temperature difference is that the moisture in food moves from the high-temperature region along the chemical potential, and finally enters the low-temperature region. This process can occur in the same food or in different foods. In the former case, moisture only moves within the food, and in the latter case, moisture must pass through the air medium, which is a slow process.

The moisture transfer caused by the difference in A_w is that water is automatically transferred from a region where A_w is high to a region where A_w is low. If a cake with a high A_w is placed in the same environment as a biscuit with a low A_w, the moisture in the cake is gradually transferred to the biscuit, so that the quality of both is affected to varying degrees.

2.6.2   Phase Transfer of Water in Food

As mentioned above, the water content of food refers to the equilibrium of moisture content in food under certain external conditions such as temperature and humidity. If the external conditions change, the moisture content of the food also changes. The change of air humidity may cause phase transfer of food moisture, and the pattern of air humidity change is closely related to the direction and strength of water phase transfer in food.

The main forms of water phase transfer in food are water evaporation and condensing.

2.6.2.1   Water Evaporation

The phenomenon in which the water in the food is lost by the transformation from the liquid phase to the gas phase is called water evaporation. A dry or semi-dried food with a low water activity can be obtained by drying or concentrating process based on water evaporation. However, for fresh fruits, vegetables, meat, poultry, fish, shellfish, and many other foods, water evaporation will cause adverse effects on food quality, such as the shrink of appearance, the change in original freshness and brittleness, and in serious cases, the lost for commercial values. Also due to the evaporation of water that promotes the activity of hydrolase in food and thus hydrolyzes high molecular substances, food quality deteriorates with a shortened shelf life.

From a thermodynamic point of view, the evaporation process of food moisture is a process of transfer-balance of water vapor formed by aqueous solution in food and water vapor in the air. Although the temperature does not differ a lot in the food from the ambient, the water vapor pressure of the food is not necessarily the same as that in the ambient, as the chemical potential of the moisture between the two phases is different. Their difference can be expressed by the following formula (2.13):

$$v\mu = \mu_F - \mu_E = R(T_F \ln p_F - T_E \ln p_E) \tag{2.13}$$

where μ is the chemical potential of water vapor, p is the water vapor pressure, T is the temperature, and the subscripts F and E represent the food and the environment, respectively. Therefore, $\Delta\mu$ is the difference between the chemical potential of water vapor in the food and water vapor in the air. Based on this, the following conclusions can be drawn:

(1) If $\Delta\mu > 0$, the transfer of water vapor in the food to the outside is a spontaneous process. In this situation, if the water vapor pressure above the aqueous food solution drops, the equilibrium state between the original food aqueous solution and the water vapor above is destroyed. To reach a new equilibrium state, some of the water in the aqueous solution in the food evaporates until $\Delta\mu = 0$. For open, unpackaged foods, especially when the relative humidity of the air is low, $\Delta\mu$ is difficult to be zero, so the evaporation of moisture in the food is continuously carried out, and the quality of the food is seriously damaged.

(2) If $\Delta\mu = 0$, the water vapor of the aqueous food solution is in a state of dynamic equilibrium with the water vapor in the air. From the net result, food does not evaporate water or absorb water, which is an ideal environment during the shelf life of food.

(3) If $\Delta\mu < 0$, the transfer of water vapor in the air to the food is a spontaneous process. In this situation, the moisture in the food does not evaporate, but absorbs water vapor in the air and becomes damp, which affect the stability of the food (A_w is increased).

The evaporation of water is mainly related to the humidity of the air and the saturation humidity difference. The saturation humidity difference refers to the difference between the saturated humidity of the air and the absolute humidity in the air at the same temperature. If the saturation humidity difference is greater, the amount of water vapor that can be contained in the air to reach saturation is greater, so that the evaporation of food moisture is large; conversely, the amount of evaporation is small.

The factors affecting the saturation humidity difference are mainly air temperature, absolute humidity, and flow rate. The saturated humidity of the air changes with temperature change, that is, the saturated humidity of the air increases when the temperature rises. When the relative humidity is constant, the temperature rises, the saturation humidity difference becomes large, and the evaporation amount of the food moisture increases. When the absolute humidity is constant, if the temperature rises, the saturated humidity increases, so the saturation humidity difference also increases, and the relative humidity decreases. Similarly, the evaporation of food moisture increases. If the temperature does not change and the absolute humidity changes, the saturation humidity difference also changes. If the absolute humidity increases and the temperature does not change, the relative humidity also increases. The saturation humidity difference decreases, and the moisture evaporation of the food decreases. The flow of air can remove more water vapor from the air surrounding the food, thereby reducing the water vapor pressure of this part in the air and increasing the saturation humidity difference, thereby accelerating the evaporation of food moisture and drying the surface of the food.

2.6.2.2 Condensation of Water Vapor

The phenomenon that water vapor in the air condenses into liquid water on the surface of the food is called condensation of water vapor. The maximum amount of water vapor per unit volume of air can be reduced as the temperature decreases. When the temperature of the air drops by a certain value, it is possible to supersaturate the original saturated or unsaturated air. In the state, when it encounters the surface of the food or the food packaging container, etc., the water vapor may condense into liquid water. If the food is a hydrophilic substance, the water vapor is coagulated, spread out, and fused with it, such as cakes and candies, which are easily wetted by the condensed water and finally adsorbs the water; if the food is a hydrophobic substance, the water vapor shrinks into small water droplets after condensation. For example, the surface of the egg and the wax paper layer on the surface of the fruit are hydrophobic substances. When the water vapor condenses on it, it cannot expand and only shrink into small water droplets.

2.7 Summary

Water is the most abundant ingredient in the food. It is extremely important to the nature of food. It causes food corruption. It is the factor that determines the rate of various chemical reactions in food. It is also a factor that causes adverse reactions during freezing. The presence of moisture in food directly affects the properties of the food. Moisture content, water activity, glass transition temperature, and molecular mobility are all important parameters for predicting food storage stabilitsy.

Questions:

1. How to theoretically explain the unique physical and chemical properties of water?
2. In what way do ions, hydrophilic substances, and hydrophobic substances in food interact with water?
3. What are the forms of water in food? What are their characteristics?
4. What are the relationships and differences between moisture content and water activity?
5. The isothermal adsorption curves of different substances are different. What are the factors that affect the shape of the curve?
6. How to explain the food storage process with the function of Tg?

7. What is Mm, and what is the relationship between Mm and food stability?
8. Compare the advantages and disadvantages of water activity and molecular mobility in predicting food stability?
9. Explain the relationship between A_w in food and lipid oxidation?
10. Explain the following terms: Bound or fixed water; Chemically compound water or constituent water; Vicinal water; Multilayer water; Hydrophobic interaction; Hydrophobic hydration; Clathrate hydrate; Water activity; Moisture sorption isotherm; Hysteresis of moisture sorption isotherm; Glass transition temperature; Molecular mobility.

Bibliography

1. Belitz, H.D., Grosch, W., Schieberle, P.: Food Chemistry, Springer, Berlin, Heidelberg (2009)
2. Beuchat, L.R., Komitopoulou, E., Beckers, H., Betts, R.P., Bourdichon, F., Fanning, S., Joosten, H.M., Ter Kuile, B.H.: Low-water activity foods: increased concern as vehicles of foodborne pathogens. J. Food Prot. **76**(1), 150–172 (2013)
3. Cheung, P.C.K., Mehta, B.M.: Handbook of Food Chemistry, Springer, Berlin, Heidelberg (2015)
4. Damodaran, S., Parkin, K.L., Fennema, O.R.: Fennema's Food Chemistry. CRC Press/Taylor & Francis, Pieter Walstra (2008)
5. Disalvo, E.A., Hollmann, A., Martini, M.F.: Hydration in lipid monolayers: correlation of water activity and surface pressure. Subcell. Biochem. **71**, 213–231 (2015)
6. Kan, J.: Food Chemistry. China Agricultural University Press, Beijing (2016)
7. Mannaa, M., Kim, K.D.: Influence of temperature and water activity on deleterious fungi and mycotoxin production during grain storage. Mycobiology **45**(4), 240–254 (2017)
8. Schiraldi, A., Fessas, D., Signorelli, M.: Water activity in biological systems—a review. Pol. J. Food Nutr. Sci. **62**(1), 5–13 (2012)
9. Sereno, A.M., Hubinger, M.D., Comesana, J.F., Correa, A.: Prediction of water activity of osmotic solutions. J. Food Eng. **49**(2–3), 103–114 (2001)
10. Starzak, M., Peacock, S.D., Mathlouthi, M.: Hydration number and water activity models for the sucrose-water system: a critical review. Crit. Rev. Food Sci. Nutr. **40**(4), 327–367 (2000)
11. Subbiah, B., Blank, U.K.M., Morison, K.R.: A review, analysis and extension of water activity data of sugars and model honey solutions. In: Food Chemistry, vol. 326 (2020)
12. Syamaladevi, R.M., Tang, J., Villa-Rojas, R., Sablani, S., Carter, B., Campbell, G.: Influence of water activity on thermal resistance of microorganisms in low-moisture foods: a review. Compr. Rev. Food Sci. Food Saf. **15**(2), 353–370 (2016)
13. Zhang, L., Sun, D.-W., Zhang, Z.: Methods for measuring water activity (A(w)) of foods and its applications to moisture sorption isotherm studies. Crit. Rev. Food Sci. Nutr. **57**(5), 1052–1058 (2017)

Dr. Lichao Zhao is a young professor and doctoral supervisor at South China Agricultural University. He graduated from Ocean University of China and South China Agricultural University, and was a visiting scholar at Rutgers University in the United States. He is the deputy director of the Youth Working Committee Branch in the university, and a member of the Youth Working Committee of the Guangdong Food Institute. He is mainly engaged in research on the relationship between food nutrition, food safety, and health, covering natural product chemistry, comprehensive utilization of agricultural by-products, rapid detection and prevention and control of foodborne pathogens in special conditions, biotransformation mechanisms of active substances in the body, and health food processing. His teaching activity is mainly on food chemistry, functional food science, and other courses, and he has participated in the compilation of many textbooks, and is the deputy editor of the twenty-first century curriculum textbook "Food Chemistry". He has presided over or participated in more than 50 various projects, and published more than 50 research papers.

Dr. Mingyue Song, associate professor and master supervisor in College of Food Science, South China Agricultural University, graduated from Northwest A&F University and the University of Massachusetts Amherst in Food Science with a doctorate degree. In 2017, he joined the "Young Talents" plan and currently is the deputy director of the Department of Food Science. His teaching activity is related to the undergraduate courses "Food Chemistry", "Introduction to Functional Foods", and postgraduate courses "Information Retrieval and Document Writing". His main research areas are focused on the separation, extraction, and structure identification of functional food factors, their activity evaluation and mechanism of action, absorption and metabolism, and their interaction with intestinal flora. In recent years, he has presided over several projects from the National Natural Science Foundation of China, the Natural Science Foundation of Guangdong Province and enterprises, and published more than 20 SCI papers.

Chapter 3
Proteins

Xinhuai Zhao, Yu Fu, and Chunli Song

Abstract In this chapter, proteins and amino acids will be introduced according to their chemical structures, classification as well as those important physicochemical and biochemical properties. It has been elucidated in detail about the mechanisms and the influencing factors involved in protein denaturation and in governing the functional properties of proteins. This chapter also introduces the functionality evaluations of proteins and their applications in the food industry, and the main methods used in protein modification. After this, readers should acquire the physicochemical and nutritional changes of proteins during food processing and storage and the controlling techniques for these changes.

Keywords Proteins · Amino · Physicochemical properties · Biochemical properties · Functional propertics · Protein modification

Proteins are highly complex biopolymers and are made up of carbon, hydrogen, oxygen, nitrogen, sulfur, phosphorus, and certain metal irons (such as Zn and Fe) at the elemental level. Natural proteins have molecular weights in the range from 10^4 to 10^5 Da, whereas some proteins have molecular weights up to 10^6 Da. Proteins are major components of a cell/organism, accounting for more than 50% of the dry weight of cells. Proteins provide essential nutrients for the growth and sustainability of life. Several proteins, such as enzymes and hormones, are regarded as biocatalysts to regulate the growth, digestion, metabolic activities, and secretion, whereas several proteins take part in energy transfer in the body, such as insulin, hemoglobin, and growth hormone. Certain proteins are also recognized as essential substances in the biological system. Immunoglobulin could take an important part in the immune

X. Zhao (✉)
School of Biology and Food Engineering, Guangdong University of Petrochemical Technology, Maoming 525000, China
e-mail: zhaoxh@gdupt.edu.cn

Y. Fu
College of Food Science, Southwest University, Chongqing 400715, China

C. Song
College of Food and Bioengineering, Qiqihar University, Qiqihar 161006, China

49

system of the body. However, several proteins like trypsin inhibitors are regarded as anti-nutritional factors. Overall, proteins contribute to food processing for the textural and flavor attributes of foods.

Amino acids are the basic structural units of proteins. There are 20 or 18 kinds of amino acids in food proteins. The amino acids link together via the amide bonds to build native proteins. Generally, proteins can be classified into three categories according to their chemical compositions, namely simple, conjugated, and derived proteins. Among the three types of proteins, simple proteins refer to these proteins only consist of amino acids, whereas conjugated proteins are these proteins that also contain the non-protein constituents or prosthetic groups. The so-called derived proteins are these proteins derived from simple or conjugated proteins by enzymatic or chemical methods. Proteins can also be classified according to their functional properties, including structural proteins, bioactive proteins, and food proteins. In most textbooks, the first protein classification is generally adopted and further subdivided according to their solubility.

To fulfill protein requirements for the human, it is necessary to develop food proteins via new technologies and make full use of the existing protein resources. Therefore, it is essential for the food scientists to understand the physical, chemical, and biological properties of proteins as well as the potential effects of food processing and storage on protein properties.

3.1 Amino Acids

3.1.1 Chemical Structures and Classification of Amino Acids

Native amino acids (except for praline) consist of at least a carboxyl and amino groups, together with a side chain (i.e. residual) R group. The amino group is bonded directly to the α-carbon atom of amino acids. Therefore, amino acids are generally referred to as α-amino acids. Amino acids may have both L- and D-configuration (certain bacteria), but only the amino acids in L-configuration can be used by humans.

$$
\begin{array}{c}
\text{O} \\
\| \\
\text{R}-\text{CH}-\text{C}-\text{OH} \\
| \\
\text{NH}_2
\end{array}
$$

The R represents different residual groups of amino acids

Amino acids are usually classified into four groups, based on their different residual groups.

(1) Non-polar amino acids having hydrophobic R group. The hydrophobicity of R group increases with an increasing number of C atoms in the hydrocarbon chain. Examples of non-polar amino acids include alanine, glycine,

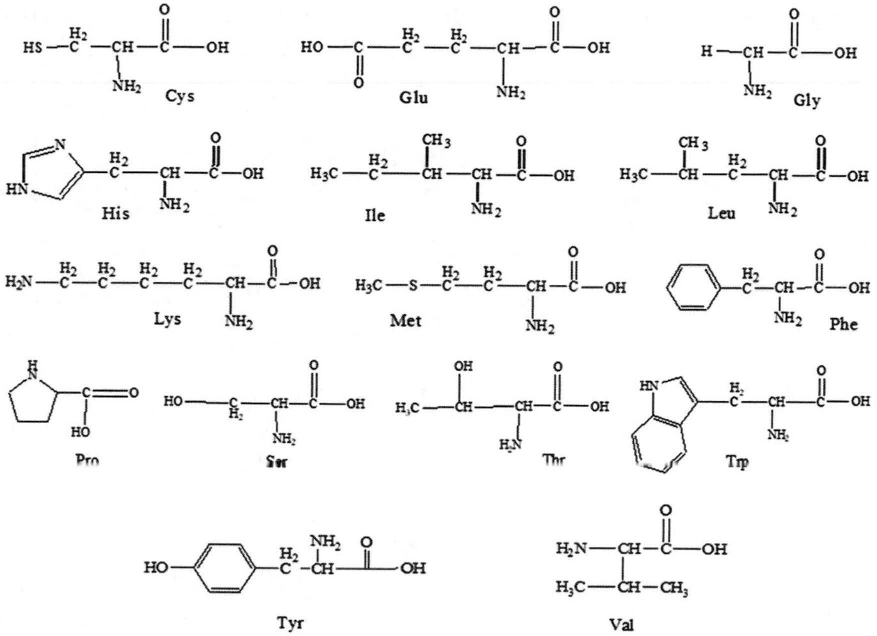

Fig. 3.1 Molecular structure of amino acids at neutral pH

leucine, isoleucine, valine, proline, tryptophan, phenylalanine, and methionine. It should be noted here that proline is an α-subamino acid.

(2) Polar but uncharged amino acids. The R group is polar but ionizable and can form hydrogen bonds with other polar groups. Typical polar but uncharged amino acids are serine, threonine, tyrosine, cysteine, asparagine, glutamine, and glycine. However, the residual groups of tyrosine and cysteine are ionizable under strongly alkaline conditions, so tyrosine and cysteine have polarity greater than the other five polar amino acids. Cysteine is usually present in the form of cystine. Both asparagine and glutamine can be hydrolyzed to remove the amide group and thus converted to respective aspartic and glutamic acids.

(3) Basic amino acids include lysine, arginine, and histidine. The residual groups of those amino acids are amino or imino groups, so they are polar and positively charged at the pH values below their pKa values.

(4) Acidic amino acids include glutamic and aspartic acids containing a carboxyl group in their side chain. The carboxyl group is dissociated to lose protons, and thus negatively charged.

There are also other derivatives of common amino acids with special structure, such as hydroxyproline and 5-hydroxylysine found in collagen, methyl histidine and, α-N-methyllysine in animal muscle protein, while we are not introduced to other amino acids here.

The chemical structures of these amino acids are shown in Fig. 3.1. Several physicochemical constants of these amino acids are given in Tables 3.1, 3.2, and 3.3.

Table 3.1 Primary α-amino acids that occur in proteins and their solubility and melting point

Name	Three letters abbreviation	One letter abbreviation	Molecular weight	Solubility in water at 25 °C (g/L)	Melting point (°C)
Alanine	Ala	A	89.1	167.2	279
Arginine	Arg	R	174.2	855.6	238
Asparagine	Asn	N	132.2	28.5	236
Aspartic acid	Asp	D	133.1	5.0	269–271
Cysteine	Cys	C	121.1	0.05	175–178
Glutamine	Gln	Q	146.1	7.2	185--186
Glutamic acid	Glu	E	147.1	8.5	247
Glycine	Gly	G	75.1	249.9	290
Histidine	His	H	155.2	41.9	277
Isoleucine	Ile	I	132.2	34.5	283–284
Leucine	Leu	L	131.2	21.7	337
Lysine	Lys	K	146.2	739.0	224
Methionine	Met	M	149.2	56.2	283
Phenylalanine	Phe	F	165.2	27.6	283
Proline	Pro	P	115.1	1620.0	220–222
Serine	Ser	S	105.1	422.0	228
Threonine	Thr	T	119.1	13.2	253
Trytophan	Trp	W	204.2	13.6	282
Tyrosine	Tyr	Y	181.2	0.4	344
Valine	Val	V	117.1	58.1	293

Table 3.2 Specific optical rotation (°) of amino acids (the medium and temperature are not shown here)

Name	$[\alpha]_D(H_2O)$	Name	$[\alpha]_D(H_2O)$
Alanine	+14.7	Lysine	+25.9
Arginine	+26.9	Methionine	+21.2
Aspartic acid	+34.3	Phenylalanine	−35.1
Cystine	−214.4	Proline	−52.6
Glutamic acid	+31.2	Serine	+14.5
Glycine	0	Threonine	−28.4
Histidine	−39.0	Trytophan	−31.5
Isoleucine	+40.6	Tyrosine	−8.6
Leucine	+15.1	Valine	+28.8

Table 3.3 pKa and pI values of ionizable groups in free amino acids and proteins at 25 °C

Name	pK$_{a1}$ (α-COOH)	pK$_{a2}$ (α-NH$_3^+$)	pK$_{aR}$ (Side chain)	pI
Alanine	2.35	9.69		6.02
Arginine	2.17	9.04	12.48	10.76
Asparagine	2.02	8.80		5.41
Aspartic acid	1.88	9.60	3.65	2.77
Cysteine	1.96	10.28	8.18	5.07
Glutamine	2.17	9.13		5.65
Glutamic acid	2.19	9.67	4.25	3.22
Glycine	2.34	9.60		5.98
Histidine	1.82	9.17	6.00	7.59
Isoleucine	2.36	9.68		6.02
Leucine	2.30	9.60		5.98
Lysine	2.18	8.95	10.53	9.74
Methionine	2.28	9.21		5.74
Phenylalanine	1.83	9.13		5.48
Proline	1.94	10.6		6.30
Serine	2.20	9.15		5.68
Threonine	2.21	9.15		5.68
Trytophan	2.38	9.39		5.89
Tyrosine	2.20	9.11	10.07	5.66
Valine	2.32	9.62		5.96

3.1.2 Physicochemical Properties of Amino Acids

(1) Optical activity

Amino acids (except for glycine) have at least one chiral center at the α-carbon atom (as shown in Fig. 3.1) and hence exhibit optical activity. The data in Table 3.2 list the optical properties of amino acids. The direction and magnitude of the optical rotation depend not only on the nature of these residual R groups but also on the pH, temperature, and other factors of the aqueous solution. The optical properties of amino acids can be used in their quantitative measurement and qualitative identification.

(2) UV absorption and fluorescence

Amino acids have no absorption in the visible region but have absorption near 210 nm in the ultraviolet region (Fig. 3.2). Tyr, Trp, and Phe contain aromatic rings thereby have strong ultraviolet absorption (Fig. 3.2). The three amino acids show maximum wavelengths of 278, 279, and 259 nm, with molar extinction coefficients of 1,340, 5,590, and 190 mol/cm, respectively. These coefficients are often used as reflectors

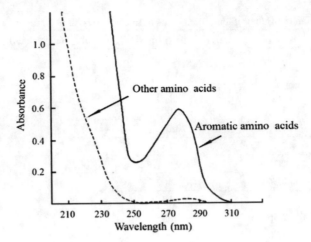

Fig. 3.2 Ultraviolet absorbance spectra of some amino acids

to monitor these three aromatic amino acids. The bound Tyr and Trp have maximum absorption near 280 nm and are responsible for quantitative analysis of proteins using ultraviolet absorption.

Tyr, Trp, and Phe can also be excited to produce fluorescence in the ultraviolet region with emission wavelengths of 304 and 348 nm at the excitation wavelength of 280 nm, respectively. The three aromatic acids also exhibit fluorescence with emission wavelengths of 282 nm at the excitation wavelength of 260 nm. Other amino acids do not exhibit fluorescence.

(3) Dissociation

All amino acids contain at least an amino and one carboxyl groups and are mainly in a form of dipolar ions or zwitterions in neutral aqueous solutions. In other words, an amino acid can either accept a proton as alkaline or dissociate a proton as an acid. In this way, a mono-amino- and mono-carboxyl amino acid after being fully protonated can be regarded as a dibasic acid with two dissociation constants corresponding to the carboxyl group (pKa_1) and protonated amino group (pKa_2). A third dissociation constant (pKa_R) occurs when the side chain of amino acid has dissociable groups such as the α-amino or α-carboxyl groups of basic or acidic amino acids. The dissociation constants and isoelectric points (pI) of these amino acids are shown in Table 3.3.

$$R-\underset{\underset{NH_3^+}{|}}{CH}-\overset{\overset{O}{\|}}{C}-OH \xrightarrow[pK_{a_1}]{-H^+} R-\underset{\underset{NH_3^+}{|}}{CH}-\overset{\overset{O}{\|}}{C}-O^-$$

$$R-\underset{\underset{NH_3^+}{|}}{CH}-\overset{\overset{O}{\|}}{C}-OH \xrightarrow[pK_{a_2}]{-H^+} R-\underset{\underset{NH_2}{|}}{CH}-\overset{\overset{O}{\|}}{C}-OH$$

When the molecules of an amino acid are electrically neutral in solution (i.e. the net charge is zero), they do not move in the electric field. The corresponding pH environment is called the isoelectric point (pI) of the amino acid. The solubility hereof the amino acid is the lowest. The pI values of amino acids can be estimated from their pKa_1, pKa_2, and pKa_3 values using the following expressions.

For a mono-amino and mono-carboxyl amino acid, $2pI = pKa_1 + pKa_2$.

For a basic amino acid, $2pI = pKa_2 + pKa_3$.

For an acidic amino acid, $2pI = pKa_1 + pKa_3$.

The subscript numbers 1, 2, and 3 refer to the α-carboxyl, α-amino, and side-chain ionizable groups, respectively.

The isoelectric point property can be used to selectively separate an amino acid from a mixture of amino acids. In addition, when amino acids are combined to form proteins, the isoelectric properties of proteins are also affected by the dissociation of amino acids.

(4) Hydrophobicity

Hydrophobicity of an amino acid can be defined as the free energy change arising from transferring 1 mol amino acid from ethanol to water phases. In the case of neglecting the change in activity coefficients, the free energy change can be expressed as below.

$$\Delta G^o = -RT \ln(\frac{S_{ethanol}}{S_{water}})$$

where, $S_{ethanol}$ and S_{water} represent the solubility of the amino acid in ethanol and water (mol/L), respectively.

As is true of all other thermodynamic parameters, ΔG° also has an additive function. If an amino acid molecular has multiple groups, its ΔG° can be the sum of the free energy changes when transferring multiple groups in the amino acid from ethanol to water phases. That is, $\Delta G^\circ = \Sigma \Delta G^\circ$.

An amino acid molecule can be divided into two parts, one is a glycine group and the other is a residual group (R group). For example, Phe can be considered as below, and the free energy change of transferring Phe from ethanol to water phases can then be considered as below. Thus, the hydrophobicity of the side chains can be determined by subtracting $\Delta G^\circ_{glycine}$ from ΔG°.

Table 3.4 Hydrophobicity of amino acid side chain at 25 °C determined using Tanford method

Amino acid	ΔG° (ethanol→water) (kJ/mol)	Amino acid	ΔG° (ethanol→water) (kJ/mol)
Ala	2.09	Leu	9.61
Arg	3.1	Lys	6.25
Asn	0	Met	5.43
Asp	2.09	Phe	10.45
Cys	4.18	Pro	10.87
Gln	−0.42	Ser	−1.25
Glu	2.09	Thr	1.67
Gly	0	Trp	14.21
His	2.09	Tyr	9.61
Ile	12.54	Val	6.27

Benzyl group   Glycyl group

$$\Delta G^\circ = \Delta G^\circ_{(side\,chain)} + \Delta G^\circ_{(glycine)}, \text{ or } \Delta G^\circ_{(side\,chain)} = \Delta G^\circ - \Delta G^\circ_{(glycine)}$$

The hydrophobicity of these residual groups can be determined by measuring the solubility of each amino acid in two different media (Tanford method, see Table 3.4). The more positive the value of an amino acid is, the more hydrophobic it behaves, where presumably the R group tends to be located inside of proteins. Otherwise, the more negative the value is, the more hydrophilic the amino acid possesses. Thereby, this hydrophilic R group tends to be distributed on the surface of protein molecules. However, an exception is Lys. Lys has a positive hydrophobicity value but is a hydrophilic amino acid due to its four methylene groups. Adsorption coefficients of amino acids are proportional to their hydrophobicity values, and hydrophobicity is commonly used to predict the adsorption behavior of amino acids on a carrier of hydrophobic compounds.

3.1.3 Chemical Reactions of Amino Acids

These functional groups such as amino, carboxyl, and residual groups in amino acid molecules can undergo various chemical reactions.

3.1.3.1 The Reactions of Amino Groups

(1) Reaction with nitrite

α-NH_2 of amino acids would react quantitatively with HNO_2 to yield nitrogen and hydroxyl acids. The generated N_2 is measured by the volume while its level is directly proportional to the amount of amino acids.

$$R-CH(NH_2)-C(=O)-OH + HNO_2 \longrightarrow R-CH(OH)-C(=O)-OH + H_2O + N_2$$

Different from α-NH_2, the ε-NH_2 has a weak reaction with HNO_2. The α-NH_2 of Pro does not react with HNO_2. Other amino acids such as Arg, His, and Trp do not interact with HNO_2 as their side chains form a cyclic structure with the nitrogen atom when being treated with HNO_2.

(2) Reaction with aldehydes

The α-NH_2 of amino acids reacts with an aldehyde compound to yield a product known as Schiff-base compound, which is an intermediate product of a non-enzymatic browning reaction (i.e. the Maillard reaction).

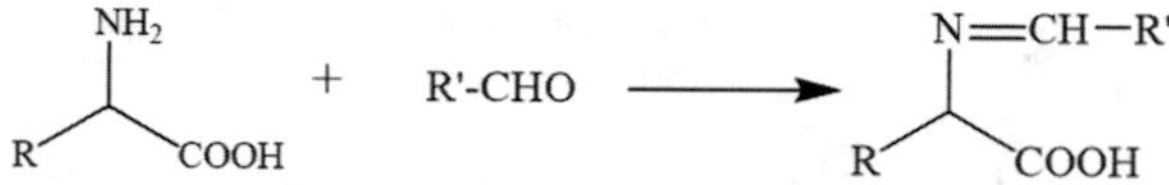

(3) Acylation

Acylation of amino acids (α-NH_2 group) by benzyloxyformyl chloride under alkaline conditions is utilized in the synthesis of peptides.

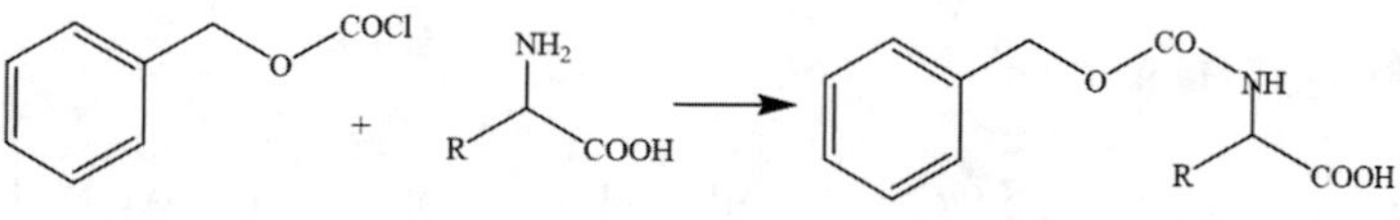

(4) Alkylation

Derivatization of amino acids (ε-NH_2 group) by dinitrofluorobenzene yields stable yellow compounds, which can be used to label N-terminal amino acid residues and free ε-NH_2 groups present in peptides and proteins.

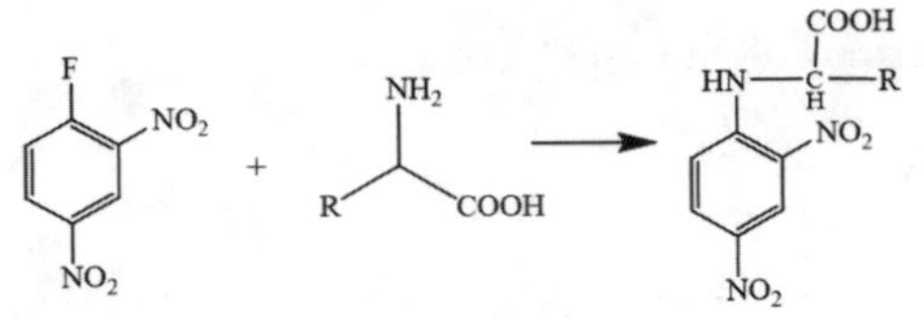

3.1.3.2   The Reactions of Carboxyl Groups

(1)   Esterification

Methyl or ethyl esters of amino acid can be obtained in anhydrous methanol or ethanol in the presence of dry HCl.

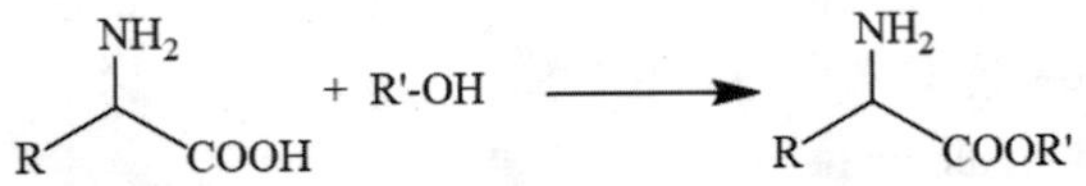

(2)   Decarboxylation

Escherichia coli over-expressing the glutamate decarboxylase can catalyze decarboxylation of glutamic acid, which can be used for the analysis of glutamic acid.

3.1.3.3   The Reactions Involving Both Amino and Carboxyl Groups

(1)   Forming peptide bonds

The connecting reaction between the carboxyl group of amino acid and the amino group of another amino acid leads to covalently linking of the two amino acids via an amide bond (also known as a peptide bond), which is the basis of protein synthesis.

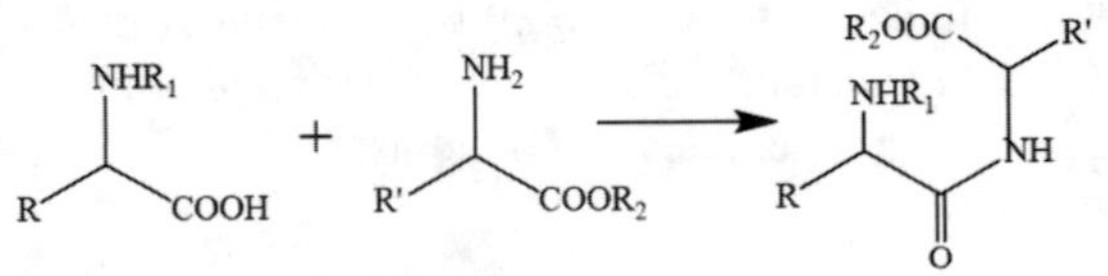

(2) Reaction with ninhydrin

Under weak alkaline conditions, ninhydrin can react with amino acids, forming a final purple product known as Ruhemann's purple, which has a maximum absorbance at 570 nm. This reaction can be used to quantify amino acids as well as proteins. Pro is an exception as it has no α-amino group, and its reaction product has a maximum absorbance at 440 nm and thereby gives yellow color.

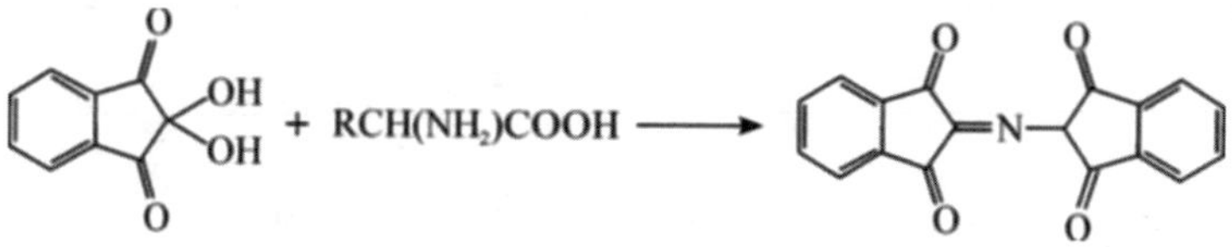

3.1.3.4 Reaction of the Side Chain Groups

The side chain group (R group) influences the chemical properties of an α-amino acid. A functional group in the side chain will add many reactions that the α-amino acid can take part in. Phenolic reaction is one of the most popular assays for side chain group of amino acids. The principle of the Folin-phenol assay is the reduction of the Folin-phenol reagent in the presence of phenolic compounds, which results in the production of molybdenum–tungsten blue. Besides the phenolic reaction, those having chemical groups that form bonds with sulfhydryls (–SH) are the most common reactions for proteins. Sulfhydryls, also called thiols, exist in proteins in the side chain of cysteine amino acids. Sulfhydryl groups are often transformed into disulfide bonds (–S–S–) within or between polypeptide chains in the presence of an oxidizing agent. Typically, –S–S– group is reversible to reduced sulfhydryl groups (–SH) in the presence of a reducing agent. This conversion has an important influence on the functional properties of proteins.

$$-SH + -SH \rightarrow -S-S-$$

In the identification of amino acids or proteins, some side chain groups show important chemical reactions, which are listed in Table 3.5.

3.1.4 Preparation of Amino Acids

(1) Protein hydrolysis

Amino acids can be produced by acidic, basic, or a series of enzymatic hydrolysis of native proteins. Separation of single amino acid from a mixture can be performed by

Table 3.5 Several important color reactions of amino acids and proteins

Reaction	Reagent	Amino Acid/Group/Bond	Color
Millons's reaction	Mercuric nitrate, mercurous nitrate	Phenol group (Tyr)	Red
Xanthoproteic reaction	Concentrated nitric acid	Benzene ring/Tyr, Trp	Yellow but orange with alkaline addition
Hopkins–Cole reaction	Glyoxylic acid	Trp/Indole ring	Purple
Ninhydrin reaction	Hydrated ninhydrin	α-, ε-amino groups	Blue or purple
Ehrlich reaction	p-dimethylaminobenzaldehyde	Indole ring (Trp)	Blue
Sakaguchi reaction	α-naphthol, sodium hypochlorite	Arg	Intense red
Sullivan reaction	1,2-naphthoquinone 4-sulfonate, sodium sulfite, sodium thiosulfate, sodium cyanide	Cystine, Cys	Red

isoelectric precipitation and crystallization. A combination of two or more of these methods is often needed to get pure compounds. However, strong acid or alkaline catalyzed destroys some amino acids in proteins such as Trp, Asn, and Glu.

(2) Chemical synthesis

Chemical synthesis leads to the production of both L- and D-amino acids. However, pharmaceutical and medical applications often need chiral pure L- or D-amino acids. Several enzymatic processes thus are used to convert the D/L-racemate into the pure isomer. The best known and industrial operated process is the enzymatic conversion of D/L-Met and Trp.

(3) Microbial fermentation

Amino acids can be produced via microorganism fermentation by utilizing inexpensive by-products as hydrocarbon sources. In fact, microbiological method has been successively employed to prepare Glu and Lys. This method is a promising way in the future.

3.2 Proteins and Peptides

3.2.1 Protein Structures

Proteins are macromolecules composing amino acids as structural units. The rotation of the peptide bonds in protein molecules leads to different protein configuration or conformation. Therefore, the spatial structures of proteins are very complicated and usually are described at the three structural levels.

(1) Primary structure

The primary structure of a protein denotes the linear sequence in which the constituent amino acids are linked via peptide bonds, listing the amino acids starting at the amino-terminal end through to the carboxyl-terminal end. The primary structure of several proteins has already been determined, such as insulin, hemoglobin, cytochrome C. Milk proteins like α-a_{S1}-, α-a_{S2}-, and β-$_{A2}$-caseins are also known to us. The total number of amino acid residues in the sequences varies; for example, a few proteins are illustrated by several dozens of amino acid residues, whereas a majority of proteins contain as many as several hundred (100–500) residues. Some uncommon proteins may have amino acid residues of up to thousands. However, the primary structures of some proteins are not yet fully determined.

In general, primary structures of proteins to a large extent determine their basic properties. Interactions between atoms of the molecular backbone (primary structure) have an influence on the next local folded structures such as secondary and tertiary structures. It is theoretically possible to create vast amino acid sequences as the proteins are built from a set of amino acids. For example, when a protein is constructed from 100 amino acid residues by all 20 amino acids, the possible sequences of the protein might statistically be up to 20^{100} (equally to 10^{130}). However, only about, 10^4–10^5 proteins are synthesized in nature, while only thousand proteins among these proteins have been isolated and characterized.

(2) Secondary structure

Secondary structure refers to the periodic spatial arrangement of amino acid residues of protein molecules at certain segments of the polypeptide chain. In proteins, three types of helical structures, namely α-helix, π-helix, and γ-helix are found, while the α-helix is the major form. Another common structural feature found in proteins is extended sheet-like structure, which mainly includes β-strand and β-bend. In addition, random coil structure is also a dominant structure for some proteins, which is characterized by the absence of symmetrical axis or symmetrical surfaces. Hydrogen bonds play an important role in conformational stability in the secondary structure of proteins.

The α-helical structure (right-handed α-helix) is an ordered and the most stable conformation in proteins. Each helical rotation involves 3.6 amino acid residues, with each residue extending the axial length by 0.15 nm. The axial length occupied per rotation is 0.54 nm. The apparent diameter of the helix is 0.6 nm. In this structure, each backbone N–H group in the polypeptide chain is hydrogen-bonded to the C=O group of its neighboring loop. Therefore, the hydrogen orientation of the α-helix is the same as the dipole orientation. Pro is an exception due to its two attributes that hinder α-helix formation. On one hand, ring structure formed by propyl side chain to the amino group restricts the rotation of N–C_α. On the other hand, there is no hydrogen available at the nitrogen atom. Thus, the peptide chains rich in Pro cannot form α-helix; alternatively, they tend to assume a random coil structure. One example is β-casein, in which about 17% of total amino acids are Pro. Therefore, β-casein shows a disordered or random structure.

The β-sheet structure is a structural feature in a cylindrical or nearly cylindrical shape and is a more extended form than the α-helical structure. Conversion from α-helix to β-sheet is usually observed when α-helix type proteins are heated. In this β-sheet configuration, the polypeptide chains interact with each other through hydrogen bonding, involving the backbone carbonyl group and amide proton on neighboring strands. In general, depending on N → C directional orientations of the strands, two forms of β-sheet structure are found in proteins, namely parallel β-sheet and anti-parallel β-sheet. Neighboring strands run in the same orientation from the N- to C- terminal ends in parallel β-sheet structures, whereas an anti-parallel β-sheet may alternate the orientation of N-terminal end in the peptide chain. In the two forms, the side chains of successive amino acid residues in a strand alternately locate themselves above and below the plane sheet.

Another common secondary structure found in proteins is the β-bend or β-turn, which can be regarded as a special helix form with "zero distance". This structure confers polypeptide chain folding back on itself, and the stability of the bend is maintained via hydrogen bonds.

(3)   Tertiary structure

The tertiary structure of a protein refers to the spatial arrangement attained when a polypeptide chain with a secondary structure is further folded by various forces to form a compact three-dimensional form. From an energetics viewpoint, the formation of tertiary structure involves the optimization of various interactions such as hydrogen bonds, electrostatic interactions, disulfide bonds, and van der Waals forces between various groups in protein. In most globular proteins, the R groups of the polar amino acids are generally located on the water accessible surface, while the R groups of the non-polar amino acids are inevitably buried in the interior of proteins. However, the non-polar amino acids of certain lipoproteins mainly attach to the molecular surface of the protein particles.

(4)   Quaternary structure

Quaternary structure refers to the spatial arrangement of proteins with two or more polypeptide chains in a special way. This arrangement may lead to the formation of biological properties. The peptide chains of these quaternary complexes are generally referred to as subunits, which are the same or different and have their own primary,

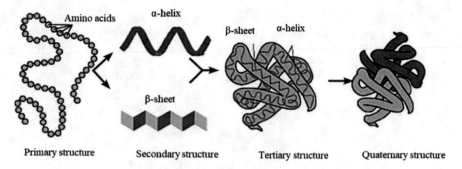

Fig. 3.3 Schematic diagram of structural levels of proteins

secondary, and tertiary structures. The interaction between peptide chains is primarily dominated by non-covalent interactions such as hydrogen bonding and hydrophobic interaction. When the hydrophobic amino acid content of a protein is more than 30%, it exhibits a greater tendency to form an oligomeric structure than those contain fewer hydrophobic amino acid residues.

The levels of protein structure are represented in Fig. 3.3.

3.2.2 The Forces Involved in the Structural Stability of Proteins

As mentioned earlier, the secondary structure of the proteins is primarily maintained via hydrogen bonds between various amino acid residues. Hydrogen bond, electrostatic interactions, hydrophobic interaction, and van der Waals interaction make contribution to protein folding of the tertiary and quaternary structures. The characteristics of the forces involved in the structural stability of proteins are given in Table 3.6. The strong driving force (bond energy) involved in the structural stability of observed proteins is only the covalent bonds, mainly disulfide bonds, whereas other forces have lower bond energy than covalent bonds. Any change in the environment of the proteins certainly would influence their folding pattern. When major structural changes appear in the secondary, tertiary, and quaternary structures, protein denaturation occurs.

The forces involved in structural stability of proteins can also be depicted in Fig. 3.4.

Table 3.6 Forces involved in structural stability of proteins

Type	Bond energy (kJ/mol)	Interaction distance (nm)	Functional groups from side chains	Reagents/conditions for weakening forces	Force enhancement
Covalent bonds	330–380	0.1–0.2	Disulfide moiety	Cysteine, Na_2SO_3, CH_3CH_2SH, etc.	–
Hydrogen bonds	8–40	0.2–0.3	Amide, carboxyl, and phenolic groups	Guanidine, urea, detergent, acid, heating	Cooling
Hydrophobic interaction	4–12	0.3–0.5	Aliphatic and aromatic side chains	Organic solvents, surfactants	Heating
Electrostatic interaction	42–84	0.2–0.3	Carboxyl and amino groups	High or low pH, salt solution	–
van der Waals interaction	1–9	–	Permanent, induced, and instantaneous dipples	–	–

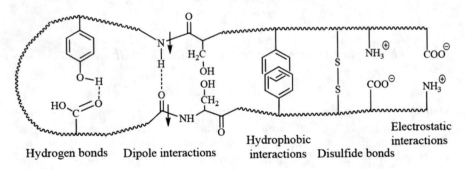

Hydrogen bonds   Dipole interactions   Hydrophobic interactions   Disulfide bonds   Electrostatic interactions

Fig. 3.4 Schematic diagram of the forces involved in structural stability of proteins

3.2.3 Protein Classification

There are a number of ways of classifying proteins. Generally, proteins can be classified into three categories based on their chemical compositions, namely simple, conjugated, and derived proteins. Proteins also can be classified according to their solubility as water-soluble, salt-soluble, alkaline (acid)-soluble, and alcohol-soluble proteins. The solubility characteristics and major sources of various common proteins are shown in Table 3.7, and the related factors affecting protein solubility can be referred to in the relevant textbooks.

Table 3.7 Protein classification and typical proteins

Proteins		Characteristics	Presence	Typical examples
Simple proteins	Albumins	Soluble in water, diluted salt, acid/alkaline solutions, and precipitated by saturated ammonium sulfate as well as heat coagulation	Animal cells and body fluids	Albumin, α-lactalbumin, ovalbumin
	Globulins	Soluble in dilute salt, acid/alkaline solutions but insoluble in pure water, and precipitated in half-saturated ammonium sulfate, also almost coagulation when heated	Animal cells and body fluids	Serum globulin, β-lactoglobulin, glycinin, myosin, lysozyme
	Glutenins	Soluble in dilute acid/alkaline solutions but not in pure water, ethanol, and neutral salt solution	Plant seeds	Gluten, glutenin
	Prolamines	Soluble in dilute acid/alkaline and 66–80% ethanol–water solution, but insoluble in pure water and salt solution. Contain large amounts of Pro and Glu but only small amounts of Lys	Plant seeds	Gliadin, zein
	Scleroproteins	Usually insoluble in water, salt, diluted acid/alkaline solution, resistance to protease	Animal tissues	Collagen, elastin, keratin

(continued)

Table 3.7 (continued)

Proteins		Characteristics	Presence	Typical examples
	Histones	Soluble in pure water but insoluble in ammonia, and precipitated in aqueous phosphotungstic acid solution (acid or neutral pH)	Animal cells	Thymus histone, erythrocyte histone, nuclear protein
	Protamines	Soluble in dilute acid and pure water but insoluble in ammonia, and precipitated in aqueous phosphotungstic acid solution (acid or neutral pH). Contain large amounts of the Arg	Mature germ cells	Fish protamine
Conjugated proteins	Nucleoproteins	Nucleic acid such as ribonucleic acid and deoxyribonucleic acid	Animal and plant cells	Thymus histone, viral protein
	Phosphoproteins	Containing phosphate group and can be attacked by phosphatase	Animal cells and body fluids	Casein, egg yolk phosphoprotein
	Metalloproteins	Containing metals like iron and copper, also having some pigments	Plant/animal cells and their body	Hemoglobin, myoglobin, cytochrome c, catalase
	Glycoprotein	Containing saccharide groups	Animal cells	Serum glycoprotein, ovomucoid
Derived proteins	Primary derivatives	Initially denatured protein; acid- or base-denatured protein		Rennet-coagulated casein
	Secondary derivatives	Decomposition products of proteins with significant property changes		Peptides

3.2.4 Physical and Chemical Properties of Proteins

3.2.4.1 Acid–Base Properties of Proteins

As we know, proteins contain ionizable amino and carboxyl groups in N- and C-terminal ends, respectively, that is, proteins can be considered as ampholytes. In addition, the side chains of proteins also may have ionizable groups. Proteins should exist as polyvalent ions. The dissociation and the resulted total charge of a protein are depending on these ionizable groups as well as the pH value of the solution. Typically, a protein carries a charge of zero at a certain pH known as isoelectric point (pI). At its isoelectric pH (pI), a protein does not possess any charge and thus will not move in an applied electric field and exhibits the lowest solubility. When the pH > pI, a protein holding a net negative charge moves toward the anode. Similarly, when the pH < pI, a protein has a net positive charge and thereby moves toward the cathode.

3.2.4.2 Protein Hydrolysis

Under the action of acid/alkaline/enzyme, proteins receive a cleavage in peptide bonds, resulting in the formation of a serial of intermediate products with various chain lengths such as poly-peptides, oligo-peptides, and di-peptides. Surely, the final products are amino acids.

Alkaline hydrolysis can damage Cys and Arg, while acid hydrolysis can destroy Trp. Alkaline hydrolysis also induces racemization of amino acids. However, enzymatic hydrolysis can avoid amino acid loss, due to its mild reaction conditions. One protease is unable to convert proteins into free amino acids. In general, several proteases are used. In addition, the enzymatic hydrolysis rate is slower than the chemical hydrolysis.

3.2.4.3 Color Reaction

Biuret reaction is an important reaction used to detect the presence of proteins; however, this reaction is not specific for proteins. In an alkaline environment, the nitrogen of the peptide bonds can donate its lone pair of electrons to Cu^{2+}, forming a stable and violet-colored complex (Fig. 3.5). At least two peptide linkages must be present due to the coordination of Cu^{2+} with –CONH– group. This means that di-peptides and free amino acids are unable to give the test. In a neutral environment, protein or polypeptides and ninhydrin reagent give blue or purple color. Of course, the ninhydrin reagent can also react with ammonium salt and amino acids.

In addition, some color reactions are based on specific reactions of certain amino acids. An example of these reactions is the presence of phenolic groups of Tyr or thiol

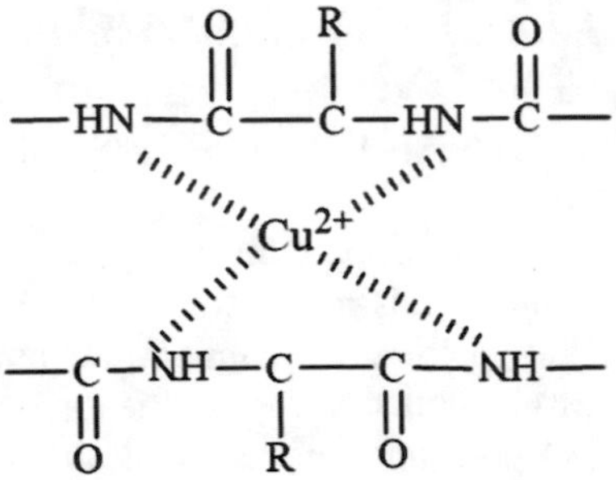

Fig. 3.5 Formation of
protein–Cu^{2+} complex by the
so-called biuret reaction

groups of His. These reactions can also be used in the qualitative and quantitative analyses of proteins.

3.2.4.4 Hydrophobicity

Proteins also have their hydrophobicity. Hydrophobicity of a protein can be represented by averaging the hydrophobicity of each amino acid, if its amino acid composition is known. That is, the hydrophobicity values of proteins can be determined by dividing the sum of the hydrophobicity values of amino acids by the number of amino acid residues.

$$\Delta \bar{G}^o = \frac{\Sigma \Delta G^o}{n}$$

Hydrophobicity of the proteins is one of the major factors governing physicochemical properties of proteins such as structure, interfacial, and fat-binding properties, and others. The hydrophobicity can be used to reflect the interaction between protein and water or other chemicals. The relationship between hydrophobicity, surface tension, and emulsifying activity index of several proteins is shown in Fig. 3.6.

3.2.5 Peptides

Peptides are short chains of amino acid monomers linked via amide (peptide) bonds. Peptides have lower molecular weights than proteins. Peptides are distinguished depending on the length of peptide chain that formed. The shortest peptide, consisting of two amino acids, is called di-peptides, followed by tri-peptides, tetra-peptides, and so on. On the other hand, a polypeptide chain consists of less than 10 amino acid residues usually called oligo-peptide. Likewise, poly-peptide is the general term for the peptides containing 10 or more amino acid residues. Therefore, the properties of peptides are different from amino acids or proteins. The amount of peptides in foods is very lower than proteins, and their functional properties are usually ignored.

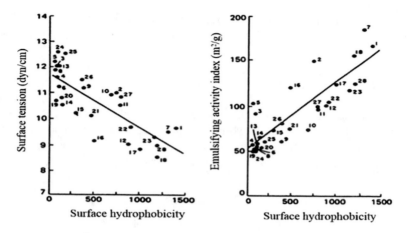

Fig. 3.6 Relationship between surface hydrophobicity. of proteins and the interfacial property of the emulsions prepared by corn oil and proteins (0.2%, w/w)

3.2.5.1 Physiochemical Properties of Peptides

(1) Dissociation

Peptides like amino acids and proteins have their dissociable functional groups. pK values and isoelectric points of peptides can be determined. The dissociation of peptides is also influenced by peptide composition and the aqueous environment they are exposed such as pH and ionic strengths. However, the dissociation of peptides is not as intensive explored as that of proteins or amino acids.

(2) Solubility, viscosity, and osmotic pressure of peptides

Peptides with small molecular weights generally have high solubility, which also shows small changes at various pH values. In comparison with proteins, peptides remain soluble at higher concentrations and over a wide pH range, making them ideal for the processing of some acidic foods. In addition, small peptides have similar behavior than amino acids in trichloroacetic acid (commonly used in protein precipitation); that is, small peptides are soluble in 3% or more of trichloroacetic acid solution, while large peptides are precipitated by 3% trichloroacetic acid.

The viscosity of peptide solution is significantly lower than that of protein solution. Peptides do not facilitate to form gels. The osmotic pressure of the peptide solution is lower than that of the free amino acid and thereby is beneficial for gastrointestinal absorption. Some small peptides have even better absorption than free amino acids, which is important to food nutrition.

(3) Chemical properties of peptides

The chemical properties of peptides are generally like those of proteins and amino acids, and thus can undergo those reactions that most amino acids carry out. The biuret reaction can also be used for quantitative determinations of peptides. It should be noted that this reaction cannot be used to distinguish polypeptide and proteins.

3.2.5.2 Bioactive Peptides

Bioactive peptides play vital roles in human health and nutrition. The tetra- and tri-peptide gastrins are the most favorite bioactive peptides to promote gastric acid secretion. Glutathione takes part in redox reaction and can scavenge free radicals and hydrogen peroxide to exert protective functions in the body.

$$ROOH + 2\,GSH \xrightarrow{Enzyme} ROH + GSSG + H_2O$$

$$H_2O_2 + 2\,GSH \xrightarrow{Enzyme} GSSG + 2\,H_2O$$

Some bioactive peptides can be obtained by selective degradation of native proteins, e.g. enzymatic hydrolysis. The explored physiological functions of the polypeptides are summarized below.

(1)   Mineral binding peptides enhance mineral absorption in human. Caseinphosphopeptide is derived during casein hydrolysis with trypsin. Caseinphosphopeptide contains a plurality of phosphoserines, which enable its calcium-binding activities. Calcium and caseinphosphopeptide form soluble complexes to allow calcium solubility or prevent the formation of insoluble calcium phosphate. The enhancement of the intestinal calcium paracellular absorption is thus observed.

(2)   Certain hydrolysates from the native proteins show anti-hypertensive effect. These peptides from soybean and casein have been found to have significant anti-hypertensive effect and thus are referred to anti-hypertensive peptides. Inhibition on angiotensin-converting enzyme by these peptides leads to anti-hypertension.

(3)   Several peptides also show immuno-modulatory effects and can stimulate the proliferation of human immune lymphocytes and enhance the phagocytic ability of macrophages. These peptides can promote pathogen resistance in both humans and animals.

(4)   Several peptides also have anti-microbial action. Nisin is an anti-microbial peptide produced by the subsp. *Lactis* in *Lactococcus lactis*. Nisin consists of 34 amino acids and has strong inhibition on many of the Gram-positive bacteria. Hence, Nisin is not as resistant as common antibiotics due to its rapid degradation by chymotrypsin in the digestive tract.

(5)   Other bioactive peptides. The ratio of the branched-chain amino acids to the aromatic amino acids in mixtures of the amino acids or oligo-peptides is expressed as Fischer ratio. Peptides with high Fischer ratio, useful for patients with liver diseases, are also explored for nutritional and therapeutic applications.

3.3 Protein Denaturation

Protein molecules are formed by amino acids that are linked together in specific orders and balanced by intra-molecular and inter-molecular forces. Finally, they form spatial structures, namely primary, secondary, tertiary, and quaternary structures. Hence, protein conformation is the result of many actions. However, this conformation is unstable and there will be some changes to varying extents in the secondary, tertiary, and quaternary structures of proteins under acidic, alkaline, heat, organic solvent, or radiation treatments, in which the process is called protein denaturation. Therefore, protein denaturation does not involve changes in the sequences of amino acid connection, namely the primary structures of proteins.

Protein denaturation can affect the structures, physiochemical, and biological properties of proteins, generally including:

(1) Exposure of the hydrophobic groups inside protein molecules; protein solubility thus decreases in water;
(2) Loss of certain bioactivity of proteins such as the loss of enzyme or immune activity;
(3) More peptide bonds of proteins are exposed, which are, therefore, more susceptible to hydrolysis by proteases;
(4) Changes in the water-binding capacity of proteins;
(5) Changes in the viscosity of protein dispersion system;
(6) Loss of crystallization capacity of proteins.

The extent of protein denaturation can be evaluated by measuring the changes in some properties of proteins such as optical, sedimentation, viscosity, electrophoretic, thermodynamic properties, etc. In addition, protein denaturation can also be studied by immunological methods, e.g. via the well-known enzyme-linked immunosorbent assay (ELISA).

The denaturation of native proteins is sometimes reversible. When denaturation factors are removed, proteins will return to their native structures, namely protein renaturation. In general, proteins are more likely to undergo reversible denaturation under mild conditions, while irreversible denaturation will occur under those severe conditions. When disulfide bonds that stabilize protein conformation are destroyed, it is difficult for denatured proteins to renaturate.

The factors that cause protein denaturation include physical (e.g. temperature, pH), chemical, chemical reagent, mechanical processing, etc. No matter what factor causes protein denaturation, from the viewpoint of proteins themselves, protein degeneration is analogous to a physical change process, which does not involve any chemical reaction.

3.3.1 Physical Denaturation

(1)   Heating treatment

Heating treatment is a commonly used process during food processing, which is the most common factor responsible for protein degeneration. Proteins undergo drastic changes at a certain temperature known as denaturation temperature. Protein denaturation with thermal treatment leads to a considerable extension of protein deformation. For instance, native serum protein is ellipsoidal (length: width ratio of 3:1). After thermal denaturation, this ratio is increased to 5.5:1. The molecular shape of the protein is significantly extended.

For the chemical reactions, temperature coefficients are usually in a range of 3–4. However, for protein denaturation with thermal treatment, the temperature coefficients may be up to 600. This property is important for food processing. High-temperature instantaneous sterilization and ultra-high temperature sterilization technology thus employ high temperature to greatly enhance the speed of protein denaturation and inactivation of bioactive proteins or microorganisms in a short time, whereas there is a slight increase in chemical reaction speed in terms of other nutrients to ensure less loss of nutrients.

Thermal denaturation of proteins is related to protein compositions, concentrations, water activity, pH, and ionic strengths. When proteins have more hydrophobic amino acids, they are more stable than those with more hydrophilic amino acids. Thus, bioactive proteins are stable in dry state and show strong resistance to temperature change; however, they are liable to denaturation in the moist and thermal state.

(2)   Freezing treatment

Low-temperature treatment can also cause denaturation of some proteins. L-Threonine cystinase is stable at room temperature, but unstable at 0 °C. Soybean 11S proteins and milk protein will aggregate and precipitate when they are cooled or frozen. Some enzymes (e.g. oxidases) can be activated at relatively low temperatures.

Low-temperature protein denaturation may be caused by changes in hydration environment of proteins. The force balance to maintain protein structures is destroyed, and hydration layer of some chemical groups is also destroyed. The interaction between these groups may cause protein aggregation or subunit rearrangement. It is also possible that the salt effect of the frozen system induces protein denaturation. In addition, concentration increase caused by freezing may lead to the increased exchange reactions of intra-molecular and inter-molecular disulfide bonds in proteins, thus leading to protein denaturation.

(3)   Mechanical treatment

Due to the shearing force, both kneading and whipping can extend protein molecules and destroy α-spiral structure, which then leads to protein denaturation.

The greater the shearing rate, the greater the degree of protein denaturation. For example, when a 10–20% whey protein solution at pH 3.5–4.5 and 80–120 °C is subjected to a shearing rate of 8,000–10,000 s^{-1}, it forms a protein-based fat substitute. Mechanical denaturation of proteins is also used in the production of salad dressing and ice cream.

(4) Static high-pressure treatment

High static pressure treatment can also lead to protein denaturation. Although native proteins have relatively stable conformation, spherical protein molecules are not rigid. Some void spaces show certain flexibility and compressibility, invariably inside the protein molecules. The protein molecules, therefore, will be deformed (i.e. denaturation) under high pressure. At normal temperatures, denaturation of proteins occurs at pressures ranging from 100 to 1,000 MPa. Sometimes, high pressure can lead to protein denaturation or enzyme inactivation, which is restored when the high pressure is removed.

High static pressure treatment neither causes the destruction of nutrients, color, and flavor of foods nor form harmful compounds. High-pressure treatment of meat products can lead to lysis of muscle fibers in muscle tissues, thereby resulting in improved quality for meat products.

(5) Electromagnetic radiation

The effect of electromagnetic waves on protein structure is related to the used wavelength and energy of electromagnetic wave. Visible light has little effect on protein conformation due to its long wavelength and low energy. Using high-energy electromagnetic wave such as UV, X-ray, and γ-ray can affect protein conformation. High-energy rays absorbed by aromatic amino acids can induce conformational changes of proteins and even chemical changes of amino acid residues such as destruction of covalent bonds, ionization, and free radicalization. Therefore, high-energy rays can not only lead to protein denaturation but also may affect the nutritional values of proteins.

Radiation preservation of foods has little effect on proteins as the used radiation dosage is relatively low. On the other hand, the dissociation of water can reduce the dissociation of other substances in foods.

(6) Interfacial effect

Irreversible protein degeneration occurs when proteins are adsorbed at gas–liquid, liquid–solid, or liquid–liquid interfaces. The water molecules at the gas–liquid interface have higher energy than those molecules in bulk solution and can interact with protein molecules, which gives rise to the increase in molecular energy of proteins. When some chemical effects (bonds) in protein molecules are broken, the structure stretches a little. Finally, water molecules enter internal protein molecules, which further leads to the extension of protein molecules. The hydrophobic and hydrophilic residues of proteins, respectively, thus arrange to the different polarity of two phases (air/oil and water), which eventually lead to protein denaturation.

Protein molecules have porous structures, which can be easily absorbed in the interface. If protein structure is tight, stabilized by disulfide bonds, or absence of obvious hydrophobic and hydrophilic regions, proteins cannot be absorbed in the interface easily, so interfacial protein denaturation is difficult.

3.3.2 Chemical Denaturation

(1) Acidic and alkaline factors (pH)

Most proteins are stable within a specific pH range. However, if they are in extreme pH conditions, dissociation of these dissociable groups within protein molecules such as amino and carboxyl groups will occur and thereby generating strong intra-molecular electrostatic interactions, which then lead to stretching and denaturation of protein. If accompanied by heating treatment, the denaturation rate of proteins will be even higher. In some cases, however, proteins (e.g. enzymes) can be restored to their native structures, when pH is adjusted to its original range after acid–base treatment.

Proteins are more stable at their isoelectric points than at any other pH value. At neutral pH, proteins do not have too much net charge and intra-molecular electrostatic repulsive force is relatively small, most proteins are, therefore, relatively stable in the neutral condition.

(2) Salts

Metal ions such as Ca^{2+} and Mg^{2+} may be the constituent parts of proteins and play an important role in protein conformation. Therefore, the removal of Ca^{2+} and Mg^{2+} will reduce the stability of protein molecules toward heat and enzyme treatments. It is easy for Cu^{2+}, Fe^{2+}, Hg^{2+}, Pb^{2+}, and Ag^{2+} to interact with the -SH groups of protein molecules or to convert disulfide bonds into -SH groups, which changes the force that stabilizes protein structure and thereby leads to protein denaturation. Hg^{2+} and Pb^{2+} can react with the His and Trp residues of proteins and bring protein denaturation.

For various anions, their influences on the structural stability of proteins follow the series: $F^- < SO_4^{2-} < Cl^- < Br^- < I^- < ClO_4^- < SCN^- < Cl_3CCOO^-$. At high concentration, anions have stronger impact on protein structure than cations. Generally, chloride ions, fluorine ions, and sulfate ions are structure stabilizers of proteins, whereas thiocyanate and trichloroacetic acid are structure destabilizers of proteins.

(3) Organic solvents

Most organic solvents can bring about protein denaturation as they reduce the dielec-tric constant of solution and increase intra-molecular electrostatic force of protein molecules. Or else, they may destroy or increase hydrogen bonds within protein molecules and change the original force that stabilizes protein conformation. In addition, they can enter into the hydrophobic regions of proteins and destroy the hydrophobic interaction of protein molecules. They may give rise to the changed protein structure and denaturation.

At low concentrations, organic solvents have little effect on protein structure, and some of them even have stabilizing effects. However, at high concentrations, all organic solvents can cause denaturation of proteins.

(4)   Organic compounds

High concentrations of urea and guanidine salts (4–8 mol/L) will break down the hydrogen bonds in protein molecules, leading to protein denaturation. Surfactants such as sodium dodecyl sulfonate can destroy the hydrophobic regions of proteins and promote the stretching of protein molecules and therefore are powerful denaturing agents.

(5)   Reducing agents

With -SH groups, mercaptoethanol ($HSCH_2CH_2OH$), cysteine, and dithiothreonol can reduce disulfide bonds existing in protein molecules and thus change the original conformation of proteins, resulting in irreversible degeneration of proteins.

$$HSCH_2CH_2OH + {-}S{-}S{-}Pr \rightarrow {-}S{-}SCH_2CH_2OH + HS{-}Pr$$

For food processing, protein denaturation is generally favorable, but in some cases such as enzyme separation and milk concentration, it should be avoided. At this time, excessive denaturation of proteins will lead to enzyme inactivation or precipitation formation, which is an undesirable change.

3.4   Functional Properties of Proteins

The functional properties of proteins are these physiochemical properties beneficial to the characteristics of foods, apart from their nutritional values, such as gelation, solubility, foamability, emulsification, viscosity, etc. The functional properties of proteins affect the sensory quality of foods, especially in terms of texture, and they also play an important role in the physical characteristics of food products, food processing, or storage, which can generally be divided into three categories.

(1)   Hydration properties, which depend on the interaction between proteins and water, including water adsorption and retention, wettability, swelling, adhesion, dispersibility, and solubility.

(2)   Structural properties (interaction between protein molecules) such as precipitation, gelation, texturization, dough formation, etc.

(3)   Surface properties, involving the role of proteins in two phases with different polarity, mainly including foaming and emulsification.

According to some functions of proteins in food sensory quality, the fourth property, sensory property, can be classified, which involves the turbidity, color, flavor binding, chewiness, and smoothness of proteins in foods.

Functional properties of proteins are not only independent and completely different from each other but also related to each other. For example, gelation involves both the interaction between protein molecules (formation of three-dimensional spatial network structure) and the interaction between protein molecules and water molecules (water retention). Viscosity and solubility of proteins are also related to the interaction between proteins and water.

The functional properties of proteins are the results of synergistic action of many related factors. The physiochemical properties (molecular sizes, shapes, chemical compositions, and structures) of proteins as well as many external factors have an impact on the functional properties of proteins. Overall, the factors affecting the functional properties of proteins can be divided into three aspects: (1) the inherent properties of proteins, (2) environmental conditions, and (3) food processing.

In general, certain functional property of protein is not the result of certain physicochemical property, so it is difficult to explain how the physicochemical property of proteins plays a role in the functional properties. However, some physicochemical constants of proteins are well correlated with other functional properties (Table 3.8).

Proteins are not only important nutritional components but also their functional properties are incomparable and indispensable, compared with some other food components, and are crucial for the quality of some foods. The functional properties of proteins required in common foods are shown in Table 3.9.

Table 3.8 Contribution of hydrophobicity, charge density, and structure of proteins to functional properties

Functional properties	Hydrophobicity	Charge density	Structure
Solubility	No contribution	With a contribution	No contribution
Emulsification	Contribution by surface hydrophobicity	Generally, no contribution	With a contribution
Foaming	Contribution by total hydrophobicity	No contribution	With a contribution
Fat binding	Contribution by surface hydrophobicity	Generally, no contribution	No contribution
Water binding	No contribution	With a contribution	In doubt
Thermal coagulation	Contribution by total hydrophobicity	No contribution	With a contribution
Dough formation	Slight contribution	No contribution	With a contribution

Table 3.9 Functional properties of proteins required in various foods

Foods	Functional properties
Beverages	Solubility, thermal stability, viscosity at different pH values
Soup, sauce	Viscosity, emulsification, water binding
Dough baking products (bread, cake, etc.)	Molding and forming viscoelastic film, cohesion, thermal denaturation and gelation, emulsification, water absorption, foaming, and browning
Dairy products (cheese, ice cream, dessert, etc.)	Emulsification, fat retention, viscosity, foaming, gelation, coagulation
Egg	Foaming, gelation
Meat products (sausages, etc.)	Emulsification, gelation, cohesion, absorption, and retention of water and fat
Meat substitutes (histochemical plant proteins)	Absorption and retention of water and fat, insolubility, hardness, chewiness, cohesion, thermal denaturation
Food coating	Cohesion, adhesion
Confectionery (milk chocolate, etc.)	Dispersion, emulsification

3.4.1 Protein Hydration

Most foods are hydrated systems. Physiochemical and rheological properties of each component are not only affected by water but also by water activity. Protein conformation is largely related to the interaction between proteins and water. In addition, hydration process of proteins is also involved in the application of protein concentrate or isolate produced from different raw materials in foods. The ability of proteins to absorb and retain water affects not only the viscosity and other properties of proteins but also the quality and the quantity (directly related to the product cost) of food products. Therefore, it is very useful to study the hydration and rehydration of proteins.

Protein hydration has arisen from the interaction of various polar groups on the surface of protein molecules with water molecules. In general, about 0.3 g/g of water can bind firmly to proteins, while about 0.3 g/g of water may bind loosely to protein. Due to the different contents of amino acids, different proteins have various water-binding capacities. Polar groups of amino acids have the stronger binding capacities to water, thus ionized amino acids and protein salts have relatively higher binding capacities. The binding capacities of different amino acid residues to water are shown in Table 3.10.

Protein concentration, medium pH, temperature, ionic strength, and the presence of other components can affect protein–protein and protein–water interactions. The total water-binding amount of proteins increases with an increase in protein concentration, but proteins usually show minimal hydration at their isoelectric points.

Table 3.10 Hydration capacities of amino acid residues (mol water/mol residues)

Amino acid residue		Hydration capacity	Amino acid residue		Hydration capacity
Polar residue	Asn	2	Ionic residue	Asp	6
	Gln	2		Glu	7
	Pro	3		Tyr	7
	Ser, The	2		Arg	3
	Trp	2		His	4
	Asp (unionized)	2		Lys	4
	Glu (unionized)	2	Nonpoloar	Ala	1
	Tyr	3		Gly	1
	Arg (unionized)	4		Phe	0
	Lys (unionized)	4		Val, Ile, Leu, Met	1

After animal slaughter, the water-binding capacities of muscle tissues are the worst during rigor mortis, which is caused by the decrease of muscle pH from 6.5 to 5.0 (close to isoelectric point), resulting in the decreased tenderness and poor quality of meat products. In general, the water-binding capacities of proteins decrease with the increase of temperature, which is due to the destruction of hydrogen bonds formed between proteins and water by heating, the reduction of the interaction between proteins and water, the denaturation and aggregation of proteins during heating, the reduced surface area of proteins, as well as the reduced water-binding effectiveness of polar amino acids (see Fig. 3.7). However, the heating treatment also sometimes improves the water-binding capacities of proteins. Proteins with very compact structures can undergo subunit dissociation and molecular extension during heating treatment, which exposes some previously covered peptide bonds and polar groups to protein surface and thus improves water-binding capacity. Or else, protein gelation can occur when proteins are heated. The yielded three-dimensional network can hold a large amount of water and thereby improves the water-binding capacities of proteins. Ions present in the protein system also have an impact on the water-binding capacities of proteins, which is the result of competitive effects of water–salt–protein. In usual, low salt concentration can increase the water-binding capacities of proteins (salting-in), while high salt concentration decreases the water-binding capacities of proteins (salting-out) and may even cause protein dehydration.

For some monomer proteins, their water-binding capacities can be estimated by an empirical formula and amino acid compositions. The calculated results are in good agreement with the experimental results. For some proteins composed of multiple subunits, the calculated values are generally greater than the experimental values.

Fig. 3.7 Water-binding
capacities of muscle protein
at different temperatures and
pH values

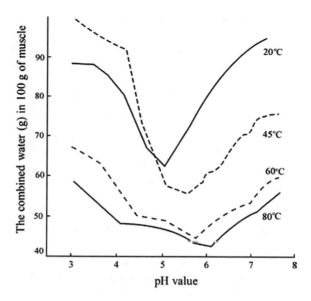

$$\text{Water-binding capacity (g water/g protein)} = f_C + 0.4 f_P + 0.2\, f_N.$$

where, f_C, f_P, and f_N represent the percentage of charged, polar, and non-polar (i.e. hydrophobic) amino acid residues in protein molecules, respectively. It can be seen from the coefficients that the charged amino acids contribute most to water-binding capacity but the non-polar amino acids have the least influence on water-binding capacity.

Water-binding capacity plays an important role in the texture of various foods, especially meat products and dough. Other functional properties of proteins, such as the gelation and emulsification, are also related to protein hydration. During food processing, protein hydration is usually measured or reflected by water-holding capacity or water retention. Water-holding capacity refers to the ability of proteins to retain (or bind) water in their tissues. The retained water includes adsorbed water, physical retained water, and hydrodynamic water. Water-holding capacity can affect tenderness, juiciness, and softness of foods. Therefore, water-holding capacity is of great significance to food quality.

3.4.2   Protein Solubility

As organic macro-molecules, proteins exist in the dispersed (colloidal) state in water. Accordingly, proteins do not have true solubility in water. The protein amount dispersed in water is usually measured using an empirical index called protein solubility. Protein solubility is very important to protein ingredients, which determines their extraction, separation, and purification. Denaturation degree of proteins can also be reflected by assessing their solubility changes. In addition, the application of proteins in acidic beverages is directly governed by their solubility.

The commonly used indices for protein solubility include protein dispersibility index (PDI), nitrogen solubility index (NSI), and water soluble nitrogen (WSN).

$$PDI\ (\%) = \frac{The\ weight\ of\ water\ dispersed\ protein}{The\ weight\ of\ total\ protein} \times 100$$

$$NSI\ (\%) = \frac{The\ weight\ of\ water\ soluble\ nitrogen}{The\ weight\ of\ total\ nitrogen} \times 100$$

$$WSN\ (\%) = \frac{The\ weight\ of\ water\ soluble\ nitrogen}{The\ sample\ weight} \times 100$$

Some conditions such as pH, ionic strength, temperature, and solvent type can affect protein solubility. Protein solubility is usually the lowest at a pH value near isoelectric point. When pH values are higher or lower than the isoelectric points, the net charges of proteins are negatively or positively charged, and their solubility levels thus increase (Fig. 3.8). Protein solubility is the lowest at isoelectric point, but this behavior is different between proteins. Casein and soy protein isolate are almost insoluble at their isoelectric points, while whey protein concentrate is still very soluble at isoelectric point. The proteins with solubility varying greatly with pH values can be conveniently extracted and separated by changing pH values. Proteins with solubility varying slightly with pH value need to be separated and extracted by other methods.

Salts have different impacts on protein solubility. When the concentration of neutral salt is 0.1–1 mol/L, protein solubility in water can be increased (salting-in).

Fig. 3.8 NSI of several proteins at various pH values

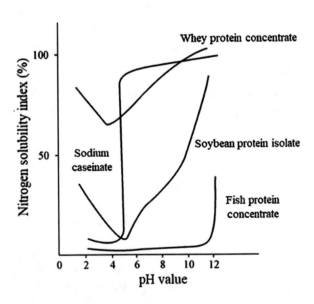

When the concentration of neutral salt is greater than 1 mol/L, protein solubility in water can be reduced and even precipitation can occur (salting-out). When salting-in or salting-out happens, the mathematical relationship between solubility and salt ionic strength follows the two formulas below.

$$Salt-in: S = S_0 + k\mu^{0.5}$$
$$Salt-out: \log S = \log S_0 - k\mu$$

Due to their reduction of dielectric constant of organic solvents, electrostatic repulsion between protein molecules is weakened by organic solvents such as acetone, ethanol, etc. The attraction between protein molecules is relatively increased, so protein aggregation and even precipitation can occur. In other words, organic solvents reduce protein solubility.

Protein solubility is irreversibly decreased after heating treatment of proteins. Protein extraction and purification process may lead to a certain degree of protein insolubility, such as defatted soybean powder. Protein concentrate and isolate thus have nitrogen solubility index varying from 10 to 90%. Generally, when other conditions are fixed, protein solubility increases with temperature at the range of 0–40 °C; as the temperature is further increased, stretching and especially denaturation of protein molecules occur, and protein solubility is eventually decreased (Table 3.11).

It is generally believed that proteins with higher initial solubility can rapidly dispersed in large quantities in the system, so a good dispersion system is obtained, which is conducive to the diffusion of protein molecules to the air or water–oil interface and improvement of other functional properties of proteins.

Table 3.11 The relative changes of solubility of some proteins after processing

Protein	Treatment	Solubility	Protein	Treatment	Solubility
Serum protein	Native	100	Albumin	Native	100
	Heating	27		80 °C, 15 s	91
β-lactoglobulins	Native	100		80 °C, 30 s	76
	Heating	6		80 °C, 60 s	71
Soybean protein isolate	Native	100		80 °C, 120 s	49
	100 °C, 15 s	100	Rapeseed protein isolate	Native	100
	100 °C, 30 s	92		100 °C, 15 s	57
	100 °C, 60 s	54		100 °C, 30 s	39
	100 °C, 120 s	15		100 °C, 60 s	14
				100 °C, 120 s	11

3.4.3 *Viscosity*

The viscosity of a fluid can be illustrated by two plates with relative movement. The plates are filled with fluid, and the relative movement of plates occurs under external force (F). If the fluid has a high viscosity, the plates will move very slowly, whereas the plates will move very fast. Therefore, viscosity is a measure of the resistance of fluid against motion. Generally, viscosity coefficient μ is used to represent the viscosity level of a kind of liquid. It is the numerical ratio of shear force (τ) and shear rate (γ), while shear rate is the ratio of movement velocity (υ) and the distance (d) of two plates with relative movement.

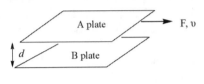

$$\tau = \mu \times \gamma = \mu \times \frac{\upsilon}{d}$$

Newtonian fluid (ideal fluid) has a fixed μ value. That is, μ value does not change along with the change of shear stress or rate. However, the dispersion systems composed of macro-molecules (including solution, emulsion, suspension liquid, gels, etc.) do not have the property of Newtonian fluid. The μ values of the dispersion systems will change as fluid shear rate or the stress changes, and their relationship in numerical value varies.

$\tau = m \times \gamma^{n}$ (m is consistency coefficient, n is flow index, $n < 1$)

There are many factors influencing the viscosity of proteins, including sterilization, pH shift, proteolysis, and the presence of inorganic ions. The main factor influencing viscosity characteristics of protein fluid is the apparent diameters of dispersed protein molecules or particles, which varies with the following parameters: (1) the inherent characteristics of protein molecules, such as sizes, volumes, structures, charges, and protein concentrations; (2) the interaction between proteins and solvents (water) molecules; and (3) the interactions between protein molecules.

The μ value of a protein solution will be decreased with the increased velocity. This phenomenon is called "shear thinning". The reasons are as follows: (1) protein molecules gradually move in the same direction so that protein molecules are orderly arranged, and thereby the friction resistance generated by liquid flow is reduced; (2) protein hydration environment deforms in the direction of motion; and (3) hydrogen bonds and other weak bonds are broken, resulting in the dissociation of protein aggregates and network structures as well as the reduction of protein volume. In summary, shear thinning can be explained by a reduction in the apparent diameters of protein molecules or particles in the direction of motion.

The breakdown of weak forces in protein molecule usually slowly occurs as protein solution flows. Therefore, the apparent viscosity of protein solution decreases with the increase of time before it reaches equilibrium. When the shear stops, the original aggregates may or may not reform. If protein aggregates can be reformed, the viscosity coefficient will not be decreased. The system is thixotropic. Both whey protein concentrate and soybean protein isolate can produce a thixotropic system.

There is no simple relationship between protein viscosity and solubility. The insoluble proteins obtained by thermal denaturation do not produce high viscosity after dispersion in water. However, for whey proteins with good solubility but poor water absorption and swelling ability, they also cannot form high viscosity dispersion system in water. For those proteins (such as soy protein, sodium caseinate) with great initial water absorption capacities, their dispersions have high viscosity, so there is a positive correlation between water adsorption capacity of proteins and dispersion viscosity.

The viscosity and consistency of protein system are main functional properties of liquid foods, such as beverage, broth, and soup, which affect food quality and processing like transportation, blending, heating, and cooling.

3.4.4 Protein Gelation

The scientific terms gelation, association, aggregation, polymerization, precipitation, flocculation as well as coagulation of proteins are used to reflect the changes of protein molecules at different levels of aggregation. These terms have certain differences. Protein association refers to the changes at the subunit or molecular level. Protein polymerization or aggregation generally refers to the generation of larger protein polymers. Protein precipitation refers to all precipitation reactions because of partial or total loss of protein solubility. Protein flocculation refers to the disordered aggregation reaction without protein denaturation. Protein coagulation is the disordered aggregation of the denatured proteins, and more, protein gelation is the orderly aggregation reaction of the denatured proteins.

Gel is the product formed after protein gelation. Protein gels have ordered three-dimensional network structure, contain other components, and play important role in food texture (e.g. meat). Proteins can not only form a semi-solid viscoelastic texture but also have the functions of water holding, fat stabilization, and adhesion. For some protein foods such as tofu and yogurt, protein gelation is the basis of quality formation.

The network structure of protein gels is the result of balanced protein–protein interaction, protein–water interaction as well as attraction and repulsive force between adjacent peptide chains. Electrostatic attraction and protein–protein interaction (including hydrogen bonds and hydrophobic interaction, etc.) are conducive to the proximity of protein peptide chains, while electrostatic repulsion and protein–water interaction are reasons for the separation of protein peptide chains. In most cases, heat treatment is necessary for protein gelation (protein denaturation and extension of

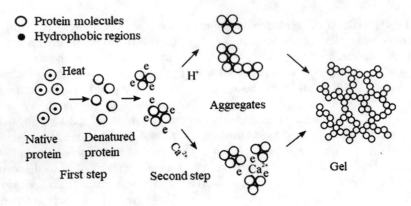

Fig. 3.9 Schematic diagram of gelation process of soybean protein

peptide chains), followed by cooling (formation of hydrogen bonds between peptide chains).

When protein gels are formed, the addition of a small amount of acid or Ca^{2+} can improve the speed and strength of protein gels. Sometimes, proteins can form gels without heating. Some proteins via addition of Ca^{2+} salt, appropriate enzymatic hydrolysis, and addition of alkaline solution, followed by adjustment of pH value to isoelectric point may also yield protein gels. The role of calcium ions is to form the so-called salt bridges.

Generally, the process of protein gelation can be divided into two steps: (1) conformation change or partial extension of protein molecules, and the occurrence of denaturation and (2) the denatured protein molecules gradually aggregate to form a network structure that can hold water and other substances in an orderly manner (Fig. 3.9).

According to the used pathway of gel formation, gels are generally divided into heat-induced gels (such as gels formed by heating egg albumin) and non-heat-induced gels (gel formed by adjustment of pH, addition of bivalent metal ions, or partial protein hydrolysis). According to the thermal stability of protein gels, they can be classified into thermal reversible gel (such as gelatin that can form solution when heated and restore to gel after cooling) and thermal irreversible gel (such as egg white protein and soy protein; gel state is formed by heat treatment never change). The thermally reversible gels are stable mainly through the formation of hydrogen bonds between protein molecules. The thermally irreversible gels mostly involve the formation of disulfide bonds between protein molecules. Once disulfide bonds are formed, they are not easy to break again and cannot be damaged by heating.

When proteins form gels, protein molecules may arrange in two different ways (Fig. 3.10). The gel formed by orderly cluster arrangement is transparent or translucent, such as the gels yield from serum protein, lysozyme, egg albumin, soybean globulin, etc. The gel formed by free aggregation is opaque, such as myosin gel formed under high ionic strength and gels formed by whey protein and β-lactoglobulin. Among common protein gels, these two different ways may simultaneously exist

Fig. 3.10 Schematic diagram of network structure of protein gels. **a** Orderly aggregation; **b** Free aggregation

and are affected by gelling conditions (such as protein concentration, pH, ion type, ion strength, heating temperature, heating time, etc.).

Protein gelation is usually produced by protein solutions, but insoluble proteins or protein dispersions in brine can also form gels. Therefore, protein solubility is not a necessary but only conducive factor for protein gelation.

Gelation is a very important functional property of proteins and plays an important role in preparation of many foods such as dairy products, gels, and various heated minced meat and fish products. Protein gelation can be used to form solid elastic gels, improves water/fat holding capacity, and also contributes to emulsification and foaming stability of food ingredients. Gelation is one of the most important functions of proteins and one of the most frequently considered indices in food processing.

3.4.5 Protein Texturization

Proteins are the basis of texture or structure of many foods such as animal muscle. However, some proteins in nature do not have the corresponding texture and chewing properties such as the isolated soluble plant proteins or milk proteins. Therefore, there are certain limitations in application of these protein ingredients in food processing. However, these proteins can currently be processed via the so-called protein texturization to form films or fibrous products with chewing properties and good water-holding properties and maintain good properties after hydration or heating treatment in the future. The texturized proteins can be used as meat substitutes or substitutes in foods. In addition, protein texturization can also be used to reorganize some by-products of animal proteins.

There are three common methods for protein texturization.

(1) Thermal protein coagulation and film formation

When the water of concentrated soybean protein solution is evaporated on a smooth hot metal surface, proteins show thermal coagulation and form hydrated protein film. Soybean protein solution kept at 95 °C for a few hours thus yields a thin layer film of proteins. These protein films are textured proteins with stable structure against heating treatment and normal chewing properties. Yuba, a traditional soybean product, is processed by the above method.

If protein solution (such as the ethanol solution of zein) is uniformly coated on the surface of a smooth object, proteins can also form a uniform film (protein film) by interaction with each other after solvent evaporates. Protein films have various mechanical strengths and barrier functions against water, oxygen, and other gases, which can be used as edible packaging materials.

(2)   Thermoplastic protein extrusion

Thermoplastic extrusion method is to let protein mixture pass through a cylinder with the function of rotating screw. Under the action of high pressure, high temperature, and strong shear, solid materials can be transformed into a viscous state and then quickly pass through the cylinder and enter atmospheric environment. The moisture evaporates rapidly, resulting in the formation of a highly swelling and dried porous structure. The final product is called textured protein (commonly known as puffed protein). After water adsorption, the product has a fibrous elastic structure with chewing properties and is stable under sterilization conditions. It can be used as meat substitute and filler for meatballs, hamburgers, etc. This method is the most commonly used method for protein texturization at present.

Despite the absence of muscle fiber structure, thermoplastic protein extrusion results in the texturized protein with similar textural feature. The texturized soybean flour shows uniform texture and fibrous laminar structure characteristics (Fig. 3.11).

(3)   Protein fiber formation

This is another way to form texturized proteins, sharing same production principles of synthetic fibers. Protein solution with high concentration is prepared at pH > 10. Due to the increase of electrostatic repulsion, protein molecules are dissociated and fully extended. After degasification and clarification, protein solution passes through a nozzle with many holes under high pressure. At this time, the

Fig. 3.11 Microstructure of texturized soybean protein

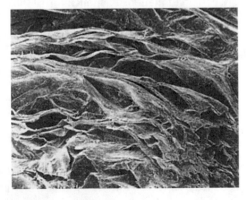

stretched protein molecules are oriented along the outflow direction and extended in a parallel way, and thereby orderly arranged. When the liquid from the nozzle enters an acidic solution containing NaCl, due to both isoelectric point and salt-out effect, the occurred protein coagulation induces the formation of hydrated protein fiber through hydrogen, ionic, and disulfide bonds, and other forces. By rotating the roller, protein fibers are stretched to increase their mechanical resistance and chewiness and reduce their water holding capacity. Then, through the heating of roller to remove part of the water, both fiber adhesion and protein toughness are improved. Finally, through a series of processing, such as seasoning, adhesion, cutting, and molding, artificial meat, or protein product analogous to meats can be formed.

Among three methods of protein texturization, thermoplastic protein extrusion is quite economical, simple, and without rigorous requirement of raw materials. Thermoplastic extrusion is not only suitable for raw materials with low protein contents (such as defatted soybean powder) but also for the raw materials with high protein contents. Protein fiber formation can only be used for the processing of protein isolates as they have higher protein contents.

3.4.6 Dough Formation

Wheat, barley, and rye have the same characteristics. In the presence of water, gluten protein in endosperm can form a strong cohesive and viscoelastic paste (dough) through mixing, kneading, and other treatments. Wheat flour has the strongest ability to form paste. After dough formation, gluten in wheat flour and other components such as starch, sugar and polar lipids, non-polar lipids, and soluble proteins are conducive to the formation of gluten three-dimensional network structure as well as the final texture of bread, which are entrapped in this three-dimensional structure.

Gluten protein is mainly composed of glutenin and gliadin, which account for more than 80% of the total protein in flour. The characteristics of dough are directly related to glutenin and gliadin. First, the contents of these dissociable amino acids in glutenin and gliadin are low, so they are insoluble in neutral water. Second, they contain many glutamine and hydroxyl-containing amino acids, so it is easy for them to form intermolecular hydrogen bonds. Gluten thus has a strong ability to absorb water and adhesion property, which is also related to hydrophobic interaction. Finally, glutenin and gliadin contain -SH groups, which can form disulfide bonds, so they are tightly bound together in dough to make it tough. When flour is kneaded, protein molecules are stretched, while disulfide bonds are formed and hydrophobic interaction is enhanced. Gluten protein is then transformed into three-dimensional and viscoelastic network structure. Starch particles and other components are entrapped in this structure. If the reducing agent is added to destroy these disulfide bonds, the cohesive structure of dough is destroyed. When oxidant like $KBrO_3$ is added to promote the formation of disulfide bonds, it is conducive to the elasticity and toughness of dough.

The proper balance between gluten and gliadin is important to bread making. The molecular weight of glutenin is as high as 1×10^6 Da, and glutenin contains many disulfide bonds (intra-chain and inter-chain). The molecular weight of gliadin is only 1×10^4 Da, with only intra-chain disulfide bonds found in gliadin molecules. Gluten determines the elasticity, adhesion, and strength of dough, while gliadin determines the fluidity, extensibility, and expansibility of dough. The strength of bread is related to glutenin. High content of glutenin will inhibit the expansion of residual CO_2 during the fermentation process and bulging of dough.

If the content of gliadin is too high, it will lead to excessive expansion. As a result, the resulting gluten will be easily broken and permeable, and dough will collapse. The addition of polar lipids to dough was beneficial to the interaction between glutenin and gliadin, and the network structure of gluten was improved. However, addition of neutral lipids was unfavorable. Addition of globulin to dough is generally not conducive to dough structure, while addition of denatured globulin can eliminate adverse effects.

When dough is kneaded, the three-dimensional network structure of gluten protein will not be well formed, if kneading strength is insufficient. As a result, the dough strength is insufficient. Excessive kneading can also cause breakage of some disulfide bonds in gluten, resulting in a decrease of dough strength. When dough is baked, water released by gluten is absorbed by the gelatinized starch molecules, but gluten retains nearly half amount of water. Gluten protein is fully stretched during kneading, which cannot be further extended during baking.

3.4.7 Emulsifying Property

Many daily foods are regarded as protein-stable emulsions. The formed dispersions include water-in-oil (W/O) and oil-in-water (O/W) types. Milk, ice cream, margarine, cheese, mayonnaise, and minced meat are the most common water–oil dispersion system, where proteins stabilize this emulsion system. Proteins are adsorbed at the interface between oil droplet and water phase to produce anti-coagulable physical and rheological properties (such as electrostatic repulsion and viscosity). The most important role of soluble proteins is that they could spread to the oil–water interface and adsorb at the interface. Part of proteins will contact the interface and the hydrophobic (or hydrophilic) amino acid residues are oriented to oil (or water) phase, which reduces free energy of system. The rest of the proteins partially will be unfolded and spontaneous adsorbed on the interface to exhibit corresponding interface properties. It is generally believed that the greater the hydrophobicity of proteins, the higher protein concentration adsorbed on the interface. Therefore, it will lead to smaller interfacial tension and stable emulsion system.

Globulins such as serum protein and whey protein have a relatively stable structure and higher surface hydrophilicity, so are not good emulsifiers. Casein is a good emulsifier due to its structural characteristics (random coil) and the relative separation of hydrophilic and hydrophobic regions in the peptide chain. Soy protein isolate, meat, and fish proteins also have good emulsifying properties.

The emulsion system is thermodynamically unstable, and the interaction between fat globules will inevitably lead to emulsion instability. The result is the complete separation of oil and water phases. In addition, the instability O/W system can be coalescence, flocculation, and stratification. Coalescence refers to the rupture of membrane between fat globules, which leads to the formation of large fat globules. Flocculation refers to the process of aggregation between fat globules without rupture of membranes. Stratification refers to the floating of fat globules due to their density less than the continuous (i.e. water) phase. All kinds of emulsion instability can occur independently or simultaneously.

Protein solubility is positively correlated with emulsifying properties. Generally, insoluble proteins have no influence on emulsion formation. Therefore, the improvement of protein solubility will be conducive to improved emulsification performance. For example, when NaCl exists in minced meat (0.5–1 mol/L), emulsifying capacity of proteins can be improved due to salt-in effect of NaCl. However, once emulsion is formed, adsorption of insoluble proteins on the membrane will promote the stability of fat globules. The pH of solution also affects emulsification. Gelatin and ovalbumin at pI have good emulsification performance. Most of the other proteins like soybean protein, peanut protein, casein, myofibril, and whey protein have better emulsification performance when they are not at pI. At this time, the dissociation of side chains in amino acids will generate greater electrostatic repulsion, which contributes to stability and avoidance of droplet aggregation. At the same time, it is helpful for protein dissolution and water binding, which also improve the stability of protein membrane.

Heating can reduce the viscosity of protein membrane adsorbed in the interface, and thus decrease the stability of emulsion. However, if protein gelation is generated by heating, viscosity and hardness of the membrane can be improved; emulsion stability is thus improved. For example, gelation of myofibrillar protein is beneficial to the stability of emulsion systems, e.g. sausages, which can improve water holding capacity and fat retention of meat products, and enhance adhesion between components at the same time.

Low molecular weight surfactants are generally adverse to emulsion stability of proteins, as they will be completely absorbed at the interface, leading to weakened protein adsorption at the interface, reduction in viscosity of protein membrane, and consequently reduced emulsion stability.

The emulsifying properties of proteins are generally determined by emulsifying activity index (EAI), emulsifying capacity (EC), and emulsifying stability index (ESI), which reflect the ability of proteins to help form emulsion systems and stabilize them. These indices reflect that (1) proteins can reduce the interface tension to help emulsion formation and (2) the emulsion system is stabilized by increasing viscosity of the adsorption membrane, steric resistance, and other factors. There is no correlation between the ability of proteins to form emulsified dispersion system and the ability to stabilize the emulsified dispersion system. The emulsifying properties of some proteins are shown in Tables 3.12 and 3.13.

Table 3.12 Emulsifying activity index of some proteins (ionic strength of solution = 0.1 mol/L)

Protein	Emulsifying activity index		Protein	Emulsifying activity index	
	pH 6.5	pH 8.0		pH 6.5	pH 8.0
Egg albumin	–	49	Whey protein	119	142
Lysozyme	–	50	β-lactoglobulin	–	153
Yeast protein	8	59	Sodium caseinate	149	166
Hemoglobin	–	75	Bovine serum albumin	–	197
Soy protein	41	92	Yeast protein (88% acylation)	322	341

Table 3.13 Emulsifying capacity **and emulsifying stability of some proteins**

Protein source	Type	EC/ (g/g)	ES (24 h)/%	ES (14 d)/%
Soybean	Protein isolate	277	94	88.6
	Soybean flour	184	100	100
Egg	Protein powder	226	11.8	3.3
	Liquid protein	215	0	1.1
Milk	Casein	336	5.2	41.0
	Whey protein	190	100	100

The interaction between proteins and lipids is helpful to the formation and stability of emulsion but it may also have adverse effects. When proteins are extracted from lipid-rich raw materials, the formation of emulsion may decrease extraction efficacy and protein purity.

3.4.8   Foaming Property

Foam refers to the dispersion system formed by gas dispersed in the continuous liquid phase or semi-solid phase. Typical food examples include ice cream, beer, and so on. In a stable foam system, continuous phase of elastic thin layer separates each bubble, and the diameter of bubble ranges from 1 μm to several cm.

Food foam has several characters: (1) the incorporation of a large amount of gas; (2) there is a large surface area between gas and continuous phases; (3) relatively high solute concentration at the interface; (4) rigid or semi-rigid and elastic films that can expand; (5) opaque foam that can reflect light. The main difference between a foam and an emulsion is whether the dispersed phase is gas or fat and that gas takes up a larger percentage of the volume in the foam system. Foam has a large interfacial area, while the interfacial tension is much higher than the emulsified dispersion system. Foam thus is more unstable and prone to rupture. At this point, the roles of proteins are to adsorb at the gas–liquid interface to reduce the interfacial tension, and at the same

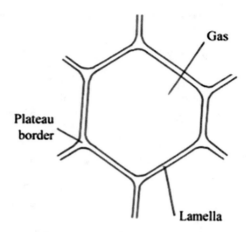

Fig. 3.12 Structure diagram of foam

time to render rheological characteristics and stability of the formed membrane to increase the strength, viscosity, and elasticity of membrane against adverse external effects. The typical structure of foam is shown in Fig. 3.12, in which the properties of lamella have a very important influence on foam stability.

The method for foam formation includes: (1) gas goes through the porous disperser and then passes into protein solution to generate corresponding gas bubbles; (2) in the presence of abundant gas, gas bubbles can be generated by mechanically stirring or oscillating protein solution; (3) under high pressure, gas is dissolved in solution, followed by sudden release of the pressure. The gas expands to form foam. In the process of foam formation, proteins are rapidly diffused and adsorbed at the gas–liquid interface in the first place, and then molecular structure of proteins is rearranged after entering the interface layer. Among them, diffusion process is a decisive factor.

The reasons for foam instability are listed here. (1) The thin layer of foam is drained due to gravity, pressure difference, evaporation, etc., which reduces the thickness of thin layer and eventually leads to foam rupture. (2) Due to the different sizes of bubbles, the pressure of gas in the small bubbles is higher, while the pressure in the large bubbles is lower. Therefore, gas transfers from the small bubbles to the large bubbles through continuous phase, resulting in the decrease of the total foam area, which is a spontaneous process to reduce the free energy of surface. At this point, interfacial expansion will lead to an increase in interfacial tension. To reduce the increased tension, protein molecules (carrying water molecules together) will migrate from the low-tension region to high-tension region, reducing the thickness of the thin layer in original region (Marangoni effect) and thus reducing foam stability. (3) The thin layer that separates bubbles ruptures. The thickness and strength of thin layer are reduced due to drainage and other factors, while the diameter of foam increases through agglomeration, eventually leading to rupture of bubble. Foam drainage correlates with thin layer rupture. The rupture of thin layer increases foam drainage, and enhanced drainage promotes thickness and strength decreases of thin layer, which is equivalent to a vicious cycle.

Table 3.14 Intrinsic properties that affect the foaming properties of proteins

Solubility	Rapid diffusion to the gas–liquid interface
Hydrophobicity	The polar zone and the hydrophobic zone are independently distributed, which can reduce the interfacial tension
Flexibility of the peptide chain	It is beneficial to extension and deformation of protein molecules at the interface
Interactions between peptide chains	It is beneficial to the interaction of protein molecules to form a good viscoelastic and stable adsorption membrane
Dissociation of groups	It is beneficial to the repulsion between bubbles, but high charge density is also adverse to protein adsorption on the membrane
Polar group	The combination of water and interaction between protein molecules are beneficial to the stability of the adsorbed membrane

Factors influencing the foaming properties of proteins include the following nine points:

(1) Intrinsic properties of proteins. A protein with good foaming property should be the protein that can quickly diffuse to the gas–liquid interface, and easy to be adsorbed, expanded, and rearranged at the interface to form a viscoelastic adsorption membrane through the interaction between molecules. β-Casein with a loose, free and curly structure is such a protein. Lysozyme, by contrast, is a tightly wound globulin that has disulfide bonds in multiple molecular sites and thus has a poor foaming property. The relationship between physical and chemical properties of proteins and the foaming properties are shown in Table 3.14.

Proteins with good foaming ability are generally very poor in foaming stability, while proteins with poor foaming ability are relatively good in foaming stability, for foaming ability and foam stability of proteins are two different molecular properties. Foaming ability depends on rapid diffusion of protein molecules, while the decrease of interfacial tension and distribution of hydrophobic groups are mainly determined by solubility and hydrophobicity of protein, and softness of peptide chains. Foaming stability is mainly determined by rheological properties of protein solutions, such as protein hydration in adsorption membranes, protein concentration, membrane thickness, and appropriate molecular interactions of proteins. Usually, egg albumin is the best protein foaming agent. Other proteins such as serum protein, gelatin, casein, gluten, soy protein, etc., also have good foaming properties.

(2) Salts. Salts affect the solubility, viscosity, stretching, and depolymerization of proteins as well as foaming properties. For example, NaCl increases expansion but decreases foaming stability. Ca^{2+} can improve foaming stability by forming salt bridges with the carboxyl groups of proteins.

(3) Sugars. Sugars usually inhibit foam expansion of proteins, but it can improve the viscosity of protein solutions. Hence, foaming stability is improved.

(4) Lipids. When low concentration lipids contaminate protein solution, the foaming ability of proteins will be seriously impaired. Polar lipids can be also adsorbed at the gas–water interface, interfere with protein adsorption, affect the interaction between proteins adsorbed, and thus decrease the foam stability.

(5) Protein concentration. Proteins at 2–8% concentration can reach the maximum degree of expansion, where the liquid phase has the best viscosity and the film has the appropriate thickness and stability. When protein concentration exceeds 10%, the viscosity of protein solution is too high, which affects the foaming ability, resulting in smaller bubbles and stiffened bubbles.

(6) Mechanical treatment. The formation of foam by mechanical treatment requires appropriate agitation, which leads to protein extension. However, stirring strength and time must be moderate. Excessive agitation will cause protein flocculation and reduce the degree of expansion and foam stability of proteins, for the flocculated proteins cannot be properly adsorbed at the interface.

(7) Heating treatment. Heating treatment is generally adverse to foam formation, as it causes gas expansion, viscosity reduction, and bubble rupture. However, proper heat treatment of some compact proteins before foaming is beneficial, since it can cause the unfolding of protein molecules and facilitate their adsorption at the gas–liquid interface. If heating treatment can lead to protein gelation, foaming stability will be greatly improved.

(8) pH. When pH is close to pI, the foam system stabilized by protein is very stable. This is due to that, the repulsive force between protein molecules is very low, which promotes protein–protein interaction and protein adsorption on the membrane. A sticky adsorption membrane thus is formed with enhanced foaming power and stability. The foaming ability of proteins at pH other than pI is generally better, but their stability is generally poor.

To evaluate the foaming property of proteins, one method is used to evaluate their ability to encapsulate gas (i.e., foaming power), while another method is to evaluate the foam life (i.e. foam stability). Foaming power increases with protein concentration, so data comparison at a single concentration is inaccurate. The values of foaming power under three different conditions are usually compared (Table 3.15).

Foam stability is generally a reflector of rupture extent of the foam sample placed for a period, or the drainage speed of the foam sample at different times. Foam stability can be measured by the time required to drain 1/2 liquid volume from the

Table 3.15 Comparison of foaming power values of the three proteins

Proteins	Maximum foaming power (concentration 20–30 g/L)	The concentration (g/L) with 1/2 maximum foaming force	Foaming power at 10 g/L
Gelatin	228	0.4	221
Caseinate	213	1	198
Soy protein isolate	203	2.9	154

foam sample after foam rupture or by the changes of foam volume at different times. Despite any methods, foam stability depends on protein concentration.

3.4.9 Binding of Flavor Compounds

Aldehydes, ketones, acids, phenols, and the fatty decomposition products present in foods may produce off-flavor. These substances can also bind to proteins or other food ingredients. They can be released in food processing and finally perceived by consumers, thereby affecting the sensory quality of foods. However, the interaction between proteins and flavor compounds also has desirable aspects. For example, they can confer texturized plant proteins with meat flavor.

The interaction between proteins and flavor compounds includes physical and chemical binding. The force involved in the physical binding is mainly van der Waals force, which is reversible binding with energy change near 20 kJ/mol. The force involved in chemical binding (usually irreversible binding) includes hydrogen bonds, covalent bonds, electrostatic force, etc. The energy change is more than 40 kJ/mol. It is generally believed that there are some identical but independent binding sites in the structure of proteins, which result in their binding by interacting with flavor compounds (F) as below.

$$\text{Proteins} + n\text{F} \leftrightarrow \text{Proteins-F}_n$$

The Scatchard model describes the binding of proteins to flavor compounds. Where, V is the amount of flavor compounds (mol/molprotein) when flavor compounds are to proteins to achieve equilibrium, L is the amount of free flavor compounds (mol/L), K is the equilibrium constant of binding (L/mol), and n is the total number of binding sites for flavor compounds in 1 mol of proteins.

$$\frac{V}{[L]} = K(n - V)$$

For proteins composed of single peptide chain, a good result can be obtained by using this model. However, for proteins composed of polypeptide chains, the binding amount of flavor compounds per mole of proteins will decrease with the increase of protein concentration. This is due to that, the interaction of proteins reduces the effectiveness of proteins to bind with flavor compounds, e.g. some sites are hidden. The binding constants of flavor compounds to some proteins are shown in Table 3.16.

The binding of flavor compounds to proteins is affected by environmental factors. Water can improve the binding of polar volatile compounds to proteins but it does not affect the binding of non-polar substances, as water can increase the diffusion rate of polar substances. High salt concentration weakens the hydrophobic interaction of proteins, causing protein unfolding and increasing their binding to carbonyl compounds. Casein binds more carbonyl compounds at neutral or alkaline pH values

Table 3.16 Binding constants of flavor compounds of some proteins

Protein		The flavor compounds bound	N (mol/mol)	K (L/mol)	$\Delta G°$ (kJ/mol)
Serum albumin		2-heptanone	6	270	−13.8
		2-nonanone	6	1800	−18.4
β-lactoglobulin		2-heptanone	2	150	−12.4
		2-nonanone	2	480	−15.3
Soy protein	Native	2-heptanone	4	110	−11.6
		2-octanone	4	310	−14.2
		2-nonanone	4	930	−16.9
		Nonanal	4	1094	−17.3
	Partial denaturation	2-nonanone	4	1240	−17.6

than it does under acidic conditions, which is associated with the non-ionization of amino groups. Hydrolysis of proteins generally reduces their ability to bind flavor compounds (especially excessive hydrolyzed proteins), which is related to the destruction of the primary structure or binding sites of proteins. Thermal protein denaturation causes expanding of protein molecules, leading to an increase in the binding capacity of flavor compounds. The presence of lipids promotes the binding and retention of various carbonyl volatiles. When proteins are vacuum freeze-dried, 50% of volatiles initially bound to proteins can be released due to vacuum.

3.4.10 Binding of Other Compounds

In addition to binding to water, lipids, and volatile compounds, proteins can still bind to metal ions, pigments, dyes, and others. Proteins are also able to bind to a few mutagenic and other active compounds. This binding can produce a detoxification effect or toxic enhancement effect. Sometimes, this binding can still lead to nutritive loss of proteins. The binding of proteins to metal ions facilitates absorption of some essential minerals such as iron and calcium, and the binding of protein to pigments can be used for quantitative analysis of proteins. Isoflavones, which bind to soybean proteins, ensure their beneficial effect.

3.5 Protein Changes During Food Processing and Storage

Food processing may induce several beneficial effects. Enzyme inactivation of the oxidases can prevent potential oxidation, while microorganism inactivation can enhance food preservability. Food processing also can convert some substances (flavor precursors) into desired flavor compounds. However, nutritional and functional properties of proteins during food processing and storage may undergo some changes. These changes even bring safety issues to the processed foods.

3.5.1 Heat Treatment

Sterilization is widely employed for most foods and has some impacts on protein functionalities. Suitable sterilization conditions are thus recommended. Pasteurization at 72 °C for bovine milk induces inactivation for most enzymes, with a slight effect on whey proteins, flavor, and nutritional value. However, higher pasteurization temperature results in protein aggregation, casein dephosphorylation, and whey protein denaturation, and thus shows significant effect on milk quality. During meat processing, myofibril proteins and myosinogen will be aggregated at 80 °C, while the –SH groups in myofibril proteins are converted into –S-S- groups. At 90 °C, myofibril proteins will yield H_2S, and proteins may react with reducing sugars to undergo the Maillard reaction.

Heat treatment (especially using mild conditions) is beneficial to food proteins. Blanching or stemming treatments lead to enzyme inactivation and thus inhibit the production of off-color or off-flavor by the enzymatic oxidation. The anti-nutritional factors and toxins in plant foods are mostly damaged or inactivated during heat treatment. What is more, suitable heat treatment ensures partly unfolding of proteins and exposure of buried amino acids and thus increases hydrolysis and digestion of proteins. Suitable heat treatment also induces the production of some flavor substances to improve the sensory quality of foods.

However, excessive heat treatment is adverse to proteins, as amino acids have some reactions like deamination, desulfuration, and decarboxylation, which result in amino acid loss and decreased nutritional values. When the processed foods have both proteins and reducing sugars, the lysine residues in proteins can react with reducing sugars to form one product of the Maillard reaction, the Schiff bases. This product is not digestible in the body. The carbonyl compounds, generated from non-reducing sugars at high temperature or from lipid oxidation, also can react with proteins via the Maillard reaction. More importantly, under higher temperature together with long treatment time, proteins in the absence of reducing agents can undergo undesired changes in peptide bonds, resulting in decreased enzymatic digestion and bio-availability. Besides, several carcinogenic/mutagenic products can be formed in meats at high temperature (190–200 °C), due to amino acid pyrolysis.

3.5.2 Freezing Treatment

Foods stored at lower temperature (e.g. freezing treatment) have delayed microorganism growth, inhibited enzyme activity, and decreased rates for chemical reactions. Freezing treatment has no influence on nutritional values of proteins, but usually exerts clear effect on their functional properties. Meat foods after freezing and thawing treatments obtain a damage effect on tissues and cell membranes, and the water–protein interaction is irreversibly replaced by the adverse protein–protein interaction. Meat foods thus show poor quality and lower water retention. Freezing treatment of bovine milk induces casein precipitation after thawing treatment, which brings a worse sensory quality.

Protein denaturation under freezing conditions is controlled by the employed freezing rate. In general, a quick-freezing rate ensures the formation of smaller ice crystals, which results in weaker mechanical action on the cells and less protein denaturation. It is widely recommended using quick freezing treatment for foods to ensure minimum flavor and texture loss.

3.5.3 Dehydration Treatment

Food dehydration reduces food mass and especially water activity and thus has beneficial effect on food storage. However, it also brings some adverse impacts on proteins.

Hot-air drying now is less used in the food industry, as this technology results in poor rehydration, rigid texture, and lower flavor quality for the dehydrated meats and fishes. Vacuum drying has less impact on meat quality than hot-air drying. In this case, the applied partial pressure of oxygen as well as temperature is kept at lower levels and thereby leads to slow rates in oxidation and the Maillard reaction, respectively. Drum drying has been used in the production of milk powder; however, this technology is also mostly replaced by spray drying technology, as it usually results in decreased protein solubility and burnt flavor in the milk powder. Vacuum freeze-drying is widely used in the present, as this technology ensures proteins with slight denaturation but good rehydration. Vacuum freeze-drying has no effect on nutritional values and bioavailability of proteins and is especially suitable for bioactive proteins like enzymes and probiotics.

The most used drying technology is spray drying. Liquid foods are atomized into small droplets and heated by the hot air to evaporate water very quickly. This technology has little impact on proteins and thus is used in the production of protein ingredients or food rich in proteins.

3.5.4 Radiation Treatment

Low-dose radiation cannot damage protein structure and therefore has no impact on nutritional values of amino acids or proteins. However, high-dose radiation induces water dissociation to form the most active free radicals, OH. These radicals thus can react with proteins, resulting in many reactions such as deamination and cross-linking. Consequently, proteins have altered functional properties.

3.5.5 Alkaline Treatment

When alkaline treatment is used in food processing, especially in the condition of strong alkaline combined with temperature, some adverse reactions related to proteins will occur. Nutritional proteins may be seriously damaged. For example, proteins may undergo dephosphorization (phosphatized Ser) or desulfurization (Cys) to produce the active dehydroalanine residues (Fig. 3.13).

After that, dehydroalanine residues can react with lysine and cysteine residues to the respective lysinoalanine and lanthionine residues (Fig. 3.14). Proteins thus have decreased nutritional values and digestibility.

Alkaline treatment at a temperature higher than 200 °C may lead to racemization of amino acids (Fig. 3.15). Native L-amino acids are thus partly converted into D-amino acids, which results in decreased bioavailability as these D-amino acids have no biological values in the body. Alkaline treatment of proteins also destroys several amino acids like Arg, Ser, Thr, and Lys.

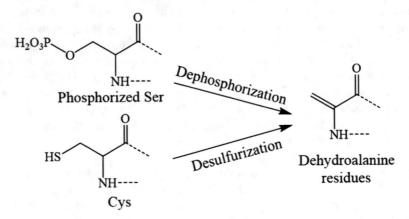

Fig. 3.13 Two pathways for the formation dehydroalanine residues

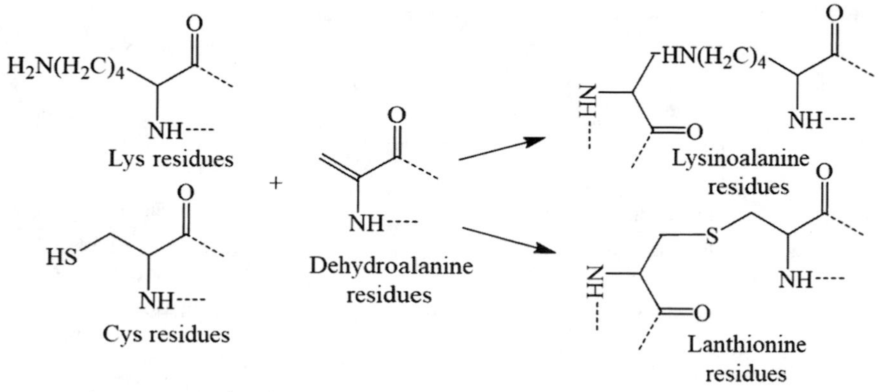

Fig. 3.14 The cross-linking of Lys and Cys residues of proteins with the formed dehydroalanine residues

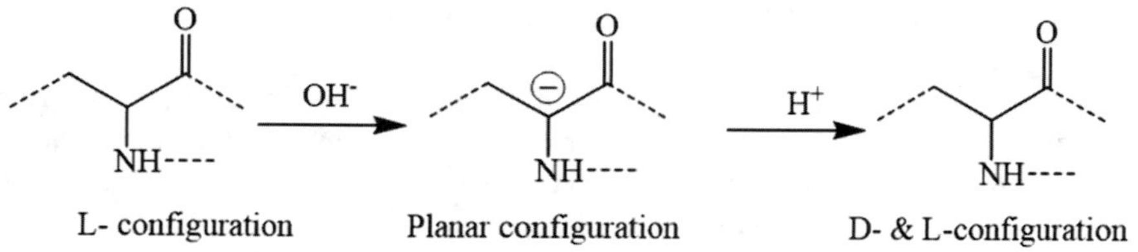

Fig. 3.15 Racemization of amino acids under alkaline condition

3.5.6   Other Protein Reactions

3.5.6.1   The Reaction of Proteins with the Free Radicals from Lipid Oxidation

Lipid (unsaturated fatty acids) oxidation leads to the formation of alkoxy and peroxyl (LOO·) radicals. These radicals in turn can react with proteins (Pr), resulting in protein cross-linking.

$$LOO· + Pr → LOOPr· \text{ and } LOOPr· + O_2 → LOOPrOO· → \cdots\cdots$$

$$\text{Or } LOO· + Pr → LOOH + Pr· \text{ and } Pr· + Pr → Pr\text{-}Pr· → Pr\text{-}Pr\text{-}Pr·$$

Protein cross-linking brings decreased nutritional values. Malondialdehyde, one product from lipid oxidation, is regarded as an important factor inducing adverse changes in protein functionalities.

Oxidized lipids have been revealed to destroy amino acids, especially when water content in the reaction system is low. The damaging extents of amino acids are given in Table 3.17.

Table 3.17 Amino acid loss in the five proteins arisen from oxidized lipids

Reactants		Reaction conditions		Amino acid loss (%)
Proteins	Lipids	Times	Temperatures (°C)	
Cytochrome c	linolenic acid	5 h	37	His, 59; Ser, 55; Pro, 53; Val, 49; Arg, 42; Met, 38; Cys, 35
Trypsin	linoleic acid	40 min	37	Met, 83; His, 12
Lysozyme	linoleic acid	8 d	37	Trp, 56; His, 42; Lys, 17; Met, 14; Arg, 9
Casein	Ethyl linoleate	4 d	60	Lys, 50; Met, 47; Ile, 30; Phe, 30; Arg, 19; His, 28; Thr, 27; Ala, 27
Albumen	Ethyl linoleate	24 h	55	Met, 17; Ser, 10; Lys, 9; Leu, 8; Ala, 8

3.5.6.2 The Reaction Between Proteins and Nitrites

Nitrates exist in soil, water as well as plant and animal foods. Nitrates can be inverted into nitrites under reducing condition (e.g. by microorganisms). Nitrates in vegetables may be converted to nitrite during normal storage. Vegetable pickling or rotting process also leads to nitrate conversion.

Nitrites set nonnegligible safety risk to the human, as they can react with other substances to form nitroso compounds. The well-known N-nitroso amines are formed by nitrites and secondary amines and are considered as the most carcinogenic compounds in foods. Nitrites react with free or bound amino acids like Pro and Trp in foods especially for processed meat products (Fig. 3.16). Other secondary amines, such as those generated from the Maillard reaction, also can react with nitrites. Other amino acids like Cys, Arg, and Tyr also have potential to react with nitrites.

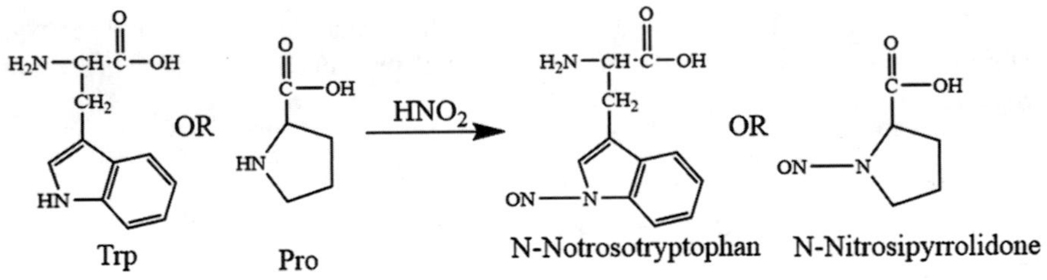

Fig. 3.16 The two nitrosamines generated from two amino acids

3.5.6.3 The Reaction of Proteins with Sulfites and Other Reducing Agents

Sulfites have a reducing property and are able to break the –S-S- bonds in the proteins, yielding S-sulfonate derivatives.

$$\boxed{P_1}\text{-S-S-}\boxed{P_2} + SO_3^{2-} \longrightarrow \boxed{P_1}\text{-S-SO}_3^- + \boxed{P_2}\text{-S}^-$$

If other reducing agents such as Cys or β-mercaptoethanol are used, the S-sulfonate derivatives will be converted into Cys residues. Under acidic or alkaline pH, the S-sulfonate derivatives will decompose to disulfides. Clearly, the S-sulfonation does not decrease the bioavailability of Cys; however, this treatment induces protein unfolding and finally property changes, due to increased electrostatic repulsion and disulfide breakage.

3.5.6.4 Oxidation of Amino Acid Residues

Several oxidants such as H_2O_2 and hypochlorites, and those oxidative substances formed during food processing and storage all are potential to induce oxidation of sensitive amino acids (Met, Cys, Trp, His, and Tyr). The S-containing amino acids in acidic conditions can be oxidized to several products like sulfinic acids, sulfoxides, and sulphones (Fig. 3.17). Both sulfoxides and sulphones are biological unavailable. In the presence of peroxidases (e.g. horseradish peroxidase) and H_2O_2, Tyr residues in proteins can be converted into the dityrosine residues (Fig. 3.18), leading to protein cross-linking. In general, oxidized amino acids have lower bioavailability and even are harmful to the body.

Under mild, acidic, and oxidative conditions, for example, using performic acid (HCO_3H), dimethylsulfoxide (DMSO) or N-bromosuccinimide (NBS), Trp can be

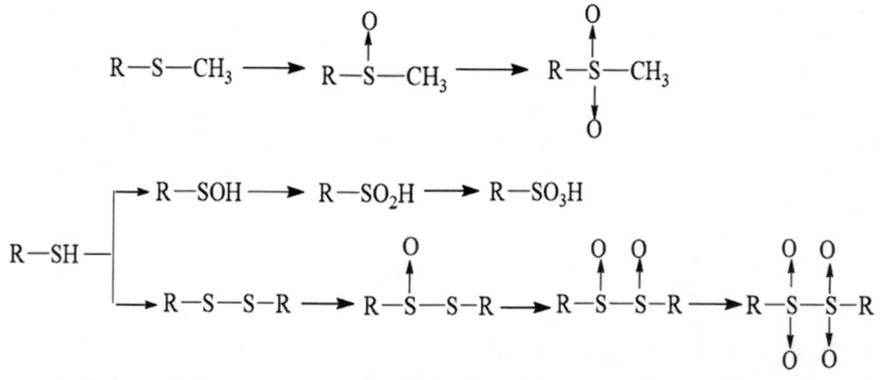

Fig. 3.17 The oxidative products of S-containing amino acids

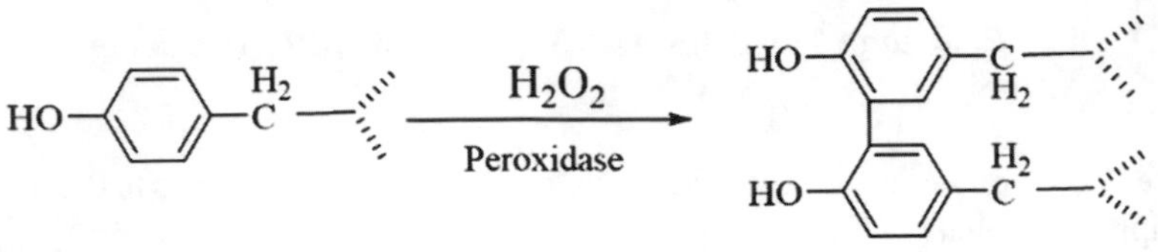

Fig. 3.18 Enzymatic cross-linking of the Tyr residues

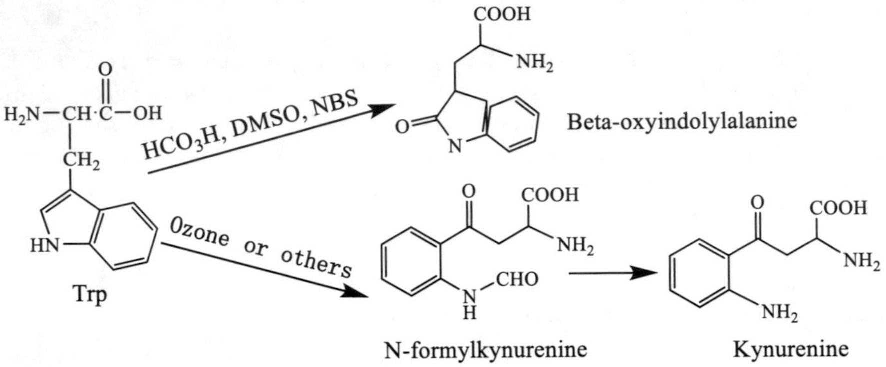

Fig. 3.19 Trp oxidation and resultant products

oxidized to β-oxyindolylalanine; however, using stronger oxidants such as ozone and H_2O_2 lead to the formation of other decomposition products (Fig. 3.19). Kynurenine has high toxicity and is carcinogenic in animals.

The products from phyto-oxidation of riboflavin also can induce oxidation of amino acids. Sensitive amino acids include Tyr, Cys, Trp, ad His. Oxidation for the S-containing amino acids and Trp usually occurs following decreased reaction rate consistently as Met > Cys > Trp.

3.5.6.5 Protein Cross-Linking

Free –NH_2 (mainly from Lys residues) can react with aldehydes to form condensation products, the Schiff bases. Malonaldehyde is one product of lipid oxidation. One malonaldehyde molecule can react with two –NH_2 or protein molecules as below, which leads to protein cross-linking. In addition, glutaraldehyde has similar ability to induce protein cross-linking.

$$OHC\frown CHO + 2\ Pr\text{-}NH_2 \longrightarrow Pr\text{-}N=CH\text{-}CH=CH\text{-}NH\text{-}Pr + 2\ H_2O$$

Protein cross-linking alters protein properties significantly, such as solubility and water holding capacity. The cross-linked proteins even show resistance to protease-induced hydrolysis. Malonaldehyde is also one of the oxidative products from lipids

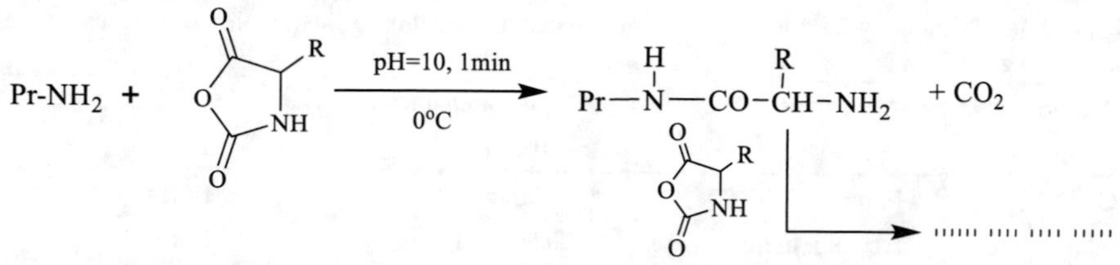

Fig. 3.20 Elongation of the side chain of proteins by carbonylated dehydrated anhydrides

in the body. Malonaldehyde thus reacts with proteins, while the resultant products are accumulated in the body. As the age increases, lipofuscin is yielded and regarded as a biomarker of aging.

3.5.6.6 Elongation Reaction of the Side Chain

When carbonylated dehydrated anhydrides react with proteins, $-NH_2$ in the side chain (e.g. Lys) will be elongated as shown in Fig. 3.20. The generated iso-peptide bonds are also bioavailable. Thus, this reaction can be used to change amino acid composition or fortify the essential amino acids.

3.5.6.7 Acrylamide Formation

Several high-temperature processed foods such as fried potato chips show lower contents of acrylamide, whilst those low-temperature processed foods only contain a few acrylamides. In general, acrylamide is regarded with two formation approaches in high-temperature foods, via lipid decomposition and the reaction between amino acid Asp and reducing sugar glucose (Fig. 3.21). These approaches have been verified using model reaction systems. Acrylamide is thus regarded as a potential toxic substance generated in food processing.

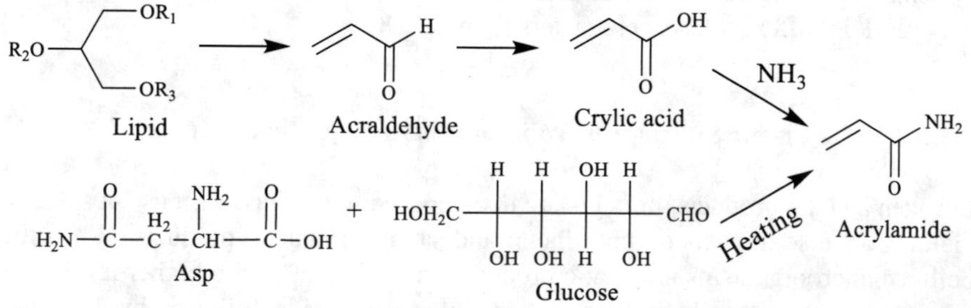

Fig. 3.21 Potential formation approaches of acrylamide in the high-temperature processed foods

Table 3.18 The aldehydes and flavor characteristics formed during the reaction of various amino acids and glucose

Amino acids	Aldehydes	Flavor characters	
		100–150 °C	180 °C
Gly	Formaldehyde	Caramel	Burnt
Ala	Acetaldehyde	Caramel	Burnt
Val	2-Methyl propanal	Cake	Chocolate
Leu	3-Methyl butanal	Bread, chocolate	Burnt cheese
Ser	Glycolaldehyde	Maple sugar	Burnt
Phe	Hyacinthine	Rosebush	Caramel
Met	Thioformaldehyde	Potato, cabbage	Potato
Pro		Corn	Bread
Ser		Bread, butter	
Arg		Bread, popcorn	Burnt
Lys		Potato	Fried potato
Asp			Caramel
Glu		Caramel	Burnt
Ile	2-Methyl butyraldehyde	Cake, moldy	Burnt cheese
Thr	2-Hydroxyl propanol	Maple sugar, chocolate	Burnt

3.5.7 Effect of Protein Reactions During Food Processing and Storage on Sensory Quality of Foods

3.5.7.1 The Flavor Substances Generated from the Maillard Reaction

Strecker reaction, one reaction involved in the Maillard reaction, is an important way to generate flavor substances for processed foods. The Maillard reaction is essential to sensory quality of bread. However, excessive reaction extent leads to a negative effect on sensory quality, for example, the burnt flavor. In the model reaction systems containing various amino acids and glucose, different reaction temperatures may result in different flavor characters (Table 3.18).

3.5.7.2 The Flavor Substances from the Rotten Aquatic Products

In rotten aquatic products, proteins are decomposed to produce various substances, yielding adverse impacts on both flavor and safety of aquatic products. When the polluted microorganisms in aquatic products are proliferated into higher levels, the secreted proteases thus hydrolyze proteins into amino acids, followed by deamination and decarboxylation of amino acids. These reactions lead to the formation of amine compounds with lower molecular weights and unacceptable off-flavor. For example, Glu, Lys, His, and Trp are converted into γ-aminobutyric acid, cadaverine, histamine, and indole, respectively. Both cadaverine and histamine have high toxic effects on the body and may induce allergic food poisoning. Thus, it is recommended to control cadaverine and histamine levels in various aquatic products. Cadaverine and histamine levels also can be used to reflect the freshness of aquatic products.

3.5.7.3 Flavor Substances Formed During Milk Processing

Fresh milk usually contains acetone, acetaldehyde, butyric acid, methyl sulfides, and other volatiles. Heating treatment of milk leads to the formation of desired flavor. In usual, bovine milk and related products contain dimethyl sulfide as the most important flavor substance, which is generated from Met as flavor precursor (Fig. 3.22). Other S-containing flavor substances are generated from the Cys in the whey proteins. Overall, protein decomposition is considered important to milk flavor quality.

During cheese ripening, protein or amino acids are the main resources to form flavor substances. The main reactions involved in the flavor formation are decarboxylation, deamination, desulfurization, Strecker degradation, and others. The formed compounds include amines, alcohols, aldehydes, S-containing compounds, and others.

3.5.8 Chemical Modification of Proteins

The side chains of proteins have several active chemical groups. Thus, some external chemical groups can be introduced into protein molecules via various chemical reactions. Attachment of these external chemical groups in the side chains of proteins brings modified structure and especially property changes. However, some side reactions accompanied with the modifying reactions may be unacceptable or harmful. Several chemical modifications used for the chemical groups in protein side chains are summarized below.

(1) Hydrolysis. This reaction is also regarded as a conversion of chemical groups, for example, using a hydrolysis reaction to convert Gln and Asn to Glu and Asp, respectively.

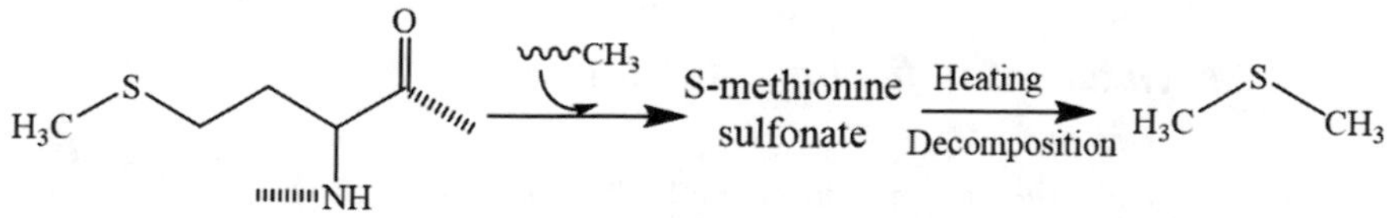

Fig. 3.22 The formation of dimethyl sulfide during milk heating

Table 3.19 The chemical reactions and groups involved in chemical modification of proteins

Chemical groups/bonds	Chemical modification	Chemical groups/bonds	Chemical modification
-NH$_2$ groups	Acylation, alkylation	Carboxyl groups	Esterification, amidation
-S-S- bonds	Oxidation, reduction	-SH groups	Oxidation, alkylation
Sulfide groups	Oxidation, alkylation	Phenolic groups	Acylation, electrophilic substitution
Imidazolyl groups	Oxidation, alkylation	Indolyl groups	Oxidation, alkylation

(2)   Alkylation. This reaction introduces alkyl (e.g. carboxyl methyl) groups into protein side chains. These alkyl groups may be attached to –OH, –NH$_2$, and –SH groups of the side chains.

(3)   Acylation. Various carboxyl groups also can be introduced into side chains. Organic acids of lower molecular weights, dicarboxylic acids, and long-chain fatty acids all can be used to modify proteins. The main reaction sites are –OH and –NH$_2$ groups of the side chains.

(4)   Phosphorylation. Using phosphorus oxychloride or polyphosphates, proteins can be conjugated with phosphate groups. Main reaction sites involve –OH and –NH$_2$ groups of the side chains.

Table 3.19 shows the main chemical groups and reaction types involved in chemical protein modifications.

When proteins are modified with these reactions, their property changes depend on the introduced chemical groups. The introduction of several ionic groups like carboxyl methyl, dicarboxylic acid, and phosphate groups brings enhanced intramolecular electrostatic repulsion and greater unfolding, and thereby alters protein solubility. The introduction of carboxyl and phosphate groups also increases the Ca-sensitivity of protein. Hydrolysis of amide groups in protein side chains, in general, results in increased protein solubility and improved foaming and emulsifying properties. If proteins are introduced with non-polar groups such as long-chain fatty acids, protein hydrophobicity is enhanced. Modified proteins thus have changed interface properties. Acylation of proteins with long-chain fatty acids improves emulsifying properties.

3.5.9   Enzymatic Modification of Proteins

Enzymatic modification of proteins usually has no safety consideration, as this approach does not alter chemical structures of amino acids. Several enzymatic reactions may be used in protein modification, especially, enzymatic hydrolysis, the plastein reaction, and enzymatic cross-linking.

3.5.9.1 Limited Enzymatic Hydrolysis

Enzymatic hydrolysis of proteins with higher hydrolysis extent yields smaller peptides and free amino acids. Most functional properties of proteins are thus destroyed totally. Modified proteins only have higher solubility at a wide pH range. Using limited hydrolysis and specific proteases, modified proteins can be endowed with improved emulsifying and foaming activity but damaged gelation. Limited hydrolysis leads the exposure of hydrophobic groups, modified proteins have decreased solubility. Limited protein hydrolysis has important application in cheese production, as rennin's effect on caseins ensures the formation of cheese curds.

Enzymatic hydrolysis of proteins may result in the formation of bitter peptides, which have potential impact on sensory quality of foods. Bitter intensity of bitter peptides is highly depending on amino acid compositions of proteins and the used proteases. In general, the proteins with average hydrophobicity larger than 5.85 kJ mol^{-1} are easily to produce bitter peptides, while using the non-specific proteases other than the specific proteases also may generate more bitter peptides.

3.5.9.2 Plastein Reaction

Th plastein reaction of proteins contains several reaction stages. First, proteins are enzymatic hydrolyzed by proteases to yield protein hydrolysates (or peptides). Second, protein hydrolysates are treated with proteases for protein re-synthesis (i.e. plastein reaction), yielding the plasteins. Protein hydrolysis is done with normal conditions, while plastein reaction is done under higher substrate concentration. In general, plastein reaction involves three chemical reaction and interaction; that is, transpeptidation, condensation, and hydrophobic interaction. If extrinsic amino acids are added into reaction system, amino acid compositions of the plasteins are thus changed. Thus, plastein reaction has potential to alter protein nutritional values or bio-functions. Plastein reaction has been used to fortify proteins with essential amino acids, or increase bio-activities such as anti-oxidation, inhibition on angiotensin-converting enzyme, anti-coagulation, and others.

3.5.9.3 Enzymatic Cross-Linking

Transglutaminase (TGase) can catalyze the acyl transfer reactions between the Lys and Gln residues of proteins and induce the formation of new covalent bonds, i.e. ε-(γ-glutamyl) lysine isopeptide bonds. Under mild reaction conditions, TGase results in both intra-molecular and inter-molecular protein cross-linking. Modified proteins are thus protein polymers with different extents of polymerization. TGase modification thus is a widely used enzymatic protein cross-linking. Overall, many proteins including caseins, whey proteins, soybean proteins, cereal proteins, and meat proteins have been modified using TGase to improve the quality attributes of foods like

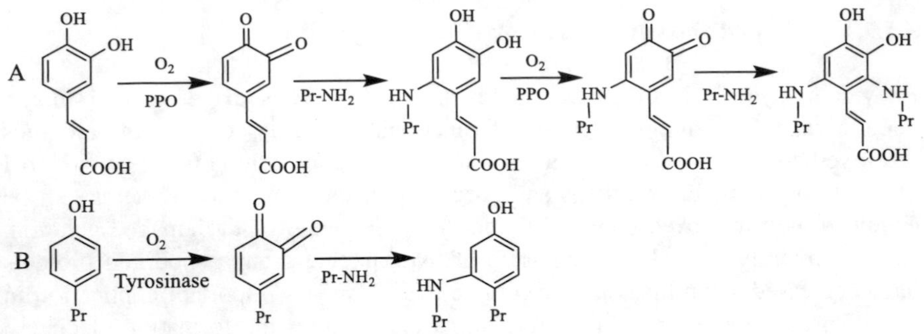

Fig. 3.23 The polyphenoloxidas (PPO)-induced protein cross-linking. Approach A, in the presence of caffeic acid; Approach B, in the absence of phenolic compounds of lower molecular weights; Pr, protein

yoghurt and bread. TGase-induced cross-linking of β-casein leads to decreased emulsifying activity but enhanced stability for the resultant casein-fat emulsion. Suitable addition of TGase to cross-link cereal proteins improves bread quality; however, higher TGase usage brings worse bead quality. TGase usage in milk processing may decrease yogurt syneresis, via the induced protein cross-linking.

Tyr residue in proteins also can undergo cross-linking through the catalysis action of peroxidases. The tyrosinases from mushrooms have potential to cross-link lysozyme, casein, and ribonuclease, in the presence of phenolic compounds (i.e. cross-linkers) of lower molecular weights (cross-linkers). Tyr in proteins also can be used as a cross-linker. The proposed two chemical mechanisms are depicted in Fig. 3.23.

The transglutaminase (TGase, EC 2.3.2.13) from the liver of guinea pig is very expensive and thus mostly used in biological researches. The TGase from *Streptoverticilliummobaraense* is composed of 331 amino acid residues and has molecular weight about 40 kDa and a Cys residue in its active center. Catalysis activity for microbial TGase is independent of calcium ions, while many metal ions (except for Cu^{2+}, Zn^{2+}, Pb^{2+}) have no or slightly effect on TGase activity. Suitable pH values and temperature for TGase catalysis are 5–8 and 37–50 °C.

TGase belongs to one of the acyl transferases and can induce acyl transfer and other reactions (Fig. 3.24). The Gln residues (γ-amide groups) of proteins are acyl donors while the Lys residues (ε-NH$_2$ groups) of proteins are acyl receptors. Protein cross-linking thus occurs via forming the intra-molecular or inter-molecular ε-(γ-glutamyl) lysine isopeptide bonds (Reaction I, Fig. 3.27). Protein cross-linking can be applied for the reaction system containing homologous protein even heterogeneous proteins, resulting in changed functional properties in rheology, gelation, emulsification, and hydration. TGase-induced protein cross-linking has no adverse impact on nutritional values of proteins. Other amine compounds including the NH$_2$-containing saccharides can be used as acyl receptors to replace the Lys residues. In these cases, amine compounds or saccharides can be conjugated into protein molecules at the Gln residues (Reaction II, Fig. 3.27). Several studies have used this approach to connect

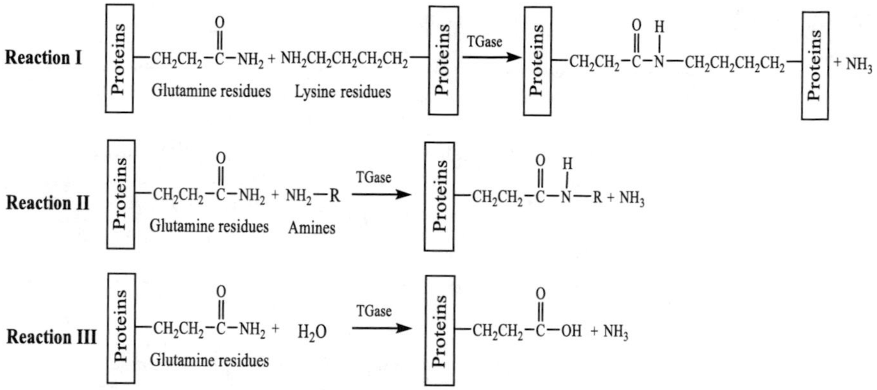

Reaction I

$$\text{Proteins}-CH_2CH_2-\overset{O}{\overset{\|}{C}}-NH_2 + NH_2CH_2CH_2CH_2CH_2-\text{Proteins} \xrightarrow{\text{TGase}} \text{Proteins}-CH_2CH_2-\overset{O}{\overset{\|}{C}}-\overset{H}{\overset{|}{N}}-CH_2CH_2CH_2CH_2-\text{Proteins} + NH_3$$

Glutamine residues   Lysine residues

Reaction II

$$\text{Proteins}-CH_2CH_2-\overset{O}{\overset{\|}{C}}-NH_2 + NH_2-R \xrightarrow{\text{TGase}} \text{Proteins}-CH_2CH_2-\overset{O}{\overset{\|}{C}}-\overset{H}{\overset{|}{N}}-R + NH_3$$

Glutamine residues   Amines

Reaction III

$$\text{Proteins}-CH_2CH_2-\overset{O}{\overset{\|}{C}}-NH_2 + H_2O \xrightarrow{\text{TGase}} \text{Proteins}-CH_2CH_2-\overset{O}{\overset{\|}{C}}-OH + NH_3$$

Glutamine residues

Fig. 3.24 The three reactions catalyzed by transglutaminase

soybean and milk proteins or protein hydrolysates with glucosamine or oligochitosan, resulting in the formation of glycated proteins or peptides. This interesting protein glycation (TGase-type protein glycation) also has protein cross-linking and is clearly different from the Maillard-type protein glycation. Moreover, TGase also has weak ability to catalyze the hydrolysis of the Gln residues to form Glu residues (Reaction III, Fig. 3.27). However, this hydrolysis only occurs in the absence of acyl receptors. Overall, TGase has important application in the food industry and can be used to cross-link and glycate food proteins for property modification.

3.6   The Main Food Proteins

Food proteins mostly come from edible animals and plants. The animal-derived proteins (such as those from meats, milks, eggs, and fishes) and cereal proteins are regarded as traditional edible proteins and are important food components or ingredients (Table 3.20). In Asia countries, traditional edible proteins also include soybean proteins. Meanwhile, the so-called new protein resources are also proposed as potential proteins for the food industry. The new protein resources include single cell proteins (SCPs), leaf proteins, and alga proteins and are the main targets for the research and development of new protein resources in the future.

3.6.1   Soybean Proteins

Soybean proteins are mainly composed of globulins and other minor proteins. These globulins are soluble in the water at the pH value far away from their isoelectric point (near pH 4.5), or soluble in salt solutions. Based on their essential amino acid

Table 3.20 Main food proteins and their applications

Resources	Proteins	Applications
Cereals	Glutes, corn proteins	Breakfast, bakery products, whipped toppings
Eggs	Whole egg, albumen, lipovitellinin	Various applications including emulsification, foaming, cohesion–adhesion, gelation
Fishes	Muscles, collagen/gelatin	Gelation, comminuted meat products
Animals	Muscles, collagen/gelatin, the serum proteins from porcine or bovine blood	Gelation, emulsification, water holding
Milks	Whole milk, skimmed milk, caseinate, whey protein powder	Various applications including emulsification, cohesion–adhesion, thickening
Oil seeds	The protein products from soybean, peanut, sesame, and others, various protein concentrates and isolates	Bakery products, protein beverages, meat analogs, or substitutes

compositions, soybean proteins have nutritional value close to that of the animal proteins and are rich in higher lysine but lack of the S-containing amino acids. Overall, soybean proteins can be regarded as one of these protein resources with both good nutritional and functional properties and have potential application in various processed foods.

Based on their sedimentation coefficients, soybean proteins can be classified into four different parts. The soybean proteins, extracted with water and ultracentrifuged at fixed conditions, are grouped into the so-called 2S, 7S, 11S, and 15S globulin parts. Here, the symbol S represents the Svedburg unit. The 2S globulin part contributes to about 20% of the total soybean proteins and mainly contains protease inhibitors, cytochrome C, allantoicase, and two other proteins. The dominant 7S globulin part contributes to about 37% of the total soybean proteins and mainly is composed of β-amylase, hemagglutinin, lipoxygenases, and 7S globulins. 11S globulins (about 1/3 of the total soybean proteins) are the major components of the 11S globulin part. However, detailed compositions of the 15S globulin part are still unknown in the present. The 15S globulin part contributes to about 10% of the total soybean proteins and might be the polymers of soybean globulins. In brief, the 7S and 11S globulin parts contribute to about 70% of total proteins, and thereby are regarded as the most important fractions in soybean proteins. What is more, composition classification of soybean proteins is strictly dependent on the used conditions, for condition change induces protein disaggregation or aggregation. For example, when the used salt concentration (or ionic strength) is decreased from 0.5 to 0.1 mol/L, 11S globulins will be aggregated into the 9S globulins.

When soybean is extracted with solvents to prepare edible oil, the left by-product is defatted soybean flour. Defatted soybean flour mainly contains proteins and carbohydrates and is now used to prepare various soybean protein products, which can be applied in food processing as protein ingredients (Table 3.21).

(1) The defatted soybean flour (DSF) is yielded from the dehulled soybean by a solvent extraction using the so-called 6# solvent (mainly containing hexane). DSF is also subjected to an important treatment (i.e. flash evaporation), aiming to remove residual solvent and to inactivate the anti-nutritional factors in DSF. In usual, a half of DSF are proteins with less loss in functional properties.

More importantly, flash evaporation of the extracting solvent ensures less protein denaturation and higher nitrogen solubility index for DSF. DSF also has higher

Table 3.21 Soybean protein products and their applications in various foods

Functional properties	Main mechanisms	Foods	Product types
Solubility	Protein hydration, pH effect	Beverages	F, C, I, H
Water adsorption and binding	Water binding via hydrogen bond interaction, water entrapment	Meat products, bread, cake	F, C
Viscosity	Thickening, water binding	Soup	F, C, I
Gelation	Formation protein network via various interactions, water entrapment	Meat products, cheese	C, I
Cohesion–adhesion	Hydrophobic interaction, hydrogen bonding	Meat, bakery products, pasta products	F, C, I
Elasticity	Disulfide bonds, gel deformation	Meat and bakery products	I
Emulsification	Interfacial adsorption and emulsion stability	Sausages, cake, soup	F, C, I
Fat binding	Hydrophobic interaction	Meat products	F, C, I
Flavor binding	Hydrophobic and other interactions, flavor entrapment	Meat analogs or substitutes, bakery products	C, I H
Foaming	Interfacial adsorption, film formation	Desserts, dressings	I, W, H
Color control	Bleaching with lipoxygenase	Bread	F

Notes C, soybean protein concentrate; F, defatted soybean flour; H, hydrolyzed soybean proteins; I, soybean protein isolate; W, soybean whey protein

lipoxygenase activity and thereby can be applied in the bleaching treatment of wheat flour.

(2) The soybean protein concentrate (SPC) is usually prepared as follows. DSF is extracted with pH 4.5 water or ethanol–water, or first wet-heated and then extracted with water. These treatments lead to the removal of the soluble oligosaccharides from DSF and result in higher protein content (about 70%) for SPC. Protease inhibitors and some proteins in DSF are also lost in the soybean whey fraction. SPC thus has lower protease inhibitor levels together with lower protein recovery. In usual, about 2/3 raw proteins are recovered.

(3) The soybean protein isolate (SPI) is usually prepared using the alkali-extraction and acid-precipitation procedures. DSF is extracted with diluted alkaline solution, and the resultant supernatant (protein solution) is acidified into isoelectric point (pH 4.5) to precipitate proteins. After that, protein precipitate is neutralized and spray-dried to obtain SPI. SPI is free of cellulose and anti-nutritional factors and has higher protein content (about 90%) with good properties in dispersion, emulsification, gelation, and thickening.

3.6.2   Milk Proteins

Milk proteins mainly are composed of caseins and whey proteins. Caseins contribute to about 80% of total milk proteins and are characterized as α_{s1}-casein, α_{s2}-casein, β-casein, and κ-casein. Whey proteins contribute to about 20% of total milk proteins, including β-lactoglobulin, α-lactoalbumin, and other minor components such as immunoglobulins, lactoferrin, lysozyme, lactoperoxidase, and other peptides or proteins.

In the milk, milk proteins exist mostly in a format of protein aggregates. Caseins can form casein micelles up to a level of 10^{14}/mL milk. Casein micelles usually show particle sizes of 30–300 nm. Some larger casein micelles even have a size larger than 600 nm. Colloidal phosphate calcium plays an important role in the formation and stability of casein micelles. The structure and involved interactions for casein micelles or sub-micelles are proposed as those shown in Fig. 3.25.

Caseins can be separated from the milk using acid precipitation (pH 4.6) or rennin treatment. The resultant products thus show property differences. Casein products (especially sodium caseinate) are important protein ingredients in the food industry. Sodium caseinate has good dispersion and stability at pH > 6, is a commercial protein additive with very good emulsification, water-binding, thickening, foaming, and gelation. Both whey protein concentrates and isolate (WPC and WPI) also are good protein ingredients and have special application in infant foods, and more, whey protein products can be used in acidic foods with good dispersion and solubility, compared with casein products. Whey protein products also can be applied in the preparation of mimic fat. Table 3.22 summarizes the main chemical features of commercial milk protein products.

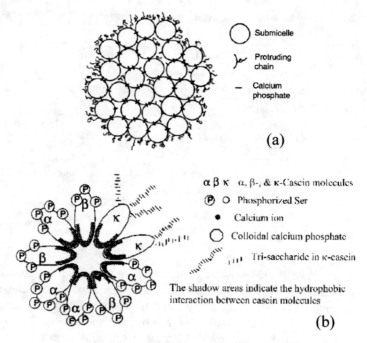

Fig. 3.25 The structures of casein micelle (**a**) and interaction involved in structure formation (**b**)

Table 3.22 Main chemical features of commercial milk protein products

Milk protein products	Preparation methods	Contents(%, dry basis)			
		Protein	Ash	Lactose	Fat
Caseins	Acid precipitation	95	2.2	0.2	1.5
	Rennin treatment	89	7.5	–	1.5
	Co-precipitation	89–94	4.5	1.5	1.5
Whey protein concentrate	Ultrafiltration	59.5	4.2	28.2	5.1
	Ultrafiltration + reverse osmosis	80.1	2.6	5.9	7.1
Whey protein isolate	Spherosil S method	96.4	1.8	0.1	0.9
	Vistec method	92.1	3.6	0.4	1.3

Milk proteins also have several important bio-functions (Table 3.23), which can provide desired health benefits to the body.

Table 3.23 Relationship between surface hydrophobicity

Milk proteins	Important bio-functions
α, β, κ-Caseins	Metal (Cu, Fe, Ca) carrier, precursors of active peptides
α-lactalbumin	Ca carrier, immuno-modulation, anti-cancer effect
β-lactoglobulin	Retinol carrier, fatty acid-binding, potential anti-oxidants
Immunoglobulins	Immune function
Lactoferrin	Anti-bacterial, anti-virus, immuno-modulative, anti-oxidative, anti-cancer, and Fe-binding functions
Lactoperoxidase	Anti-bacterial activity
Lysozyme	Anti-bacterial activity, synergized with immunoglobulins and lactoferrin
Glycopeptides	Anti-bacterial and anti-virus effect

3.6.3 Meat Proteins

Animal meats or muscles are important dietary foods of the human and are also main resources of food proteins. The proteins in animal muscles are usually classified as three groups, that is, myofibrillar, sarcoplasmic, and stromal proteins, comprising about 55, 30, and 15% of the total muscle proteins, respectively. Sarcoplasmic proteins can be extracted by water and diluted salt solutions from muscle tissues, while myofibrillar proteins are only extractable using concentrated salt solutions. Stromal proteins are insoluble proteins.

Myofibrillar proteins comprise myosin, actin, and other proteins. Myosin has a molecular weight of about 500 kDa and an isoelectric point around 5.4, possesses ATPase activity, and will be coagulated at a temperature up to 50–55 °C. Actin monomer has a molecular weight of about 43 kDa and an isoelectric point around 4.7 and can bind with myosin to form actomyosin. In myofibrillar proteins, muscle contraction is the result of the myosin–actin interaction.

Sarcoplasmic proteins are mainly composed of myoglobin, albumin, and other proteins. Myoglobin molecule contains one Fe^{2+}. It has an isoelectric point around 6.8 and makes critical contribution to meat color or quality. However, myoglobin is very sensitive to condition change (especially oxidative condition), as Fe^{2+} is easily converted into Fe^{3+}, which results in an undesirable brown color in meat. Other albumin proteins such as myogen also have instable properties, for example, they may undergo denaturation around a temperature of 50 °C.

Stromal proteins mainly comprise collagen and elastin. Collagen is chemically rich in glycine, proline, and hydroxyl praline and has both intra-molecular and inter-molecular cross-linking. Cross-linking extent of collagen increases as animal age increases. Higher cross-linking extent induces stable properties for collagen but worse tenderness for meat quality. However, long-time heating may convert collagen into gelatin. Gelatin is soluble in hot water can produce heat-reversible gels and thereby has various applications in the food industry. Elastin has no hydroxyl proline

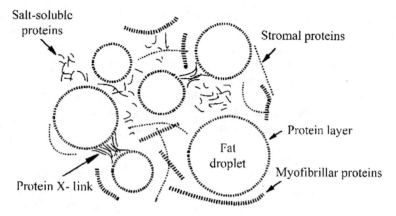

Fig. 3.26 The roles of muscle proteins in meat products for emulsion

and tryptophan but is rich in proline, glycine, and valine. Elastin is resistant to both pepsin and trypsin digestion but can be hydrolyzed by elastase.

In meat, especially comminuted meat products, all proteins make contribution to the formation and stability of the emulsion (Fig. 3.26). Longer chopping time ensures a thick protein film around fat droplets and thus brings emulsion stability. Moreover, myosin has the best emulsifying properties than other meat proteins.

Tenderness is a critical reflector of meat quality, especially for meat texture. Plant-derived proteases such as papain, bromelin, and ficin can be used in meat tenderization, via enzymatic hydrolysis of meat proteins. These proteases are usually injected into animal muscles after animal slaughter. In general, bromelin has better action on collagen but shows slightly weaker action on elastin and myofibrillar proteins. Papain exhibits highly action on myofibrillar proteins but less action on collagen. Ficin, however, has good hydrolysis effect on both myofibrillar and stromal proteins

3.6.4 Egg Proteins

The eggs especially hen eggs have been used as dietary foods for a very long time. Whole egg comprises the edible egg white and yolk and the inedible egg shell. The egg white is mainly composed of globulin, a few carbohydrates, and negligible lipids. The carbohydrates in the egg white exist in two states: free or bound with proteins, while glucose is the most abundant carbohydrate. The main components of the egg yolk are proteins and lipids, while carbohydrate content is very low. What is more, most of the lipids are bound with proteins, which results in a critical influence on the functional properties of the yolk proteins.

The egg yolk can be used as emulsifier in the food industry. Emulsifying properties of the egg yolk mostly depend on its lipoproteins. Among the three lipoproteins (vitellin, phosvitin, and α-/β-lipoproteins) in the egg yolk, lipoproteins have better emulsification. Low-density lipoprotein in the egg yolk is also better in emulsifying properties than bovine serum albumin at same protein concentration. The addition of lipolipids has no impact on the emulsifying properties of low-density lipoprotein. In general, good emulsification of the low-density lipoprotein is due to its lipids–proteins complexes, which has high binding capacity to fat.

The egg white can be used as foaming agent with better performance than caseinate. Among these proteins in the egg white, they are detected with decreased foaming activities as the order of ovomucin, ovoglobulin, ovotransferrin, ovalbumin, ovomucoid, and lysozyme. Both ovomucoid and lysozyme have stable structures due to inter-molecular disulfide bonds; therefore, they are difficult in molecule deformation and have lower abilities in interface adsorption. The egg white is also important gelling agent in the food industry, as it can produce heat-irreversible gels easily. The gel strength and turbidity of the resultant gels are governed by medium pH and ionic strength. Heat-denatured ovalbumin has similar molecular conformation than the native one. Under a pH value near isoelectric point or higher ionic strength, the denatured ovalbumin performs a random aggregation via hydrophobic interaction; however, under a pH value far away isoelectric point and lower ionic strength, the denatured ovalbumin performs an ordered linear aggregation, as electrostatic repulsion between protein molecules prevents the occurrence of the random aggregation (Fig. 3.27).

Medium pH and ionic strength also have impacts on gelling properties of ovalbumin. Ovalbumin gels have higher gel strength under mild pH and ionic strength but give worse transparency under higher ionic strength and a pH value near isoelectric point (Fig. 3.28).

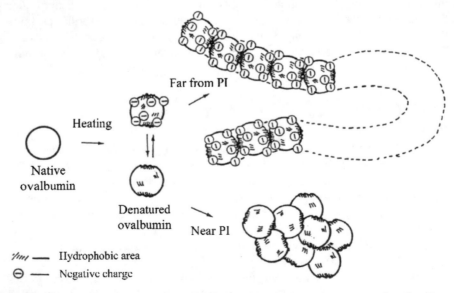

Fig. 3.27 Proposed model for the heat denaturation and formation of aggregates of ovalbumin

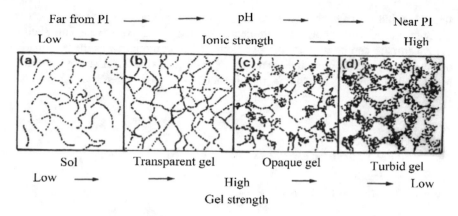

Far from PI $\longrightarrow$ $\longrightarrow$ pH $\longrightarrow$ $\longrightarrow$ Near PI

Low $\longrightarrow$ $\longrightarrow$ Ionic strength $\longrightarrow$ $\longrightarrow$ High

| Sol | Transparent gel | Opaque gel | Turbid gel |

Low $\longrightarrow$ $\longrightarrow$ High $\longrightarrow$ $\longrightarrow$ Low

Gel strength

Fig. 3.28 Proposed model for the gel networks produced with heated ovalbumin

3.6.5 Cereal Proteins

Cereals including wheat, rice, corn, barley, and oat are traditional plant foods. Cereal proteins are classified into four kinds: albumins, globulins, gliadins, and glutenins (Table 3.24). Protein contents of these cereals vary with cultivated varieties, plant regions, growth conditions, and others. In general, gliadins and glutenins are major protein in cereals, accounting for 85% of total proteins. Nutritional values of cereals are thus highly controlled by gliadins and glutenins. Gliadins have lower content in lysine. Thus, the first limited amino acid in cereal proteins is lysine.

Gliadins and glutenins have poor solubility in water and, in usual, are regarded as two components of the well-known gluten in the food industry. Gliadins are composed with a single peptide chain with molecular weight of about 30–60 kDa, have intra-molecular disulfide bonds, and are usually divided into α-, β-, γ-, and ω-gliadins. Glutenins are composed of the so-called low and high molecular weight (LMW and HMW) subunits, with molecular weights of about 31–48 and 97–136 kDa, respectively. A glutenin molecule usually has about 3–5 HMW and 15 LMW subunits. Glutenins also have disulfide bonds between these subunits.

The disulfide bonds have a critical effect on gluten properties, for they govern solubility of both gliadins and glutenins. When reducing agents are added to wheat

Table 3.24 Relationship between surface hydrophobicity

Cereals	Albumins	Globulins	Gliadins	Glutenins
Wheat	5	10	69	16
Rice	5	10	5	80
Corn	4	2	55	39
Sorghum	8	8	52	32

flour, they induce an exchange reaction for the disulfide bonds. Thus, protein solubility increases while dough strength decreases. However, oxidative agents such as benzoperoxide can be used to improve flour and dough quality, as these oxidative agents are able to promote the formation of disulfide bonds. Active soybean flour also has similar effect. In addition, the interaction between gluten and starches may result in decreased starch retrogradation.

3.6.6   New Protein Resources

3.6.6.1   Single Cell Proteins (SCPs)

The proteins from several microorganisms and microalgae are regarded as SCPs and have potential application as food proteins. Compared with traditional food proteins, SCPs have some advantages in their production. SCP production is not dependent on climate and region conditions and can be done on industrial scale. The cells grow at higher rate, thus bringing higher protein production. Organic residues are also potential substrates for cell growth. In general, these proteins from algae, yeasts, and bacteria are the most important SCPs. The main chemical compositions of these organisms are given in Table 3.25.

The yeasts contain about 50% proteins (on dry basis), which lack the S-containing amino acids. Moreover, the yeasts have higher nucleic acid contents. Excessive yeast intake might induce higher uric acid level and metabolic disturbance. The bacteria contain about 75% proteins (on dry basis) that are also poor in the S-containing amino acids, and lipids in bacteria are mainly composed of saturated fatty acids. Both yeasts and bacteria thus cannot be used as food proteins directly. In usual, it is recommended to treat the cells with several steps, to remove the undesired cell wall, nucleic acid, and ash. The refined SCPs thus have similar chemical compositions to SPI. Fortification of SCPs with the S-containing amino acids leads to improved nutritional values.

Two microalgae chlorella and spirulina have application potential as new protein resources. They contain proteins, about 50–60% of dry basis. These proteins are rich in essential amino acids except for the S-containing amino acids.

Table 3.25 Relationship between surface hydrophobicity

Components	Algae	Yeasts	Bacteria	Molds
Nitrogen	7.5–10	7.5–8.5	11.5–12.5	5–8
Lipids	7–20	2–6	1.5–30	2–8
Ash	8–10	5.0–9.5	3–7	9–14
Nucleic acid	3–8	6–12	8–16	Very changeable

3.6.6.2 Leaf Proteins

Leaves cereal grasses and legumes contain 2–4% of proteins. After mechanical expression, the leaf juices are subjected to a treatment for the elimination of anti-nutritional factors, followed by a heat treatment to coagulate crude proteins. The dried protein products contain about 60% proteins, 10% lipids, 10% ash, and other substances (such as vitamins and colorants). Decolorized leaf protein products have better sensory quality and can be used in cereals for lysine fortification.

3.6.6.3 Fish Proteins

The proteins from those non-commercial fishes also can be used as potential dietary proteins for the human. The fresh fishes are recommended with water and lipid removal, which leads to decreased off-flavor arisen from the auto-oxidation of unsaturated fatty acids. If fishbone and entrails are also removed, fish protein products will have protein contents larger than 95%.

Fish proteins as one of those animal proteins have similar amino acid compositions to egg and milk proteins. However, fish proteins usually have poor dispersion, solubility, and water binding. With special treatments such as hydrolysis on fish proteins, an improvement in functional properties can be achieved.

3.7 Summary

Amino acids are basic structural elements of proteins. Proteins possess primary, secondary, tertiary, and quaternary structures, via covalent connection of amino acids and various forces. Proteins have stable spatial conformation and show various functional properties in food systems. However, condition changes may induce altered spatial conformation and forces and thereby will lead to protein denaturation and especially property changes. Food processing brings protein denaturation and various reactions for proteins (like cross-linking, decomposition, and hydrolysis) or amino acids (like oxidation and isomerization). These changes or reactions will yield negative or positive effects on protein nutrition, property, and safety as well as food quality.

Proteins are key components in foods and make contribution to nutritional values and quality attributes of foods. Proteins in some cases are critical factors to control food texture. Proteins after their hydration may form protein dispersion in water and thus provide beneficial properties for processed foods. The important foods from animals, plants, and soybean usually have good functional properties in gelation, emulsification, foaming, texturization, viscosity, and others. Proteins also can be endowed with modified properties via various structural modifications.

Questions

1. Please give a scientific definition for these technical terms:
   Hydrophobicity of amino acids, simple proteins, complex proteins, protein structure, denaturation, functional property, shear-thinning, cross-linking, Plastein reaction

2. Based on the chemical classification of amino acids, summarize the main reactions for the $-NH_2$ groups.

3. Please list the main results arisen from protein denaturation, the normal denaturation approaches, and the involved mechanisms.

4. Please explain the potential benefits of using the UHT technology in liquid foods, from a chemical kinetics point of view.

5. Please summarize these functional properties of proteins, the chemical mechanisms involved, and their application in the food industry.

6. Please explain different roles of wheat proteins in dough formation.

7. Please summarize interface properties of food proteins and explain the roles of milk and muscle proteins in dairy and meat products.

8. Please summarize the adverse reactions of proteins subjected to strong alkaline and high-temperature treatments.

9. Please summarize main protein modification and the beneficial effects of using enzymatic protein modification.

10. Please read a paper reporting single cell proteins (SCPs) extraction and separation or describe how bioactive proteins are separated from milk.

Bibliography

1. Belitz, H.D., Grosch, W., Schieberle, P.: Food Chemistry. Springer, Berlin, Heidelberg (2009)
2. Becalski, A., Lau, P.Y.L., Lewis, D., Seaman, S.W.: Acrylamide in foods: occurrence, sources, and modeling. J. Agric. Food Chem. **51**(3), 802–808 (2003)
3. Damodaran, S., Paraf. A.: Food Proteins and Their Applications. Marcell Dekker, Inc., New York (1997)
4. Damodaran, S., Parkin, K.L., Fennema, O.R.: Fennema's Food Chemistry. CRC Press/Taylor & Francis, Pieter Walstra (2008)
5. Fox, P.E., Mcsweeney, P.L.H.: Dairy Chemistry and Biochemistry. Blackie Academic & Professional, London (1998)
6. Friedman, M.: Chemistry, biochemistry, nutrition, and microbiology of lysinoalanine, lanthionine, and histidinoalanine in food and other proteins. J. Agric. Food Chem. **47**(4), 1295–1319 (1999)
7. Hall, G.M.: Methods of Testing Protein Functionality. Blackie Academic & Professional, London (1996)
8. Nakai, S., Modler, H.W.: Food Protein: Processing Applications. Wiley-VCH Inc., New York (2001)
9. Phillips, G.O., Williams, P.A.: Handbook of Food Proteins. Woodhead Publishing Ltd., Cambridge (2011)
10. Udenigwe, C.C., Rajendran, S.R.C.K.: Old products, new applications? Considering the multiple bioactivities of plastein in peptide-based functional food design. Curr. Opin. Food Sci. **8**, 8–13 (2016)

11. Yada, R.Y.: Proteins in Food Processing. Woodhead Publishing Ltd., Cambridge (2018)
12. Zhao, X.H., Xu, H.H., Jiang, Y.J.: Food Proteins. China Science Publishing & Media Ltd., Beijing (2009)

Dr. Xinhuai Zhao is professor and doctoral supervisor at School of Biology and Food Engineering, Guangdong University of Petrochemical Technology and College of Food Science, Northeast Agricultural University, China, engages in the research of food science, especially in food chemistry. His main research interests include food protein, phytochemicals, and food health. He worked at Northeast Agricultural University and Guangdong University of Petrochemical Technology, serving as the editorial board member of three peer-reviewed journals and as a reviewer for many scientific journals. He has published "Food Chemistry" through Chemical Industry Press, Beijing, "Dairy Chemistry" through Science Press, Beijing, and "Food Protein" through Huaxia Yingcai Foundation, and published more than 300 research papers.

Dr. Yu Fu is currently a professor at College of Food Science, Southwest University, China. Dr. Fu obtained his Ph.D. degree in Food Science from Aarhus University. Thereafter, he worked as a postdoctoral fellow at University of Aberdeen and University of Copenhagen. He was also a visiting scientist at University of Manitoba. He has published over 30 SCI-indexed papers and 5 book chapters. He received several awards, including "PhD student of the year of EFFoST", "Chinese government self-financed student abroad of China Scholarship Council", "the third prize of Natural Science and Technology of Harbin", etc. His research field includes food proteins and peptides.

Dr. Chunli Song works as a professor at Faculty of Food and Bioengineering, Qiqihar University, Heilongjiang Province, China. She obtained a Ph.D. degree from Northeast Agricultural University, Harbin, China. She also has been a visiting scholar at University of Missouri. Her work focuses on food protein chemistry and is specialized in the structure-function properties of the modified food proteins.

Chapter 4
Carbohydrates

Jie Pang and Fusheng Zhang

Abstract Carbohydrates occupy a large portion of food materials and they show different properties according to their unique structures. In this chapter, different categories of carbohydrates including monosaccharides, oligosaccharides, and polysaccharides are, respectively, introduced according to their structures, properties, and applications. Some characteristic phenomena and reactions in the food industry related to carbohydrates are explained in detail as well, such as mutarotation, Maillard reaction, isomerization, etc. Meanwhile, representatives of carbohydrates that are frequently applied in food processing are shown with their unique physiochemical properties and structures, which will influence their industrial application and help to deal with the need for food manufacturing such as fructose, starch, pectin, cellulose, etc. Finally, it should be acquired from this chapter that a comprehensive introduction to carbohydrates in food and understand its application in the food industry.

Keywords Monosaccharides · Oligosaccharides · Polysaccharides · Maillard reaction · Amylose · Gelatinization and retrogradation

4.1 Overview

Carbohydrates are the biggest class of compounds in nature. They are the main products of the photosynthesis in green plants. The content of carbohydrates in plants can reach more than 80% of the dry weight. Liver sugar and blood sugar in animals are also carbohydrates, accounting for about 2% of dry weight in animals.

The molecular composition of carbohydrates can generally be expressed in the general formula of $C_n (H_2O)_m$. It seems that such substances are composed of carbon and water, so they are named carbohydrates. But this term is not accurate,

J. Pang (✉)
College of Food Science, Fujian Agriculture and Forestry University, Fuzhou 350002, China
e-mail: pang3721941@163.com

F. Zhang
College of Food Science, Southwest University, Chongqing 400715, China

© The Author(s), under exclusive license to Springer Nature Singapore Pte Ltd. 2021
J. Kan and K. Chen (eds.), *Essentials of Food Chemistry*,
https://doi.org/10.1007/978-981-16-0610-6_4

as the ratio of hydrogen to oxygen in organic compounds such as formaldehyde (CH_2O), acetic acid ($C_2H_4O_2$) is 2:1, but they are not carbohydrates. In some other organic compounds such as rhamnose ($C_6H_{12}O_5$) and deoxyribose ($C_5H_{10}O_4$), the ratio of hydrogen to oxygen does not conform to the general formula of 2:1, but they do be carbohydrates. It is more scientific and reasonable to assume carbohydrates as saccharide, but it is still used today due to the long history. According to the chemical structure of carbohydrates, it should be defined as polyhydroxyaldehydes or polyhydroxyketones, their derivatives, and condensates.

According to the degree of hydrolysis, carbohydrates are divided into three categories: monosaccharides, oligosaccharides, and polysaccharides. Monosaccharides are the simplest carbohydrates that can no longer be hydrolyzed into smaller units of sugar. Based on the number of carbon atoms in a monosaccharide molecule, it can be further divided into three kinds of sugars, namely, Triose, Tetrose, Pentose, Hexose, and so on. According to the characters of the carbonyl group in the monosaccharide molecule, it also can be divided into aldoses and ketoses. The most important and common monosaccharides in nature are glucose and fructose.

Oligosaccharides are compounds that can hydrolyze to produce 2–10 monosaccharides. According to the number of monosaccharides produced after hydrolysis, oligosaccharides can be divided into disaccharides, trisaccharides, tetrasaccharides, pentasaccharides, etc. Among them, disaccharides are the most important, such as sucrose, lactose, maltose, etc. They can also be divided into reducing oligosaccharides and non-reducing oligosaccharides according to their different reductive properties.

Polysaccharides refer to sugars with a degree of monosaccharide polymerization greater than 10, such as starch, cellulose, glycogen, etc. According to the composition, polysaccharides can be divided into homogeneous polysaccharides and heteropolysaccharides. Homogeneous polysaccharides refer to polysaccharides consisting of the same glycosyl groups, such as cellulose and starch. Heteropolysaccharides refer to polysaccharides consisting of two or more different monosaccharide units, such as hemicellulose, pectin, mucopolysaccharides, etc. According to the different non-sugar groups, polysaccharides can be divided into pure polysaccharides and compound polysaccharides, mainly including glycoproteins, glycolipids, lipopolysaccharides, glucidamin, etc. In addition, according to the different functions of polysaccharides, they can be divided into structural polysaccharides and active polysaccharides.

Carbohydrates are the main source of energy needed by living organisms to maintain life activities. They are the basic material for synthesizing other compounds and the main structural component of living organisms. About 80% of the total energy consumed by humans is supplied by carbohydrates, which is the source of life for humans and animals. In food, carbohydrates serve not only as nutritional compounds but also as sweetener. Polysaccharides are widely used as thickener and stabilizer in food system. In addition, carbohydrates are precursors of flavor and color in food processing, which play an important role in the sensory quality of food.

4.2   Monosaccharides and Oligosaccharides

4.2.1   Structure of Monosaccharides and Oligosaccharides

4.2.1.1   Structure of Monosaccharides

In chemical structure, except acetone sugar, other monosaccharide contains chiral carbon atoms (Asymmetric carbon atoms) in the molecule, so most of monosaccharides have optical isomers. Natural monosaccharides are always D-type, and there are only two natural L-type in food, that is, L-arabinose and L-galactose. Fischer's formula is the most representative of the straight-chain configuration of monosaccharides. Common monosaccharides can be regarded as derivatives of D-glyceraldehyde. The structure of D-aldose derived from C_3 to C_6 is expressed by Fisher formula in Fig. 4.1.

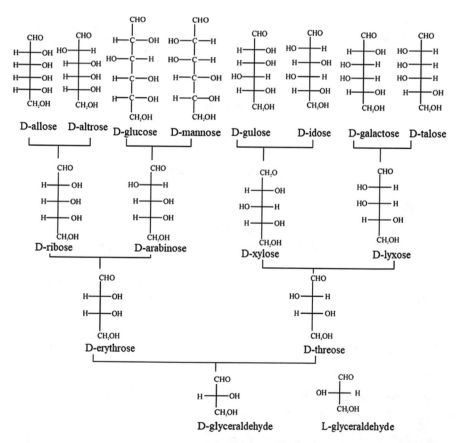

Fig. 4.1   Structural formula of D-aldose (C_3–C_6)

Monosaccharides that contain five or more carbon atoms may have ring structure besides straight chain structure, especially in aqueous solution where they usually exist in a cyclic structure (intramolecular semiacetal or semiketal configuration), that is, carbonyl group in monosaccharide molecule reacts with an alcoholic group of itself to form five-membered furan sugar ring and more stable six-membered pyran sugar ring. Due to the addition of a chiral carbon atom to the ring structure, there are two additional configurations, namely α-type and β-type. Figure 4.2 shows the formation of D-glucose ring structure. The ratio of the two configurations is

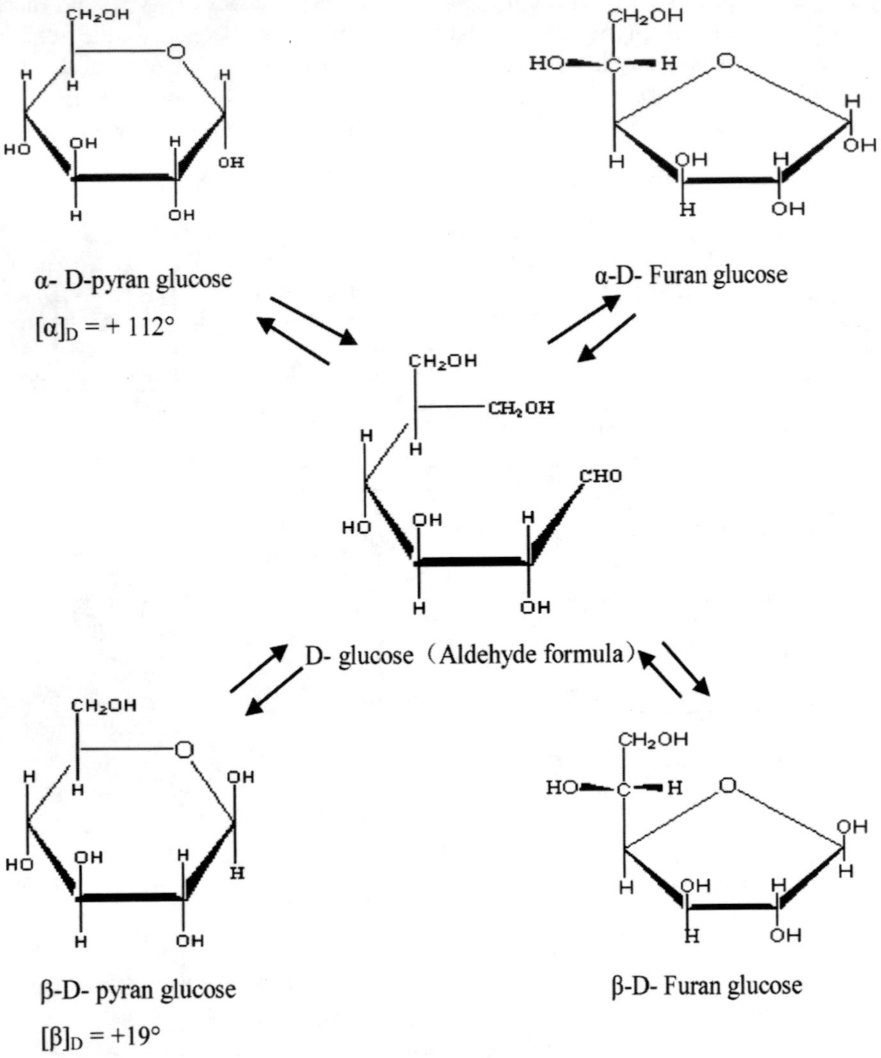

α- D-pyran glucose

$[\alpha]_D = +112°$

α-D- Furan glucose

D- glucose (Aldehyde formula)

β-D- pyran glucose

$[\beta]_D = +19°$

β-D- Furan glucose

Fig. 4.2 The equilibrium state of D-glucose (Haworth representation)

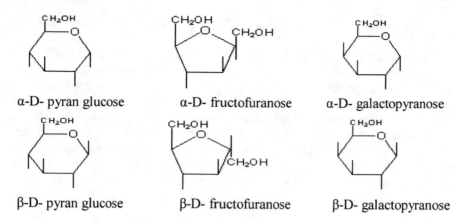

α-D- pyran glucose   α-D- fructofuranose   α-D- galactopyranose

β-D- pyran glucose   β-D- fructofuranose   β-D- galactopyranose

Fig. 4.3 Annular configurations of several monosaccharides

Fig. 4.4 Chair style of pyran
ring

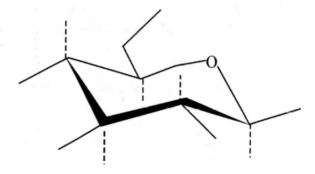

α-type: β-type = 37: 63 at 20 °C and in equilibrium. Pure D-glucose belongs to α-D-pyran glucose. When it is made into aqueous solution, the specific rotation decreases gradually from initial +112° to +52.7° due to the conversion of α-type to β-type. This phenomenon is called mutarotation. The higher the temperature, the faster the speed of mutarotation.

Haworth style that was put forward by W.N. Haworth is the most common way to write monosaccharide ring structure (Fig. 4.3). Natural sugar rings are not actually planar structures. Pyranoglucose has two different conformations—chair style or boat style, and most of hexose exists in chair style, as shown in Fig. 4.4.

4.2.1.2   Structure of Oligosaccharides

Oligosaccharides are formed by the dehydration between the hemiacetal hydroxyl groups of C_1 in aldose (or C_2 in ketose) and the hydroxyl groups of another monosaccharide, which meanwhile yields glycoside bond. Therefore, oligosaccharides are composed of monosaccharides bound by glycoside bonds. Glycoside bonds have two configurations, α-type and β-type. The binding sites occur at 1→2, 1→3, 1→4, 1→6, etc.

Oligosaccharides are usually named systematically. The configuration of monosaccharide residues and glycoside bonds are represented by the prescribed symbols D or L and α or β, respectively. The position and orientation of carbons linked by glycoside bonds are represented by Arabic numbers and arrows (→), and "O" stands for the substitution located at the original hydroxy group. For example, the systematic name of maltose is 4-O-α-D-pyran-glucosyl (1→4)-D-pyran-glucose; the systematic name of lactose is β-D-galactosyl (1→4)-D-pyran glucose; and the sucrose is α-D-pyran glucosyl (1→2)-β-D-fructofuranose glycoside (Fig. 4.5). In addition, as the customary names are simple and convenient to use and have been

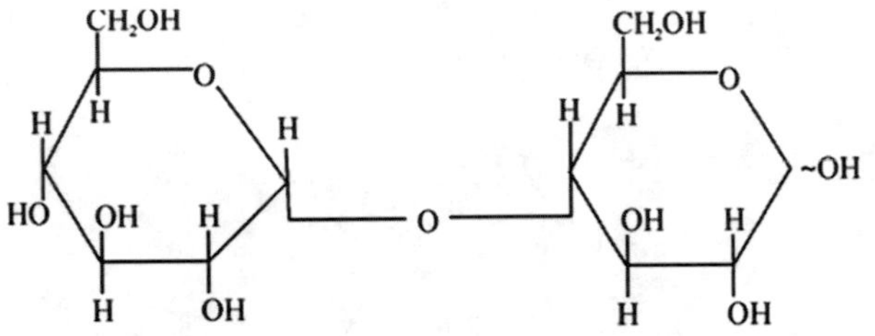

Maltose (*α-D*-pyran-glucosyl-(1→4)-*D*-pyran-glucose)

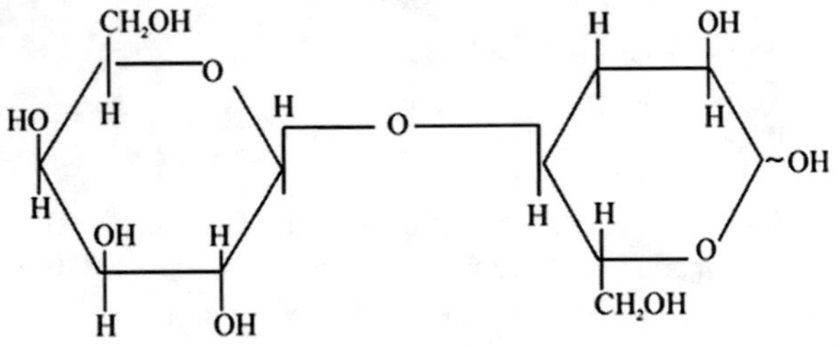

Lactose (*β-D*-galactosyl-(1→4)-*D*-pyran-glucose)

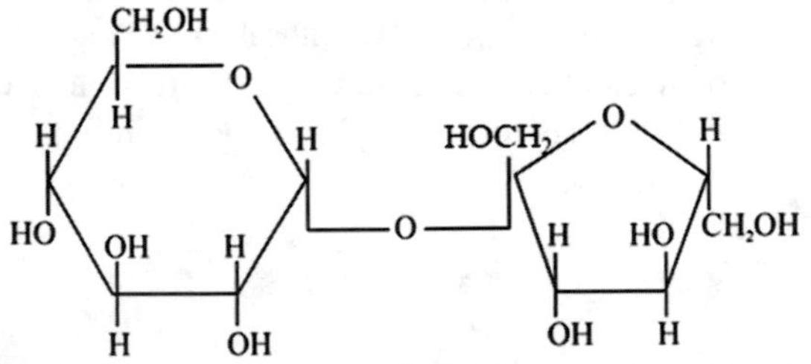

Sucrose (*α-D*-pyran-glucosyl-(1→2)-*β-D*-fructofuran glycoside)

Fig. 4.5 Structural formulas of several oligosaccharides

used for a long time, they are still frequently used, such as sucrose, lactose, trehalose, raffinose, and so on.

4.2.2 Physical Properties of Monosaccharides and Oligosaccharides

4.2.2.1 Sweetness

Sweet taste is an important property of sugar. The intensity of sweet taste is expressed by sweetness. However, the sweetness cannot be determined quantitatively by physical or chemical methods, only by sensory comparison method. Sucrose (non-reducing sugar) is usually used as the reference substance, the sweetness is defined as 1.0 for 10 or 15% sucrose aqueous solution at 20 °C, while the sweetness of other sugars is compared with sucrose. For example, the sweetness of fructose is 1.5, and glucose is 0.7. This sweetness is relative, and it is listed in Table 4.1 of the sweetness of some common monosaccharides.

The sweetness of sugar is related to its molecular structure, molecular weight, molecular state of existence, and external factors, that is, the higher the molecular weight, the smaller the solubility, the smaller the sweetness. In addition, alpha and beta forms of sugar also affect the sweetness of sugar. For D-glucose, if the sweetness of alpha form is set at 1.0, the sweetness of beta form is about 0.666. Crystalline glucose is in alpha form, and part of it is transformed into beta form after dissolving in water, so the newly dissolved glucose solution is the sweetest. On the contrary, if the sweetness of beta-type fructose is 1.0, the sweetness of alpha-type fructose is 0.33; the crystalline fructose is beta-type, after dissolution, part of it becomes alpha-type, and when it reaches equilibrium, its sweetness decreases.

Table 4.1 Specific sweetness of monosaccharides

Monosaccharide	Sweetness
Sucrose	1.00
α-D-Glucose	0.70
β-D-Fructose	1.50
α-D-Galactose	0.27
α-D-Mannose	0.59
α-D-Xylose	0.50
Maltose	0.5
Lactose	0.4
Maltitol	0.9
Sorbitol	0.5
Xylitol	1.0

High-quality sugar should have the characteristics of appropriate sweetness, quick sweetness response, and rapid disappearance. The commonly used monosaccharides basically meet these requirements, but there are still differences. For example, compared with sucrose, fructose has a faster sweetness response, a faster speed of reaching the highest sweetness and a shorter duration; while glucose has a slower sweetness response, a slower speed of reaching the highest sweetness, and a lower sweetness, but it has a cool feeling.

When different kinds of sugar are mixed, they have a synergistic effect on sweetness. For example, when sucrose is combined with fructose syrup, its sweetness can be increased by 20–30%. The sweetness of 5% glucose solution is only half of sucrose in the same concentration, but when the solution contains 5% glucose and 10% sucrose, the sweetness is equivalent to that of 15% sucrose solution.

Oligosaccharides, except sucrose, maltose, and other disaccharides, can be used as a kind of sweeteners with low calorific value and low sweetness. They are widely used in food, especially as functional food sweeteners.

4.2.2.2 Optical Rotation

Optical rotation refers to the property of a substance that makes the vibration plane of linear polarized light rotate to the left or right. Its optical rotation direction is expressed by different symbols, namely, D- or (+), L- or (−).

Except acetone sugar, the other monosaccharides have chiral carbon atoms in their molecular structure, so they all have optical rotation, which is an important index for the identification of sugar. The specific rotation of sugar refers to the angle at which the polarized light rotates 1 mL solution containing 1 g sugar when the transmittance layer is 0.1 m. It is usually expressed by $[\alpha]_{\lambda}^{t}$. Among them, t is the temperature at the time of measurement, and λ is the wavelength of the polarized light. When determining specific rotation, sodium light is usually used and denoted by symbol D. Table 4.2 shows the specific rotation of several monosaccharides.

When sugar is first dissolved in water, its specific rotation is in dynamic change, but it tends to stabilize after a certain period. This phenomenon is called mutarotation.

Table 4.2 Specific optical rotation values of various sugars at 20 °C (sodium light)

Monosaccharide	Specific rotation $[\alpha]_D^{20}/(°)$
D-Glucose	+52.2
D-Fructose	−92.4
D-Galactose	+80.2
L-Arabinose	+104.5
D-Mannose	+14.2
D-Arabinose	−105.0
D-Xylose	+18.8

Table 4.3 Solubility of sugar

Sugar	20 °C		30 °C		40 °C		50 °C	
	C	S	C	S	C	S	C	S
Fructose	78.94	374.78	81.54	441.7	84.34	538.63	86.94	665.58
Sucrose	66.6	199.4	68.18	214.3	70.01	233.4	72.04	257.6
Glucose	46.71	87.67	54.64	120.46	61.89	162.38	70.91	243.76

Note C, concentration (%); S, solubility (g/100 g water)

This is due to the conformational transformation of sugar. Therefore, when determining the rotatory power of sugar with variable optical activity, the sugar solution must be kept for a period (24 h) before determination.

4.2.2.3 Solubility

The multiple hydroxyl groups in the monosaccharide molecule make it soluble in water, especially in hot water, but not in organic solvents such as ether and acetone. At the same temperature, the solubility of various monosaccharides is different, of which fructose is the most soluble, followed by glucose. Temperature has a decisive influence on the dissolution process and dissolution rate. Generally, the solubility increases with the increase of temperature. The solubility of fructose, glucose, and sucrose at different temperatures is shown in Table 4.3.

The solubility of sugar is also closely related to the osmotic pressure of its aqueous solution. At a certain concentration, the osmotic pressure increases with the increase of the concentration. The high osmotic pressure property of high concentration sugar is used to preserve jams and preserves, which requires high solubility of sugar. Sugar can inhibit the growth of yeast and mold only when the concentration exceeds 70%. At 20 °C, the highest concentrations of sucrose, glucose, and fructose are 66%, 50%, and 79%, respectively, so only fructose has good food preservation at this time, while sucrose or glucose alone cannot meet the requirements of preservation and quality. The concentration of fructose syrup varies with the content of fructose. When fructose occupies 42%, 55%, and 90%, the concentration of fructose syrup is 71%, 77%, and 80%, respectively. Therefore, fructose syrup with higher fructose content has better preservation performance.

4.2.2.4 Hygroscopicity, Moisture Retention, and Crystallinity

Hygroscopicity refers to the absorption of water by sugar under the condition of high air humidity. Moisture retention property refers to the property that sugar keeps moisture at low air humidity. These two properties are of great significance for maintaining the softness, elasticity, storage, and processing of food. Different sugars have different hygroscopicities. Among all sugars, fructose has the strongest hygroscopicity, followed by glucose. Therefore, the use of fructose or high fructose syrup to produce bread, cakes, soft sweets, condiments, and other foods has a good effect. However, due to its strong hygroscopicity and moisture retention, it cannot be used to produce hard candy, crispy candy, and crispy biscuits.

Sucrose and glucose are easy to crystallize, while fructose and high fructose syrup are difficult to crystallize. Sucrose crystals are coarse while glucose crystals are fine. Starch syrup is a mixture of glucose, oligosaccharide, and dextrin, which cannot crystallize and can prevent the crystallization of sucrose. In the production of candy, it is necessary to make use of the differences in the crystalline properties of sugar. For example, when the saturated sucrose solution is supersaturated due to the evaporation of water, the sucrose molecules will be arranged neatly and recrystallized in the presence of sudden temperature change or crystal seeds, which can be used to produce ice sugar and other products. Another example is that when hard candy is produced, sucrose cannot be used alone; otherwise, when the water content is less than 3% after boiling, sucrose crystals will appear after cooling down, which will make hard candy fragmented and thus hard candy products will not be transparent and tough. When hard candy is produced, if a proper amount of starch syrup (DE value 42) is added, no crystals can be formed and hard candy of various shapes can be made. This is ascribed that starch syrup does not contain fructose and has less hygroscopicity and good candy preservation. At the same time, the dextrin in starch syrup can increase the stickiness, toughness, and strength of candy, so that candy is not easy to crack. In addition, the addition of other substances, such as milk, gelatin, will also prevent the formation of sucrose crystallization in the process of candy making. Furthermore, high sugar concentration is required in the production of preserves. If sucrose is used, crystallinity is easy to occur, which not only affects the appearance but also reduces the anti-corrosion effect. Therefore, the non-crystallinity of fructose or fructose syrup can be used to replace sucrose properly, and the quality of products can be greatly improved.

4.2.2.5 Viscosity

The viscosity of sugar decreases with the increase of temperature, but the viscosity of glucose increases with the increase of temperature. The viscosity of monosaccharide is lower than that of sucrose, and the viscosity of oligosaccharide is higher than that of sucrose. The viscosity of starch syrup decreases with the increase of conversion degree.

The viscosity characteristics of syrup have practical significance for food processing. For example, in a certain viscosity range, the sugar paste boiled from syrup can have plasticity, to meet the needs of stringing and forming in the candy process. When stirring cake protein, adding boiled syrup is to use its viscosity to wrap and stabilize the bubbles in the protein.

4.2.2.6   Osmotic Pressure

Like other solutions, monosaccharide aqueous solutions have the characteristics of lower freezing point and higher osmotic pressure. The osmotic pressure of sugar solution is related to its concentration and molecular weight, that is, osmotic pressure is proportional to the molar concentration of sugar; at the same concentration, the osmotic pressure of monosaccharide is twice that of disaccharide. For example, fructose or fructose syrup has good antiseptic effect because of its high osmotic pressure. Oligosaccharides have smaller osmotic pressure due to their high molecular weight and low water solubility.

4.2.2.7   Fermentability

Sugar fermentation is of great significance to food. Yeast can ferment glucose, fructose, maltose, sucrose, mannose, etc. to produce alcohol while producing carbon dioxide, which is the basis of wine production and bread loosening. But the fermentation speed of various sugars is different. The speed order for most yeast fermentation is: glucose > fructose > sucrose > maltose. Lactic acid bacteria can ferment lactose to produce lactic acid in addition to the above-mentioned sugars. However, most oligosaccharides cannot be directly fermented by yeasts and lactic acid bacteria. They must be hydrolyzed to produce monosaccharides before they can be fermented.

In addition, due to the fermentability of sucrose, glucose, and fructose, other sweeteners can be used instead of sugars in the production of certain foods to avoid food deterioration or soup turbidity caused by microbial growth and reproduction.

4.2.3   Chemical Reactions of Monosaccharides and Oligosaccharides

Simple monosaccharides and most oligosaccharides can undergo esterification, etherification, acetalization, and some additional reactions of carbonyl groups, as well as some special reactions, as they have carbonyl and hydroxyl groups. Several food-related and important reactions will be introduced here.

4.2.3.1   Maillard Reaction

Maillard reaction, is also called carbonyl ammonia reaction, refers to the reaction of carbonyl group and amino group through condensation and polymerization to produce melanin-like substance. Since the reaction was first discovered by French biochemist Louis-Camille Maillard in 1912, it was named after him. The final product of Maillard reaction is colored substance with complex structure, which deepens the color of the reaction system. Therefore, this reaction is also called "browning reaction". This browning reaction is not caused by enzymes, so it belongs to non-enzymatic browning reaction.

Almost all foods contain carbonyl and amino groups, among which carbonyl derives from sugars or aldehydes and ketones produced by oxidative rancidity of oils, and amino derives from proteins, so carbonyl ammonia reaction may occur. Therefore, the phenomenon of food color deepening caused by carbonyl ammonia reaction is common in food processing. For example, the golden color produced by baking bread, the brown–red color produced by roasting meat, the brown color produced by smoking; brewing food such as the brown color of beer, and the brown–black color of soy sauce and vinegar are all related with Maillard reaction.

I. Mechanisms of Maillard reaction

Maillard reaction process can be divided into three stages: initial stage, intermediate stage, and final stage. Each stage includes several reactions.

1. Initial stage

The initial stage includes carbonyl ammonia condensation and molecular rearrangement.

(1) Carbonyl ammonia condensation

The first step in carbonyl ammonia reaction is the condensation reaction between free amino groups in amino compounds and free carbonyl groups of carbonyl compounds (Fig. 4.6). The initial product is an unstable imine derivative called Schiff base, which is then cyclized to N-glucosylamine.

In the reaction system, sulfite can form an additive compound with aldehyde in the presence of sulfite. The product can condense with R-NH$_2$, but the condensation product cannot further produce Schiff base and N-glucosylamine (Fig. 4.7). Therefore, sulfite can inhibit the browning of carbonyl ammonia reaction.

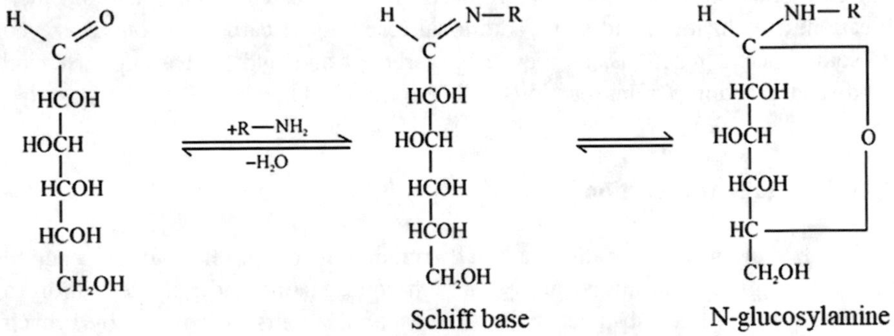

Fig. 4.6 Carbonyl ammonia condensation reaction formula

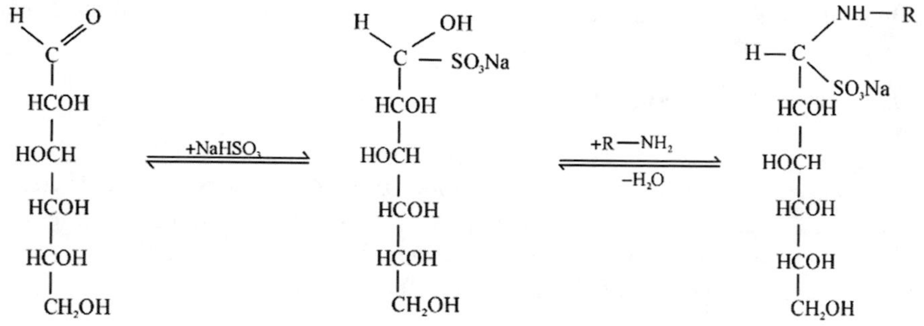

Fig. 4.7 Additional Reaction Formula of Sulfite with Aldehyde

The carbonyl ammonia condensation reaction is reversible. Under dilute acid conditions, the product is easily hydrolyzed (Fig. 4.8). In the process of carbonyl ammonia condensation, the pH value of the reaction system decreases due to the

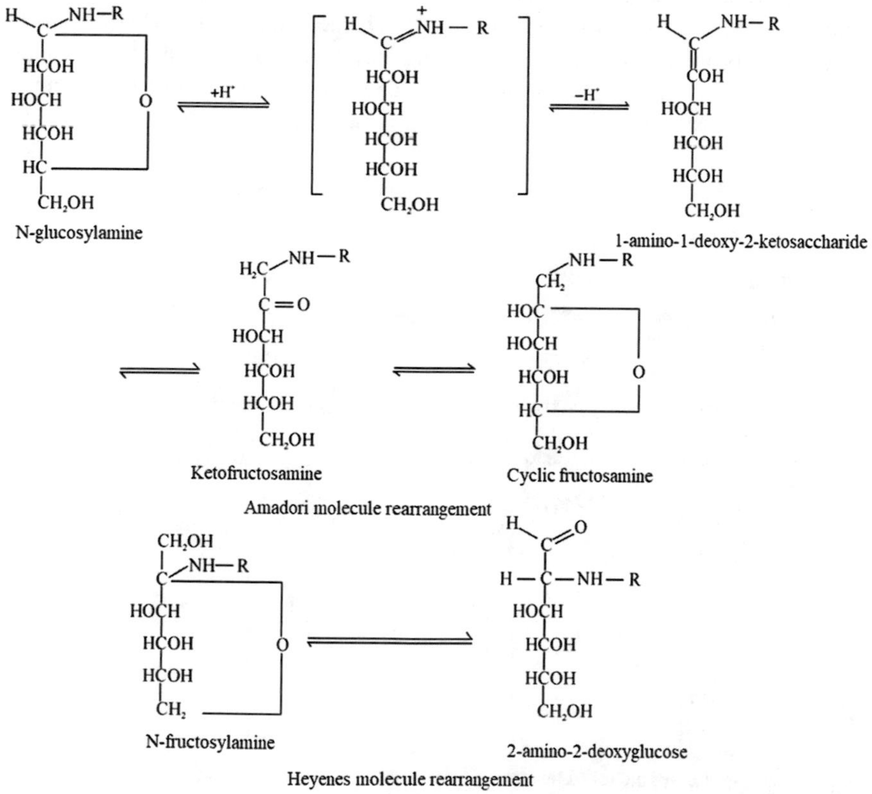

Fig. 4.8 Molecular rearrangement in Maillard reaction

gradual decrease of free amino group, so it is advantageous for carbonyl ammonia reaction under alkaline conditions.

(2) Molecular rearrangement

Under the catalysis of acid, N-glucosylamines form 1-amino-1-deoxy-2-ketosaccharides, namely monofructosamines through Amadori molecule rearrangement. In addition, ketosacharides can also form ketosylamines with amino compounds, while ketosylamines can be isomerized to 2-amino-2-deoxyglucose through the rearrangement of Heyenes molecule rearrangement (Fig. 4.8).

2. Intermediate stage

The rearrangement product fructosamines may be further degraded through multiple pathways to produce various carbonyl compounds, such as hydroxymethylfurfural (HMF), redutones, etc. These compounds can also react further.

(1) Dehydration of fructosylamine to hydroxymethylfurfural

When the pH is less than 5, the amine residue ($R-NH_2$) is removed first, and then dehydrated to 5-hydroxymethylfurfural (Fig. 4.9). The accumulation of HMF is closely related to the browning rate, and the browning may occur soon after the accumulation of HMF. Therefore, the accumulation of HMF can be measured by spectrophotometer to monitor the occurrence of browning reaction in food.

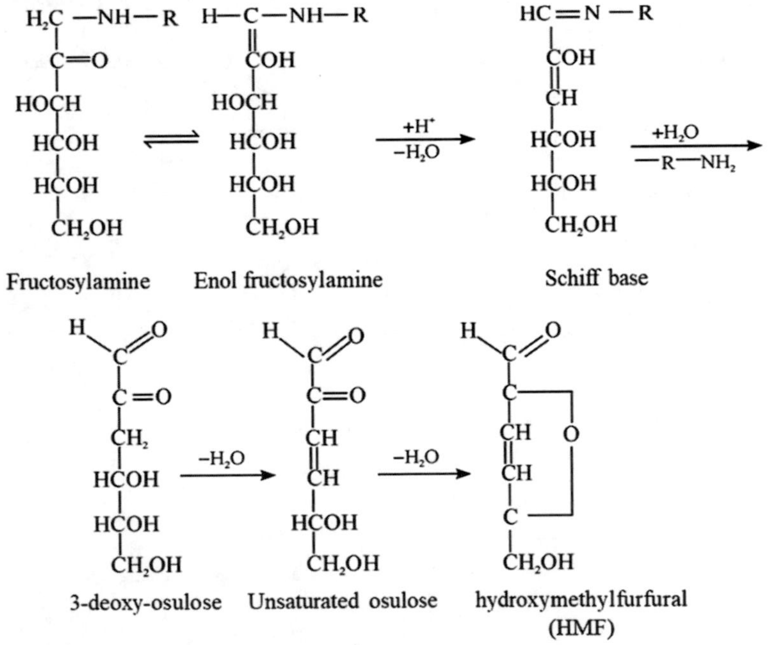

Fig. 4.9 Reaction of fructosylamine dehydration to hydroxymethylfurfural

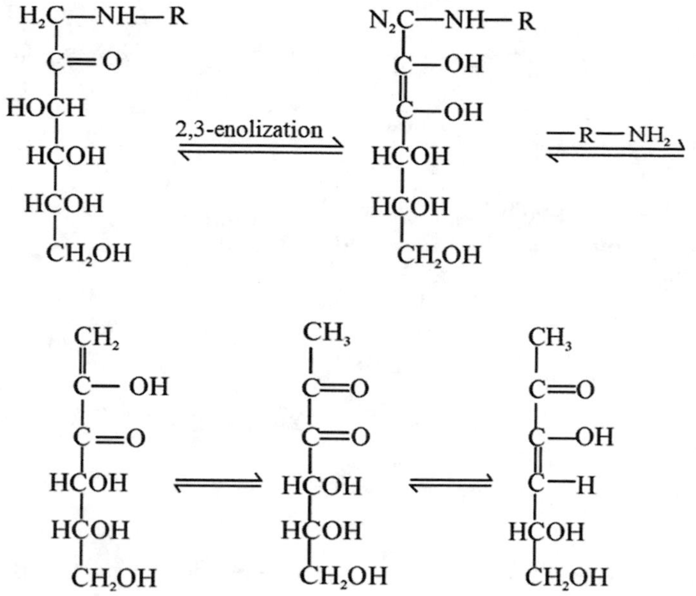

$$
\begin{array}{ccc}
\begin{array}{l}
\mathrm{H_2C-\!NH-\!R}\\
|\\
\mathrm{C}\!=\!\mathrm{O}\\
|\\
\mathrm{HOCH}\\
|\\
\mathrm{HCOH}\\
|\\
\mathrm{HCOH}\\
|\\
\mathrm{CH_2OH}
\end{array}
&
\xrightarrow[\text{2,3-enolization}]{\quad\quad}
&
\begin{array}{l}
\mathrm{N_2C-\!NH-\!R}\\
|\\
\mathrm{C-\!OH}\\
\|\\
\mathrm{C-\!OH}\\
|\\
\mathrm{HCOH}\\
|\\
\mathrm{HCOH}\\
|\\
\mathrm{CH_2OH}
\end{array}
&
\xrightarrow[\quad]{-R-NH_2}
\end{array}
$$

$$
\begin{array}{ccccc}
\begin{array}{l}
\mathrm{CH_2}\\
\|\\
\mathrm{C-\!OH}\\
|\\
\mathrm{C}\!=\!\mathrm{O}\\
|\\
\mathrm{HCOH}\\
|\\
\mathrm{HCOH}\\
|\\
\mathrm{CH_2OH}
\end{array}
&
\rightleftharpoons
&
\begin{array}{l}
\mathrm{CH_3}\\
|\\
\mathrm{C}\!=\!\mathrm{O}\\
|\\
\mathrm{C}\!=\!\mathrm{O}\\
|\\
\mathrm{HCOH}\\
|\\
\mathrm{HCOH}\\
|\\
\mathrm{CH_2OH}
\end{array}
&
\rightleftharpoons
&
\begin{array}{l}
\mathrm{CH_3}\\
|\\
\mathrm{C}\!=\!\mathrm{O}\\
|\\
\mathrm{C-\!OH}\\
\|\\
\mathrm{C-\!H}\\
|\\
\mathrm{HCOH}\\
|\\
\mathrm{CH_2OH}
\end{array}
\end{array}
$$

Fig. 4.10 Fructosylamine rearrangement reactions

(2) Fructosylamine removes amine residues and rearranges to produce reductones

In addition to the 1,2-enolation of Amadori molecular rearrangement during the above reaction process, 2,3-enolation can also occur, and finally reductone compounds are generated. The chemical properties of reduced ketones are more active. They can be further dehydrated and then condensed with amines. They can also be cracked into smaller molecules such as diacetyl, acetic acid, acetone aldehyde, etc. (Fig. 4.10).

(3) Interaction of amino acids with dicarbonyl compounds

In the presence of dicarbonyl compounds, amino acids can decarboxylate and deaminate to form aldehydes and carbon dioxide, and their amino groups are transferred to dicarbonyl compounds and react further to form various compounds (flavor components, such as aldehydes, pyrazines, etc.). This reaction is called Strecker degradation reaction (Fig. 4.11). Isotope tracing has proved that 90–100% of the carbon dioxide produced in carbonylation is from amino acid residues rather than sugar residues. Therefore, even though the Strecker degradation reaction is not the only way to produce carbon dioxide in the browning reaction system, it is also the main one.

(4) Formation of other reaction products of fructosamines

In the intermediate stage of Maillard reaction, fructosamines can generate various heterocyclic compounds, such as pyridine, benzopyridine, benzopyrazine, furan, and

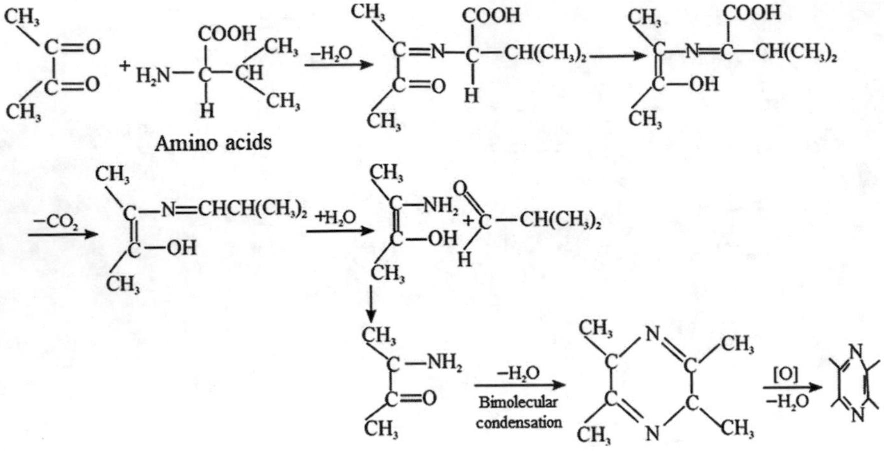

Fig. 4.11 Strecker degradation Reaction

pyran, in addition to reductones (Fig. 4.12), so the reaction in this stage is a complex one.

In addition, the products of Amadori rearrangement reaction can be oxidized and cracked to form amino-substituted carboxylic acids (Fig. 4.13). Therefore, ε-carboxymethyllysine can be used as an index of Maillard reaction process in this reaction system.

3. Final Stage

In the final stage of carbonyl ammonia reaction, on the one hand, polycarbonyl unsaturated compounds (such as reductones) are cracked to produce volatile compounds. On the other hand, condensation and polymerization are carried out to produce melanin-like substance, thus completing the whole Maillard reaction.

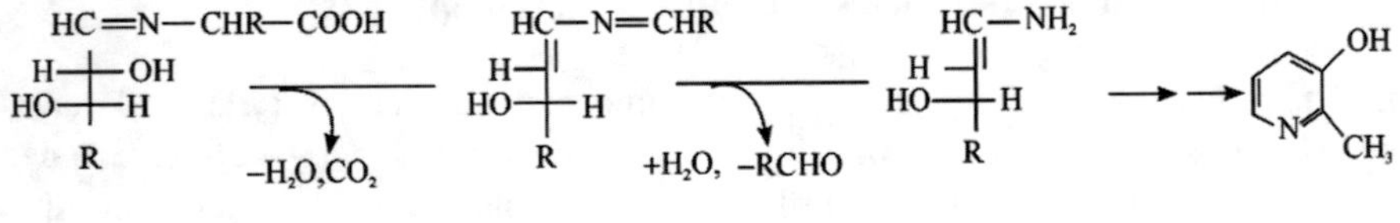

Fig. 4.12 Formation of pyridine compounds in Maillard reaction

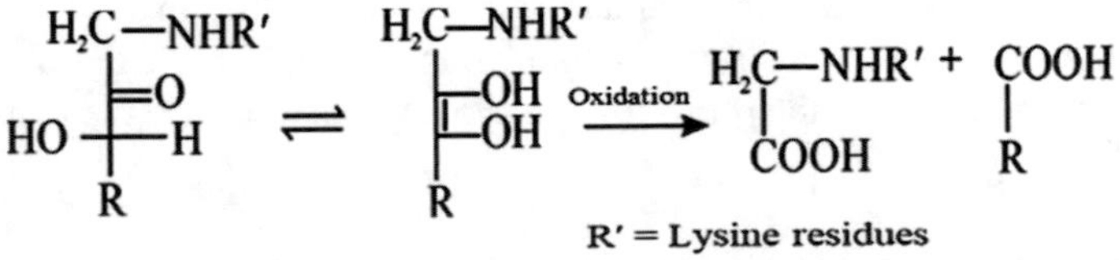

R' = Lysine residues

Fig. 4.13 Oxidative cracking of the product of Amadori rearrangement reaction

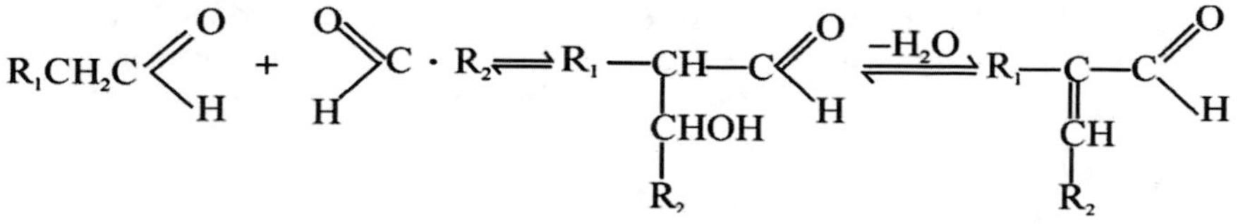

$$R_1CH_2C\diagup^{O}_{H} + \diagdown^{O}_{H}C \cdot R_2 \rightleftharpoons R_1-\underset{\underset{R_2}{|}}{\underset{CHOH}{|}}CH-C\diagup^{O}_{H} \xrightarrow{-H_2O} R_1-\underset{\underset{R_2}{|}}{\underset{CH}{\parallel}}C-C\diagdown^{O}_{H}$$

Fig. 4.14 Alcohol-aldehyde condensation reaction

(1)   Alcohol-aldehyde condensation

Alcohol-aldehyde condensation is the process of the self-phase condensation of two molecules of aldehydes and further dehydration to produce unsaturated aldehydes (Fig. 4.14).

(2)   The polymerization reaction for producing melanoidin-like substances

After the intermediate reaction, the products include furfural and its derivatives, dicarbonyl compounds, reductones, aldehydes produced by the degradation of Strecker and the cracking of sugar, etc. These products further condensate and polymerize to form complex macromolecule pigments.

In a word, there are many carbonyl ammonia reaction products in food system, which have an important impact on food flavor, color, and so on.

II.   Factors affecting carbonyl ammonia reaction

Maillard reaction mechanism is very complex, not only related to the types of monosaccharides and amino acids involved but also affected by temperature, oxygen, water, and metal ions. Controlling these factors can stimulate or inhibit non-enzymatic browning, which has practical significance for food processing.

1.   The effects of substrates

For different reducing sugars, Maillard reaction rate follows: ribose > arabinose > xylose in pentose; galactose > mannose > glucose in hexacarbohydrate; and the browning rate of pentose is about 10 times than that of hexacarbohydrate; and aldose > ketose, monosaccharide > disaccharide. It is noteworthy that some unsaturated carbonyl compounds (such as 2-hexenal) and alpha-dicarbonyl compounds (such as glyoxal) have higher reactivity than reducing sugars.

For different amino acids, the Maillard reaction rate of ε-NH$_2$ amino acid is much faster than that of α-NH$_2$ amino acid. Therefore, it can be predicted that the loss of lysine is greater in Maillard reaction. For α-NH$_2$ amino acids, the shorter the length of carbon chain, the stronger the reactivity.

2.   The effect of pH

Maillard reaction can occur in both acid and alkali environments, but the reaction speed increases with the increase of pH value when it is above 3. Besides, adding acid to reduce the pH value before the drying of egg powder and adding sodium carbonate to restore the pH value when the egg powder is dissolved again can effectively inhibit the browning of egg powder.

3. **Moisture**

Maillard reaction rate is proportional to the concentration of reactants, and it is difficult to proceed under completely dry conditions. Browning is easy to proceed with when water content is 10–15%. In addition, browning is also related to fat. When water content exceeds 5%, lipid oxidation and browning are accelerated.

4. **Temperature**

Maillard reaction is greatly affected by temperature, the 10 °C change of temperature leads to a three to five times change of browning speed. Therefore, food processing should avoid high temperature for a long time and low temperature is suitable for storage.

5. **Metal ions**

Iron and copper can promote Maillard reaction, so the metal ions should be avoided in food processing. Calcium can bind with amino acids to form insoluble compounds, which can inhibit Maillard reaction, and Mn^{2+}, Sn^{2+}, etc. can inhibit Maillard reaction.

6. **Air**

The presence of air affects Maillard reaction. Vacuum or inert gas can reduce the oxidation of fats and carbonyl compounds, as well as their reaction with amino acids. In addition, although the removal of oxygen did not affect the carbonyl reaction at the early stage of Maillard reaction, it could affect the formation of pigments at the later stage of the reaction.

For many foods, to increase color and aroma, it is necessary to use proper browning reactions in processing. For example, the production of tea, the baking of cocoa beans and coffee, the heating sterilization of soy sauce, and so on. However, for certain foods, browning reaction can cause their color deterioration, so it should be strictly controlled, such as high-temperature sterilization of dairy products, vegetable protein drinks.

4.2.3.2 Caramelization Reaction

When sugars, especially monosaccharides, are heated to a high temperature above the melting point (generally, 140–170 °C or more) without the presence of amino compounds, due to the dehydration and degradation of sugars, browning reaction also occurs. This reaction is called caramelization reaction, also known as Caramel action.

Different monosaccharides have different reaction rates according to their different melting points. The melting points of glucose, fructose, and maltose are 146 °C, 95 °C, and 103 °C, respectively. Thus, the caramelization reaction of fructose is the fastest.

Different pH values of sugar solutions have different reaction speeds. The higher the pH value, the quicker the caramelization reaction, where 10 times of reaction speed is observed when the pH value is 8 compared with 5.9.

There are two main types of caramelization products: one is the dehydration product of sugar, caramel (or sauce color); the other is the pyrolysis product of sugar, volatile aldehydes or ketones, etc. For some foods, such as baked and fried food, proper caramelization can make the products have pleasant color and flavor. In addition, caramel pigments, which are used as food colorants, are obtained by this reaction.

I.   The formation of caramel

Caramelization can occur when carbohydrates are heated under anhydrous conditions or treated with dilute acid at high concentrations. Glucose can produce glucose anhydride with right optical rotation (1,2-dehydration-α-D-glucose) and left optical rotation (1,6-dehydration-β-D-glucose). The specific rotatory value of the former is +69° and that of the latter is −67°. It is easy to distinguish between them. Yeast can only ferment the former. Fructose can form fructose anhydride (2,3-dehydration-β-D-fructofuranose) under the same conditions.

The process of caramel formation from sucrose can be divided into three stages. Initially, sucrose melts. When heated to about 200 °C, after foaming for about 35 min, sucrose loses a molecule of water and forms isosucrose anhydride (Fig. 4.15), which has no sweetness and mild bitterness. This is the initial reaction of sucrose caramelization.

After isosucrose anhydride is produced, the foaming stops temporarily and restarts later, which is the second stage of caramel formation, lasting about 55 min, longer than the first stage. During this period, the water loss reaches 9%. The product was caramel anhydride, with the average molecular formula, $C_{24}H_{36}O_{18}$.

$$2C_{12}H_{22}O_{11} - 4H_2O \rightarrow C_{24}H_{36}O_{18}$$

Caramel anhydride shows a bitter taste and is soluble in water and ethanol with a melting point at 138 °C. After 55 min of foaming in the intermediate stage, the third stage of foaming begins and further yields caramene.

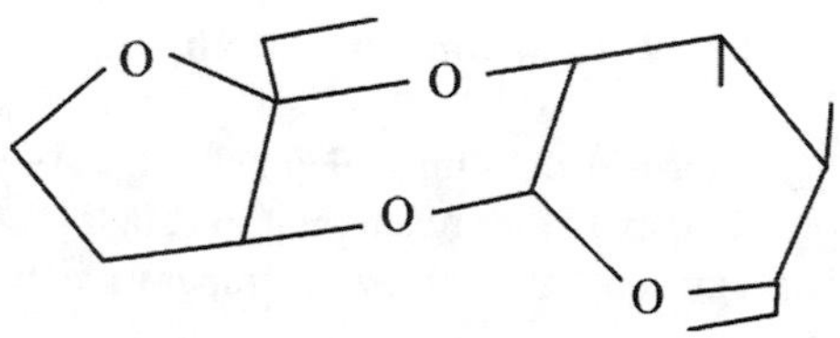

Fig. 4.15 Structural formula of isosucrose anhydride

$$3C_{12}H_{22}O_{11} - 8H_2O \rightarrow C_{36}H_{50}O_{25}$$

The melting point of caramene is 154 °C, which is soluble in water. If heated again, a dark, insoluble substance of high molecular weight is produced, which is called caramelin with an average formula of $C_{125}H_{188}O_{80}$. The structure of these complex pigments is still unclear, but they have the following functional groups: carbonyl, carboxyl, hydroxyl, and phenol.

The raw materials for producing caramel pigments are sucrose, glucose, maltose, or molasses. High temperature and weak alkaline conditions can increase the rate of caramel reaction. Catalysts can accelerate the reaction and produce different types of caramel pigments. At present, there are three commercial caramel pigments on the market. The first is acid-resistant caramel pigments produced with sucrose catalyzed by ammonium bisulfite, which catalyze the cleavage of sucrose glycoside bonds, and ammonium ions participate in Amadori rearrangement. It can be used in cola drinks, other acidic drinks, baking foods, syrups, candies, and seasonings. The solution of this pigment is acidic (pH 2–4.5), and it contains colloidal particles with negative charges. The second is made by the heating of sugar and ammonium salt to produce caramel pigments with red–brown colloidal particles with positive charge. The pH of the aqueous solution is 4.2–4.8, which can be used for baking food, syrup, pudding, etc. The third is the red–brown colloidal particles produced by direct pyrolysis of sucrose with slightly negative charge. The pH of the aqueous solution is 3–4, which can be used in beer and other alcoholic beverages. The isoelectric point of caramel pigments plays an important role in food manufacturing. For example, in a beverage with pH 4–5, if caramel pigments with isoelectric point pH 4.6 are used, flocculation, turbidity, and even precipitation will occur.

Phosphate, inorganic acid, alkali, citric acid, fumaric acid, tartaric acid, malic acid, etc. have catalytic effects on the formation of caramel.

II.   Formation of furfural and other aldehydes

Another type of change in sugar under intense heat is cracking and dehydration to form some aldehydes. For example, when monosaccharides are heated under acidic conditions, they are mainly dehydrated to form furfural or furfural derivatives. They can be polymerized or reacted with amines to produce dark brown pigments. When sugar is heated in alkaline, it will first tautomerize to form enol sugar, and then break to form formaldehyde, pentose, glycolaldehyde, tetracarbon sugar, glyceraldehyde, pyruvic aldehyde, etc. These aldehydes can produce dark brown substances through complex condensation, polymerization, or carbonylation of ammonia.

4.2.3.3   Interaction with Alkali

Monosaccharide is unstable in alkaline solution and prone to isomerization, decomposition, and other reactions. The stability of monosaccharides in alkaline solution has a great relationship with temperature, which is quite stable at lower temperature.

With the increase of temperature, monosaccharides will undergo isomerization and decomposition reactions very quickly. The degree of these reactions and the products formed are affected by many factors, such as the type and structure of monosaccharides, the type and concentration of alkali, the temperature and time of reaction, etc.

I. Isomerization

When monosaccharides are treated with dilute alkali solution, enol intermediates are first formed, and C_2 loses chirality. Changes of hydrogen atom at C_1 hydroxyl of glucose enol intermediate yield different monosaccharide isomers. If the hydrogen atom is added to C_2 from left side, the hydroxyl on C_2 will be on the right side and D-glucose will still be obtained. But if hydrogen atom on C_1 hydroxyl is added to C_2 from right side, the hydroxyl on C_2 will be transferred to left side, and the product is D-mannose. Similarly, hydrogen atoms on C_2 hydroxyl groups can also be transferred to C_1 and D-fructose can be produced. Therefore, the equilibrium mixture of D-glucose, D-mannose, and D-fructose can be obtained by the conversion of enol intermediate under the action of dilute alkali. Similarly, the same equilibrium mixture can be obtained by treating D-fructose or D-mannose with dilute alkali (Fig. 4.16).

The enolation of monosaccharides occurs not only in the formation of 1,2-enol but also in 2, 3-enol and 3,4-enol with the increase of alkali concentration, thus forming

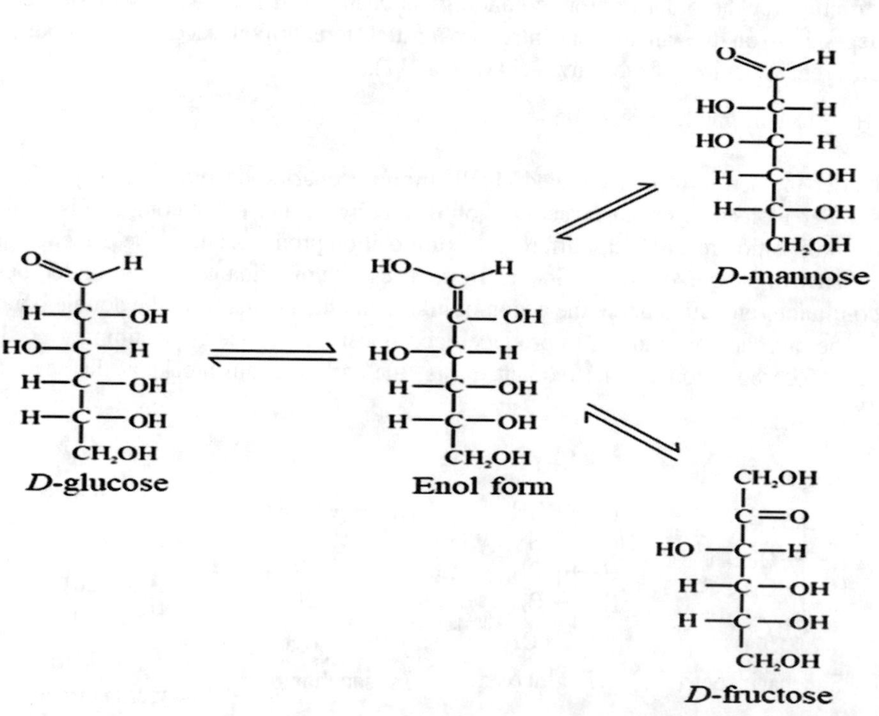

Fig. 4.16 Isomerization of glucose

other hexanol and hexone sugars. However, under the action of weak alkali, enolation usually stops at the stage of forming 2, 3-enol.

Before the application of enzymatic method, the glucose solution or starch syrup was treated by alkali reaction to produce fructose syrup. The conversion rate of fructose in this method was only 21–27%, and the loss of sugar was about 10–15%. At the same time, colored substances were produced, which made the refining process very difficult. Since the use of isomerase in 1957, three generations of high fructose syrup (HFS) products have been produced in food industry. In the first generation of HFS products, D-glucose is 52%, D-fructose is 42%, high carbon sugar is 6%, and solid content is about 71%. In the second generation of HFS products, D-glucose is 40%, D-fructose is 55%, high carbon sugar is 5%, and solid content is about 77%; In the third generation of HFS products, D-glucose is 7%, D-fructose is 90%, high carbon sugar is 3%, and solid content is about 80%.

II. Formation of saccharin acid

When monosaccharides react with alkali, with the increase of alkali concentration, heating temperature or heating time, intramolecular redox reaction and rearrangement of monosaccharides occur, resulting in carboxylic acid compounds. This carboxylic acid compound is called saccharic acid compound. Its chemical composition is not different from that of the original monosaccharide, but the molecular structure (or the order of atomic bonding) is changed. It has a variety of isomers, depending on the alkali concentration, and different monosaccharides produce different structures of saccharin acid (Fig. 4.17).

III. Decomposition reaction

Under the action of concentrated alkali, monosaccharide decomposes to produce smaller molecules of sugar, acid, alcohol, aldehyde, and other compounds. This decomposition reaction has different decomposition products due to the presence or absence of oxygen or other oxidants. In the presence of oxidants, hexose undergoes continuous enolation under the action of alkali and then splits from the double bond in the presence of oxidants to produce decomposition products containing 1, 2, 3, 4, or 5 carbon atoms. If no oxidant exists, the carbon chain breaks at the second

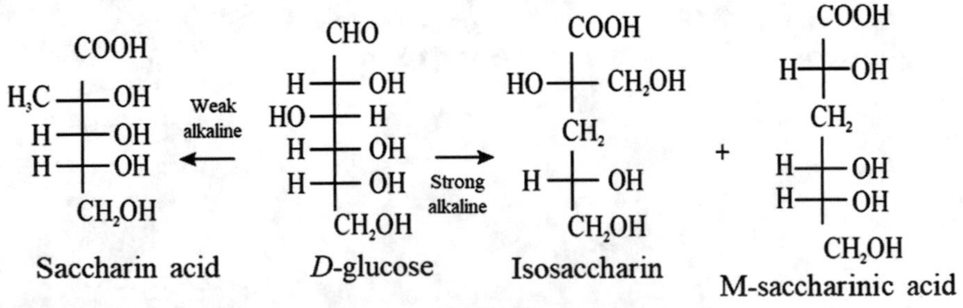

Fig. 4.17 Formation reaction of saccharin acid compound

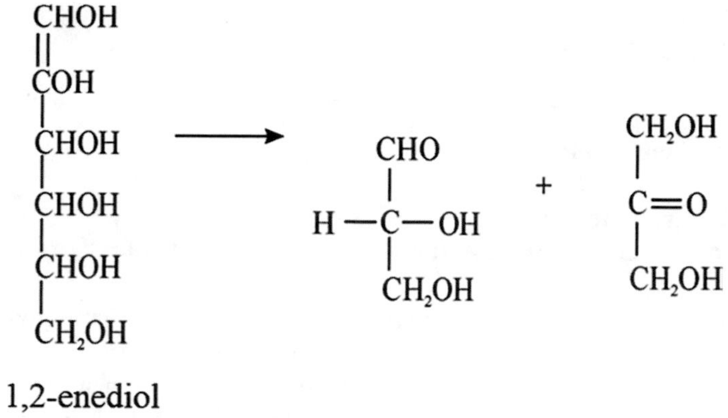

$$\begin{array}{l} CHOH \\ \parallel \\ COH \\ | \\ CHOH \\ | \\ CHOH \\ | \\ CHOH \\ | \\ CH_2OH \end{array} \longrightarrow \begin{array}{l} CHO \\ | \\ H-C-OH \\ | \\ CH_2OH \end{array} + \begin{array}{l} CH_2OH \\ | \\ C=O \\ | \\ CH_2OH \end{array}$$

1,2-enediol

Fig. 4.18 Decomposition of monosaccharides under non-oxidative conditions

single bond away from the double bond. The specific reaction formula is shown in Fig. 4.18.

4.2.3.4   Interaction with Acids

The effect of acid on sugar varies with the type, concentration, and temperature of acid. Very weak acidity can promote the conversion of monosaccharide α and β isomers; in dilute acid and heating conditions, monosaccharides can also undergo intermolecular dehydration reaction, thus condensation to glycosides (the inverse reaction of glycoside hydrolysis), yielding products including disaccharides and other oligosaccharides, which is also called complex reaction. In addition to the formation of disaccharides with α- or β-1,6 bonds, there are also trace amounts of other disaccharides. For example, the compound reaction of two-molecule glucose mainly combines the α-1,6 bond and the β-1,6 bond to form isomaltose and gentian disaccharide. In food industry, acid hydrolysis of starch is used to produce glucose. Due to the complex reaction of glucose, about 5% isomaltose and gentian disaccharide are produced, which not only affects the yield of glucose but also affects the crystallization and flavor of glucose.

When sugar and strong acid are heated together, furfural is produced by dehydration. For example, pentose produces furfural, hexose produces 5-hydroxymethyl furfural, and hexose is more prone to this reaction than hexose. Furfural and 5-hydroxymethyl furfural can react with some phenolic substances to form colored condensates, which can be used to identify carbohydrates. For example, resorcinol plus hydrochloric acid turns red when exposed to ketose, while turns light yellow when exposed to aldose. This reaction is called Sellwaneffs' test, which can be used to identify ketose and aldose.

The dehydration reaction of sugars is related to pH. The experiments show that the production of 5-hydroxymethyl furfural and colored substances is low at pH 3.0.

At the same time, the formation of colored substances increased with the increase of reaction time and concentration.

4.2.3.5   Oxidation and Reduction of Sugar

I.   **Oxidation reaction**
(1)   **Oxidation of Toulon reagent and Ferrin reagent (alkaline oxidation)**

Monosaccharides contain free carbonyl or keto groups. Since keto groups can be converted into aldehydes in dilute alkali solutions, monosaccharides have aldehyde generality and can be oxidized to acids and reduced to alcohols. Therefore, aldose and ketose can be oxidized by weak oxidants such as Toulon reagent or Ferrin reagent, the former produces silver mirror, the latter produces brick red precipitation of cuprous oxide, and the aldehyde group of sugar molecule is oxidized to carboxyl group.

$$C_6H_{12}O_6 + Ag(NH_3)^+_2OH^- \rightarrow C_6H_{12}O_7 + Ag\downarrow$$

Glucose or Fructose   Gluconic acid

$$C_6H_{12}O_6 + Cu(OH)_2 \rightarrow C_6H_{12}O_7 + Cu_2O\downarrow$$

Red precipitation

Sugars that can be oxidized by these weak oxidants are called reducing sugars, so fructose is also reducing sugars. This is mainly due to the keto-enol tautomerism of fructose in dilute alkali solution and the continuous conversion of keto group to aldehyde group, which reacts with oxidant.

(2)   **Bromine water oxidation (acid oxidation)**

Bromine water can oxidize aldose to form gluconic acid, and gluconic acid can easily lose water when heated to obtain gamma and delta esters. For example, D-glucose is oxidized to D-gluconic acid and D-gluconic acid–delta-lactone (GDL) by bromine water. The former can form calcium gluconate with calcium ions as the ingredient of oral calcium, while the latter is a mild acidic agent suitable for meat and dairy products. However, ketose cannot be oxidized by bromine water as it does not cause isomerization of sugar molecules under acidic conditions, so this reaction can be used to distinguish aldose from ketose.

D-glucose   D-gluconic acid-δ-lactone   D-gluconic acid-γ-lactone

(3) Nitric acid oxidation

The oxidation of dilute nitric acid is stronger than that of bromine water. It can oxidize aldehyde group and primary alcohol group of aldoses to form binary acid with the same carbon number. Galactose is oxidized to form galactose diacid, which is insoluble in acidic solution, while other hexanol oxidized to form binary acid can be dissolved in acidic solution, so the reaction can be used to identify galactose and other hexanol sugar.

D-glucose   D-glucaric acid   Lactone

(4) Periodate oxidation

Like other compounds with two or more hydroxyl or carbonyl groups on adjacent carbon atoms, carbohydrates can also be oxidized by periodate, resulting in the breaking of carbon–carbon bonds. The reaction is quantitative and consumes a mole of periodate for every break of a carbon–carbon bond. Therefore, this reaction is now one of the important means to study the structure of carbohydrates.

(5) Other

In addition to the oxidation reactions described above, ketose is cracked at the keto group under the action of strong oxidant to produce oxalic acid and tartaric acid. Monosaccharides react with strong oxidants to produce carbon dioxide and water. Under the action of oxidase, only the primary alcohol group of the sixth carbon atom is oxidized to carboxyl group to form glucuronic acid, while the aldehyde group is not oxidized.

2. **Reduction reaction**

Aldehydes or ketones in monosaccharide molecules can also be reduced to polyols. NaHg and $NaBH_4$ are commonly used as reducing agents. For example, D-glucose is reduced to sorbitol, xylose is reduced to xylitol, and D-fructose is reduced to a mixture of mannitol and sorbitol.

Sorbitol, mannitol, and other polyols exist in plants. Sorbitol is non-toxic, has slight sweetness and hygroscopicity, and its sweetness is 50% of sucrose. It can be used as a moisturizer for food, cosmetics, and medicines. The sweetness of xylitol is 70% of sucrose, which can replace sucrose as sweetener for diabetic patients.

4.2.4 Important Oligosaccharides in Food and Their Properties

Oligosaccharides are found in a variety of natural foods such as vegetables, grains, legumes, milk, honey, etc. The most common and important oligosaccharides in foods are disaccharides such as sucrose, maltose, and lactose. Most of the oligosaccharides other than this, due to their significant physiological functions, are functional oligosaccharides, which have received much attention in recent years.

4.2.4.1 Disaccharide

Disaccharides can be regarded as compounds formed by the dehydration of two molecules of monosaccharides, all soluble in water, sweet, optically active, and crystallizable. According to its reducing properties, disaccharides are further divided into reducing disaccharides and non-reducing disaccharides.

(a) **Sucrose (saccharose or cane sugar)**

Sucrose is a non-reducing sugar in which C_1 of α-D-glucose and C_2 of β-D-fructose are bound by glycosidic bonds. In nature, sucrose is widely distributed in the fruits, roots, stems, leaves, flowers, and seeds of plants, especially in sugar cane and sugar beet. Sucrose is the most demanding energy extract in humans and is the most important energy sweetener in the food industry.

Pure sucrose is a colorless and transparent monoclinic crystal with a relative density of 1.588 and a melting point of 160 °C. When heated to the melting point, it forms a glassy solid and further a brown caramel when heated to above 200 °C.

Sucrose is easily soluble in water, and its solubility increases with increasing temperature; when KCl, K_3PO_4, NaCl, etc. are present, its solubility also increases, and when $CaCl_2$ is present, it decreases. Sucrose is difficult to dissolve in organic solvents such as ethanol, chloroform, and ether.

Sucrose is dextrose, and the specific optical rotation of its aqueous solution is +66.5°. When it is hydrolyzed, an equal amount of glucose and fructose mixture is obtained. The specific optical rotation at this time is −19.9°, that is, the optical rotation direction of the hydrolysis mixture is changed, so the hydrolysate of sucrose is referred as inverted sugar.

Sucrose does not have a reducing property, has no mutarotation phenomenon, and has no osazone reaction, but can be combined with an alkaline-earth metal hydroxide to form a sucrose salt. This property is used industrially to recover sucrose from waste molasses.

(b)   Maltose

Maltose is a disaccharide composed of two molecules of glucose combined by α-1, 4 glycosidic bonds and has two isomers of α-maltose and β-maltose. Maltose is a transparent needle-like crystal, soluble in water, slightly soluble in alcohol, and insoluble in ether. Its melting point is 102 °C–103 °C, the relative density is 1.540, the sweetness is 1/3 of sucrose, and the taste is cool and soft. The specific optical rotation of α-maltose is +168°, while of β-maltose, is +112°; and at the equilibrium of the mutarotation, is +136°. Maltose is reductive and can form glycocalyx with excess phenylhydrazine.

Maltose is present in malt, pollen, nectar, tree honey and the petioles, stems, and roots of soybean plants. Under the action of amylase (i.e. maltase) or saliva, starch hydrolysis can obtain maltose. Maltose is formed when the dough is fermented and the sweet potato is steamed. The main component of the sugar contained in the wort used in beer production is maltose.

(c)   Lactose

Lactose is the main sugar component in mammalian milk. Milk contains 4.6–5.0% lactose and human milk contains 5–7%. Lactose is produced by the binding of one molecule of β-D-galactose to another molecule of D-glucose via a β-1, 4-glycosidic bond. Lactose is a white solid at room temperature, with weak solubility. Its sweetness is only 1/6 of sucrose and has reducibility. It can form strontium and has optical rotation with specific optical rotation of +55.4°, which is commonly used in food industry and the pharmaceutical industry.

Lactose helps the metabolism and absorption of calcium in the body, while it can also cause lactose intolerance in people who lack lactase in the body. Only when it is hydrolyzed into monosaccharides, it can be used as energy.

(d)   Cellobiose and trehalose

Cellobiose is a basic structural component of cellulose. It has no free state in nature. It is a typical β-glucoside by binding of two molecules of glucose to β-1, 4 glycosidic bonds. A half-acetal hydroxyl group is still retained in the molecule, so it has a reducing property, can be rotated, and can be formed into a quinone, which is a reducing disaccharide. Cellobiose is a colorless crystal with a melting point of 225 °C.

Trehalose, formerly known as glutinous sugar, is a disaccharide formed by the combination of two glucose molecules with a 1, 1 glycosidic bond. There are three isomers, namely trehalose (α, α), isotrehalose (β, β), and new trehalose (α, β). Trehalose is widely found in plants such as seaweed, fungi, and ferns, as well as in yeast and in the blood of invertebrates and insects. Trehalose has a non-specific protective effect on biological tissues and biomacromolecules and thus can be used as a protective agent for industrially unstable drugs, foods, and cosmetics.

4.2.4.2 High Fructose Syrup (HFS)

High fructose syrup (HFS), also known as high fructose corn syrups or isomerized syrup, is an isomerization of glucose solution obtained by enzymatic hydrolysis of starch, resulting in a mixed sugar syrup composed of fructose and glucose formed by glucose isomerase to transform part of glucose into fructose.

HFS is divided into three products with fructose content of 42%, 55%, and 90% according to the amount of fructose contained therein, and the sweetness is 1.0, 1.4, and 1.7 times of sucrose respectively.

As a new type of edible sugar, fructose syrup has the greatest advantage of containing a considerable amount of fructose, and fructose has many unique properties, such as synergy of sweetness, cold and sweetness, high solubility and high osmotic pressure, hygroscopicity, moisture retention and crystallinity, superior fermentability and processing and storage stability, significant browning reaction, etc., and these properties are more prominent with the increase of fructose content. Therefore, it is now widely used as a substitute for sucrose in the food field.

4.2.4.3 Other Oligosaccharides

(a) **Raffinose**

The raffinose, also known as honey trisaccharide, is the main component of soy oligosaccharides together with stachyose. It is another oligosaccharide widely found in the plant kingdom except sucrose. It is found in cottonseed, sugar beet and legume seeds, potatoes, various cereal grains, honey, and yeast are more abundant (Fig. 4.19).

The raffinose is α-D-galactopyranosyl ($1 \rightarrow 6$)-α-D-glucopyranose ($1 \rightarrow 2$)-β-D-fructofuranose. Pure raffinose is white or light yellow, long needle-like crystal. The crystal usually has five molecules of crystal water. The melting point of the raffinose with crystal water is 80 °C, and the temperature without crystal water is 118–119 °C. The raffinose is easily soluble in water and has a sweetness of 20–40% of sucrose. It is slightly soluble in ethanol and insoluble in petroleum ether. Its hygroscopicity is the lowest of all oligosaccharides, and it does not absorb water and agglomerate even in an environment with a relative humidity of 90%. The raffinose is a non-reducing sugar, which is not very active in Maillard reaction and has good thermal stability.

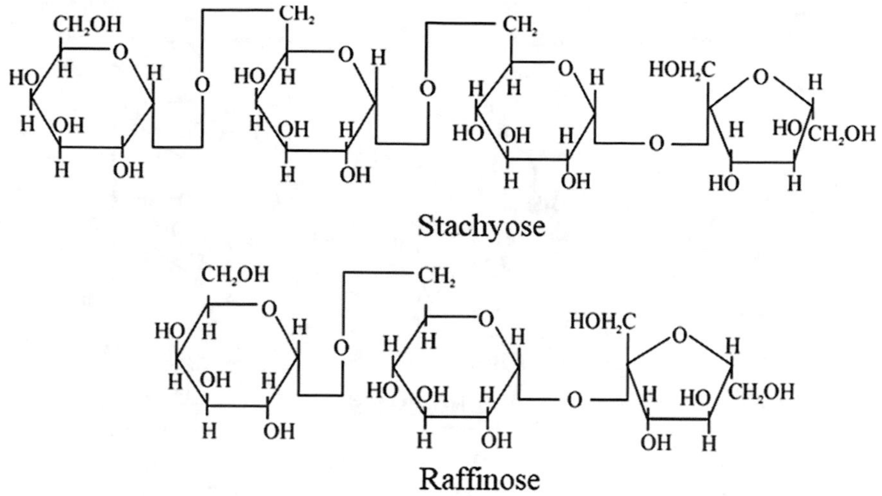

Fig. 4.19 Structure of stachyose and raffinose

There are two main methods for industrial production of raffinose. One is extracted from beet molasses and the other is extracted from detoxified cottonseed.

(b) Oligo-fructose (fructooligosaccharide)

Oligo-fructose, also known as oligofructose or trichome trisaccharide oligosaccharide, refers to a mixture of 1-kestose, fungitetraose and 1F-fructofuranosylnystose by linking 1–3 fructose groups via β-$(2 \rightarrow 1)$ glycosidic bonds on the fructose residue of sucrose molecules. Its structural formula can be expressed as G–F–Fn (G is glucose, F is fructose, n = 1–3) and is a linear heterooligosaccharide composed of fructose and glucose (Fig. 4.20).

Oligofructose is found in natural plants such as Jerusalem artichoke, asparagus, onions, bananas, tomatoes, garlic, honey, etc. Oligofructose has a variety of physiological effects, such as effective proliferative factors of bifidobacteria, water-soluble dietary fiber, low-calorie sweeteners that are difficult to digest, the improvement of intestinal environment, and the prevention for dental caries. Nowadays, oligofructose is obtained by moderately enzymatic hydrolysis of Jerusalem artichoke, and sucrose is used as raw material. By β-D-fructofuranosidase with fructose, a β-$(1 \rightarrow 2)$ glycosidic bond is combined with 1–3 fructose on the sucrose molecule. The β-D-fructofuranosidase is mostly produced by *Aspergillus oryzae* and *Aspergillus niger*.

The viscosity of oligofructose is like that of sucrose, and its sweetness is lower than that of sucrose. However, oligofructose has an obvious effect of inhibiting starch aging and is applied to starchy foods with good effect.

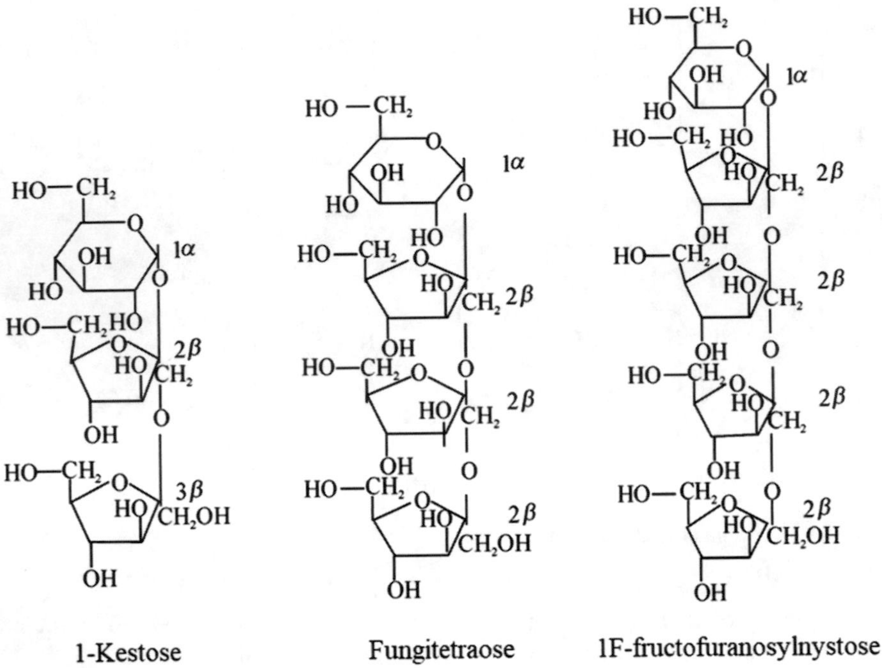

Fig. 4.20 Structure of oligofructose

(c) **Xylo-oligosaccharide**

Xylo-oligosaccharide is an oligosaccharide formed by 2–7 xyloses linked by β-(1→4) glycosidic bonds (Fig. 4.21), mainly composed of xylobiose and xylotriose. There are many xylan-rich plants in nature, such as corn cobs, sugar cane, and cottonseed. The xylan can be obtained by alkali hydrolysis, acid hydrolysis, or thermal hydrolysis. The sweet taste of xylooligosaccharides is like that of sucrose. It has unique acid resistance, heat resistance, and non-decomposability. It also has the effects of improving the intestinal environment, preventing dental caries,

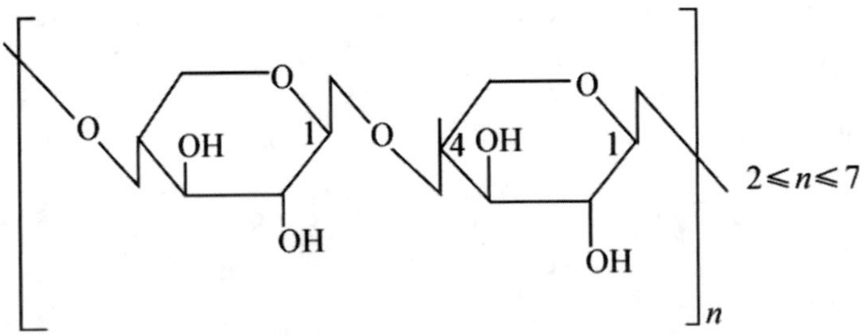

Fig. 4.21 Structure of xylooligosaccharide

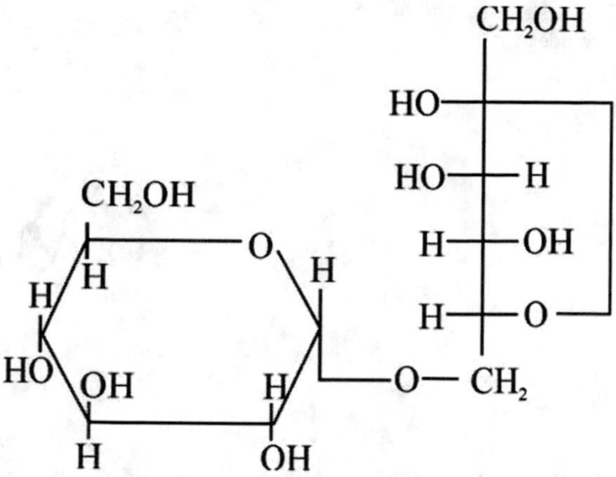

Fig. 4.22 Structural formula
of isomaltulose

promoting the body's absorption of calcium and metabolism in the body independent of insulin. Therefore, xylooligosaccharides are one of the most promising functional oligosaccharides and are widely used.

(d) **Isomaltulose**

Isomaltulose, also known as palatinose, has a structure of 6-O-α-D-glucopyranosyl-D-fructose (Fig. 4.22).

Isomaltulose has a sweet taste like sucrose and has a sweetness of 42% of sucrose. At room temperature, its solubility is relatively small, 1/2 of sucrose, but its solubility increases sharply with increasing temperature and can reach 85% of sucrose at 80 °C. The crystal of isomaltulose contains one molecule of water. It is the same as fructose. It is an orthorhombic crystal with optical rotation. The specific optical rotation is + 97.2°, and the melting point is 122–123 °C. It has no hygroscopicity and strong acid hydrolysis and is not used for fermentation of most bacteria and yeast to prevent dental caries, so it is used in acidic foods and fermented foods.

Isomaltulose was first discovered in the process of sugar beet production. At present, it is industrially used to convert sucrose as a raw material by α-glucosyltransferase.

(e) **Cyclodextrin**

The cyclodextrin is a closed-loop oligosaccharide composed of α-D-glucose combined with α-1,4 glycosidic bonds, and the degree of polymerization is 6, 7, and 8 glucose units, respectively, which are called α-, β-, γ-cyclodextrin. Industrially, a glucosyltransferase (EC.2.4.1.19) produced by *Bacitlusmacerans* is used to transform starch to cyclodextrin (Fig. 4.23).

The cyclodextrin structure has high symmetry and has a cylindrical three-dimensional structure, and the cavity depth and inner diameter are both 0.7–0.8 nm; the glycosidic oxygen atom in the molecule is coplanar, and the primary hydroxyl group on the hydrophilic group glucose residue C_6 on the molecule is arranged on the

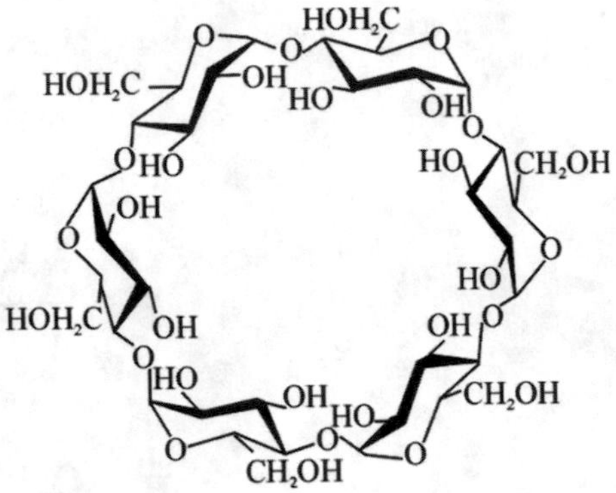

Fig. 4.23 Structure of the cyclodextrin

outer side of the ring, and the hydrophobic group C–H bond is arranged on the inner wall of the cylinder to make the middle empty hole hydrophobic. In view of its molecular structural characteristics, cyclodextrin can easily contain fat-soluble substances such as essential oils, flavors, etc. in its internal space, Therefore, it can be used as a microencapsulated wall material as a protective agent for volatile odor components, a modified embedding agent for bad odors, a moisturizer for food cosmetics, an emulsifier, a foaming accelerator, a stabilizer for nutrients and pigments, etc. Among them, β-cyclodextrin has the best application effect. In addition, it can also be used to embed the analyte to improve analysis sensitivity.

4.3 Polysaccharides

4.3.1 Structure of Polysaccharide

Polysaccharides refer to sugars with a degree of monosaccharide polymerization greater than 10. In nature, the degree of polymerization of polysaccharides is more than 100, the degree of polymerization of most polysaccharides is 200–3000 and that of cellulose is 7000–15000. Polysaccharides have two structures: one is straight chain and the other is branched chain. They are macromolecule compounds formed by monosaccharide molecules through 1,4 and 1,6 glycoside bonds. Like proteins, polysaccharides structures can be divided into primary, second, third, and fourth levels. The primary structure of polysaccharides refers to the order in which glycoside bonds connect monosaccharide residues in the linear chain of polysaccharides. The secondary structure of polysaccharides refers to the various polymers formed by hydrogen bonding between the polysaccharide skeleton chains. It only relates to the conformation of the main chain of polysaccharides but does not involve the spatial arrangement of side chains. The formed regular and coarse spatial conformation based on the primary and secondary structures of polysaccharides is the tertiary structure of polysaccharides. However, it should be noted that in the primary and secondary structures of polysaccharides, irregular and larger branching structures will hinder the formation of tertiary structures, while external disturbances, such as changes in solution temperature and ionic strength, also affect the pattern of tertiary structures. The fourth-order structure of polysaccharides refers to the aggregates formed by non-covalent bonds between polysaccharide chains. This aggregation can occur between the same polysaccharide chains, such as the hydrogen bond interaction between cellulose chains, or between different polysaccharide chains, such as the interaction between the polysaccharide chains of *Xanthomonas flavus* and the unreplaceable regions in the galactomannan skeleton.

4.3.2 Properties of Polysaccharide

4.3.2.1 Polysaccharides Solubility

Polysaccharides have many free hydroxyl groups, so they have strong hydrophilicity and are easy to hydrate and dissolve. In food system, polysaccharides can control water movement, and water is also an important factor affecting the physical and functional properties of polysaccharides. Therefore, many functional properties and textures of food are related to polysaccharides and water.

Polysaccharide is a substance with relatively high molecular weight. It does not significantly reduce the freezing point of water. It is a freezing stabilizer (not a cryoprotectant). For example, when starch solution is frozen, a two-phase system is formed, one is crystalline water (ice) and the other is a glassy substance composed of 70% starch molecule and 30% non-frozen water. Non-frozen water is a component of highly concentrated polysaccharide solution. For its high viscosity, the movement of water molecules is limited. When most polysaccharides are in the freeze-concentrated state, the movement of water molecules is greatly restricted. Water molecules cannot move to the active position of ice crystal nucleus or crystal nucleus growth, thus inhibiting the growth of ice crystal, providing freezing stability. The structure and texture of food products are effectively protected from damage, thereby the quality and storage stability of products improved.

Except for highly ordered crystalline polysaccharides, which are insoluble in water, most polysaccharides cannot crystallize and are easy to hydrate. In food industry and other industries, water-soluble polysaccharides and modified polysaccharides are called colloids or hydrophilic colloids.

4.3.2.2 Viscosity and Stability of Polysaccharide Solution

Polysaccharides (hydrophilic colloids or gels) mainly have the functions of thickening and gelling. In addition, they can control the flow properties and texture of liquid foods and beverages and change the deformability of semi-solid foods. In food

Fig. 4.24 Irregular cluster
polysaccharide molecules

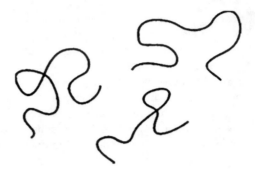

products, generally 0.25–0.5% concentration of glue can produce great viscosity and
even form gel.

The viscosity of polymer solution is related to the size, morphology, and conformation of molecules in solvents. Generally, polysaccharide molecules are in a disordered random coil state in solution (Fig. 4.24). However, the actual state of most
polysaccharides deviates from that of strict random coils, forming compact coils. The
properties of coils are related to the composition and connection of monosaccharides.

Linear polymers in solution occupy a large space when they rotate and have high
collision frequency and friction with each other, so they have high viscosity. The
solution viscosity of linear polymers is very high, even when the concentration is
very low, the solution viscosity is still very high. Highly branched polysaccharides
occupy a much smaller volume than straight-chain polysaccharides with the same
relative molecular weight, so the collision frequency is low, and the solution viscosity
is relatively low.

For linear polysaccharides with one kind of charge, the electrostatic repulsion of
the same charge causes chain extension, increases the chain length and the occupied
volume of the polymer, so the viscosity of the solution is greatly increased. Generally, the non-charged straight-chain homogeneous polysaccharide molecules tend to
associate and form partial crystallization, as the non-charged polysaccharide chains
collide with each other easily to form intermolecular bonds, thus forming an association or partial crystallization. For example, amylose dissolves in water under heating
conditions. When the solution is cooled, the molecules immediately aggregate and
precipitate, and this process is called aging.

Polysaccharide solutions generally have two kinds of liquidity. One is pseudoplastic and the other is thixotropic. Linear polymer molecular solutions are generally pseudoplastic. The higher the molecular weight, the greater the pseudoplasticity. Large pseudoplasticity is called "short flow", and its taste is non-sticky. Small
pseudoplasticity is called "long flow", and its taste is viscous.

Most of the hydrophilic colloidal solutions decrease with the increase of temperature. Therefore, by using this property, higher content of hydrophilic colloids can
be dissolved at high temperature, and the solution will play a thickening role after
cooling down. Except for xanthan gum solution, the viscosity of xanthan gum solution
remains basically unchanged in the range of 0–100 °C.

4.3.2.3 Gel

In many food products, some polymer molecules (such as polysaccharides or proteins) can form sponge-like three-dimensional network gel structure (Fig. 4.25). Continuous three-dimensional network gel structure is composed of polymer molecules through hydrogen bond, hydrophobic interaction, Van der Waals effect, ion bridging, entanglement, or covalent bond formation. The porous inside is filled with liquid phase, which is composed of low molecular weight solute and some polymer water solution.

Gels have both solid and liquid properties, making them semi-solid with viscoelastic properties, showing partial elasticity and partial viscosity. Although the polysaccharide gel contains only 1% polymers and contains 99% moisture, it can form a strong gel, such as sweet gel, jelly, imitated fruit block, etc.

Different gels have different uses. Their selection criteria depend on the desired viscosity, gel strength, rheological properties, pH of the system, processing temperature, interaction with other ingredients, texture and price, etc. In addition, the expected functional characteristics must also be considered. Hydrophilic colloids have multi-functional uses. They can be used as thickeners, crystallization inhibitors, clarifiers, film formers, fat substitutes, flocculants, foam stabilizer, sustained-release agents, suspension stabilizer, water-absorbing swelling agents, emulsion stabilizers, capsules, etc. These properties are often used as the basis for the selection of uses.

Fig. 4.25 Schematic diagram of typical three-dimensional network gel structure

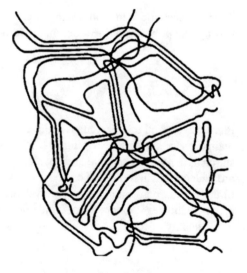

4.3.2.4 Hydrolysis of Polysaccharides

Polysaccharides are easier to hydrolyze than proteins in food processing and storage. Therefore, relatively high concentrations of edible gum are often added to avoid the decrease of the viscosity of the system due to the hydrolysis of polysaccharides.

The hydrolysis of oligosaccharides or polysaccharides, catalyzed by acids or enzymes, is accompanied by a decrease in viscosity. The degree of hydrolysis depends on the strength of acid or the activity of enzyme, time, temperature, and the structure of polysaccharide.

4.3.2.5 Flavor Binding Function of Polysaccharides

Macromolecular carbohydrates are a kind of good flavor fixatives, and Arabic gum is the most widely used one. Arabic gum can form a film around the flavor substance particles, which can prevent water absorption, the loss due to evaporation and chemical oxidation. Mixtures of Arabic gum and gelatin are used as wall materials of microcapsules, which is great progress in fixing food flavor components.

4.4 Main Polysaccharides in Food

The polysaccharides in foods mainly include starch, pectin, cellulose, hemicellulose, hydrophilic polysaccharide glue, modified polysaccharides, etc.

4.4.1 Starch

Plants, roots, and tubers contain abundant starch. Starch and starch products are the main diets for humans, providing 70–80% of calories to humans. Starch and modified starch have unique chemical and physical properties and functional properties. They are widely used in foods and can be used as adhesives, turbid agents, dusting agents, film formers, foam stabilizers, preservatives, gelling agents, polishes, water-holding agents, stabilizers, texture agents, and thickeners. Starches play a very important role in the quality of food.

4.4.1.1 Starch Structure

Starch is a high polymer composed of D-glucose through α-1, 4 and α-1, 6 glycosidic bonds, which can be divided into amylose and amylopectin. In the native starch granules, the two starches are present at the same time, and the relative amounts are different depending on the source of the starch (Table 4.4).

Table 4.4 Amylose content in different varieties of starch

Starch source	Amylose content	Starch source	Amylose content
Rice	17	Oat	24
Glutinous rice	0	Light pea	30
Common corn	26	Wrinkled pea	75
Waxy corn	0	Potato	22
High amylose corn	70–80	Sweet potato	20
Sorghum	27	Cassava	17
Waxy sorghum	0	Green beans	30
Wheat	24	Broad bean	32

Amylose is a linear macromolecule formed by the linkage of D-glucose through α-1, 4 glycosidic bonds (Fig. 4.26), and the degree of polymerization is about 100–6000, generally 250–300. The amylose molecule is not a fully linear molecule, but

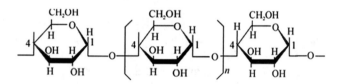

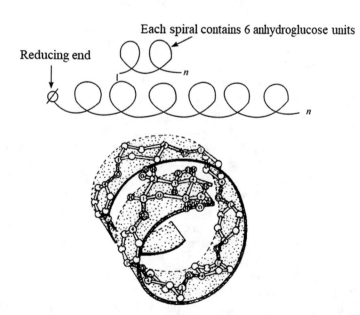

Fig. 4.26 Structure of amylose

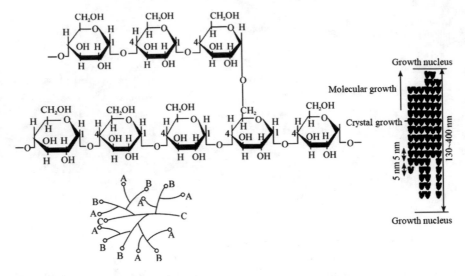

Fig. 4.27 Structure of amylopectin

a hydrogen bond between the hydroxyl groups in the molecule to distort the entire chain into a helical structure with a helical pitch of six glucose residues. Amylopectin is a macromolecule formed by the linkage of D-glucose through α-1, 4 and α-1, 6 glycosidic bonds (Fig. 4.27). The structure has branches, that is, each amylopectin molecule consists of one main chain and several side chains that are attached to the main chain.

The main chain is generally referred to as a C chain, and the side chain is further divided into an A chain and a B chain. The A chain is an outer chain, which is linked to the B chain via an α-1, 6 glycosidic bond, and the B chain is linked to the C chain via an α-1, 6 glycosidic bond. The number of the A chain and the B chain is approximately equal, and inside the A chain, the B chain, and the C chain, the linkage is α-1, 4 glycosidic bonds. Each branch contains an average of about 20–30 glucose residues. The branches are 11–12 glucose residues apart, and each branch is also curled into a helical structure. Therefore, the amylopectin molecule is approximately spherical and has a "twig" shape. The amylopectin molecule has a degree of polymerization of about 1,200–3,000,000, generally more than 6,000. It is much larger than the amylose molecule and is one of the largest natural compounds.

The starch granules are formed by radially arranging amylose and/or amylopectin molecules and have a structure in which alternating layers of crystalline and amorphous regions are present. Branches of amylopectin clusters (B-chain and C-chain) exist in a helical structure that is stacked together to form several small crystalline regions (microcrystal bundles) (Fig. 4.28), which are branched on the side chain of glucose. Residues are formed by hydrogen bond association in parallel, and there are mainly three crystal forms, namely, types A, B, and C (but when the complex forms a complex with an organic compound, it will exist in a V-type structure). This

Fig. 4.28 Structure of the
microcrystal beam

also means that the whole amylopectin molecule is not involved in the formation of the microcrystal bundle, but a part of the chain participates in the formation of the microcrystal bundle, and some of the chains do not participate in the formation of the microcrystal bundle, but become the non-starch particles. At the same time, the amylose also mainly forms an amorphous zone. The crystallization zone constitutes a compact layer of starch granules, the amorphous zone constituting a sparse layer of starch granules, and the compact layer and the sparse layer alternately arranged to form starch granules. The crystal structure is only a small part of the starch granules, and most of them are amorphous.

Observing the starch granules under a polarizing microscope, a black polarizing cross, or Maltese cross, can be seen, dividing the starch granules into four white areas (Fig. 4.29), and the intersection of the polarized crosses is in the starch granules. As well, the polarizing cross is situated in the grain core (umbilical point). This phenomenon is called birefringence, indicating that the starch granules have a crystal structure, which also indicates that the starch molecules in the starch granules are radially aligned and ordered.

At the same time, when the starch granules are observed under a microscope, a ring-like fine line like a tree ring around the umbilical point (referred to as a wheel pattern, Fig. 4.30) can be found, which has a screw-shell shape and a different density of the grain. The ringing of potato starch granules is most obvious, and the ring of cassava starch granules is also very clear, but the grain starch granules have almost no ring pattern.

The shape of the starch granules is generally divided into three types: circular, polygonal, and oval (Fig. 4.31), which vary from starch origin. For example, the large size of potato starch and sweet potato starch is oval, while the small particles are round; the rice starch and corn starch granules are mostly polygonal; the broad bean starch is oval and close to the shape of kidney; the mung bean starch and pea starch granules are mainly round shape and oval.

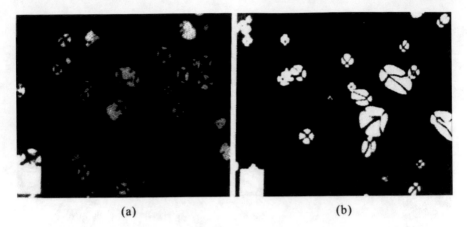

Fig. 4.29 Polarized microscopic morphology of wheat (**a**) and potato (**b**) starch granules

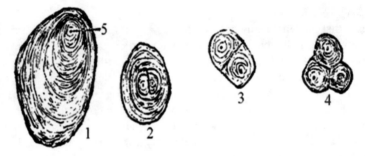

Fig. 4.30 Ring structure of potato starch granules. 1-simple starch granules; 2-semi-complex starch granule; 3, 4-complex starch granules; 5-grain core

The size of different starch granules varies greatly, and the size of one kind of starch granules also has a difference (see Table 4.5). The shape and size of starch granules are affected by factors such as seed growth conditions, maturity, endosperm structure, and relative proportions of amylose and amylopectin and have a great influence on the properties of starch.

4.4.1.2 Physical Properties of Starch

Starch is a white powder. Due to the strong water absorption and water holding capacity of the hydroxyl groups present in the starch molecule, the water content of the starch is relatively high, about 12%, but is also related to the source of the starch.

Pure amylopectin is easily dispersed in cold water, while amylose is the opposite. Natural starch is completely insoluble in cold water, but when heated to a certain temperature, natural starch will swell. Amylose molecules diffuse from starch

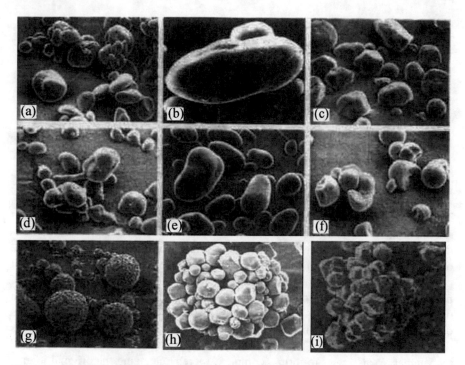

Fig. 4.31 Scanning electron micrograph of some starch granules **a, b** wheat; **c** corn; **d** high amylose corn; **e** potatoes; **f** cassava; **g** rice; **h** buckwheat; **i** Amaranth seeds

Table 4.5 Size of several starch granules (μm)

Starch type	The size of granule	Average particle sizes
Potato	5–100	65
Sweet potato	5–40	17
Rice	3–8	5
Corn	5–30	15
Wheat	2–10, 25–35	20
Green beans	8–21	16
Broad bean	20–48	32

granules into water to form a colloidal solution, while amylopectin remains in the starch granules. When the temperature is high enough and constantly stirred, the amylopectin will also swell and form a stable viscous colloidal solution. When the colloidal solution is cooled, the amylose is recrystallized and precipitated and can no longer be dispersed in hot water, while the degree of recrystallization of amylopectin is very small. The aqueous starch solution is dextrorotatory, with the specific optical rotation from $+201.5°$ to $+205°$, and the average specific gravity is about 1.5–1.6.

Fig. 4.32 Iodine–starch
complex

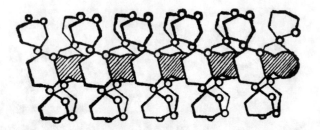

Starch and iodine can form a color complex, sensitively. The complex formed by amylose and iodine is brownish blue, and by amylopectin and iodine, is blue–violet. The complex color of dextrin and iodine is determined by dextrinmolecules in decreasing manner, from blue, purple, orange to colorless. This color reaction is related to the molecular size of amylose. Short amylose with a degree of polymerization of 4–6 does not develop color with iodine, and short amylose with a degree of polymerization of 8–20 is red with iodine. Amylose molecules with a degree of polymerization greater than 40 are dark blue with iodine. Although the degree of polymerization of the amylopectin molecule is large, the degree of polymerization of the branched side chain portion is only 20–30, so it is magenta with iodine. This color reaction is not a chemical reaction. In aqueous solution, amylose molecules exist in a spiral structure. Each helix adsorbs an iodine molecule and is connected by Van der Waals forces to form a complex, thereby changing the original iodine. The iodine molecule acts like a shaft running through the amyloid molecular helix (Fig. 4.32). Once the helix is extended, the bound iodine molecules are released. Therefore, the hot starch solution is stretched by the spiral structure, and the iodine does not appear dark blue. After cooling, it is dark blue due to the restoration of the spiral structure.

Pure amylose can quantitatively bind iodine, and each gram of amylose can bind 200 mg of iodine. This property is usually used for the determination of amylose content.

In addition to complexing with iodine to form a complex, amylose can form a complex like starch–iodine manner with fatty acids, alcohols, surfactants, etc.

4.4.1.3 Hydrolysis of Starch

The starch will undergo hydrolysis reaction under the catalysis of inorganic acid or enzyme, which are called acid hydrolysis method and enzymatic hydrolysis method, respectively. The hydrolysate of starch differs depending on the catalytic conditions and the type of starch, but the final hydrolysate is glucose.

(a) Acid hydrolysis

The acid hydrolysis of the glycosidic linkages of the starch molecules is random and initially produces large fragments. The degree of hydrolysis of starch is different, and the molecular size of the hydrolyzed product is also different, which may be purple dextrin, red dextrin, leuco dextrin, maltose, glucose. Different sources of starch have a different degree by acid hydrolysis. Generally, potato starch is easier to hydrolyze than corn starch, wheat, sorghum, and other cereal starches; and rice starch is more difficult to hydrolyze. Amylopectin is more susceptible to hydrolysis than amylose. The difficulty degree for glycosidic acid hydrolysis follows α-1, 6 > α-1, 4 > α-1, 3 > α-1, 2; and α-1, 4 glycosidic bond hydrolysis rate is greater than that of β-1, 4 glycosides. The crystalline zone is more difficult to hydrolyze than the amorphous zone. In addition, the acid hydrolysis reaction of starch is also related to temperature, substrate concentration, and inorganic acid species. Generally, the catalytic hydrolysis by hydrochloric acid and sulfuric acid is great.

Industrially, hydrochloric acid is sprayed into the uniformly mixed starch, or the stirred aqueous starch is treated with hydrogen chloride gas; and the mixture is then heated to obtain the desired degree of depolymerization, followed by acid neutralization, washing, and drying. The product is still granulated but very easily broken. This starch is called acid-modified or thin-boiling starches. This process is called thinning. The gelation of the acid-modified starch is improved, the gel strength is increased, and the viscosity of the solution is decreased. Further modification of starch with acid is to produce dextrin, purple dextrin, red dextrin, colorless dextrin, etc.

(b) Enzymatic hydrolysis

The enzymatic hydrolysis of starch is called saccharification in the food industry, and the amylase used is also called saccharification enzyme. Enzymatic hydrolysis of starch generally undergoes three steps of gelatinization, liquefaction, and saccharification. The amylase used for the hydrolysis of amylase mainly includes α-amylase (liquefaction enzyme), β-amylase (invertase, saccharification enzyme), and glucoamylase.

α-amylase is an endonuclease that cleaves both amylose and amylopectin molecules from the inside of the α-1, 4 glycosidic bond at any position, and the glucose residue at the reducing end of the product is α-configuration. α-amylase does not catalyze the hydrolysis of α-1, 6 glycosidic linkages, but circumvent and continues to catalyze the hydrolysis of α-1, 4 glycosidic linkages. In addition, α-amylase is incapable of catalyzing the hydrolysis of α-1, 4 glycosidic bonds in maltose molecules, so its hydrolysis products are mainly α-glucose, α-maltose, and very small dextrin molecules.

β-amylase can catalyze the hydrolysis of α-1, 4 glycosidic bonds from the reducing tail of starch molecules, cannot catalyze the hydrolysis of α-1, 6 glycosidic bonds, and cannot circumvent and continue to catalyze the hydrolysis of α-1, 4 glycosidic bonds. Thus, the β-amylase is an exozyme and the hydrolysate is β-maltose and β-limiting dextrin.

Glucose amylase starts from the non-reducing tail to catalyze the hydrolysis of starch molecules. The reaction can occur on α-1,6, α-1,4, α-1,3 glycosidic bonds, which can catalyze the hydrolysis of any glycosidic bond in starch molecules. Glucose amylase belongs to the external enzyme, and the final product is all glucose.

Some debranching enzymes specifically catalyze the hydrolysis of 1→6 linkages of amylopectin, producing many linear molecules of low relative molecular mass, one of which is an isoamylase and another is a pullulanase.

4.4.1.4 Starch Gelatinization

Raw starch molecules are closely arranged by many intermolecular hydrogen bonds, forming bundles of micelles, and the gap between them is small, even small molecules such as water are difficult to penetrate. The raw starch with micelle structure is called β-starch. After heating in water, β-starch destroys the weak hydrogen bond in the micelle of starch crystallization zone with the increase of heating temperature, and some micelles are dissolved to form the voids, so that the water molecules are immersed inside. A part of the starch molecules is hydrogen-bonded, as the micelles are gradually dissolved. The voids are gradually enlarged, and the starch granules are hydrated, with the volume expanded by several tens of times, and the micelles of the raw starch granules disappear. This phenomenon is called starch swelling. With continuous heating, the micelles in the crystallization zone all collapse, and the starch molecules form a single molecule, and surrounded by water (hydrogen bond), and become a solution state, as the starch molecules are in chain or branched. Finally, they are involved, resulting in a viscous paste solution. This phenomenon is called starch gelatinization (Fig. 4.33), and the starch in this state is called α-starch.

The gelatinization can be divided into three stages: ① reversible water absorption stage. When moisture enters the amorphous part of the starch granules, starch acts through hydrogen bonds and water molecules. The volume of the particles expands slightly, with no obvious change in appearance, and the crystal structure inside of the starch grain is not changed. At this time, it is cooled and dried, and it can be

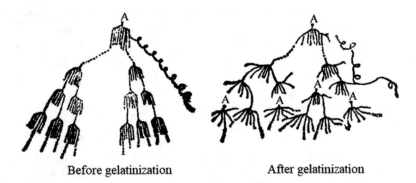

Before gelatinization   After gelatinization

Fig. 4.33 Schematic diagram of molecular morphology before and after starch gelatinization

restored, and the phenomenon of birefringence is unchanged; ② irreversible water absorption stage. With the increase of temperature, water enters the gap of starch microcrystal beam, along with irreversible water absorption. Particle volume expands and hydrogen bond between starch molecules is destroyed as molecular structure is stretched. At the same time, crystal structure and birefringence begin to disappear; ③ disintegration of starch granules. As all the starch molecules enter the solution, the viscosity of the system reaches the maximum, and the phenomenon of birefringence completely disappears.

Starch gelatinization generally has a temperature range, and the temperature at which the birefringence phenomenon begins to disappear is called the onset of gelatinization temperature, and the temperature at which the birefringence phenomenon completely disappears is called the complete gelatinization temperature. The temperature at which the gelatinization starts and the temperature at which the gelatinization is completed usually indicate the gelatinization temperature of the starch (Table 4.6).

Starch gelatinization, viscosity of starch solution, and properties of starch gel depend not only on the type of starch, the temperature of heating but also on the type and number of other components coexisting, such as sugar, protein, fat, organic acid, water, and substances such as salt.

The gelatinization temperature of various starches is different. The starch with higher amylose content has higher gelatinization temperature; even the same starch has different gelatinization temperature due to different particle sizes. In general, the gelatinization temperature of small granular starch is higher than large granular starch.

High concentrations of sugar will reduce the rate of starch gelatinization, the peak viscosity, and the strength of the gel formed. Disaccharides are more effective than monosaccharides in increasing the gelatinization temperature and lowering the viscosity peak, usually sucrose > glucose > fructose. Sugar reduces gel strength by plasticizing and interfering with the formation of binding regions.

Lipids, such as triacylglycerols, and lipid derivatives, such as monoacylglycerols and diacylglycerol emulsifiers, also affect the gelatinization of starch, that is, fats that form complexes with amylose delay the swelling of the particles, as in the low-fat-content white bread, usually 96% of the starch is completely gelatinized. However, the addition of fat to the gelatinized starch system, if no emulsifier is present, has no effect on the maximum viscosity value that can be achieved, but reduces the temperature at which the maximum viscosity is reached. For example, during gelatinization

Table 4.6 Gelatinization temperature of several starches

Starch	S	C	Starch	S	C
Japonica rice	59	61	Corn	64	72
Glutinous rice	58	63	Buckwheat	69	71
Barley	58	63	Potato	59	67
Wheat	65	68	Sweet potato	70	76

Note S, Start gelatinization temperature (°C); C, Complete gelatinization temperature (°C)

Fig. 4.34 Formation of inclusion complexes of fatty acids and amylose

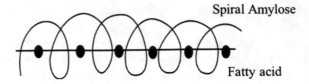

Spiral Amylose

Fatty acid

of a corn starch–water suspension, the maximum viscosity is reached at 92 °C, and if 9–12% fat is present, the maximum viscosity will be reached at 82 °C. The addition of a fatty acid with 16–18 carbon atoms or its monoacylglycerol increases the gelatinization temperature, and the temperature for the maximum viscosity also increases, while the strength of the gel decreases. The fatty acid or fatty acid component of monoacylglycerol can form an inclusion complex with helical amylose or form a clathrate with the longer side chain of amylopectin (Fig. 4.34) due to the high hydrophobicity inside the starch helix. Externally, the formation of the lipid–starch complex interferes with the formation of the binding zone and effectively prevents water molecules from entering the starch granules.

Due to the neutral character of starch, low concentrations of salt have little effect on gelatinization or gel formation, but potato amylopectin containing some phosphate groups and artificially produced ionized starch are affected by salt. For some salt-sensitive starch, depending on the conditions, the salt can increase or decrease the rate of expansion. The order of promotion of starch gelatinization by different ions is: $Li^+>Na^+>K^+>Rb^+$; $OH^- >$ salicylic acid $> SCN^- > I^- > Br^- > Cl^- > SO_4^{2-}$. In addition, compounds capable of destroying hydrogen bonds, such as urea, guanidinium salts, dimethyl sulfoxide, etc., can also gelatinize starch at normal temperature, wherein dimethyl sulfoxide dissolves before the starch has swollen, so it can be used as a solvent for starch.

Acids are commonly found in many starch thickened foods. Therefore, most foods have a pH range of 4–7, and such acid concentrations have little effect on starch swelling or gelatinization. At a pH of 10.0, the rate of swelling of the starch is significantly increased, but this pH is normally beyond the range of food. At low pH, the peak viscosity of the starch paste is significantly reduced, and the viscosity drops rapidly during cooking, for at low pH the starch is hydrolyzed, resulting in a non-thickened dextrin. In starch-thickened acidic foods, in order to avoid acid thinning, cross-linked starch is generally used.

In many foods, the interaction between starch and protein has an important impact on the texture of the food. For example, when wheat starch and gluten protein are in the dough, heated in the presence of water, starch gelatinization and protein denaturation occur, so that the baked food forms a certain structure. However, the nature of the interaction between starch and protein in food systems is still unclear.

4.4.1.5 Starch Aging (Retrogradation)

After the gelatinized α-starch is left at room temperature or below, it becomes opaque and even precipitates. This phenomenon is called starch aging. This is due to that the gelatinized starch molecules are automatically arranged in sequence at low temperatures, and the hydrogen bonds between adjacent molecules are gradually restored to form dense, highly crystallized, starch microcrystal bundles. Therefore, in a certain sense, the aging process can be regarded as the reverse process of gelatinization, but aging does not completely restore the starch to the structural state of raw starch (β-starch), which is less crystalline than raw starch. The starch after aging loses affinity with water and is not easy to interact with amylase, so it is not easily digested and absorbed by the human body, which seriously affects the texture of the food. For example, the staling of the bread loses freshness, and the viscosity of the rice soup decreases or precipitates. It is the result of starch aging. Therefore, the control of starch aging is of great importance in the food industry.

Starch from different sources, the degree of aging is not the same. This is due to that the aging of starch is related to the ratio of amylose and amylopectin. Generally, amylose is more susceptible to aging than amylopectin. The more amylose, the faster the aging. Amylopectin hardly ages as its branched structure hinders the formation of hydrogen bonds in the microcrystal bundle.

When the water content of starch is 30–60%, it is easy to age; when the water content is less than 10% or very high, aging does not likely happen. The optimum temperature for aging is about 2–4 °C, and no aging occurs when it is greater than 60 °C or less than −20 °C. It is also less susceptible to aging under conditions of partial acidity (pH = 4 or less) or partial alkali.

To prevent starch aging, the gelatinized α-starch can be quickly dehydrated at a high temperature of 80 °C or cooling dried below 0 °C to rapidly remove moisture (water content preferably down to 10% or less). In this way, it is impossible for the starch molecules to move and close to each other to become a fixed α-starch. After the α-starch is added with water, since the water is not immersed in the α-starch, the starch molecules are covered, and it is easy to gelatinize without heating. This is the principle of preparing convenient foods, such as instant rice, instant noodles, biscuits, puffed foods, etc.

Gelatinized starch is not easy to age in the presence of monosaccharides, disaccharides, and sugar alcohols, as they can interfere with the intermolecular association of starch, and the water absorption itself is strong enough to capture the water in the starch gel, so that the swollen starch becomes stable state. Surfactants or surface-active polar lipids delay the aging of starch due to the inclusion of amylose with it. In addition, some macromolecular substances such as protein, hemicellulose, vegetable gum, etc. also have a slowing effect on the aging of starch.

4.4.1.6   Resistant Starch

As people find that some starch cannot be digested and absorbed in the human small intestine, it has been challenged the idea that starch can be completely digested and absorbed in the small intestine. A new type of starch classification method has emerged, and the currently recognized classification following Englyst and Baghurst et al., where starch is classified into three categories according to the bioavailability of starch in the small intestine: One type is ready digestible starch (RDS), which refers to starch molecules that can be rapidly digested and absorbed in the small intestine, usually α-starch, such as hot rice, boiled sweet potatoes, and vermicelli; the second type is slowly digestible starch (SDS), which refers to starch that can be completely digested and absorbed in the small intestine but is slower. It mainly refers to some ungelatinized starches such as raw rice, noodles, etc.; the third category is the so-called resistant starch (RS). In fact, this classification is based on a single starch molecule, which means that some foods may contain the above three categories or two of them.

Resistant starch, referred to as RS, also known as anti-starch, resistant starch, anti-digestive starch, anti-enzymatic starch or anti-enzymatic starch. Prof. Litian Zhang, Honorary President of China Starch Association, has authorized this kind of starch as anti-digestive starch which defines its specific functionality. The European Anti-Digestive Starch Association in 1992 defined it as a general term for starch and its degradation products that are not absorbed by the healthy normal human small intestine. At present, anti-digestive starch is further divided into four kinds according to the shape and physicochemical properties of anti-digestive starch: RS1, RS2, RS3, RS4.

a.   RS1 is called physical embedded starch, which means that starch granules are difficult to be contacted with enzymes due to the barrier function of cell walls or the isolation of proteins, and thus are not easily digested. The physical action such as smashing and grinding during processing and chewing during feeding can change its content. It is commonly found in foods such as cereals and beans that are lightly milled.

b.   RS2 refers to anti-digestive starch granules, which are starches with a certain particle size, usually raw potatoes and bananas. After physical and chemical analyses, it is considered that RS2 has a special conformation or crystal structure (B-type or C-type of X-ray diffraction pattern) and is highly resistant to enzymes.

c.   RS3 is aged starch, which is mainly formed after the gelatinized starch is cooled. Starch polymer is commonly used in cooked and cold-cooked rice, bread, fried potato chips, and other foods. This kind of anti-digestive starch is divided into two parts, RS3a and RS3b. RS3a is agglomerated amylopectin, and RS3b is agglomerated amylose, where RS3b has the strongest anti-enzymatic activity.

d.   RS4 is a chemically modified starch, which is genetically modified or chemically induced by molecular structural changes and the introduction of some chemical functional groups, such as acetyl starch, hydroxypropyl starch, heat-modified starch and phosphorylated starch, etc.

Table 4.7 Direct/total ratio and anti-digestive starch content of common foods

Food name	Amylose/amylopectin ratio	Resistant starch (%)
Linear corn starch (I)	70/30	21.3 ± 0.3
Linear corn starch (II)	53/47	17.8 ± 0.2
Pea starch	33/67	10.5 ± 0.1
Wheat starch	25/75	7.8 ± 0.2
Common corn starch	25/74	7.0 ± 0.1
Potato starch	20/80	4.4 ± 0.1
Waxy corn starch	<1/99	2.5 ± 0.2

It is worth mentioning that: RS1, RS2, and RS3a can still be digested and absorbed after proper thermal processing; RS3 is the most important anti-digestive starch at present, and there are many studies on such starch worldwide.

The factors affecting the aging of starch, that is, the factors affecting the content of anti-digestive starch (here mainly referred to as RS3) in food, can be divided into internal and external factors according to their properties. Internal factors refer to factors related to starch properties and food composition in food, including the composition of raw materials, the ratio of amylose to amylopectin, the size of starch granules, and the degree of polymerization of starch molecules or chain length; External factors refer to the relevant processing conditions, treatment methods, and food forms. The amylose/amylopectin ratio and anti-digestive starch content of common foods are shown in Table 4.7.

4.4.1.7 Modified Starch

To meet the need for various uses, the natural starch needs to be physically, chemically, or enzymatically treated to cause certain changes in the original physical properties of the starch, such as water solubility, viscosity, color, taste, fluidity, and the like. This treated starch is collectively referred to as modified starch. There are many types of modified starches such as soluble starch, bleached starch, cross-linked starch, oxidized starch, esterified starch, etherified starch, and phosphate starch.

(a) **Soluble starch**

Soluble starch is a starch that has been treated with slight acid or alkali. The starch solution has good fluidity when heated and forms a firm gel when condensed. Alpha-starch is a soluble starch produced by physical treatment.

The general method for producing soluble starch is to apply 40% corn starch slurry with hydrochloric acid or sulfuric acid at a temperature of 25 °C–55 °C (below the gelatinization temperature). The treatment time can be determined by the viscosity reduction, about 6–24 h. Then, the hydrolysate is neutralized with Na_2CO_3 or dilute

NaOH and then filtered and dried to obtain soluble starch. Soluble starch is used to make gum and candy.

(b)   Esterified starch

The sugar-based monomer of starch contains three free hydroxyl groups which form a starch ester with an acid or anhydride, and the degree of substitution can vary from 0 to a maximum of 3. Commonly used are starch acetate, starch nitrate, starch phosphate, and starch xanthate.

Industrially, starch acetate is prepared by treating starch emulsion with acetic anhydride or acetyl chloride under alkaline conditions. Low-substituted starch acetate (degree of substitution < 0.2, acetyl percentage < 5%) paste has weak agglomeration showing anti-aging effect and high stability. Starch triacetate (high degree of substitution) contains 44.8% acetyl group, soluble in acetic acid, chloroform, and other chloroalkane solvents. Its chloroform solution is often used to determine its viscosity, osmotic pressure, optical rotation, and so on.

Starch xanthate can be obtained by using CS_2 to act on starch, and it can be prepared by using a high degree of cross-linked starch as a raw material to make it insoluble in water. Starch xanthate can be used to remove copper, chromium, zinc, and other heavy metal ions from industrial wastewater with satisfactory results.

Oxidizing starch with N_2O_5 in a chloroform solution containing NaF gives a completely substituted starch nitrate, which is an industrially produced starch ester derivative that can be used for explosives.

Phosphoric acid is a trivalent acid, and ester derivatives formed by the action with starch are starch monoesters, diesters, and triesters. Esterification with sodium orthophosphate and sodium tripolyphosphate ($Na_5P_3O_{10}$) gives a starch monoester, which shows high viscosity, transparency, and adhesiveness. When esterified with phosphorus oxychloride ($POCl_3$), a starch monoester and a cross-linked starch phosphodiester, a triester mixture can be obtained. Starch phosphodiesters and triesters are cross-linked starches. The swelling of the cross-linked starch granules is suppressed, leaving the gelatinization difficult and the viscosity and viscosity stability increased. Starch phosphate with low esterification can improve the freeze-thaw resistance of some foods and reduce the segregation of water during freeze-thaw process.

(c)   Etherified starch

The free hydroxyl group on the starch saccharide monomer can be etherified to give an etherified starch, wherein the methyl etherification method is a common method for studying the structure of starch or other polysaccharides, that is, dimethyl sulfuric acid and NaOH or AgI and Ag_2O act on the starch, and the free hydroxyl group thereof is methoxylated, and after hydrolysis, according to the structure of the obtained methyl sugar, the glycosidic linkage between the glucose units in the starch molecule can be determined. Industrial production generally uses the former method, particularly for the preparation of a methyl etherified starch with a low degree of substitution, while that having a high degree of substitution requires repeated methylation operations.

Low-substituted methyl etherified starch has lower gelatinization temperatures, higher water solubility, and lower agglomeration properties. The methyl etherified starch with a degree of substitution of 1.0 is soluble in cold water but insoluble in chloroform. As the degree of substitution increases, the water solubility decreases and the solubility in chloroform increases.

Granular or gelatinized starch readily reacts with ethylene oxide or propylene oxide under basic conditions to form a partially substituted hydroxyethyl or hydroxypropyl etherified starch derivative. The low degree of substitution of hydroxyethyl starch has a lower gelatinization temperature, a faster rate of heat swelling, higher transparency and adhesiveness of the paste, a weaker agglomeration property, and a transparent, soft film after drying.

(d) **Oxidized starch**

Industrially, the use of NaClO to treat starch gives oxidized starch. Since the amylose is oxidized, the chain becomes distorted and thus is less likely to cause aging. Oxidized starch paste shows low viscosity, but high stability, more transparent, and good film-forming properties. It can form a stable solution in food processing, suitable for dispersant or emulsifier. Periodic acid or its sodium salt can also oxidize adjacent hydroxyl groups to aldehyde groups, which is very useful in studying the structure of sugars.

(e) **Cross-linked starch**

Starch can act with multi-functional groups, such as formaldehyde, epichlorohydrin, phosphorus oxychloride, tripolyphosphate, or the like. These reactions help to combine starch molecules together by various cross-linking bonds, yielding cross-linked starch. Cross-linked starch has good mechanical properties, and is resistant to heat, acid, and alkali condition. As the degree of cross-linking increases, it is not gelatinized even at high temperatures. In the food industry, cross-linked starches can be used as thickeners and excipients.

(f) **Branched starch**

Starch can be grafted with acrylic acid, acrylonitrile, acrylamide, methyl methacrylate, butadiene, styrene, and other synthetic polymer monomers to form a copolymer. The obtained copolymer has the properties of a second type of polymer (natural starch and synthetic polymer) and varies with the percentage of grafting, the grafting frequency, and the average molecular weight. The grafting percentage is the weight percentage of the graft polymer in the copolymer; the grafting frequency is the average number of glucose units between the graft chains and is calculated from the graft percentage and the average relative molecular mass of the copolymer.

The structure of the branched chain of the synthetic polymer ($CH_2 = CH_X$) on the starch chain is different, and its properties are also different. If $X = -CO_2H$, $-CONH_2$, $-CO(CH_2)_n$, $-N^+R_3Cl$, etc., the obtained copolymer is soluble in water and can be used as thickener, absorbent, sizing agent, adhesive, and flocculant. If X

= –CN, –CO$_2$R, phenyl, etc., the resulting copolymer is insoluble in water but can be used in resins and plastics.

4.4.2 Pectins

Pectin is one of the plant cell wall components and exists in the intercellular layer between adjacent cell walls, acting to bind cells together. Pectin substances are widely present in plants, especially in fruits and vegetables, which give fruits and vegetables a hard texture.

4.4.2.1 Chemical Structure and Classification of Pectin Substances

The main chain of the pectin molecule is a polymer of 150–500 α-D-galactopyranosyl groups linked by α-1,4 glycosidic bonds, wherein some carboxyl groups and methanol in the galacturonic acid residues form an ester and the remaining carboxyl moiety form a salt with sodium, potassium, or ammonium ions (Fig. 4.35). The α-L-rhamnopyranose side chain is contained at a distance from the main chain, so the molecular structure of the pectin consists of a uniform region and a hair region (Fig. 4.36). The uniform region is composed of α-D-galacturonic acid groups, and the hair region is composed of highly branched α-L-rhamnogalacturonan.

The degree of methyl esterification in natural pectin substances varies greatly. The ratio of esterified galacturonic acid groups to total galacturonic acid groups is called the degree of esterification (DE), and the methoxy group content is also used to indicate the degree of esterification. The highest degree of esterification in pectin extracted from natural raw materials is 75%, and the degree of esterification in pectin products is generally 20–70%.

According to the ripening process of fruits and vegetables, pectin substances generally have three forms:

(a) **Raw pectin**. The methylated galacturonic acid chain combined with cellulose and hemicellulose is only present in the cell wall, insoluble in water, and

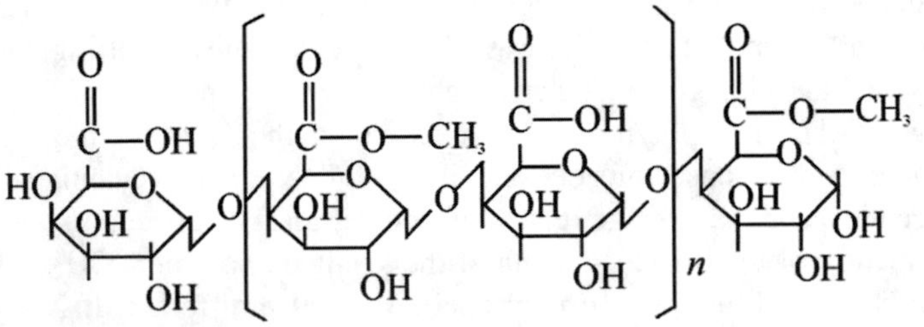

Fig. 4.35 Structure of pectin

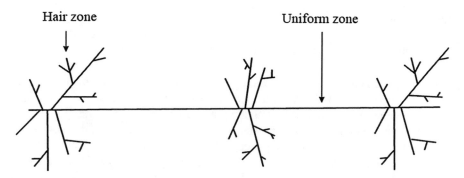

Fig. 4.36 Schematic diagram of pectin molecular structure

hydrolyzed to form pectin. It binds to cellulose and hemicellulose in immature fruit and vegetable tissues to form a firmer cell wall, making the whole tissue tough.

(b) **Pectin**. Pectin is a polygalacturonic acid chain with different degree of methylation and cation neutralization of carboxyl groups, which is present in the sap of plant cells, and its content in mature fruits and vegetables is high.

(c) **Pectic acid**. Pectic acid is a completely unmethylated polygalacturonic acid chain, which forms a water-insoluble or slightly soluble pectin acid salt in the cell juice with minerals such as Ca^{2+}, Mg^{2+}, K^+, Na^+. When the fruits and vegetables become soft ulcers, the content of pectic acid is high.

Pectin with a degree of esterification greater than 50% is usually called high-methoxyl pectin (HM). Pectin with a degree of esterification less than 50% is usually called low methoxyl pectin (LM).

4.4.2.2 Characteristics of Pectin Substances

The pectin substance can be hydrolyzed under acidic or alkaline conditions to cleave the ester group or the glycosidic bond; under the condition of high temperature and strong acid, the uronic acid residue undergoes decarboxylation.

The solubility of pectin and pectic acid in water decreases with increasing polymerization degree, and to some extent increases with the degree of esterification. Pectic acid has a low solubility (1%), but its derivatives such as methanol esters and ethanol esters are more soluble.

The solution formed by the dispersion of pectin is a high viscosity solution whose viscosity is proportional to the molecular chain length. Under certain conditions, pectin can form a gel.

4.4.2.3 Formation of Gelatin Gel

1. **Conditions and mechanism of gel formation of pectin substance**

When the pectin aqueous solution has a sugar content of 60–65%, a pH of 2.0–3.5, and a pectin content of 0.3–0.7% (depending on the pectin properties), the pectin at room temperature or even near boiling temperature can also form a gel.

During the gelation process, the water content has a great effect and excess water can prevent the pectin from forming a gel. The sugar is added to the pectin solution for dehydration, which causes the hydration layer around the pectin molecule to change, so that the water, adsorbed on the surface of the original colloidal particles is reduced, and the pectin molecules are easily combined to form chain micelles. The dehydration can accelerate the aggregation of the micelles and interweave them, forming a three-dimensional network structure without orientation. By adding a certain amount of acid in the pectin–sugar dispersion system, the H+ produced by the acid can reduce the negative charge of the pectin molecule. When the pH reaches a certain value, the pectin is close to electrical neutrality, so its solubility is reduced, so acid can accelerate the crystallization, precipitation, and agglomeration of pectin micelles, which is conducive to gel formation.

2. **Factors affecting the strength of pectin gel**
(a) **Relative molecular mass**

Under the same conditions, the greater the relative molecular mass of pectin, the stronger the gel formed. If the pectin molecular chain degrades, the gel strength formed is weaker. In addition to the very good correlation to average molecular weight, the pectin gel strength is also related to the number of points that each molecule participates in. This is as every 6–8 galacturonic acid groups form a crystallization center when the pectin solution is converted into a gel.

(b) **Pectin esterification degree**

The gel strength of pectin increases with the degree of esterification, as the crystal center of the gel network structure is located between the ester groups, and the degree of esterification in pectin also directly affects the gelation rate, which increases with increasing degree of esterification (Table 4.8).

When the degree of methyl esterification is 100%, it is called fully methylated polygalacturonic acid, and a gel can be formed if a dehydrating agent is present.

When the degree of methyl esterification is more than 70%, it is called quick-gelation pectin, and after adding sugar and acid (pH 3.0–3.4), a gel can be formed at a relatively high temperature and much easier when the temperature drops slightly. In candied jams, this gelation behavior prevents the lifting or sinking of the pulp pieces.

When the degree of methyl esterification is 50–70%, it is called slow-gelation pectin. After adding sugar and acid (pH 2.8–3.2), it can slowly form a gel at a lower temperature. The acid requirement also increases due to the increase in free carboxyl

Table 4.8 Classification and gelling conditions of pectin

Pectin type	Degree of esterification	Gelation condition	Gelation rate
High methoxy	74–77	Brix > 55, pH < 3.5	Super-fast
High methoxy	71–74	Brix > 55, pH < 3.5	Fast
High methoxy	66–69	Brix > 55, pH < 3.5	Medium speed
High methoxy	58–65	Brix > 55, pH < 3.5	Slow speed
High methoxy	40	Ca^{2+}	Slow speed
Low methoxy	30	Ca^{2+}	Fast

groups in the pectin molecule. Slow-gelation pectin is used in the production of jelly, jam, snacks, etc., and can be used as a thickener and emulsifier in juice foods. When the degree of methyl esterification is less than 50%, it is called low methoxy pectin. Even if the ratio of sugar addition and acid addition is appropriate, it is difficult to form a gel. However, its carboxyl group can react with multivalent ions (usually Ca^{2+}) to form a gel, the role of multivalent ions is to strengthen the cross-linking between pectin molecules (formation of "salt bridge"). At the same time, the presence of Ca^{2+} has a hardening effect on the texture of the pectin gel, which is the reason why the calcium salt treatment is first used in the processing of fruits and vegetables. The gelation ability of such pectin is greater affected by the degree of esterification than the relative molecular mass.

(c)   **pH value**

A certain pH value contributes to the formation of a pectin-glycol gel system. Different types of pectin have different pH values in gelation. For example, low methoxy pectin has poor sensitivity to pH change and a gel is formed in the range of pH 2.5–6.5, while a normal pectin forms a gel only in the range of pH 2.7–3.5. Inappropriate pH will not help the gelatin to form a gel, but cause hydrolysis of the pectin, especially for high methoxy pectin and under alkaline conditions.

(d)   **Sugar concentration**

Low methoxy pectin may not require the addition of sugar when forming a gel, but with the addition of 10–20% sucrose, the texture of the gel is improved.

(e)   **Temperature**

When the content of the dehydrating agent (sugar) and the pH are appropriate, the temperature has little effect on the pectin gel in the range of 0–50 °C. However, if the temperature is too high or the heating time is too long, the pectin will degrade and the sucrose will also be converted, thereby affecting the strength of the pectin.

4.4.3 Cellulose and Hemicellulose and Cellulose Derivatives

4.4.3.1 Cellulose

Cellulose is the major structural component of higher plant cell walls, usually combined with hemicellulose, pectin, and lignin, and the manner and extent of binding affect the texture of plant foods to a large extent. Cellulose is a homo-chain polymer obtained by linking β-D-glucopyranosyl units through β-1,4 glyco-sidic bonds. The degree of polymerization depends on the source of the cellulose and can generally reach 1000–14000.

The microstructure of cellulose was studied by X-ray diffraction, and it was found that cellulose is a bundle of more than 60 cellulose molecules arranged in parallel and hydrogen-bonded to each other. Although the bond of the hydrogen bond can be much weaker than that of the general chemical bond, the number of hydrogen bond between the cellulose crystallites is big, resulting in a tightly bounded microcrystal bundle and a very stable chemical property of the cellulose, such as cellulose insolubility in water, stability to dilute acids and alkalis. Cellulose does not reduce Fehling's solution, and it is not destroyed under normal food processing conditions. However, under high temperature, high pressure and acid (60–70% sulfuric acid or 41% hydrochloric acid), it can be decomposed into β-glucose.

The human body does not have the digestive enzymes to break down cellulose. When cellulose passes through the human digestive system, it does not provide nutrients and calories but does have important physiological functions. Cellulose can be used in paper, textiles, chemical compounds, explosives, film, pharmaceutical and food packaging, alcohol fermentation, feed production (yeast protein and fat), adsorbents, and clarifiers.

4.4.3.2 Hemicellulose

Hemicellulose is a type of heteropolysaccharide containing D-xylose, which is generally hydrolyzed to produce pentose, glucuronic acid, and some deoxy sugars. Hemicellulose is present in all terrestrial plants and is often found in the part of the plant that is lignified. The most important hemicellulose in food is the xylan composed of (1→4)-β-D-xylopyranosyl units, which is also a source of dietary fiber.

The crude hemicellulose can be divided into a neutral component (hemicellulose A) and an acidic component (hemicellulose B), and hemicellulose B is particularly abundant in hardwood. Both hemicelluloses have a xylan chain formed by the combination of β-D-(1→4) bonds. In hemicellulose A, there are many short-chain branches composed of arabinose in the main chain, D-glucose, D-galactose, and D-mannose. It is a typical example that arabinoxylan obtained from wheat, barley, and oat flour. Hemicellulose B contains no arabinose, which mainly contains 4-methoxy-D-glucuronic acid, so it is acidic. The structure of the water-soluble wheat flour pentamer is shown in Fig. 4.37.

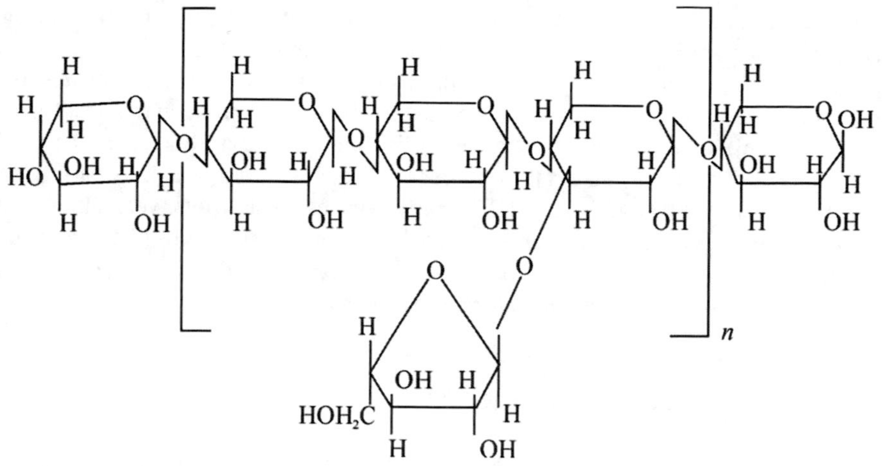

Fig. 4.37 Location of water-soluble wheat flour pentosan

Hemicellulose plays a large role in baked goods. It can improve the ability of flour to bind water, improve the quality of bread dough mixture, reduce the energy of the mixture, help the protein to enter and increase the volume of bread, and can delay the aging of bread.

Hemicellulose has a beneficial physiological effect on intestinal peristalsis, fecal volume, and fecal excretion, which can promote the elimination of bile acids and lower the cholesterol content in the blood. It has been shown that hemicellulose can reduce cardiovascular disease, colon disorders, and especially colon cancer.

4.4.3.3 Methylcellulose (MC)

The methyl group is obtained by treating the cellulose with methyl chloride under strong alkaline (sodium hydroxide) conditions to obtain methyl cellulose, and the modification is etherification. Commercial grade MC generally has a degree of substitution of 1.1–2.2, and MC with a degree of substitution of 1.69–1.92 has the highest solubility in water, and the viscosity mainly depends on the chain length of its molecule.

In addition to the properties of general hydrophilic polysaccharide gum, methylcellulose has three advantages: ① When the solution of methylcellulose is heated, the initial viscosity is lowered, which is the same as that of the general polysaccharide gum, and then the viscosity rises rapidly and forms a gel, which is converted into a solution when it cools, that is, a thermogel. This is due to that heating destroys the hydration layer outside the respective methylcellulose molecules and causes an increase in hydrophobic bonds between the polymers. Electrolytes (such as sodium chloride) and non-electrolytic agents (such as sucrose or sorbitol) can lower the temperature at which gels form, perhaps as they compete for water molecules. ②

MC itself is an excellent emulsifier, while most polysaccharides are merely emulsifying aids or stabilizers. ③ MC shows better film-forming property compared with general edible polysaccharides. Therefore, methyl cellulose can enhance the absorption and retention of water by foods, so that fried foods can reduce the absorption of oil. In some foods, it can act as a syneresis inhibitor and a filler. It can be used as texture and structural material in processed foods without gluten, as well as in frozen foods to inhibit syneresis. MC is used in sauces, meat, fruits, vegetables, and in salad dressings as thickeners and stabilizers, and it can also be used as an edible coating and a substitute fat for various foods. As it cannot be digested and absorbed by the human body, MC is a non-caloric polysaccharide.

4.4.3.4 Carboxymethylcellulose (CMC)

Carboxymethylcellulose is obtained by treating pure wood pulp with 18% sodium hydroxide to obtain alkaline cellulose and then reacted with sodium chloroacetate to form cellulose carboxymethyl ether sodium salt (cellulose-O- CH_2-CO_2-Na^+,CMC-Na), the degree of substitution (DS) of general products is 0.3–0.9, and the degree of polymerization is 500–2000. The DS for the food ingredient CMC occupying the largest market volume is 0.7.

Since CMC is composed of a negatively charged, long rigid molecular chain, it has high viscosity and stability due to electrostatic repulsion in solution and is related to degree of substitution and degree of polymerization. Carboxymethyl cellulose with a degree of substitution of 0.7–1.0 is easily soluble in water, forming a non-Newtonian fluid. Its viscosity decreases with increasing temperature. The solution is stable at pH 5–10 and has the highest stability at pH 7–9. When the pH is 7, the viscosity is the largest. When the pH is below 3, the free acid precipitate is easily formed. When a divalent metal ion is present, its solubility is lowered and an opaque liquid dispersion is formed. A gel or precipitate can be produced in the presence of a trivalent cation, so the salt tolerance of CMC is poor.

CMC-Na is widely used in the food industry. In China, it is stipulated that for the instant noodles and cans, the maximum dosage is 5.0 g/kg; and for juice milk, the maximum dosage is 1.2 g/kg. CMC-Na can stabilize the protein dispersion system, especially at the pH close to the isoelectric point of the protein. For example, egg white can be stabilized by drying or freezing with CMC-Na. CMC-Na can also improve the stability of dairy products to prevent casein precipitation. Adding CMC-Na to jam and ketchup can increase their viscosity and solid content, and make the structure of jam and ketchup soft and fine. Addition of CMC-Na to bread and cake can increase its water retention and prevent starch aging. Adding CMC-Na to the instant noodles makes it easier to control the moisture, reduces the oil absorption of the noodle cake, and increases the gloss of the noodles, where generally 0.36% of CMC-Na is used. Addition of CMC-Na to the soy sauce to adjust the viscosity of the soy sauce, so that the soy sauce has a smooth mouthfeel. CMC-Na acts similarly to sodium alginate for ice cream, but CMC-Na is inexpensive, has good solubility, and has a strong water-retention effect. Therefore, CMC-Na is often used in combination with other emulsifiers to reduce costs, and CMC-Na has synergistic effect with sodium alginate.

4.4.3.5 Microcrystalline Cellulose (MCC)

Microcrystalline cellulose (MCC) used in the food industry is a purified insoluble cellulose prepared by hydrolysis of pure wood pulp and separation of microcrystalline components from cellulose. Cellulose molecules are linear molecules composed of about 3000 β-D-glucopyranosyl units, which are very easy to associate and have long intercalations. However, the long and narrow molecular chains cannot be completely lined up, and the end of the crystallization zone is the bifurcation of the cellulose chain, which is no longer ordered, but randomly arranged. When the pure wood pulp is hydrolyzed with an acid, the acid penetrates the amorphous region with a lower density, and the molecular chains in these regions are hydrolytically cleaved to obtain a single spike crystal. Since the molecular chains constituting the spikes have a large degree of freedom of movement, the molecules can be oriented, and the crystals grow larger and larger.

There are two kinds of MCC, both showing heat and acid resistant. The first type of MCC is a powder, which is a spray-dried product. Spray drying causes agglomeration of crystallite aggregates to form a porous and sponge-like structure; and microcrystalline cellulose powder is mainly used for flavor carriers and as an anti-caking agent for cheese. The second MCC is a colloid that is dispersed in water and has similar functional properties as water-soluble gum. To produce the MCC colloid, after the hydrolysis, a large mechanical energy is applied, and the weakly bonded microcrystalline fibers are pulled apart, so that the main portion becomes a colloidal particle size aggregate with a diameter of less than 0.2 μm. To prevent recombination of aggregates during the drying process, sodium carboxymethylcellulose was added. Since carboxymethyl cellulose (CMC) provides stable negatively charged particles, the MCC is separated, preventing MCC from re-association and aiding in redispersion.

The main functions of the MCC colloid are included as follows. MCC colloid forms particularly stable foam and emulsion during high temperature processing, and it can yield a gel-like texture gel by forming a hydrated microcrystalline network structure. It helps to improve the heat resistance of pectin and starch gel, and gel adhesion. It can replace the part use of fat and control ice crystal growth. MCC is therefore able to stabilize emulsions and foams as MCC adsorbs on the interface and reinforces the interface film. So, MCC is a common ingredient in low-fat ice cream and other frozen confection products.

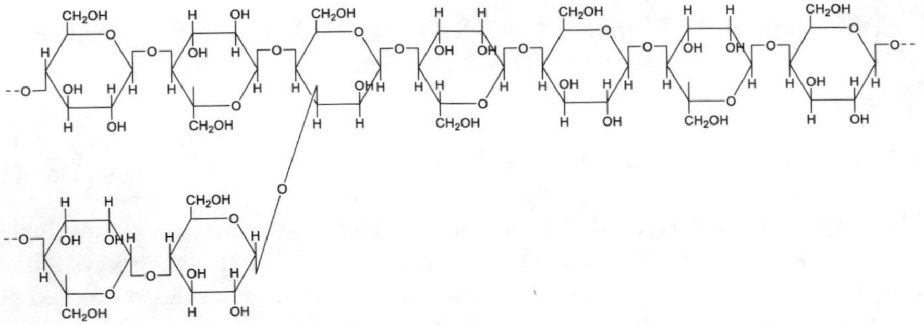

Fig. 4.38 Possible structural illustration of konjac glucomannan

4.5 Other Plant Polysaccharides

4.5.1 Konjac Gum

Konjac glucomannan is a polysaccharide composed of D-mannopyranose and D-glucopyranose linked by β-1,4 glycoside bond. The ratio of D-mannose to D-glucose is 1:1.6. At the C_3 position of D-mannose in the main chain, there are branched chains linked by β-1,3 glycoside bonds. Each 32 sugar residues have about 3 branched chains, and the branched chains are composed of several sugar residues. Each 19 sugar residues have an acetyl group, which gives it water soluble. Each 20 sugar residues contain one glucuronic acid, whose structure is shown in Fig. 4.38.

Konjac glucomannan can dissolve in water and form pseudoplastic fluid with high viscosity. After alkali treatment and acetic acid removal, it can form elastic gel and is a kind of thermal irreversible gel. When the konjac glucomannan was mixed with xanthan gum, a thermoreversible gel was formed. When the ratio of xanthan gum to konjac glucomannan is 1:1, the intensity of the gel was the largest. The melting temperature of the mixed gel is 30–63 °C, and the gel strength increased with the increase of the concentration, and decreased with the increase of the salt concentration. The interaction between helical structure of xanthan gum and konjac glucomannan is shown in Fig. 4.39.

Konjac glucomannan is applied to form a heat irreversible gel to produce a variety of foods, such as konjac cake, konjac tofu, konjac vermicelli and all kinds of bionic foods (shrimps, loin, belly slice, tendons, loach, sea cucumbers, jellyfish skin, etc.) and edible films.

4.5.2 Guar Gum and Locust Bean Gum

Guaran and locust bean gum are both galactomannan, which are important thickening polysaccharides and are widely used in food industry and other industries. Guar gum

Fig. 4.39 Interaction
between helical structure of
xanthan gum and konjac
glucomannan

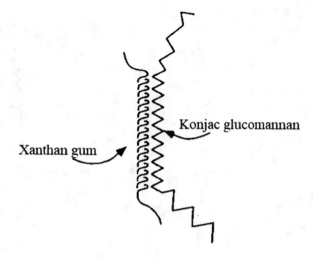

is one of the most viscous gums in all commercial gums. Its main chain is made up of β-D-mannose linked by 1,4-glycoside bond, and one side chain with (1→6) α-D-galactopyranose unit connects to each single sugar residue in the main chain (Fig. 4.40).

The branched chains of galactomannan in locust bean gum are less than that of guar gum, and the structure is irregular. Galactosyl groups in guar gum are evenly distributed in the main chain, and D-galactosyl side chains are contained in half of the main chains of pyran mannose. The locust bean gum molecule contains few galactosyl side chains and consists of a long smooth region (no side chains) and a hair region with galactosyl side chains. The ratio of mannose to galactose in guar gum is 1.6 (M/G = 1.6), while that in locust gum is 3.5, so the content of galactosyl in locust gum is very low. Due to the structural differences between guar gum and locust gum, guar gum and locust gum have different physical properties. The locust bean gum molecule has a long smooth region and can interact with other polysaccharides such as xanthan gum and carrageenan double helix to form a three-dimensional network structure of viscoelastic gel (Fig. 4.41), but guar gum and xanthan gum cannot form gel. As the less side chains of galactosyl group, the stronger synergism with other polysaccharides, the function of galactomannan is related to the content and distribution of galactose.

Guar gum is mainly used as a thickener. It is easy to hydrate to produce high viscosity, but it is often used in ice cream when it is combined with other edible

Fig. 4.40 The repetitive unit
structure of galactomannan

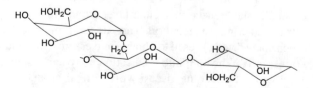

Fig. 4.41 Interaction of
Robinia bean gum molecule
with xanthan gum or
carrageenan double helix
molecules to form
three-dimensional network
structure gel

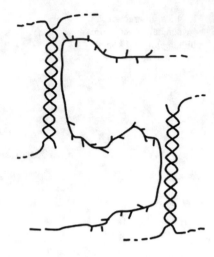

gums such as CMC, carrageenan, and xanthan gum. Locust bean gum is seldom
used alone when it is used in dairy products and frozen sweets. It is commonly used
in combination with other gums such as CMC, carrageenan, xanthan gum, and guar
gum. The dosage is generally from 0.05 to 0.25%. It can also be used in meat industry
such as fish, meat, and other seafood.

4.5.3   Arabic Gum

Arabic gum is composed of two components. One is composed of polysaccharides
without N or with a small amount of N, and the other is a protein structure with high
molecular weight. Polysaccharides are covalently bonded with hydroxyproline and
serine in the protein-peptide chain. The total protein content is about 2%, and in some
special varieties, as high as 25%. The polysaccharide molecules linked to proteins are
highly branched acidic polysaccharides with the following components: D-galactose
44%, L-arabinose 24%, D-glucuronic acid 14.5%, L-rhamnose 13%, 4-O-methyl-
D-glucuronic acid 1.5%. In the main chain of Arabic gum, β-D-galactopyranose is
linked by 1,3-glycoside bond, while the side chain is linked by 1,6-glycoside bond.

Arabia gum is easy to dissolve in water. The most unique property is high solubility
but with low viscosity. The solubility can even reach 50%. In this case, the system
is somewhat like gel. Arabic gum is not only a good emulsifier but also a good
emulsion stabilizer. This is as Arabic gum has surface activity and can form a thick,
spatially stable macromolecular layer around the oil droplets to prevent oil droplets
from gathering. Essential oils and Arabia gum are often made into emulsions, and
then spray drying is used to obtain solid essence to avoid volatilization and oxidation
of essence. When put into application, they can disperse and release flavor quickly
and will not affect the viscosity of the final products, so that they can be used for

solid beverages, cloth powder, cake powder, soup powder, etc. Another characteristic of Arabic gum is its compatibility with high-concentration sugar, so it can be widely used in the production of candy with high sugar content and low moisture content, such as toffee, pectin soft candy, and soft fruit cake. Its function in candy is to prevent sucrose crystallization, emulsification, and dispersion of fat components and prevent fat from separating out from the surface to produce "white frost".

4.6   Seaweed Polysaccharide

4.6.1   Agar

Agar is well known as a microbial medium. It comes from various seaweeds of *Rhodophyceae* and is mainly from the coast of Japan. Agar can be separated into two parts: agraose and agaropectin. The basic disaccharide repetitive units of agarose are composed of D-galactopyranose linked by 3,6-dehydrated-alpha-L-pyran galactosyl units via beta-1,4 or beta-1,3 glycoside bonds. About 1 in 10 galactose residues of the agarose chain is sulfated, as shown in Fig. 4.42.

The repetitive unit of agar gel is like that of agarose but contains 5–10% sulfate. A part of D-glucuronic acid residue and pyruvate are esters formed in the form of acetals.

The most distinctive property of agar gel is that it remains stable when the temperature greatly exceeds the initial temperature of gelation. For example, the water dispersions of 1.5% agar form a gel at 30 °C and melt at 35 °C. Agar gel is thermally reversible and is quite stable.

The applications of agar in food include preventing dehydration and shrinkage of frozen food and providing desirable texture, making it stable and desirable texture in processed cheese and butter, controlling water activity of baked and frost food and preventing from texture hardening. In addition, it is also used in canned meat products. Agar is usually used in combination with other polymers such as Tragacanth Gum, Carob Gum, or gelatin.

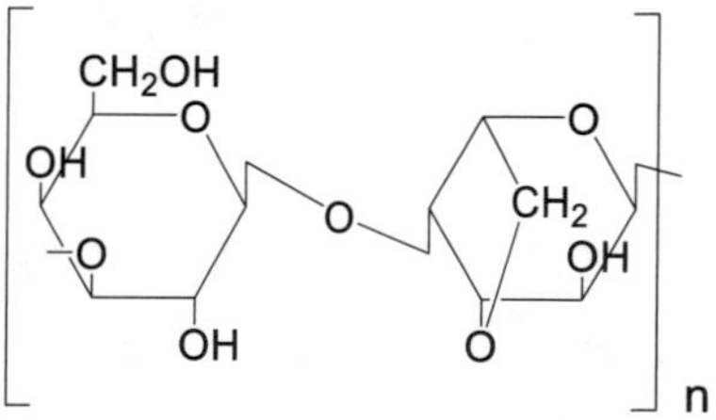

Fig. 4.42 Repeated structural units of agarose

4.6.2 Alginate

Alginate exists in the cell wall of *Phaeophycene*. Most commercial alginic acid exists in the form of sodium salt. Alginate is a linear polymer linked by 1,4-glycosidic bonds between β-D-pyran mannuronic acid (M) and α-L-guluronic acid (L) (Fig. 4.43), with a degree of polymerization of 100–1000. The ratio of D-mannuronic acid (M) to L-guluronic acid (G) varies from source to source, generally 1.5:1, which has a great influence on the properties of alginate. They are arranged in the following order:

- Mannuronic acid block-M-M-M-M-M-
- Guluronic acid block-G-G-G-G-G-
- Alternating block-M-G-M-G-M-

Gulouronic acid blocks in alginate molecular chains can easily interact with Ca^{2+}. A hole is formed between G blocks in two alginate molecular chains and combining with Ca^{2+} to form an "egg box", as shown in Fig. 4.44. The gel formed by alginate and Ca^{2+} is a thermal irreversible gel. The gel strength is related to the content of G

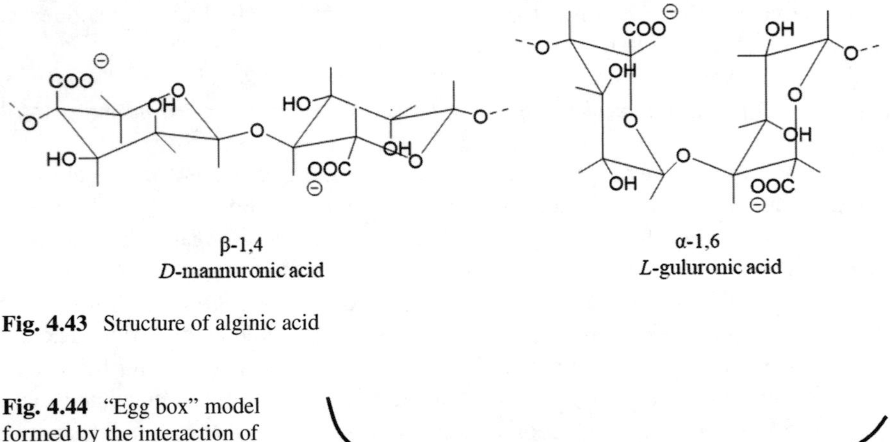

β-1,4
D-mannuronic acid

α-1,6
L-guluronic acid

Fig. 4.43 Structure of alginic acid

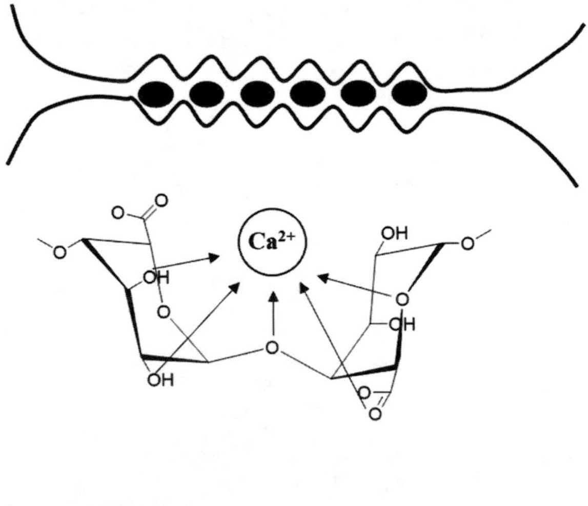

Fig. 4.44 "Egg box" model formed by the interaction of alginate salt and Ca^{2+}

block in alginate molecules and the concentration of Ca^{2+}. Alginate gel has thermal stability and less dehydration shrinkage, so it can be used to make sweet gel.

Alginates can also interact with other components of food products, such as proteins or fats. For example, alginates are liable to interact with positively charged amino acids in denatured proteins and are used in the manufacture of recombinant meat products. The synergistic gelation between alginate with high content of guluronic acid and high esterified pectin can be applied to jam, jelly, and the like. The structure of obtained gel is a thermo-reversible gel and is applied to low heat products.

Alginate can form a thermal irreversible gel with Ca^{2+}, making it widely used in food (gel food or bionic food), especially in recombinant foods such as imitation fruit, onion rings and gel candy, etc. It can also be used as thickener for soup, stabilizer for inhibiting ice·crystal growth in ice cream, yogurt, and milk.

4.6.3 Chitosan

Chitosan is a kind of non-branched chain polymer linked by N-acetyl-glucosamine through β-1,4 glycoside bonds. It is white or gray-white, slightly pearly luster, semi-transparent flaky solid, sometimes in a powder state, tasteless, and insoluble in water. Its physical and chemical properties mainly depend on the acetylation rate and the degree of polymerization. It mainly exists in the exoskeletons of crustaceans (shrimp, crab) and other animals. It contains 15–30% chitosan in soft shells such as shrimp, and 15–20% in crab. Some fungal cell wall components also contain. The basic structural unit is chitobiose, as shown in Fig. 4.45.

As there is free amino group in the molecule, chitosan is easy to form salt in acidic solution and has cationic properties. With the increase in a number of amino groups in chitosan molecule, the amino characteristics of chitosan become remarkable, which is the unique property of chitosan, thus laying the foundation for many biological and processing properties of chitosan.

Chitosan can be used as binder, moisturizer, humectant, filler, emulsifier, polishing agent, and thickening stabilizer in food industry. It can also be used as functional oligosaccharides, which can reduce cholesterol, improve immunity, and enhance the

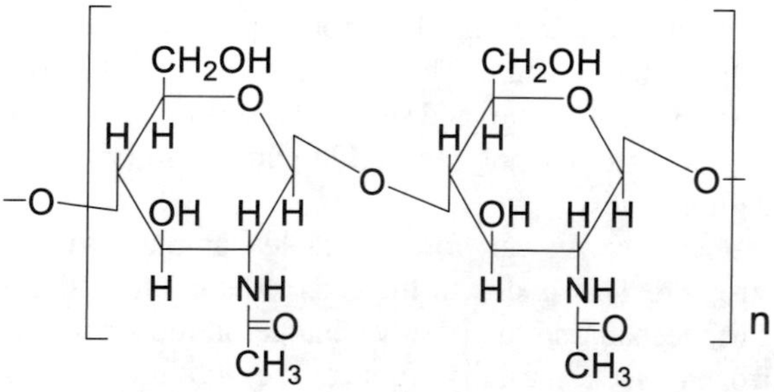

Fig. 4.45 Chemical structure of chitobiose

anti-disease and anti-infection ability of the body, especially with strong anti-cancer effect. Due to its abundant resources and high application value, it has been widely developed and used. At present, enzymatic or acid hydrolysis of shrimp skin or crab shell is commonly used in industry to extract chitosan.

At present, the modified chitosan widely used in food industry is carboxymethylated chitosan, of which N, O-carboxymethyl chitosan is used as thickener and stabilizer in food industry. It is also used as a reagent for purifying water as N, O-carboxymethyl chitosan can be precipitated by complexing with most organic ions and heavy metal ions. Moreover, N, O-carboxymethyl chitosan can be soluble in neutral pH = 7 water to form colloidal solution. It has good film-forming properties and is used for fruit preservation.

N, O-carboxymethyl chitosan was prepared with chloroacetic acid as reagent. Some technologies for direct modification of chitosan have also been developed, and the method is like that of other polysaccharide modification.

4.6.4 Carrageenan

Carrageenan is a heterogeneous polysaccharide extracted from red algae by thermal-alkali separation. It is formed by sulfated or non-sulfated D-galactose and 3,6-dehydrated-L-galactose alternately linked by α-1,3-glycoside bonds and β-1,4-glycoside bonds. Most of the sugar units have one or two sulfate groups, and the total sulfate group content is 15–40%, and the number and location of sulfate groups are closely related to the gel properties of carrageenan. There are three important types of carrageenan: kappa-, lota-, and lambda-carrageenan. Kappa- and lota-carrageenan form double thermo-reversible gel through double helix cross-linking (Fig. 4.46). Carrageenan is a random cluster structure in solution. When the solution is cooled, enough number of cross-linked regions form a continuous three-dimensional network gel structure.

Carrageenan is soluble in water as it contains sulfate anions. The less the content of sulfate group, the easier it is to change from random to helical structure. The kappa-carrageenan contains less sulfates and the gel formed is opaque. The carrageenan has a higher sulfate group content and a random cluster structure in the solution. The gel formed is transparent and elastic. By adding cations such as K^+ and Ca^{2+}, the electrostatic interaction between the sulfates and the sulfates can further strengthen the intermolecular association, which not only improves the gel strength but also improves the gelation temperature. The gel formed by kappa carrageenan is the strongest, but it is easy to release water and shrink. This can reduce the water shrinkage of carrageenan by adding other polysaccharide gums. Lambda-carrageenan is soluble but has no gelling ability.

Carrageenan can form a stable complex with milk protein and enhance gel strength. This is due to the electrostatic interaction between the sulfated anion of carrageenan and the positive charge on the surface of casein colloidal particles. In frozen sweets and dairy products, the addition of carrageenan is very low, only 0.03%.

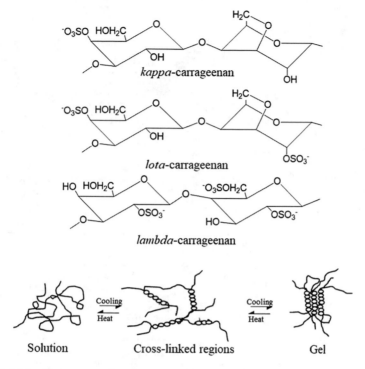

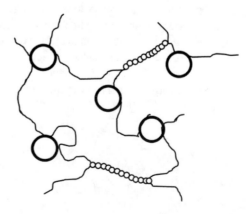

Fig. 4.46 Molecular structure of carrageenan and its formation mechanism

Low concentration of kappa-carrageenan (0.01–0.04%) interacts with kappa casein in milk protein to form weak thixotropic gel (Fig. 4.47). With this special property, cocoa particles in chocolate milk can be suspended and can also be used in ice cream and infant formula.

The traditional dessert gel is made of gelatin. Carrageenan has the characteristics of high melting point, but the gel formed by carrageenan is relatively hard.

Fig. 4.47
Kappa-carrageenan and
kappa-casein interact to form
gels

Fig. 4.48 Interaction
between kappa-carrageenan
and carob gum

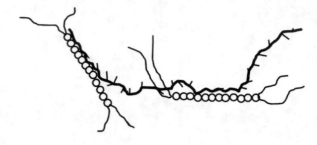

Table 4.9 Application of
carrageenan in food industry

Types of food	Food products	The function of carrageenan
Dairy	Ice cream, cheese	Stabilizers and emulsifiers
Sweet products	Instant pudding	Stabilizers and emulsifiers
Drinks	Chocolate milk	Stabilizers and emulsifiers
	A substitute for cream in coffee	Stabilizers and emulsifiers
	Sweet gel	Gelling agent
	Low heat jelly	Gelling agent
Meat	Low fat sausage	Gelling agent

The hardness of gel can change by adding galactomannan (locust bean gum) to increase the elasticity of the gel instead of gelatin, which can be used as the gel for dessert and reduce the dehydration shrinkage of the gel (Fig. 4.48). For example, the application of carrageenan in ice cream can improve the stability and foam holding capacity of the product. To soften the gel structure, some guar gum can also be added. Carrageenan can also be used in ice cream with starch, galactose mannan, or CMC. If K^+ and Ca^{2+} are added, the formation of carrageenan gel can be promoted. Adding 0.2% lambda-carrageenan or kappa-carrageenan to juice beverage can improve its texture. Carrageenan can also be used in meat products. For instance, adding kappa-carrageenan or lota-carrageenan to low-fat minced meat products can improve the taste, while carrageenan can also replace some animal fats. Therefore, carrageenan is a multifunctional food additive, which can hold water, hold oil, increase thickening, stabilize, and promote gel formation. The application of carrageenan in food industry is shown in Table 4.9.

4.7 Microbial Polysaccharides

Microbial polysaccharide is an edible gum synthesized by bacteria and fungi including mold and yeast. Flavobacterium gum, pullulan, alpha-glucan, and cyclodextrin have been used in food industry.

4.7.1 Xanthan Gum

Xanthan gum is a widely used food gum, which consists of cellulose main chain and trisaccharide side chain. The repeated unit of molecular structure is pentose, of which trisaccharide side chain is composed of two mannoses and one glucuronic acid (Fig. 4.49). The relative molecular mass of xanthan gum is about 2×10^6. The side chain of xanthan gum in solution is parallel to the main chain, forming a stable hard rod structure. When heated to above 100 °C, the hard rod can be transformed into a random coil structure. The hard rod exists in a spiral form through intramolecular association and forms a network structure through entanglement (Fig. 4.50). The xanthan gum solution has the characteristic of high vacation plasticity, shear thinning, and viscosity recovery in a wide range of shear and concentration. Its unique fluidity property is related to its structure. The natural conformation of xanthan gum polymer is hard rod, and hard rods gather together. When in shearing process, the aggregates disperse immediately and quickly gather again after shearing stops.

The viscosity of xanthan gum solution is basically unchanged in the temperature range of 28−80 °C and wide pH range of 1–11, which is compatible with high salt concentration. This is due to that xanthan gum has stable helical conformation and

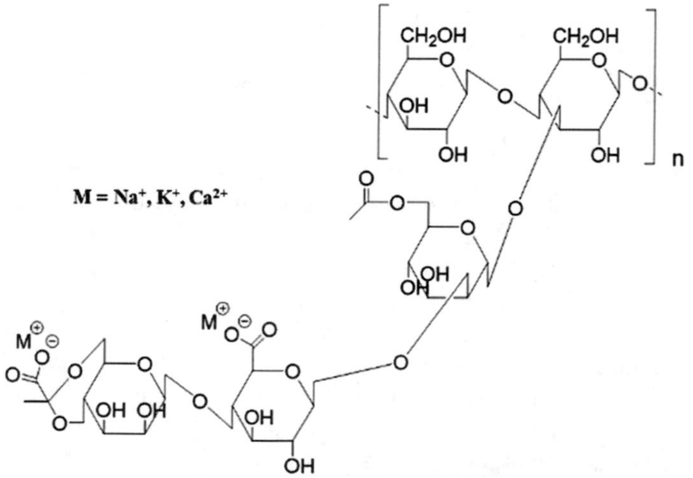

Fig. 4.49 Structure of xanthan gum

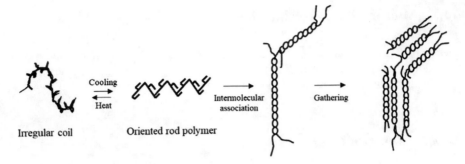

Fig. 4.50 Gelation mechanism of xanthan gum

trisaccharide side chains, which can protect the glycoside bond of main chain from breaking, so the molecular structure of xanthan gum is particularly stable.

Xanthan gum and guar gum have a synergistic effect and form a thermo reversible gel with Robinia bean gum. The gelation mechanism is the same as that of carrageenan and locust bean gum.

Xanthan gum is widely used in food industry as it has the following important properties. It can be dissolved both in cold water and hot water and shows high viscosity at low concentration, where the viscosity of the solution is basically unchanged in a wide temperature range (0–100 °C). Xanthan gum has good compatibility with salt and maintains dissolution and stability in acidic food. It has a synergistic effect with other gums and can stabilize suspensions and emulsions, exhibiting good freezing and thawing stability. These properties are inseparable from the structure of linear cellulose backbone and anionic trisaccharide side chains. Xanthan gum can improve the processing and storage properties of batter and dough, such as improving the elasticity and air holding capacity of batter and dough.

4.7.2 Pullulan

Pullulan gum, is a polymer with maltotriose as a repeating unit and linked by α-(1→6) glycoside bonds. It is a group of extrasporium polysaccharides produced by Pullulan, as shown in Fig. 4.51.

Pullulan gum is a colorless and tasteless white powder, soluble in water and forms a viscous solution after dissolving in water, which can be used as food thickener. Pullulanase can hydrolyze it into maltose. The film made of pullulan is water-soluble, oxygen-impermeable, non-toxic to human body, with the strength like nylon that is suitable for packaging of easily oxidized foods and medicines. Pullulan gum is a polysaccharide with low utilization rate in human body. It can be used to replace starch in the preparation of low-energy food and beverage.

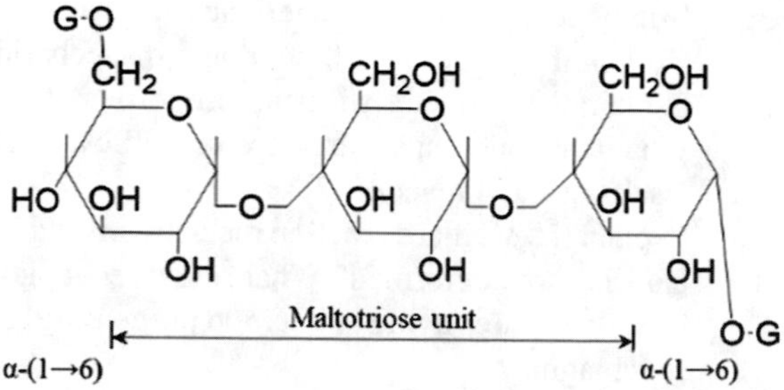

Fig. 4.51 Chemical structure of pullulan

4.7.3 Dextran

Dextran, is an extrasporium polysaccharides produced by Leunostoc mesenteroides in medium containing sucrose. It is a polysaccharide composed of α-D-pyran glucose residues linked by α-1,6 glycoside bonds.

Dextran is easily soluble in water and forms a clear viscous solution after dissolving in water. It can be used as a humectant for candy and can keep the moisture in candy and bread products. In chewing gum and fudge, it is used as a gelling agent to prevent the emergence of sugar crystallization. In ice cream, it inhibits the formation of ice crystals. In pudding mixtures, it provides an appropriate viscosity and taste.

4.8 Summary

Carbohydrates are the main source of energy needed by living organisms to maintain life activities. They are the basic materials for synthesizing other compounds and the main structural component of living organisms. In food, carbohydrates not only serve as nutritional compounds but also show many activities for food additives. Low molecular sugar can be used as sweetener, and macromolecular sugar can be widely used as thickener and stabilizer in food. In addition, carbohydrates are precursors of flavor and color in food processing, which play an important role in the sensory quality of food. There are many types of carbohydrates that exist naturally and can be processed. In addition to some known carbohydrates, many new carbohydrates will be discovered and utilized rationally.

Questions

1. Term explanation: sugar mutarotation, glycosides, oligosaccharides, sugar reduction, polysaccharides, starch gelatinization and aging, dietary fiber, Maillard reaction, caramelization, microcrystalline cellulose, cyclodextrin, fructose syrup.
2. Discuss the types of carbohydrates and their applications in food.

3. Briefly describe the characteristics and uses of the gel.
4. Explain the properties of mono and disaccharides in food applications.
5. Explain the characteristics and properties of starch.
6. What reactions can sugars have in acidic and alkaline solutions? What are the results of these reactions?
7. Explain the Maillard reaction mechanism and its influence on food processing.
8. How is caramel formed? What role does it play in food processing?
9. How does starch aging affect food processing and food quality? How to prevent starch aging?
10. Why do newly made cereals such as bread and steamed bread have the characteristics of soft internal structure, elasticity, and good taste, but with the extension of storage time, they will turn from soft to hard?
11. Why do grain foods appear golden during frying, baking, and other processing processes?
12. What kind of sugar is the candied fruit we usually eat? Why is not only used sucrose?
13. The formation mechanism and influencing factors of pectin gel?

References

1. Akkerman, R., Faas, M.M., de Vos, P.: Non-digestible carbohydrates in infant formula as substitution for human milk oligosaccharide functions: Effects on microbiota and gut maturation. Crit. Rev. Food Sci. Nutr. **59**(9), 1486–1497 (2019)
2. Almasi, H., Azizi, S., Amjadi, S.: Development and characterization of pectin films activated by nanoemulsion and Pickering emulsion stabilized marjoram (*Origanum majorana* L.) essential oil. Food Hydrocolloids **99** (2020)
3. Belitz, H. D., Grosch, W., Schieberle, P.: Food Chemistry. Springer-Verlag Berlin, Heidelberg (2009)
4. Chatterjee, C., Pong, F., Sen, A.: Chemical conversion pathways for carbohydrates. Green Chem. **17**(1), 40–71 (2015)
5. Damodaran, S., Parkin, K.L., Fennema, O.R.: Fennema's Food Chemistry. CRC Press/Taylor & Francis, Pieter Walstra (2008)
6. Kan, J.: Food Chemistry. China Agricultural University Press, Beijing (2016)
7. Kazemi, M., Khodaiyan, F., Hosseini, S.S.: Utilization of food processing wastes of eggplant as a high potential pectin source and characterization of extracted pectin. Food Chem. **294**, 339–346 (2019)
8. Li, H., Gidley, M.J., Dhital, S.: High-amylose starches to bridge the "fiber gap": Development, structure, and nutritional functionality. Compr. Rev. Food Sci Food Saf. **18**(2), 362–379 (2019)
9. Li, H., Prakash, S., Nicholson, T.M., Fitzgerald, M.A., Gilbert, R.G.: The importance of amylose and amylopectin fine structure for textural properties of cooked rice grains. Food Chem. **196**, 702–711 (2016)
10. Lund, M.N., Ray, C.A.: Control of Maillard reactions in foods: strategies and chemical mechanisms. J. Agric. Food Chem. **65**(23), 4537–4552 (2017)
11. Nooshkam, M., Varidi, M., Bashash, M.: The Maillard reaction products as food-born antioxidant and antibrowning agents in model and real food systems. Food Chem. **275**, 644–660 (2019)

12. Qin, Y., Liu, C., Jiang, S., Xiong, L., Sun, Q.: Characterization of starch nanoparticles prepared by nanoprecipitation: influence of amylose content and starch type. Ind. Crops Prod. **87**, 182–190 (2016)
13. Reynolds, A., Mann, J., Cummings, J., Winter, N., Mete, E., TeMorenga, L.: Carbohydrate quality and human health: a series of systematic reviews and meta-analyses. Lancet **393**(10170), 434–445 (2019)
14. Rodsamran, P., Sothornvit, R.: Microwave heating extraction of pectin from lime peel: characterization and properties compared with the conventional heating method. Food Chem. **278**, 364–372 (2019)
15. Xie, B.: Food Chemistry. China Science Press, Beijing (2011)
16. Yang, J.S., Mu, T.H., Ma, M.M.: Extraction, structure, and emulsifying properties of pectin from potato pulp. Food Chem. **244**, 197–205 (2018)
17. Zou, P., Yang, X., Wang, J., Li, Y., Yu, H., Zhang, Y., Liu, G.: Advances in characterisation and biological activities of chitosan and chitosan oligosaccharides. Food Chem. **190**, 1174–1181 (2016)

Dr. Jie Pang professor and doctoral supervisor at College of Food Science, Fujian Agriculture and Forestry University, China, was a visiting scholar at the Institute of Chen Xingshen Mathematics, Nankai University, and Harvard University in the United States. He is an expert who enjoys special government allowance of the State Council, is the Deputy Director of the Engineering Research Center of the Ministry of Education, a leading talent in science and technology innovation of Fujian Province, and a review member of the National Funding Committee for key projects. He is now the Dean of College of Food Science, Fujian Agriculture and Forestry University, and the Head of the postdoctoral mobile workstation for the discipline of food science and engineering, and the academic leader of the pharmacognosy discipline. He has presided over 6 general projects of the National Natural Science Foundation of China, 1 sub-task of the National Key R&D Program, and 1 national virtual simulation experiment teaching project. He won the first prize of the Provincial Science and Technology Progress (Fujiang), and the first prize and a second prize for the provincial higher education teaching achievement. He has been the Editor-in-Chief for two representative monographs and published more than 120 research papers.

Dr. Fusheng Zhang associate professor and master supervisor at College of Food Science, Southwest University, China, was a visiting scholar at Purdue University. He is a committee member of Non-thermal Processing Branch of CIFST, and a member of Konjac Association of Chinese Society for Horticultural Science. Dr. Fusheng Zhang obtained his Ph.D. degree from the College of Food Science and Nutritional Engineering, China Agricultural University, in 2011, and his main research areas are non-thermal processing technology in the fields of fruits, vegetables, and carbohydrates. Dr. Fusheng Zhang has hosted 6 national and provincial projects and published more than 60 academic papers in food science journals, and participated in compiling 6 textbooks related to food science.

Chapter 5
Lipid

Hui He and Tao Hou

Abstract In this chapter, various lipids present in food materials are illustrated. First, the composition, characteristics, and nomenclature of fat and fatty acids are introduced. The physical properties of lipid, such as crystallization behavior, melting characteristics, oil emulsification, etc., are explained in details related to their applications in food processing. The reader can acquire the evaluation methods of oil quality such as peroxide value, acid value, and iodine value, as well as the basic principles of oil processing chemistry such as enzymatic transesterification. It has been depicted that the mechanisms and influencing factors of fat oxidation, the antioxidant principle of antioxidants and the chemical changes of oils during processing and storage. Besides, many important lipids and their derivatives including saturated fatty acids, unsaturated fatty acids, lecithin, cholesterol, sterol, etc. are shown according to their structures and bioactive activities.

Keywords Triacylglycerol · Polymorphism · Emulsion · Oxidation · Hydrogenation

Fat is a familiar nutritional ingredient in food. Why are foods with high-fat content prone to deterioration in daily life? Why does the oil tend to sticky, foamy, and poor quality when it is heated for a long time? Do you worry about the risks of cardiovascular disease? Europeans have high fat intake in their diets, which induces more patients with cardiovascular diseases and obesity. It is unknown that Italians have a higher fat intake, but the incidence rate of cardiovascular diseases is significantly lower than that in other European countries. Answers to these questions will be involved in this chapter.

H. He (✉) · T. Hou
College of Food Science and Technology, Huazhong Agricultural University, Wuhan 430074, China
e-mail: hehui@mail.hzau.edu.cn

© The Author(s), under exclusive license to Springer Nature Singapore Pte Ltd. 2021　　197
J. Kan and K. Chen (eds.), *Essentials of Food Chemistry*,
https://doi.org/10.1007/978-981-16-0610-6_5

5.1   Introduction

5.1.1   *Definition and Function of Lipids*

Lipids are a wide variety of hydrophobic substances, which are insoluble in water and soluble in organic solvents. Almost all the fatty acid glycerides (i.e. triacylglycerol) belong to fat, and it is traditionally believed that solid lipid is fat, and liquid lipid is oil at room temperature. Lipid also includes a small amount of non-diacylglycerol compounds such as phospholipids, sterols, glycolipids, and other carotenoids. Due to the wide variety of lipid compounds and different structures, it is difficult to use a word to define it. The lipid compounds generally have the following common characteristics: ① they can insoluble in water and soluble in organic solvents such as ether, petroleum ether, chloroform, and acetone; ② most of them have an ester structure and most of the esters were formed by fatty acids; ③ they are produced by organisms and also can be used by organisms, which differ from the mineral oil.

However, there are some substances called lipids that do not completely accord with the above definition, such as lecithin is slightly soluble in water and insoluble in acetone, sphingomyelin, and cerebroside belonging to complex lipids are insoluble in ether.

Fat is an important component of food and one of the essential nutrients for humans. Compared with the same quality protein and carbohydrates, the fat has the highest calories, and each gram of fat provides 39.58 kJ of calories. Fat can provide essential fatty acids, act as a carrier of fat-soluble vitamins, and give the food a smooth mouthfeel, a sleek appearance, and a crispy flavor of fried foods. Plastic fat also has a modeling function. Fat is also a heat transfer medium in cooking. Additionally, lipids, which are indispensable substances for composing biological cells, have functions on lubrication, protection, and warmth preservation in organisms.

5.1.2   *Lipid Classifications*

Lipids can be classified into simple lipids, complex lipids, and derivative lipids based on their structures and compositions (Table 5.1).

Table 5.1 Lipids classifications

Main class	Subclass	Composition
Simple lipids	Acylglycerol, Waxes	Glycerol + fatty acids (about 99% of natural lipids) Long-chain fatty alcohols + long-chain fatty acids
Complex lipids	Glycerophospholipid Sphingolipids Cerebroside Gangliosides	Glycerol + fatty acids + phosphates + nitrogenous groups Sphingosine + fatty acids + phosphates + choline Sphingosine + fatty acids + sugar Sphingosine + fatty acids + carbohydrate
Derivative lipids		Carotenoids, sterols, fat-soluble vitamins, and so on

5.2   Structure and Composition of Fat

5.2.1   Structure and Nomenclature of Fatty Acids

5.2.1.1   Structure of Fatty Acids

(1)   Saturated fatty acids

The saturated fatty acids naturally presented in edible fats are mainly long-chain (carbon number >14) and linear fatty acids, while milk fat contains a certain amount of short-chain fatty acids.

(2)   Unsaturated fatty acids

The unsaturated fatty acids naturally presented in edible oils often contain one or more allyl [— $(CH = CH—CH_2)_n$ —] structural units with a methylene group (non-conjugated) between the two double bonds, and the double bond is mostly in cis-form. During the processing and storage of oils, some double bonds are converted into trans-form and appear conjugated double bonds. Some fatty acids are called essential fatty acids (EFA) because of a special physiological effect, being incapable of synthesizing by human body and indispensable to human body, such as linoleic and linolenic acid. The best source of EFA is plant oil.

Saturated fatty acids intake has a positive relationship with the incidence and mortality of coronary heart disease, while the unsaturated fatty acids can reduce blood lipid and prevent from atherosclerosis. The WHO survey shows that the Mediterranean diet is mainly dominated by olive oil, which enriched monounsaturated fatty acids. Though Mediterranean has high fat intake, the incidence rate of cardiovascular diseases is markedly lower than that in other European countries, which is due to the decrease of low-density lipoprotein cholesterin (LDL-C) by monounsaturated fatty acids. It is commonly known that monounsaturated fatty acids can improve the activity of LDL receptor and increase the elimination of LDL, followed by a decrease of LDL-C. Additionally, monounsaturated fatty acids have an antagonism against cholesterol. Polyunsaturated fatty acids also have a similar effect.

5.2.1.2   Nomenclature of Fatty Acids

(1)   Systematic nomenclature

The fatty acids are named after the parent hydrocarbon. The longest carbon chain containing a carboxyl group is the parent hydrocarbon, and if it is an unsaturated fatty acid, the parent hydrocarbon contains a double bond. The number starts from the C-terminus of the parent hydrocarbon and marks the position of the double bond.

For example: $CH_3(CH_2)_7CH=CH(CH_2)_7COOH$ is 9-octadecenoic acid.

(2)   Numerical nomenclature

n:m (n is the number of carbon atoms, m is the number of double bonds), for example, 18:1, 18:2, 18:3.

It is necessary to mark the *cis-* and *trans-* structure and position of the double bond (*c* means that *cis*-form, *t* represents *trans*-form), and the position can be numbered from the C-terminus, such as 5t, 9c-18: 2. It can also be numbered starting from the methyl terminus, labeled as 'ω-number', or 'n- number', which is the atom order of the lowest numbered double bond carbon, such as 18: 1ω9 or 18: 1 (n-9), and 18: 3ω3 or 18: 3 (n-3). However, this method is restricted to the *cis*-configuration of double bond structure. Multiple double bonds should be a five-carbon diene structure, that is, it has a non-conjugated double bond structure (such as the natural polyenoic acid). Therefore, the first double bond is positioned, followed by the position determination of the remaining double bonds. It is necessary to mark the position of the first double bond carbon; other structural fatty acids cannot be used with 'ω' or 'n'.

(3)   Common or trivial nomenclature

A great deal of fatty acids is originally derived from natural products, which are then named according to their source, such as lauric acid, palmitic acid, arachidic acid, and so on.

(4)   Abbreviations of fatty acids

As shown in Table 5.2, DHA and EPA are the main components of the functional product 'brain gold' that appeared on the market in China.

5.2.2   Structure and Nomenclature of Fat

5.2.2.1   Structure of Fat

Fats are mainly tri-esters formed by glycerol and fatty acids, namely triacylglycerols (TG).

Table 5.2 Numerical, systematic, common names, and their abbreviations of main fatty acids

Numerical name	Systematic name	Common name	Abbreviation
4: 0	Butyric acid	Butyric acid	B
6: 0	Caproic acid	Caproic acid	H
8: 0	Caprylic acid	Caprylic acid	Oc
10: 0	Decanoic acid	Capric acid	D
12: 0	Dodecanoic acid	Lauric acid	La
14: 0	Tetradecanoic acid	Myristic acid	M
16: 0	Palmitic acid	Palmtic acid	P
16: 1	Palmitoleic acid	Palmitoleic acid	Po
18: 0	Stearic acid	Stearic acid	St
18: 1ω9	9c-Octadecenoic	Oleic acid	O
18: 2ω6	9c, 12c-Octadecatrienoic	Linoleic acid	L
18: 3ω3	9c, 12c, 15c -Octadecatrienoic	Linolenic acid	α-Ln
18: 3ω6	6c, 9c, 12c -Octadecatrienoic	Linolenic acid	γ-Ln
20: 4ω6	5c, 8c, 11c, 14c -Eicosapentanoic	Arachidonic	An
20: 5ω3	5c, 8c, 11c, 14c, 17c -Eicosapentanoic	Eicosapentanoic	EPA
22: 1ω9	13c -Docosenoic	Erucic	E
22: 6ω3	4c, 7c, 10c, 13c, 16c, 19c- Docosahexanoic acid	Docosahexanoic	DHA

$$\underset{\text{Glycerol}}{\overset{\displaystyle CH_2-OH}{\underset{\displaystyle CH_2-OH}{HO-C-H}}} + \underset{\text{Fatty acids}}{3\,R_iCOOH} \longrightarrow \underset{\text{Triacylglycerol}}{\overset{\displaystyle CH_2OCOR_1}{\underset{\displaystyle CH_2OCOR_3}{R_2OCOCH}}}$$

If $R_1 = R_2 = R_3$, it is called simple glyceride. 70% of olive oil contains glycerin trioleate, when R_i is not the same, it is called mixed glyceride. Natural oils are mostly mixed glycerides. When R_1 and R_3 are different, the C_2 atom in the glycerol has chirality, and most natural oils are L-shaped. The fatty acids in natural triglycerides, regardless of whether they are saturated or not, have an even number of carbon atoms, and most of them are straight-chain fatty acids. Fatty acids with odd-numbered carbon atoms, branched chains, and cyclic structures are relatively rare.

5.2.2.2 Nomenclature of Triacylglycerol

The nomenclature of triacylglycerols are the stereospecific numbering (Sn) proposed by Hirschman and the R/S system nomenclature proposed by Cahn, due to the limited application of the latter (it does not apply to the case where the fatty acids on glycerol C_1 and C_3 are the same), so only the Sn nomenclature is introduced here. This law stipulates the writing of glycerol: the number of carbon atoms is C_1, C_2, C_3, respectively, from top to bottom, the hydroxyl group on C_2 is written on the left, and the name of triacylglycerol is as follows:

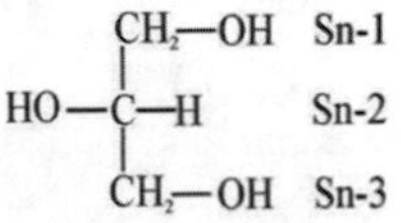

$$
\begin{array}{ll}
CH_2\!-\!OH & Sn\text{-}1 \\
HO\!-\!C\!-\!H & Sn\text{-}2 \\
CH_2\!-\!OH & Sn\text{-}3
\end{array}
$$

Glycerol

$$
\begin{array}{l}
CH_2OOC(CH_2)_{14}CH_3 \\
CH_3(CH_2)_7CH\!=\!CH(CH_2)_7COOCH \\
CH_2OOC(CH_2)_{16}CH_3
\end{array}
$$

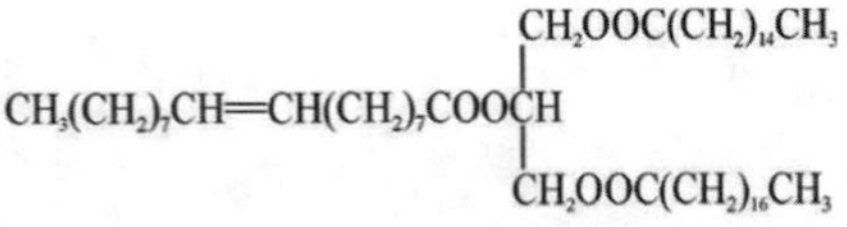

Triacylglycerol

(1)   Number naming: Sn-16:0-18:1-18:0
(2)   The abbreviation naming: Sn-POSt.
(3)   System name: Sn-glycerol-1-palmitate-2-oleate-3-stearate
      or 1-palmitoyl-2-oleoyl-3-stearoyl-Sn-glycerol.

   Sometimes the C_1 and C_3 positions are called α positions, and the C2 position is called β positions.

5.3   Physical Properties of Fat

5.3.1   Odor and Color

Pure fat is colorless and tasteless, and the slightly yellow–green color in natural oils is due to the presence of some fat-soluble pigments (such as carotenoids, chlorophyll, etc.). After the oil is refined and decolorized, the color becomes lighter. Most fats and oils are non-volatile, and a few fats and oils contain short-chain fatty acids, which can cause odor, such as milk fat. The smell of fat is mostly caused by non-fat ingredients. For example, the fragrance of sesame oil is caused by acetyl pyridinium, the fragrance of coconut oil is caused by nonyl ketone, and the pungent smell of rapeseed oil is caused by heating, which is caused by the decomposition of glucosinolate compounds.

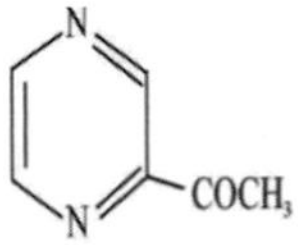

Acetyl pyridinium

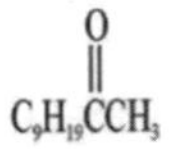

Nonyl ketone 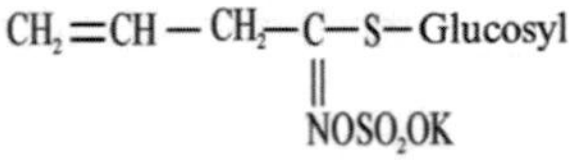   Glucosinolate compounds

5.3.2   Melting Point and Boiling Point

Since natural fats and oils are a mixture of various triacylglycerols, they have no accurate melting point and boiling point, but only a range of melting or boiling temperature. In addition, this is ascribed to the phenomenon of homogeneity of oils and fats where compounds show the same chemical composition but different crystal structures. The melting points of free fatty acids, monoacylglycerol, diacylglycerol, and triacylglycerol decrease sequentially. This is due to that their polarity decreases successively and the forces between molecules successively decrease.

The melting point of fat is generally between 40 and 55 °C. The longer the carbon chain of the fatty acid in the triacylglycerol and the higher the saturation, the higher the melting point; the melting point of the trans structure is higher than that of the *cis* structure, and the melting point of the conjugated double bond structure is higher than that of the non-conjugated double bond structure. Compared with vegetable oils, cocoa butter and grease from terrestrial animals have higher saturated fatty acid content and are often solid at room temperature; vegetable oils are mostly liquid at room temperature. Generally, when the melting point of fat is lower than the body temperature of 37 °C, its digestibility is over 96%; the higher the melting point is, the less digestible it is. The relationship between the melting point of fat and the digestibility is shown in Table 5.3.

The boiling point of fat and oil is related to the fatty acids it composes, generally between 180 and 200 °C. The boiling point increases with the growth of the fatty acid carbon chain, but the boiling point of fatty acids with the same carbon chain length does not differ a lot with molecular saturations. As the free fatty acids increase during storage and use of fats and oils, the fats and oils become easy to smoke during heating process, and the smoking point is lower than the boiling point.

Table 5.3 The relationship between the melting point and digestibility of several commonly used edible fats and oils

Fat	Melting point (°C)	Digestibility (%)
Soybean oil	−8 to −18	97.5
Peanut oil	0 to 3	98.3
Sunflower oil	−16 to 19	96.5
Cottonseed oil	3 to 4	98
Cream	28 to 36	98
Lard	36 to 50	94
Tallow	42 to 50	89
Suet	44 to 55	81
Margarine	–	87

5.3.3   Crystal Characteristics

The X-ray diffraction measurement shows that the microstructure of solid fat is a highly ordered crystal, and its structure can be obtained by periodically arranging a basic structural unit in a three-dimensional space. There are many ways to crystallize solid fats, that is, the phenomenon of polymorphism. The so-called homogeneous polycrystalline refers to substances with the same chemical composition, which can have different crystallization modes, but the same liquid phase is formed after melting. Different homopolycrystals have different stability. A metastable homopolycrystal will spontaneously transform into a stable state when it is not melted. This transition is unidirectional; and when both two homopolycrystals are stable, it can be bidirectional, which depends on the temperature; natural fats are mostly unidirectional.

The polymorphism of long carbon chain compounds is related to the different packing arrangements or the inclination angle of the hydrocarbon chain. The smallest spatial unit repeating along the chain axis in the unit cell, also known as subcell, can be used to describe the packing method. The smallest repeating unit in the fatty acid hydrocarbon chain is ethylene ($-CH_2CH_2-$).

There are seven types of packing ways of subcell in fatty acids, among which the most common are as follows.

Figure 5.1a shows the triclinic crystal stacking (T //), also known as β-type. Since the orientation of the subunit cells is the same, this stacking method has the highest stability; ordinary orthogonal packing (O⊥), also known as β′-type, is shown in Fig. 5.1b. As the subunit cell orientation at the center and the subunit cell orientation at the four vertices are different, so the stability is not as good as β-type; Fig. 5.1c shows the hexagonal type (H), also known as α-type. In this structure, the chain is randomly oriented and rotates around its long vertical axis, which shows increased disorder and lowest stability.

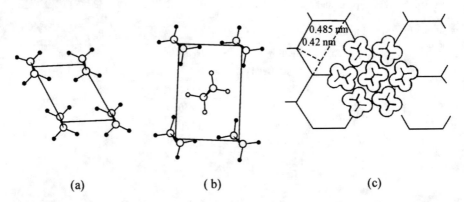

(a)   (b)   (c)

Fig. 5.1 General types of hydrocarbon subcell packing. **a** β-type (triclinic crystal stacking, T //); **b** β′-type (ordinary orthogonal packing, O⊥); **c** α-type (hexagonal type, H)

Table 5.4 Homogeneous homopolycrystal characteristics of tributylglycerol (R1 = R2 = R3)

Traits	α		β′		β
Packing ways	Regular hexagon		Orthogonal		Triclinic
Melting point	α	<	β′	<	β
Density	α	<	β′	<	β
Degree of order	α	<	β′	<	β

Table 5.5 Melting point of homopolycrystals triacylglycerol

Compounds	Melting points of homopolycrystals triacylglycerol / °C		
	α	β′	β
StStSt	55	63.2	73.5
PPP	44.7	56.6	66.4
OOO	−32	−12	4.5–5.7
PPO	18.5	29.8	34.8
POP	20.8	33	37.3
POSt	18.2	33	39
PStO	26.3	40.2	
StPO	25.3	40.2	

Table 5.4 compares the characteristics of the three crystal forms. Due to the long carbon chains, tertiary glycerols show many characteristics of hydrocarbons (Table 5.5).

Taking glyceryl tristearate as an example, when the oil gradually cools from the molten state, the α-type is formed first, which is unstable and can be transformed into β′-type and β-type under different conditions. If the α-type is heated to its melting point, it can be quickly transformed into β-type; if the temperature is kept at a few degrees above the melting point of α-type, the β′-type can be directly obtained; while the β′-type is heated to its melting point, it will melt, and be transformed into a stable β-type.

Since the Sn-1 and Sn-3 positions of the triacylglycerol are in opposite directions to the Sn-2 positions of the fatty acids, the molecules are arranged in a chair shape in the crystal lattice, as shown in Figs. 5.2 and 5.3. In the β-type arrangement, fatty acids are arranged in two staggered ways, one is double carbon chain length (DCL) method, and the other is the triple carbon chain length (TCL) method (Fig. 5.2), which are denoted as β-2 and β-3, respectively.

Generally, triacylglycerols with three same fatty acids are easy to form stable β-type, and they are arranged in β-2, while those with different fatty acids tend to stay in the β′-type state due to the different carbon chain lengths, and are arranged in β-3 manner.

Fats that are easy to crystallize into β-type are soybean oil, peanut oil, coconut oil, olive oil, corn oil, cocoa butter, and lard. Fats that are easy to crystallize into β′-type

Fig. 5.2 Distribution
manner of β type of
triacylglycerol

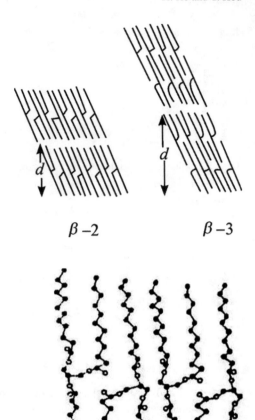

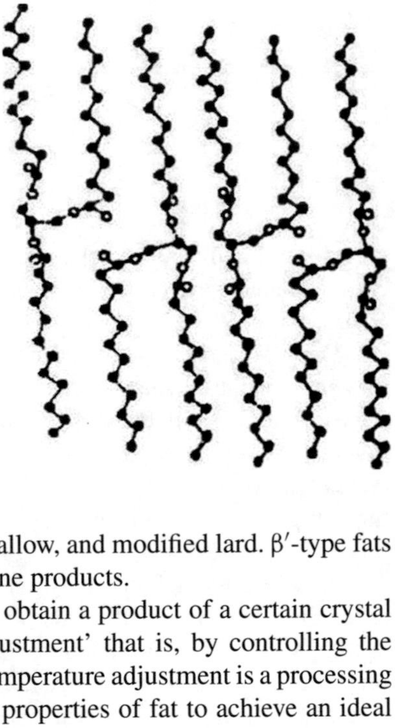

β –2 β –3

Fig. 5.3 The molecular
arrangement of glyceryl
trilaurate

are cottonseed oil, palm oil, rape oil, milk fat, tallow, and modified lard. β′-type fats
are suitable for making shortening and margarine products.

In practical applications, if it is expected to obtain a product of a certain crystal
form, it can be achieved by 'temperature adjustment' that is, by controlling the
crystallization temperature, time, and speed. Temperature adjustment is a processing
method that uses crystallization to change the properties of fat to achieve an ideal
homogeneous polycrystalline and physical state, thereby increasing the utilization
and application range of fat.

There are two main glycerides, Sn-StOSt and Sn-POSt, in cocoa butter, the raw
material for producing chocolate, which can form several homogeneous homopolycrystals: α-2 (melting point 23.3 °C), β′-2 type (melting point 27.5 °C), β-3V type
(melting point 33.8 °C), β-3VI type (melting point 36.2 °C). To obtain chocolate

with smooth appearance, fine mouthfeel and good mouth-melting (33.8 °C), the β-3V crystal of cocoa butter should be avoided to transform into β-3VI type, otherwise it will produce a rough mouthfeel and 'white frost' on the surface. Since the β-3VI type is more stable than the β-3V type, the β-3V type crystals will spontaneously transform into the β-3VI type crystals. This transformation can be inhibited by adding emulsifiers, thereby inhibiting the whitening of the chocolate surface.

5.3.4 Melting Properties

5.3.4.1 Melting

The enthalpy curve obtained by melting triacylglycerols with three same fatty acids ($R_1 = R_2 = R_3$) is shown in Fig. 5.4, the enthalpy of β-type increases with an increase in temperature. When it reaches the melting temperature, it absorbs heat but the temperature does not rise until all the solids are converted into liquid (point B), and the temperature begins to rise. The unstable α-type changes to stable β-type at E, releasing heat.

When fat melts, the volume expands except for the change in enthalpy. However, when the solid fat changes from a less stable homopolycrystal to a more stable homopolycrystal, the volume shrinks as the latter density is greater. The specific volume of the liquid oil and the solid fat can be measured by a dilatometer with the change of temperature to obtain an expansion melting curve as shown in Fig. 5.5. The instrument used in this method is simple and more practical than calorimetry. The solid begins to melt at point X and becoming liquid at point Y. At the point b of the curve is a solid–liquid mixture, where the proportion of solid fat in the mixture is ab/bc, the proportion of liquid oil is bc/ac', and the solid–liquid ratio is ab/bc at a certain temperature, also called as the solid fat index (SFI).

Fig. 5.4 Enthalpy melting curve of α and β homopolycrystal

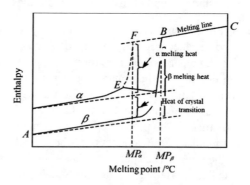

Fig. 5.5 Enthalpy or
expansion melting curve of
glyceride mixture

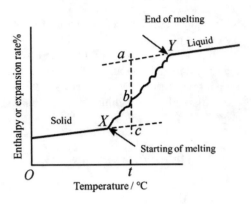

5.3.4.2   Lipid Plasticity

Lipid appears solid at room temperature is a mixture of solid fat and liquid oil, which are intertwined and cannot be separated by a usual method. This kind of fat has plasticity and can maintain a certain appearance. The plasticity of fat refers to the ability of apparent solid fat to resist deformation under a certain external force. The plasticity of lipid depends on the following conditions:

a. Solid fat index (SFI). The plasticity is best when the ratio of solid to liquid in the lipid is appropriate. If the solid fat is too much, it is too hard and the plasticity is not good; if the liquid oil is too much, it is too soft, easily deformed, and the plasticity is also not good.
b. The crystal form of fat. When the fat is in the β'-type crystal, the plasticity is the strongest. Since the β'-type introduces a large amount of small air bubbles into the product during crystallization, giving the product better plasticity and creamy cohesive properties, while the β-form contains small and large bubbles.
c. Melting temperature range. The greater the temperature difference between the start and end of melting, the more plastic the fat will be.

   Plastic fats have good spread ability and plasticity (for cake flowering) and are used in baked goods, having a crisping effect. The addition of plastic fat during the dough preparation process forms a large area of film and thin strips, which enhances the ductility of the dough. The isolation of the oil film prevents the gluten grains from sticking to each other to form a large gluten, which reduces the elasticity and toughness of the dough, and the water absorption of the dough is also reduced, thus making the product crisp. Another function of the plastic fat is to contain and maintain a certain amount of air bubbles during preparation to increase the volume of the dough. The special lipid used in the production of biscuits, cakes, and breads is called shortening, a structurally stable plastic solid fat, and it has the properties of being soft at 40 °C, not too hard at low temperatures, and not easily oxidized.

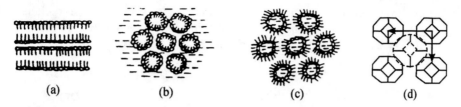

     (a)   (b)   (c)   (d)

Fig. 5.6 Liquid crystal structure of lipid. **a** Layered structure, **b** hexagonal-I structure, **c** hexagonal-II structure, **d** Cubic structure

5.3.5 Liquid Crystal State of Lipid

There are several phases for the lipid. In addition to the solid state and the liquid state, there is a phase state between the solid state and the liquid state, which is called liquid crystal or mesomorphic phase.

There are non-polar hydrocarbon chains in the liquid crystal structure of the lipid, and only weak Van der Waals' forces exist between the hydrocarbon chains. When heating the lipid, the hydrocarbon zone melts before the actual melting point is reached; and the polar group (such as the ester group and the carboxyl group) in the lipid has the inductive force, the orientation force, and even the hydrogen bonding force except for the Van der Waals' force, thus, the polar region does not melt, thereby form a liquid crystal phase. The emulsifier is a typical amphiphilic substance, and it is easy to form a liquid crystal phase.

In the lipid–water system, there are mainly three kinds of liquid crystal phases as shown in Fig. 5.6, that is, a layered structure, a hexagonal structure, and a cubic structure. The layered structure resembles a biological bilayer membrane, and a layer of water is sandwiched between two layers of ordered lipid. When the layered liquid crystal is heated, it can be converted into cubic or hexagonal-I or hexagonal-II liquid crystal. In the hexagonal-I structure, the non-polar group is toward the inside of the hexagonal column, the polar group is toward the outside of the hexagonal column, and the water is in the space between the hexagonal columns; in the hexagonal-II structure, the water is wrapped in the hexagonal column, the polar end of the oil surrounds the water, and the non-polar hydrocarbon zone faces the outside of the hexagonal column, so is as in the cubic structure. In vivo, the liquid crystal state affects the permeability of the cell membrane.

5.3.6 Emulsification of Lipid and Emulsifier

Oil and water are incompatible with each other, but under certain conditions, both can form a metastable emulsion. One of the phases is dispersed in another phase by a small droplet with a diameter of 0.1–50 μm. The former is called an internal phase or a dispersed phase, and the latter is called an external phase or a continuous phase.

The emulsion is divided into an oil-in-water emulsion (O/W, water is the continuous phase) and a water-in-oil emulsion (W/O, the oil is the continuous phase). Milk is a typical O/W emulsion, and cream is generally a W/O emulsion.

5.3.6.1 Emulsion Destabilization

The thermodynamically unstable system of emulsions loses stability under certain conditions, and delamination, flocculation, and even coalescence occur, mainly due to the following factors.

a. Gravity causes stratification: Gravity can cause stratification or sedimentation of phases with different densities.
b. Insufficient surface static charge on the surface of the dispersed phase leads to flocculation: the surface of the dispersed phase is insufficient in static charge, and the repulsive force between the droplets is insufficient, and the droplets are close to each other. At this time, the interface film of the droplet has not still been broken.
c. The interfacial film rupture between the two phases causes coalescence: the interfacial film breaks between the two phases, and the droplets are combined with the droplets, and the small droplets become large droplets, and in severe cases, they are completely phase-separated.

5.3.6.2 Emulsification by Emulsifiers

(1) Increasing the electrostatic repulsion between the dispersed phases. Some ionic surfactants can establish an electrical double layer in the oil-containing aqueous phase, resulting in an increase in the repulsive force between the small droplets, keeping the droplets stable and not flocculated, and such emulsifiers are suitable for O/W systems (Fig. 5.7).
(2) Increasing the viscosity of the continuous phase or producing a thick film with elasticity. Gelatin and many gums can increase the viscosity of the continuous phase of the emulsion. The protein can form an elastic thick film around the dispersed phase, which can inhibit the flocculation and coalescence of the

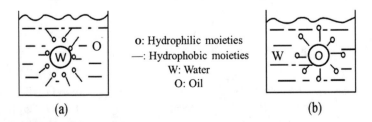

(a)   (b)

Fig. 5.7 Schematic diagram of emulsification of emulsifier. **a** W/O; **b** O/W

dispersed phase. This type of emulsifier is suitable for the O/W type system. For example, there is a casein membrane around the fat globule in the milk, which acts as emulsification.

(3) Reducing the interfacial tension between the two phases. Most emulsifiers are amphiphilic surfactants, which have a hydrophilic group and a hydrophobic group. The emulsifier is concentrated at the water–oil interface, the hydrophilic group acts with water, and the hydrophobic group acts with the oil, thereby stabilizes the emulsion by reducing the interfacial tension between the two phases.

(4) Forming a liquid crystal phase. Some emulsifiers can cause multi-molecular layer formed by liquid crystal to wrap around the oil droplets, weakening the Van der Waals attraction between the droplets and inhibiting flocculation and coalescence of the droplets. This stabilizing effect is more pronounced when the viscosity of the liquid crystal phase is much greater than that of the aqueous phase.

5.3.6.3   How to Choose Emulsifiers

The emulsifiers required for O/W and W/O systems are different and can be based on the 'hydrophilic–lipophilic balance (HLB) properties' established by the ATLAS Research Institute. The HLB value can indicate the hydrophilic lipophilic ability of the emulsifier, which can be measured by an experimental method and can also be calculated by some methods. Table 5.6 lists the applicability of different HLB values. Table 5.7 lists the HLB values and accepted daily intake (ADI) for some commonly used emulsifiers. The HLB value has algebraic additivity, that is, the HLB value of

Table 5.6 HLB values and applicability

HLB values	Applicability	HLB values	Applicability
1.5–3	Defoamer	8–18	O/W emulsifier
3.5–6	W/O emulsifier	13–15	Detergent
7–9	Wetting agent	15–18	Melting agent

Table 5.7 HLB values and ADI values for some common food emulsifiers

Emulsifier	HLB values	ADI values/(mg/kg weight)
Glycerol monostearate	3.8	Infinite quantity
Diglyceryl stearate monoester	5.5	0–25
Diacetyl succinyl monoglyceride	9.2	0–50
Sodium stearyl-2-lactate	21.0	0–20
Sorbitan tristearate	2.1	0–25
Polyoxyethylene sorbitan oleic acid monoester	15.0	0–25

the mixed emulsifier can be calculated. Usually, the mixed emulsifier has a better emulsification effect than a single emulsifier having the same HLB value.

The role of emulsifiers in food is multifaceted. For example, in addition to emulsification in ice cream, it can reduce bubbles, make ice crystals smaller, and give the ice cream a fine and smooth taste; In chocolate, it can inhibit the conversion of cocoa butter from β-3V type to β-3VI type, which inhibits the 'white frost' on the chocolate surface. When used in baked pasta foods, it can increase the volume of the product and prevent starch aging. In margarine, it can be used as a crystal modifier to adjust the consistency.

5.4 Oxidation of Lipids in Processing and Storage

5.4.1 Oxidation of Lipid

Oxidation of lipids is one of the main causes of lipids and oily food rancidity. During storage, lipid becomes rancid due to the action of oxygen, light, microorganisms, and enzymes in the air, which produces unpleasant smell and bitter taste, as well as some toxic compounds, which are collectively referred to as rancidity of lipids. The rancidity of foods during storage and processing is undesirable, but sometimes moderate oxidation of the lipid is necessary for the formation of aromas of fried foods.

The primary products of lipid oxidation are hydroperoxides, which are formed through autoxidation, photooxidation, and enzymic oxidation. ROOH is unstable and easily decomposed, and the decomposed products can be further polymerized.

5.4.1.1 Autoxidation

(1) Mechanism of automatic oxidation of lipid

Autoxidation of lipids is a free radical reaction of activated unsaturated fats with ground-state oxygen, including three stages: chain initiation, chain propagation, and chain termination. In the chain initiation stage, unsaturated fatty acids and their glycerides (RH) are easily dehydrogenated by α-methylene groups adjacent to double bonds under metal catalysis or light or heat, triggering the first of the chain reactions and yielding free radical alkyl (R•) as α-methylene hydrogen is easily removed by activation of double bonds. During the chain propagation phase, R• is cyclically produced; and during the chain termination phase, the free radicals react to form a non-radical compound.

Chain initiation (Induction period): $RH \xrightarrow{\text{Initiator}} R\cdot + H\cdot$ (5.1)

Chain propagation: $R\cdot + O_2 \longrightarrow ROO\cdot$ (5.2)

$ROO\cdot + RH \longrightarrow ROOH + R\cdot$ (5.3)

Chain termination: $R\cdot + R\cdot \longrightarrow R{-}R$ (5.4)

$R\cdot + ROO\cdot \longrightarrow ROOR$ (5.5)

$ROO\cdot + ROO\cdot \longrightarrow ROOR + O_2$ (5.6)

The first free radical is initiated usually with high activation energy, and this step is relatively slow. The activation energy of the chain propagation reaction is relatively low; thus, this step is carried out very quickly, and the reaction steps (5.2) and (5.3) can be recycled to produce a large amount of ROOH.

The oxygen in the reaction formula (5.2) is a lower energy ground-state oxygen, so-called triplet oxygen ($^3O_2$), and its electron arrangement is as shown in Fig. 5.8 and is filled in two π^* orbitals and they spin parallel. According to the Pauli incompatibility principle, this filling method is lower in energy and more stable. Since the electron spins are parallel, the total angular momentum of the electron is $2S + 1 = 2(1/2 + 1/2) + 1 = 3$(S is azimuthal quantum number), so the oxygen is called triplet oxygen. It is very difficult for the oil to react directly with $^3O_2$ to form ROOH. Since the activation energy of the reaction formula (5.7) is as high as 146–273 kJ/mol, the

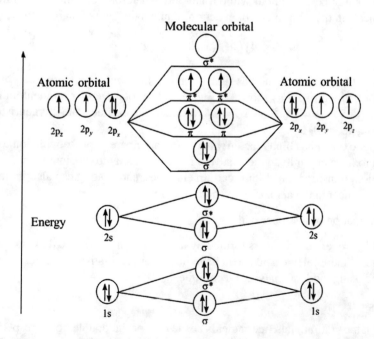

Fig. 5.8 Triplet oxygen molecular orbital

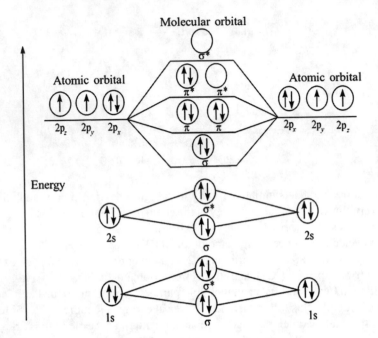

Fig. 5.9 Singlet oxygen molecular orbital

reaction cannot be carried out without any aid. Therefore, the initial generation of free radicals in the auto-oxidation reaction requires the help of an initiator.

$$RH + {}^3O_2 \longrightarrow ROOH \qquad (5.7)$$

When ${}^3O_2$ is excited (such as light), the electrons in the π^* orbit can be anti-parallel filled with spin as shown in Fig. 5.9. The total angular momentum of the electron is $2S + 1 = 2(1/2 - 1/2) + 1 = 1$, so this excited state oxygen is called singlet oxygen (${}^1O_2$).

Singlet oxygen has high reactivity and can participate in photooxidation to form hydroperoxides and initiate the first free radical in the autoxidation chain reaction. In addition, transition metal ions, certain enzymes, and heating can also initiate the first free radical in the autoxidative chain reaction.

(2) Formation of hydroperoxide

Various hydroperoxide isomers formed by the oxidation of oleic acid esters, linoleic acid esters, and linolenic acid esters have now been qualitatively and quantitatively analyzed using modern analytical techniques.

a. Oleate

The initiator first initiates free radicals at the α-C of the double bond, respectively, generating C_8 and C_{11} radicals; due to the delocalization of single electrons in the allyl structure, the single electrons at C_8 and C_{11} can also flow to C_{10} and C_9, the double bond is displaced and cis–trans isomerized to form C_{10} and C_9 radicals. Four kinds of free radicals react with $^3O_2$ to generate the corresponding ROO·and further generate four kinds of hydroperoxides. Among them, C_8 and C_{11} hydroperoxides are slightly more than C_9 and C_{10} hydroperoxides, and trans isomers account for more than 70%.

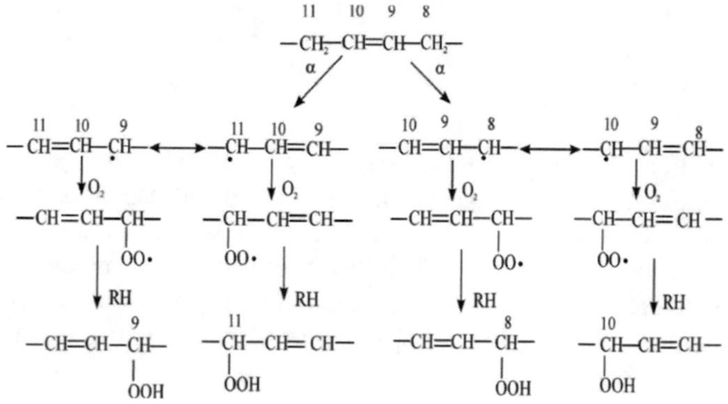

b. Linoleate

Linoleate has a pentadiene structure, and α-C_{11} is simultaneously activated by two double bonds, and the oxidation reaction rate is about 20 times faster than that of oleic acid. Therefore, a radical is first formed at C_{11}, and the radical is cis–trans isomerized to form two linoleate radicals with a conjugated double bond structure and then react with $^3O_2$ to form two kinds of ROOH.

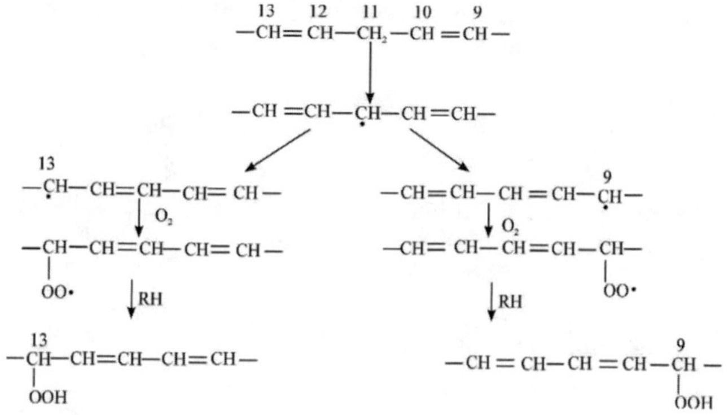

c. Linolenic

In the linolenic, there are two carbon atoms (C_{11}, C_{14}) between two double bonds, so it is easy to initiate free radicals here, and finally four hydroperoxides are formed. The oxidation rate is faster than that of linoleate.

The formation of ROOH in the process of autoxidation: free radicals are first initiated at the α-C of the double bond, and the radical resonance is stable. The double bond can be displaced and cis–trans isomerized. Participating in the reaction is $^3O_2$, and the number of kinds of ROOH generated is 2 × methylene numbers. When two or more double bonds are contained, the number of a-methylene groups is just the number of a-methylene groups that are activated by two double bonds.

5.4.1.2 Photooxidation

Some natural pigments such as chlorophyll and hemoglobin in foods are photosensitizers(Sens) that convert ground state oxygen ($^3O_2$) to excited state oxygen ($^1O_2$) when exposed to light. The highly electrophilic $^1O_2$ can directly attack any carbon atom on the double bond site of the high electron cloud density to form a six-membered ring transition state, and then the double bond shifts to form the ROOH in the trans configuration. The number of species of hydroperoxide produced was 2 × double bond number. Taking linoleate as an example, the reaction mechanism is as follows:

Since the energy of $^1O_2$ is high and highly reactive, the photooxidation reaction rate is about 1500 times faster than the autoxidation reaction rate; the comparison of the photooxidation rate with oleate, linoleate, and linolenate is 1.0: 1.7: 2.3, and their automatic oxidation rate is generally 1: 12: 25. The ROOH produced by the photooxidation reaction is further cleaved to initiate a free radical chain reaction of the autoxidation process.

5.4.1.3 Enzymatic Oxidation

The oxidation of lipid with the help of enzymes is called enzymatic oxidation. Lipoxygenase(Lox) specifically acts on polyunsaturated fatty acids with a 1,4-cis, cis-pentadiene structure (e.g., 18:2, 18:3) and dehydrogenates at the central methylene group of 1,4-pentadiene (i.e., at position ω-8) to form a radical. The isomerization then shifts the double bond position while converting to the trans configuration, forming ω-6 and ω-10 hydroperoxides with conjugated double bonds.

In addition, ketone type rancidity is also enzymatic oxidation, which is caused by the action of enzymes (such as dehydrogenase, decarboxylase, hydratase) and produced by certain microorganisms. Oxidation reactions occur mostly between the α- and β-carbon positions of saturated fatty acids and are, therefore, also referred as β-oxidation. The final products of oxidative production are keto acids and methyl ketones that have an unpleasant odor and therefore these oxidative reactions are referred to as keto-type rancidity.

5.4.1.4 Hydrolysis and Polymerization of Hydroperoxides

The hydroperoxides produced by various routes are unstable and can be degraded to produce many decomposition products. Primary, ROOH breaks at the oxygen–oxygen bond, producing alkoxy radicals and hydroxyl radicals.

1. The oxygen–oxygen bond cleavage of ROOH

Further, the alkoxy radicals break at carbon–carbon bond, which is connected to the oxygen on both sides of the carbon atom to form a compound such as aldehydes, acids, and hydrocarbons.

2. The carbon–carbon bond cleavage of RO•

Furthermore, the alkoxy radicals can also form ketones and alcohols by the following routes:

$$R_1-CH-R_2OOH \xrightarrow{R_3O^\bullet} R_1-\underset{\underset{O}{\|}}{C}-R_2COOH + R_3OH$$

$$\xrightarrow{R_4H} R_1-\underset{\underset{OH}{|}}{CH}-R_2COOH + R_4^\bullet$$

The aldehydes, ketones, alcohols, acids, etc. produced by the decomposition of hydroperoxides have an unpleasant odor, finally causing rancidity.

Small molecule compounds produced by the oxidation of lipids can also undergo polymerization to form dimers or polymers. For example, hexanal, which is produced by the oxidation of linoleic acid, can be polymerized into a cyclic trimer, triamyl trioxane with a strong smell.

$$3C_5H_{11}CHO \longrightarrow$$

5.4.1.5 Factors Influencing Lipid Oxidation

1. Composition of fatty acids and glycerides

The rate of lipid oxidation is related to the degree of unsaturation of fatty acids, the position of double bonds, and the configuration of the *cis–trans*. The chain-initiated reaction of saturated fatty acids is difficult to occur at room temperature. When unsaturated fatty acids have begun to rancid and saturated fatty acids can remain intact. While in unsaturated fatty acids, the more numbers of double bonds, the faster the rate of oxidation; the *cis* configuration is easier to oxidize than the *trans* configuration; the conjugated double bond structure is easier to oxidize than the non-conjugated double bond structure. The free fatty acid has a slightly higher oxidation rate than the bound fatty acid in the glyceride. When the free fatty acid content in the oil is more than 0.5%, the auto-oxidation rate will be significantly accelerated; the random distribution of fatty acids in glycerides is beneficial to reduce the oxidation rate (Table 5.8).

2. Oxygen

The oxidation rate of singlet oxygen is about 1500 times that of triplet oxygen. When the oxygen concentration is low, the oxidation rate is approximately proportional to the oxygen concentration; when the oxygen concentration is high, the oxidation rate is independent of the oxygen concentration. Therefore, vacuum or nitrogen-filled packaging and low-breathing materials can be used to prevent oxidative deterioration of oil-containing foods.

Table 5.8 Induction period and relative oxidation rate of fatty acids at 25° C

Fatty acid	Number of double bonds	Induction period/h	Relative oxidation rate
18:0	0		1
18:1(9)	1	82	100
18:2(9, 12)	2	19	1200
18:3(9, 12, 15)	3	1.34	2500

3. Temperature

The increase in temperature causes the oxidation reaction rate to increase. High temperature can promote the production of free radicals as well as the decomposition and polymerization of hydroperoxides. However, as the temperature rises, the solubility of oxygen decreases. Saturated fatty acids are not only stable at room temperature but also undergo significant oxidation at elevated temperatures. For example, the content of saturated fatty acids in lard is usually higher than that of vegetable oil, while the shelf life of lard is shorter than that of vegetable oil. This is due to that lard is usually obtained by smelting, and it also contains photosensitizer hemoglobin and metal ions, and undergoes the high temperature stage, which may cause the production of free radicals; while vegetable oil is obtained by extracting with organic solvent at a relatively low temperature, so the stability is better than lard.

4. Water

The relationship between the relative rate of lipid oxidation reaction and water activity is shown in Fig. 5.10. The oxidation rate is the lowest at a water activity of 0.33; the oxidation rate decreases as the water activity from 0 to 0.33. This is due to that a very small amount of water is added to the dry sample to hydrate with the catalytically oxidized metal ion. The catalytic efficiency is significantly reduced, and it can combine with hydroperoxide and prevent its decomposition; as the water activity is from 0.33 to 0.73, the oxidation rate increases as the fluidity of catalyst and the dissolved oxygen in the water increases, and the molecule swells, which leaves

Fig. 5.10 Relationship between water activity and fat oxidation rate

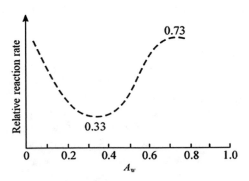

more catalytic sites exposed. As the amount of water is increased when the water activity is greater than 0.73, the oxidation rate decreases as the concentration of the catalyst is diluted.

5. Surface area

The surface area of the lipids in contact with air is proportional to the rate of oxidation of the lipids.

6. Pro-oxidants

Some divalent or polyvalent transition metal ions with suitable redox potential are effective pro-oxidants, active even at the concentration of 0.1 mg/kg. The chain initiation period is shortened and the oxidation rate is accelerated. Its catalytic mechanism may be as follows:

(1) Promote the decomposition of hydroperoxide.

$$M^{n+} + ROOH \begin{cases} M^{(n+1)+} + OH^- + RO \cdot \\ M^{(n-1)+} + H^+ + ROO \cdot \end{cases}$$

(2) Directly interacting with substrates that are not oxidized.

$$M^{n+} + RH \longrightarrow M^{(n-1)+} + H^+ + R \cdot$$

(3) Activation of oxygen molecules to produce singlet oxygen and peroxyl radicals.

$$M^{n+} + {}^3O_2 \longrightarrow M^{(n+1)+} + O_2^- \begin{cases} \xrightarrow{-e} {}^1O_2 \\ \xrightarrow{+H^+} HO_2 \cdot \end{cases}$$

Metal ions are derived from the soil in which oil crops are grown, processing and storage equipment, and the food material itself. The catalytic strengths of different metals are ranked as follows: lead > copper > brass > tin > zinc > iron > aluminum > stainless steel > silver. In addition, protoheme is also a catalyst for the oxidation of fats due to its iron content. When lard is refined, if the hemoglobin is not completely removed, the lard is rancified quickly. It should be said that the enzymes involved in enzymatic oxidation are all catalysts for the oxidation reaction.

7. Light and ray

Light and ray not only promote the decomposition of hydroperoxides but also generate free radicals. Visible light, UV, and high-energy radiation can promote oxidation. Thus, the package to decrease light exposure can retard lipid oxidation rate.

8.   Antioxidants

Adding antioxidants can decrease the rate of oil oxidation. Substances that retard
and decrease the rate of fat oxidation are called antioxidants. It will be described in
detail in Sect. 5.4.2.

5.4.1.6   The Danger of Peroxidized Lipid

Automatic oxidation of oil is a free radical chain reaction, and the high reactivity
of free radicals can lead to damage to the body, cell destruction, and human aging.
Peroxidation lipids produced during the oxidation of fats and oils lead to deterioration
of the appearance, texture, and nutritional quality of foods, and even mutagenic
substances.

(1)   Peroxidized lipid reacts almost with any ingredient in the food, reducing the
       quality of the food, for example, take the reaction of lipid peroxide with
       proteins. Hydroperoxide and its degradation products react with proteins,
       resulting in changes in food texture, decreased protein solubility (protein cross-
       linking), color change (browning), and reduced nutritional value (loss of essen-
       tial amino acid). The alkoxy radical (RO•) generated by the oxygen–oxygen
       bond cleavage of the hydroperoxide reacts with the protein (PrH) to form a
       protein radical, and the protein radical re-crosslinks.

$$\mathbf{RO \bullet + PrH \longrightarrow Pr \bullet + ROH}$$
$$\mathbf{2Pr\bullet \longrightarrow Pr - Pr}$$

The decomposition product of lipid peroxide, malondialdehyde, reacts with
ε-NH_2 of lysine in the protein to form a Schiff base, which crosslinks the
macromolecule. After the fish protein is stored in a frozen state, its decrease
in solubility is attributed to the reaction.

The aldehyde formed by the automatic oxidation of the unsaturated fatty acid
can be condensed with the amino group in the protein to form a Schiff base. If
the aldol condensation reaction is continued, a brown polymer can be formed;
as the hydrolysis occurs in the early stage of the aldol condensation reaction
(in the first or second alkyd condensation), which releases an aldehyde with
a strong odor. Therefore, the reaction not only causes discoloration but also
causes a change in flavor, as shown in Fig. 5.11.

(2)   Hydroperoxides react with almost all molecules or cells in the human body,
       destroying DNA and cellular structure. When the group of -NH_2 in enzymes
       undergoes the cross-linking reaction with malondialdehyde, they lose their

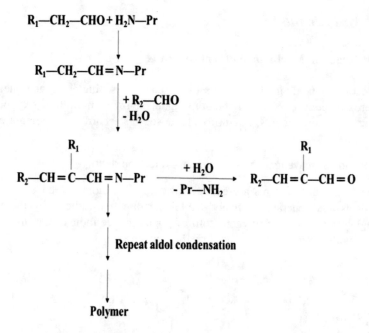

Fig. 5.11 A series of reactions between aldehydes and proteins

activity. Protein cross-linking will lose its biological function. These destroyed cellular components are phagocytized by lysosomes and cannot be digested by hydrolytic enzymes. When accumulated in the body, lipofuscin (age spots) will be produced.

(3) Lipids can produce harmful substances when oxidized at normal temperature and high temperature. Animal experiments have shown that feeding highly oxidized fat at normal temperature (10–20% in rat feed) will cause appetite decline, growth inhibition, hepatic and renal enlargement, and peroxide accumulation in adipose tissue. Feeding highly oxidized fats caused by heating can have various deleterious effects on animals. It has been reported that polar dimers produced by oxidized polymers are toxic, and cyclic cheeses formed by anaerobic thermal polymerization are also toxic. Oil that has been fried for a long time in French fries and fish fillets, or a frying oil that is repeatedly used, has an appreciable amount of carcinogenic activity.

5.4.2 Antioxidants

5.4.2.1 Reaction Mechanism of Antioxidants

Antioxidants can be divided into free radical scavengers, singlet oxygen quenchers, hydroperoxide converters, metal chelating agents, enzyme inhibitors, oxygen scavengers, enzyme antioxidants, and ultraviolet absorbers according to their antioxidant mechanism.

1. Free radical scavenger (hydrogen donor, electron donor)

Phenolic (AH_2) antioxidants are excellent hydrogen donors, which can scavenge free radicals and generate relatively stable phenolic free radicals by themselves. The single electron on the oxygen atom of phenolic free radicals can interact and conjugate with the π electronic cloud on the benzene ring.

When there is a tert-butyl group at the ortho position of the phenolic hydroxyl group, due to the steric hindrance, the attack of oxygen molecules is hindered. Therefore, the tert-butyl group reduces the possibility of further alkoxy radicals initiating free radical chain reactions, thereby having an antioxidant effect. Its mechanism of action can be expressed by the following formula:

$$ROO \bullet + AH_2 \rightarrow ROOH + AH\bullet$$
$$AH \bullet + AH\bullet \rightarrow A + AH_2$$

In addition, there are antioxidants that donate electrons, but since this antioxidant is generally a weak antioxidant, it is not commonly used.

2. Singlet oxygen quencher

$^1O_2$ easily interacts with the double bonds of the same singlet state and transforms into $^3O_2$, so carotenoids containing many double bonds are better $^1O_2$ quenchers. The mechanism of action is that the excited state of $^1O_2$ transfers energy to the carotenoids, so that the carotenoid changes from the ground state (^{1} carotenoid) to the excited state (^{3} carotenoids), and the latter can directly return to the ground state.

$$^1O_2 + {}^1 \text{carotenoid} \rightarrow {}^3O_2 + {}^3 \text{carotenoid}$$
$$^3 \text{carotenoid} \rightarrow {}^1 \text{carotenoid}$$

In addition, the $^1O_2$ quencher may also restore the photosensitizer to its ground state.

$$^1\text{carotenoid} + {}^3\text{Sen}^* \rightarrow {}^3\text{carotenoid} + {}^3\text{Sen}$$

3.   Hydroperoxide converter

Hydroperoxide is the main initial product of oil oxidation. Some compounds, such as lauric acid and stearate of thiodipropionic acid (represented by R_2S), can convert ROOH into inactive alcohol, thereby inhibiting further oil oxidation, and such substances are called hydroperoxide converter. The mechanism of action is as follows:

$$R_2S + R'OOH \rightarrow R_2S = O + R'OH$$
$$R_2S = O + R'OOH \rightarrow R_2SO_2 + R'OH$$

4.   Metal chelator

Citric acid, tartaric acid, ascorbic acid, etc. can chelate with transition metal ions, which are the co-oxidants of fats, and finally deactivate them, thereby inhibiting the oxidation of fats.

5.   Oxygen scavenger

In addition to chelating metal ions, ascorbic acid is also an effective oxygen scavenger. It has an antioxidant effect by removing oxygen from food. For example, ascorbic acid inhibits enzymatic browning, which is a deoxygenation effect.

6.   Enzyme antioxidants

Superoxide dismutase (SOD) can convert superoxide anion radical $O_2\cdot$ into $^3O_2$, the reaction is as follows:

$$2O_2^{-} + 2H^+ \xrightarrow{\text{SOD}} {}^3O_2 + H_2O_2$$
$$2H_2O_2 \xrightarrow{\text{Catalase}} 2H_2O + {}^3O_2$$

SOD, glutathione peroxidase, catalase, glucose oxidase, etc. are all enzyme antioxidants.

7.   Synergist

In actual application of antioxidants, two or more antioxidants are often used at the same time as the antioxidant effect of using several antioxidants is better than using one antioxidant alone, there are usually two synergistic mechanisms:

(1)   The synergist functions to regenerate the main antioxidants and thus play a synergistic effect. For example, BHA and BHT, both of which are phenolic

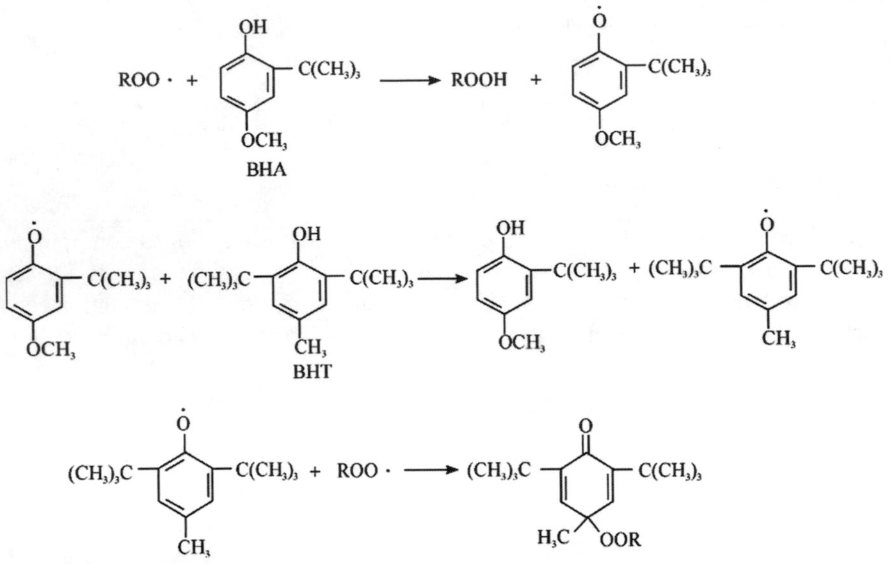

Fig. 5.12 The synergist effect of BHT to BHA

antioxidants, the former is the main antioxidant, it will first become a hydrogen donor (due to its low dissociation energy of OH bond), and BHT can only react slowly with ROO• due to steric hindrance, So the role of BHT is to regenerate BHA, as shown in Fig. 5.12.

(2) The synergist is a metal-chelating agent. For example, the combination of phenols and ascorbic acid can greatly improve the antioxidant capacity, where phenols are the main antioxidants, ascorbic acid can chelate metal ions, and ascorbic acid is also an oxygen scavenger.

5.4.2.2 Antioxidants Commonly Used in Food

Antioxidants can be divided into natural antioxidants and synthetic antioxidants according to their sources. The antioxidants allowed in China contain 14 kinds, mainly including tocopherols, tea polyphenols, propyl gallate (PG), ascorbic acid, butyl hydroxyanisole (BHA), dibutyl hydroxytoluene (BHT), etc.

1. Natural antioxidants

Many natural animal and plant materials contain some antioxidants. Due to the concern about the safety of synthetic antioxidants, natural antioxidants are becoming more and more popular. Among the natural antioxidants, phenols are still the most important category. For example, tocopherols widely distributed in nature, tea polyphenols in tea, and sesamol in sesame are all excellent natural antioxidants. In addition, there are some antioxidants in many spices, such as rosmarinic acid,

Fig. 5.13 Structures of
tocopherol

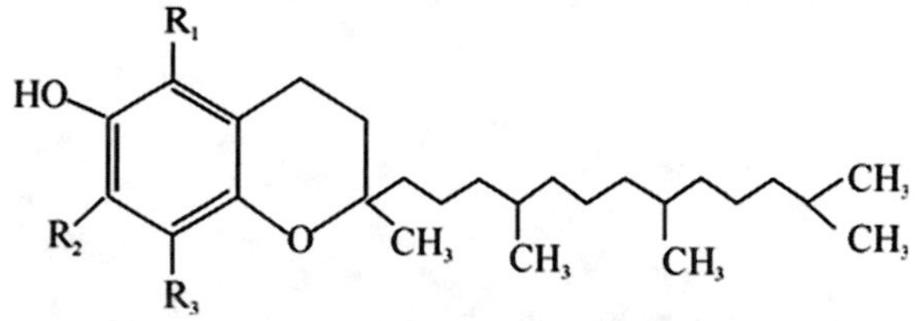

	R₁	R₂	R₃
α	CH₃	CH₃	CH₃
β	CH₃	H	CH₃
γ	H	CH₃	CH₃
δ	H	H	CH₃

gingerone in ginger. Flavonoids and some amino acids and peptides are also natural antioxidants. Some natural enzymes such as glutathione peroxidase and SOD also have good antioxidant properties, as well as the ascorbic acid, carotenoids, etc.

(1) Tocopherols. Tocopherol has a variety of structures, and the main structural forms are shown in Fig. 5.13.

In terms of antioxidant activity, the order of activity of several tocopherols is $\delta > \gamma > \beta > \alpha$. The antioxidant effect of tocopherol in animal fat is better than that in vegetable oil, but its natural distribution is high in vegetable oil. Tocopherol has the characteristics of heat resistance, light resistance, and high safety and can be used in frying oil.

(2) Tea polyphenols. Tea polyphenols are some polyphenolic compounds in tea, including epigallocatechin gallate (EGCG), epigallocatechin (EGC), epicatechin gallate (ECG), epicatechin (EC), where EGCG is the most effective in both water-bearing and oil-bearing systems.

(3) L-ascorbic acid. L-ascorbic acid is widely present in nature and can also be artificially synthesized. It is a water-soluble antioxidant and can be used in processed fruits, vegetables, meat, fish, beverages, and other foods. L-ascorbic acid acts as an antioxidant, and its effects are multifaceted: ① scavenging oxygen, for example, it is used in fruits and vegetables to inhibit enzymatic browning; ② it acts as a chelating agent and is used as a synergist in combination with phenols; ③ reducing some oxidized products, for example, L-ascorbic acid is used in meat products as a coloring aid, reducing brown metmyoglobin to red ferrous myoglobin; ④ protect the sulfhydryl group (-SH) from oxidation.

2. Synthetic antioxidants

Synthetic antioxidants are still widely used due to their good antioxidant properties and price advantages. Several of the most commonly used synthetic antioxidants are also phenols. Its structure is as follows:

OH
C(CH3)3
OCH3

OH
C(CH3)3
OCH3

OH
(H3C)3C   C(CH3)3
CH3

3-(*tert*-butyl)-4-methoxyphenol (3-BHA)   2-(*tert*-butyl)-4-methoxyphenol (2-BHA)   2,6-di-*tert*-butyl-4-methylphenol (BHT)

OH
HO   OH
COOC3H7

OH
C(CH3)3
OH

COC3H7
OH
HO
OH

propyl 3,4,5-trihydroxybenzoate (PG)   2-(*tert*-butyl)benzene-1,4-diol (TBHQ)   1-(2,4,5-trihydroxyphenyl)butan-1-one (THBP)

(1) BHA. Commercial BHA is a mixture of 2-BHA and 3-BHA. The antioxidant effect in animal fat is better than that in vegetable oil. It is easily soluble in oil and insoluble in water, with good heat resistance, and does not show color when interacting with metal ions. When used in baked food, it has a typical phenol smell. Animal experiments show that BHA has a certain degree of toxicity, and BHA also has antimicrobial effects.

(2) BHT. BHT is insoluble in water and soluble in organic reagents, with good heat resistance and stability. It is stable at ordinary cooking temperature and can be used in baked foods. It does not show color when interreacting with metal ions, with no special odor like BHA, and is low in price with strong antioxidant ability. Therefore, it is used as the main antioxidant in many countries.

(3) PG. PG shows better antioxidant performance than BHT and BHA, with good heat resistance, but it is colored when reacting with metal ions, so it is often used in combination with citric acid. As citric acid can chelate metal ions, it can not only be used as a synergist but also avoid the problem of coloring with metal ions. PG will quickly volatilize during the baking or frying process of food and can be used in canned food, instant noodles, and dried fish products.

(4) TBHQ. It is an oil-soluble antioxidant. When used in vegetable oil, it has better antioxidant effect than BHA, BHT, and PG. It does not color with iron ions and has no peculiar smell or odor.

(5) D-isoascorbic acid and its sodium salt. D-isoascorbic acid is a synthetic product and is a water-soluble antioxidant. It is used in foods such as fruits, vegetables, canned food, beer, juice, etc. Its antioxidant properties are comparable to L-ascorbic acid, but D-isoascorbic acid is easy to synthesize.

5.4.2.3 Precautions for the Use of Antioxidants

(1) Antioxidants should be added as soon as possible. As the fat oxidation reaction is irreversible, antioxidants can only play a role in hindering the oxidation of fats, delaying the time of food spoilage, but cannot change the result of the deterioration.

(2) The dosage of antioxidants should be paid attention to. One is that the dosage cannot exceed its safe dosage; the other is that when the dosage of some antioxidants is inappropriate, it will have the effect of promoting oxidation.

(3) When choosing antioxidants, attention should be paid to solubility. Only with good solubility in the system can, its antioxidant effect, be fully exerted. Therefore, fat-soluble antioxidants should be selected for grease systems, and water-soluble antioxidants should be selected for aqueous systems.

(4) In practical applications, two or more antioxidants are often used to take advantage of their synergistic effects.

5.4.2.4 Anti-Oxidation and Pro-Oxidation

Some studies have shown that the amount of some antioxidants is not completely positively correlated with antioxidation performance, and sometimes when the amount is too large, it may play a role in promoting oxidation (pro-oxidation). Take phenolic substances (AH_2) as an example, the mechanism is as follows:

(1) Low concentration of phenol can scavenge free radicals.

$$ROO \bullet + AH_2 \rightarrow ROOH + AH \bullet \text{ (scavenging peroxy free radicals)}$$
$$ROO \bullet + AH \bullet \rightarrow ROOH + A$$

(2) High concentration of phenol can promote oxidation.

As the concentration of phenol-free radicals increases, the reverse reaction of the above reaction can occur, which can promote oxidation.

$$ROOH + AH \bullet \rightarrow ROO \bullet + AH_2$$

α-tocopherol and β-tocopherol show pro-oxidation phenomena, while γ-tocopherol and δ-tocopherol do not. In animal and vegetable oils, when the concentration of α-tocopherol is not too high (less than 600–700 mg/kg of oil), it does not exhibit pro-oxidation at room temperature; when the temperature rises, the generation rate of phenolic free radicals is faster than that of the substrate auto-oxidation; that is, when the concentration of phenolic free radicals exceeds $ROO\bullet$, it may play a role in promoting oxidation. Therefore, in the food industry, it is generally advisable to control the total amount of α-tocopherol (the sum of its own content and the added amount) at 50–500 mg/kg.

Contrary to the case of α-tocopherol, low concentration of ascorbic acid (10^{-5} mol/L) can promote oxidation, especially in the co-existence with Fe^{2+} and Cu^{2+}. It is speculated that the reaction forms ascorbic acid-Fe^{2+}-oxygen complexes, which can promote the decomposition of hydroperoxides, thereby causing auto-oxidation of unsaturated fatty acids.

β-carotene has the highest antioxidant activity when the concentration is 5×10^{-5} mol/L. If the concentration is higher, the pro-oxidation effect is the dominant. Whether β-carotene is anti-oxidant also depends on the partial pressure of oxygen. When the oxygen pressure is low ($p_{O_2} < 150$ mmHg, 1 mmHg = 133.322 Pa), it has an anti-oxidation effect; when the oxygen pressure is high, it has a pro-oxidation effect.

5.5   Other Chemical Changes of Lipids During Processing and Storage

5.5.1   Hydrolysis of Lipids

The lipids can be hydrolyzed to free the fatty acids in the presence of water and under the action of heat, acid, alkali, and lipolytic enzymes. The hydrolysis reaction of lipids under the alkaline condition is called saponification reaction, and the fatty acid salts formed from this reaction is soap. Thus, the hydrolysis reaction of the lipids is widely used to produce soap in the industry.

Free fatty acids (FFAs) are not present in the adipose tissue of living animals, but they will be produced by the action of lipolytic enzymes in the body after the animals are slaughtered. As FFAs have an acidic taste and are more sensitive to oxidation than glycerol, they can lead to fast rancidity of lipids. The smelting method to obtain lipids is always under high temperature, which can inactivate lipolytic enzymes. Lipolytic enzymes are also present in vegetable oilseeds, but the amount of FFA in animal fats is less than that of unrefined vegetable oils. In the refining process of vegetable oils, FFA is neutralized by alkali addition.

During the frying process, the moisture of food enters the lipids, leading to the hydrolysis of the lipids and the release of the FFA, which cause a decrease in the smoke point of the lipids; along with the increase of the FFA content, the smoke point of the lipids decreases continuously (Table 5.9). Therefore, hydrolysis will cause a decrease in the lipid quality and the deterioration in flavor. Hydrolysis of the cream will also produce some short-chain fatty acids (C_4–C_{12}), further resulting in

Table 5.9 Relationship between the free fatty acids content in lipids and the smoke point

The free fatty acids/ %	0.05	0.10	0.50	0.60
The smoke point/°C	226.6	218.6	176.6	148.8–160.4

the production of a rancid taste. It is noteworthy that some mild hydrolyses in food processing are beneficial, such as the utilization of hydrolyses in the production of cheese and yogurt.

5.5.2 Chemical Reaction of Lipids at the High Temperature

Various chemical reactions of lipids will occur when they are cooked at a high temperature, including thermal decomposition, thermal polymerization, thermal oxidative polymerization, condensation, hydrolysis, and oxidation reactions. The long-time heating will lower the quality of the lipids, including decreasing the iodine value and the smoke point of the lipids while increasing the viscosity, the acid value, and the amount of foam in the lipids.

5.5.2.1 The Thermal Decomposition Reaction

Both saturated fats and unsaturated fats will take place thermal decomposition reactions at high temperature. The thermal decomposition reaction can be divided into thermal decomposition and thermal oxidative decomposition according to the presence of oxygen or not. The presence of metal ions such as Fe^{2+} will catalyze the thermal decomposition reaction. The thermal decomposition of saturated fats is shown in Fig. 5.14.

The thermal oxidative decomposition can occur in saturated fats at high temperatures (T > 150 °C). In that reaction, ROOH is first formed on α-, β-, or γ-carbon of the carboxyl or ester group and further decomposed into hydrocarbons, aldehydes, ketones, and other compounds. Figure 5.15 shows the thermal oxidative decomposition when oxygen attacks the β-position of saturated fats.

In addition to some low molecular weight materials, unsaturated fats mainly produce dimers at heat and anaerobic conditions. The thermal oxidative decomposition of unsaturated fats is similar with that of auto-oxidation reactions in low temperature. According to the position of the double bond, the formation and decomposition of ROOH can be predicted, but the decomposition rate of ROOH is higher at high temperatures.

5.5.2.2 The Thermal Polymerization Reaction

The thermal and thermal oxidative polymerization reactions of lipids can occur under high temperature conditions (T > 150 °C). The polymerization will result in an increase in the viscosity and lipid foam. Moreover, the thermal polymerization reaction under anaerobic conditions is a Diels–Alder reaction between polyene compounds to form a cyclic olefin. The polymerization can occur between different triglycerides (Fig. 5.16) or within the same triglyceride (Fig. 5.17).

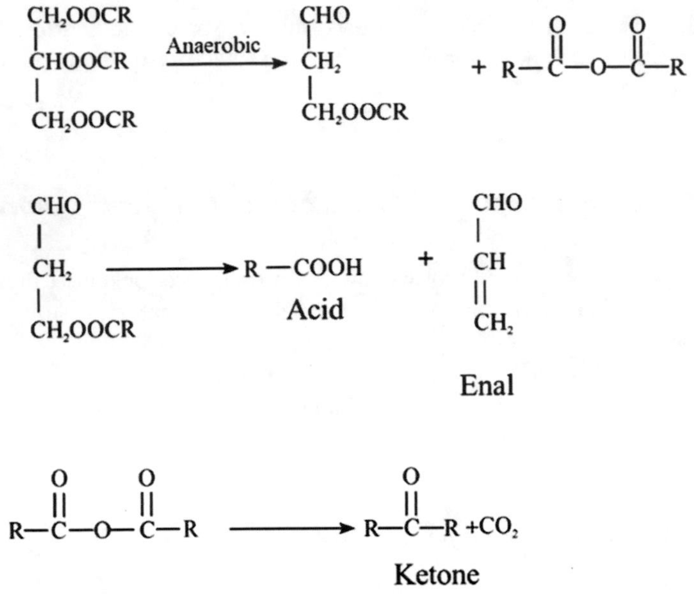

CH_2OOCR
$CHOOCR$ $\xrightarrow{\text{Anaerobic}}$ CHO CH_2 CH_2OOCR $+ R-\overset{O}{\overset{||}{C}}-O-\overset{O}{\overset{||}{C}}-R$
CH_2OOCR

CHO
CH_2 $\longrightarrow$ $R-COOH$ $+$ CHO CH CH_2
CH_2OOCR Acid

Enal

$R-\overset{O}{\overset{||}{C}}-O-\overset{O}{\overset{||}{C}}-R \longrightarrow R-\overset{O}{\overset{||}{C}}-R +CO_2$

Ketone

Fig. 5.14 The thermal decomposition reaction of saturated fats

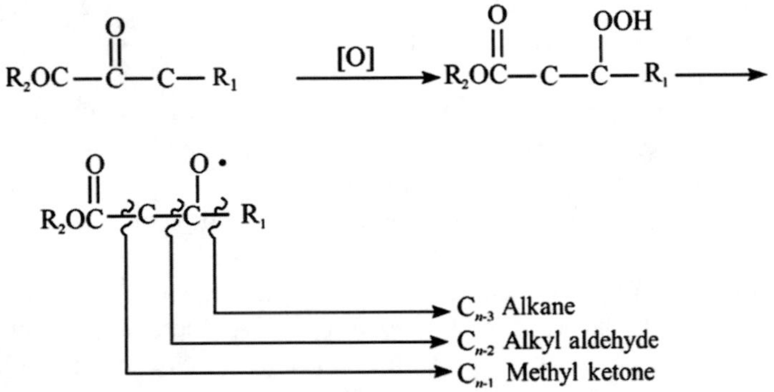

$R_2OC-\overset{O}{\overset{||}{C}}-C-R_1 \xrightarrow{[O]} R_2OC-\overset{O}{\overset{||}{C}}-\overset{OOH}{\underset{|}{C}}-R_1 \longrightarrow$

$R_2OC-\overset{O}{\overset{||}{C}}-C-\overset{O\bullet}{\underset{|}{C}}-R_1$

$\longrightarrow C_{n-3}$ Alkane
$\longrightarrow C_{n-2}$ Alkyl aldehyde
$\longrightarrow C_{n-1}$ Methyl ketone

Fig. 5.15 The thermal oxidative decomposition of unsaturated fats

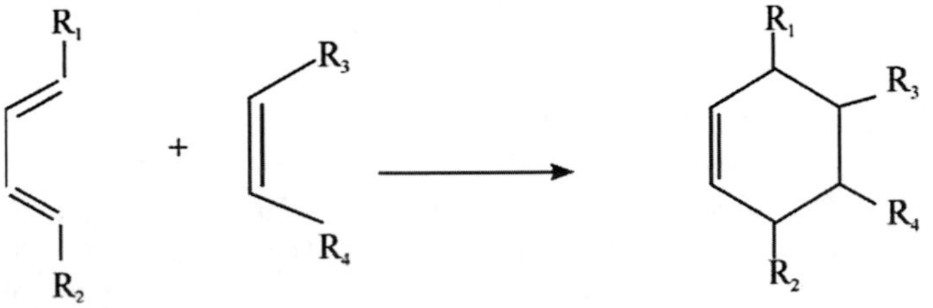

R_1
R_2 $+$ R_3 R_4 $\longrightarrow$ R_1 R_3 R_4 R_2

Fig. 5.16 The intermolecular Diels–Alder reaction

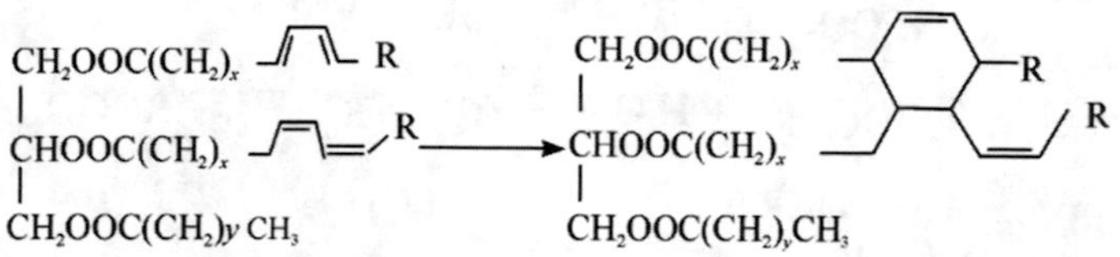

Fig. 5.17 The intramolecular Diels–Alder reaction

The thermal oxidative polymerization reaction will be occurred at high temperature conditions (200–230 °C), and the triglyceride molecules are uniformly split at the α-carbon of the double bond to generate free radicals. In addition, the free radicals will combine with each other to form some toxic dimers. These dimers can be absorbed and combine with enzymes to inactivate them in the body, leading to physiological abnormalities. The fine foam occurring in fried fish and shrimp was found to be dimers.

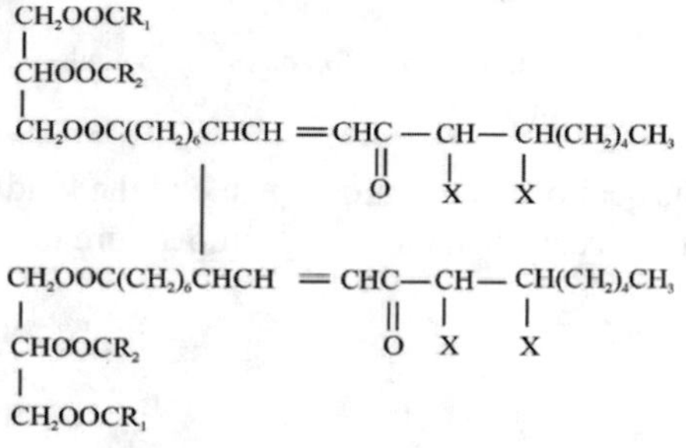

X = OH or Epoxy compound

5.5.2.3 The Condensation Reaction

The water in the food entering the oils at high temperatures, especially under frying conditions, equivalent to steam distillation, will drive away the volatile oxides in the oils and hydrolyze partially the oils at the same time, leading to the increase of the acid value and the decrease of the smoke point. The hydrolyzed products are then condensed into epoxy compounds with the larger molecular weight, as shown in Fig. 5.18.

The chemical reactions of lipids at high temperatures are not necessarily negative. The formation of aroma in fried foods is related to certain reaction products under high temperature conditions. The main components of aroma in fried food mainly are carbonyl compounds (aldehydes). For example, when glycerol trioleate is heated to 185 °C for 72 h, and water vapor is passed for 2 min every 30 min, five linear 2,4-heptadienals and lactones are found in the volatiles, which show the characteristic aroma of fried foods. However, the excessive reactions of lipids at high temperatures

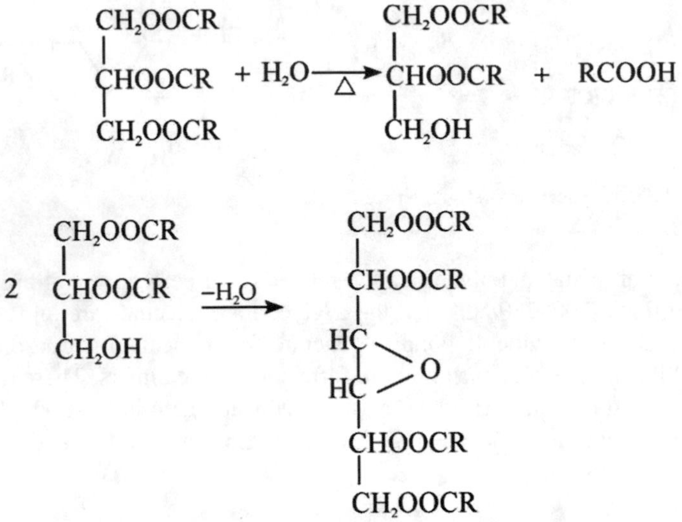

$$CH_2OOCR \mid CHOOCR \mid CH_2OOCR + H_2O \xrightarrow{\Delta} CH_2OOCR \mid CHOOCR \mid CH_2OH + RCOOH$$

$$2 \ CH_2OOCR \mid CHOOCR \mid CH_2OH \xrightarrow{-H_2O} $$

Fig. 5.18 The condensation reaction of lipids into epoxy compounds

are unfavorable for the quality and nutritional value of the lipids. In food processing, it is generally preferred to control the heating temperature to be below 150 °C.

5.5.3 Radiolysis

Irradiation can act as a mean of sterilization to extend the shelf life of food, and the radiant reaction causing the degradation of lipids is called radiolysis. The negative effect of irradiation, like heat treatments, contains the induction of chemical change.

In the radiant process of lipids, the lipid molecules absorb radiant energy to form ions and excited molecules, and the excited molecules can be further degraded. Taking the saturated fatty acid esters as an example, it is first broken at the α-, β-, and γ-position near the carbonyl group after irradiation, preferentially split at five positions (a, b, c, d, and e) near the carbonyl groups (Fig. 5.19), and the generated radiolysis products containing hydrocarbons, aldehydes, acids, ketones, and esters. The excited molecules can generate free radicals when they are decomposed. Irradiation can also accelerate the automatic oxidation of lipids at aerobic conditions, while destroy the antioxidants.

Both irradiation and heating can cause the degradation of lipids and generate similar degradation products, but the latter produces more decomposition products. Many studies have shown that the irradiation of fat-containing foods at a pasteurization dose is not dangerous.

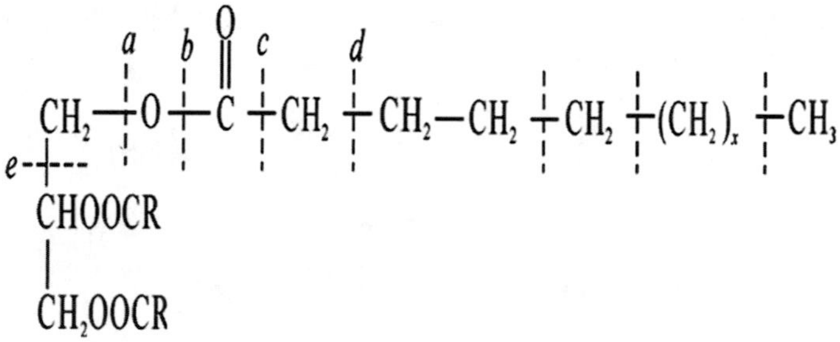

Fig. 5.19 The locations of the fractures in the radioactive reactions

5.6 Evaluation of Lipid Quality

The quality of lipids during the processing and storage periods will gradually decrease due to various chemical reactions. The oxidation of lipids is a vital factor to cause the rancidity of oil. In addition, other reactions such as hydrolysis and radiolysis can also lead to a reduction in the quality of the lipids.

5.6.1 Methods to Evaluate the Lipid Oxidations

Lipid oxidation reactions are extremely complicated because of the multitudinous oxidative products, the instability, and easy decomposable characteristic of intermediate products. Therefore, it is very important to select the evaluation index to describe lipid oxidation reactions. There is still no simple method to measure all oxidation products immediately, and it is often necessary to measure several indicators to comprehensively evaluate the degree of lipid oxidation reactions.

5.6.1.1 Peroxide Value

The peroxide value (POV) refers to the number of millimoles of hydroperoxide contained in 1 kg of lipids.

ROOH is the main primary product of lipid oxidation reactions. In the initial stage of lipid oxidation, the POV value increases with the degree of oxidation. However, the decomposition rate of ROOH will exceed the rate of its formation, and the POV value will decrease at the prolonged oxidation period. Therefore, the POV value should be finitely used to measure the initial stage of lipid oxidations. The POV value is usually determined by iodometric method:

$$ROOH + 2KI \longrightarrow ROH + I_2 + K_2O$$

The generated iodine is titrated with a solution of $Na_2S_2O_3$ to quantitatively determine the content of ROOH.

$$I_2 + 2Na_2S_2O_3 \longrightarrow 2NaI + Na_2 + S_4O_6$$

5.6.1.2 Thiobarbituric Acid

The oxidized product of the unsaturated fatty acids, aldehydes, can form colored compounds with thiobarbituric acid (TBA). For example, the colored products formed by the malondialdehyde (MDA) and TBA have maximum absorption at 530 nm, while the maximum absorption of the products formed by other aldehydes and TBA is at 450 nm, so it is necessary to measure the absorbance of the colored products at two wavelengths for the sake to measure the degree of lipid oxidations. The disadvantage of this method contains that MDA is not a product of all lipid oxidation systems, and some non-oxidized products can also develop color with TBA, such as reaction between TBA and proteins existing in food. Therefore, this method is not convenient for evaluating the oxidation of different systems, but it can still be used to compare the degree of oxidation in a single substance with different oxidation stages.

5.6.1.3 Iodine Value

The iodine value (IV) refers to the number of grams of iodine absorbed by 100 g of lipids, and the measurement of this value utilizes the addition reaction of a double bond. Since the rate of the addition reaction of iodine directly with the double bond is very slow, the iodine is first converted to iodine bromide or iodine chloride, and then an addition reaction is carried out. The amount of iodine value represents the number of double bonds of lipids, and the decrease of iodine indicates the occurrence of lipid oxidations.

The excessive IBr will turn into I_2 in the presence of KI and then titrated with $Na_2S_2O_3$ solution to obtain the iodine value.

$$IBr + KI \longrightarrow I_2 + KBr$$
$$I_2 + 2Na_2S_2O_3 \longrightarrow 2NaI + 2Na_2S_4O_6$$

5.6.1.4 Active Oxygen Method

The active oxygen method (AOM) is designed to measure the time required for POV to reach 100 (plant oils) or 20 (animal fats) at 97.8 °C under continuously injecting air with the rate of 2.33 mL/s. This method can be used to compare the antioxidant properties of different antioxidants, but it does not exactly correspond to the actual storage period of the lipids.

5.6.1.5 Schaal Method

The Schaal method is designed to periodically measure the change in POV value at 60 °C, determining the time required for the presence of oxidative rancidity of the lipids and the lipidic rancid time in sensory evaluation.

5.6.1.6 Carbonyl Group Value

The total amount of carbonyl compounds (aldehydes, ketones) produced in the decomposition of ROOH is defined as the carbonyl group value (CGV). In China, the national standard test method of CGV is 2,4-dinitrophenylhydrazine colorimetric method, and the principle is shown as follows. The phenylhydrazine formed between carbonyl compounds and 2,4-dinitrophenylhydrazine will produce quinone ion with brownish-red or wine-red color under alkaline conditions and can be measured for its absorbance at 440 nm, and the CGV value should not exceed 50 meq/kg during the frying process of edible lipids.

5.6.2 Other Methods to Evaluate the Lipid Oxidations

5.6.2.1 Acid Value

The acid value (AV) refers to the number of milligrams of KOH required to neutralize the free fatty acids in 1 g lipids. This indicator not only measures the amount of free fatty acids in the lipids but also reflects the quality of the lipids. The AV of fresh lipids is low, and the AV of edible vegetable oils must not exceed 4 in provisions of Chinese food hygiene standards.

5.6.2.2 Saponify Value

The milligrams of KOH required for the complete saponification of 1 g of lipids are defined as the saponify value (SV). The number of the SV is inversely proportional to the average molecular mass of the lipids. The SV of the general lipids is around 200, and the high SV lipids have several good properties with low melting points and easy digestibility.

5.6.2.3 Diene Value

The Diels–Alder reaction can be operated between the butene dianhydride and the lipids, so the diene value (DV) refers to the number of grams of iodine converted to maleic anhydride in the 100 g lipids. This indicator indicates the number of conjugated double bonds in the unsaturated fatty acids. Natural fatty acids always contain non-conjugated double bonds, which can be converted to conjugated double bonds by this reaction.

The quality inspection of the used fried lipids, to check whether it has deteriorated, can be achieved by measuring the insoluble matter and the smoke point in petroleum ether. The fried lipids can be considered to have deteriorated, when the insoluble matter in the petroleum ether is greater than 0.7% and the smoking point is lower than 170 °C, or the insoluble matter in the petroleum ether is greater than 1.0%, regardless of whether the smoke point is changed.

5.7 The Chemistry in Lipid Processing

5.7.1 Lipid Refining

Raw oil could be extracted from oilseed crops and animal lipid tissues using pressing, cooking, machine separation, organic solvent extraction, and some other methods. It contains phospholipid, pigment, protein, free fatty acid, some other impurity and even some poisonous contents (e.g. aflatoxin in peanut oil, gossypol in cotton seed) in raw oil. The flavor, appearance, quality, and stability of raw oil are unsatisfying and the quality, flavor, and shelf-life of raw oil could be improved by the refining process.

5.7.1.1 Degumming

Degumming is to remove gum-soluble impurity in raw oil by physical, chemical, or physical–chemical methods. Edible oil would influence food flavor if oil contains too much phospholipid since it induces foam, smoke, odor, and oxidative browning during heating process. Degumming is based on the principle that phospholipid and some proteins are oil soluble but their hydrate could not soluble in oil. Aqueous phase is separated after water or steam are bubbled into raw oil during heating process, followed by standing and layering after agitation at 50 °C, then, phospholipid and protein can be removed from the oil.

5.7.1.2 Deacidification

Free fatty acid in raw oil is more than 0.5%, and in rice bran oil, even higher than 10%. The stability and flavor would be affected by FFA in oil. Deacidification (alkali refining) is to remove the FFA using alkaline neutralization. The volume of alkali addition can be determined by oil acid value. Fatty acid salt is produced in deacidification and can be removed after aqueous phase is separated. Gum and pigment could also be removed from oil in this process.

5.7.1.3 Bleaching

Some pigments like chlorophyll and carotenoid are included in raw oil. Chlorophyll is one of the photosensitizers and could affect the stability and appearance of raw oil. It can be removed by adsorbent, and this process of eliminating pigments is called bleaching. Some other adsorbents are frequently used like active carbon and white clay. Phospholipid, fatty acid salt, and some oxidant can be absorbed by adsorbent and then removed by filtration.

5.7.1.4 Deodorization

Some useless odorous substances, mainly resulting from lipid peroxide, are included in raw oil. Oil oxidation can be suppressed by reduced pressure distillation with citric acid added to chelate transition metal ions. The method can not only remove volatile odors but also degraded non-volatile odors into volatile odors and remove them by distillation.

Although the quality of oil is improved after the refining process, some side-effect also happens such as the loss of vitamin A, vitamin E, and carotenoid, which are natural antioxidant.

5.7.2 Lipid Hydrogenation

Due to the limited resource of solid lipid, liquid oil could be transferred to solid lipid or semi-solid lipid by modification. The lipid hydrogenation is an addition reaction between double bond in unsaturated fatty acid with the catalyze of Ni and Pt. The melting point and lipid stability would increase and the color becomes lighter after this process. For example, the odor in raw fish oil disappears after hydrogenation.

Some hydrogenation products could be used as shortening and margarine in food industry. Partially hydrogenated oil could be produced under the condition of 151.99–253.3 kPa, 125–190 °C, with the aid of Ni.

5.7.2.1 Mechanism of Lipid Hydrogenation

The most common catalyst in hydrogenation is Ni, the catalyze effect of Pt is more effective than Ni, but the use of Pt in the hydrogenation process is limited due to the high price. Cu is highly efficient on linolenic acid hydrogenation in soybean oil, but it also has disadvantages such as the toxicity and the hardness to remove. Otherwise, the Ni catalyst can be recycled 50 times if there is no sulfide or the oil has been refined.

The mechanism of hydrogenation is shown in Fig. 5.20. Briefly, both liquid oil and hydrogen are absorbed by the solid catalyst. First, either end of double bond in oil combined with metal and carbon–metal compound (a) is produced and then (a) interaction with hydrogen atoms that are absorbed on the solid catalyst, and unstable products (b/c) are generated. Their only one olefinic bond carbon connected with

Fig. 5.20 Hydrogenation of oil (✳ means metal bond)

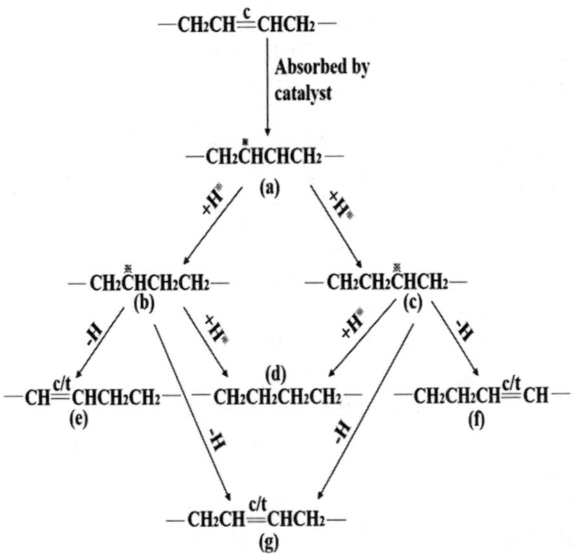

catalyst so it can spin freely. (b) and (c) then can accept another hydrogen atom to become a saturation product (d) or loss a hydrogen atom and produce a double bond again. The new double bond can be created in the original site or shift to both sides. Thus, the products of (g), (e), and (f) are produced respectively. All these products have *cis*- and *trans*-isomerization so the products of hydrogenation contain *trans*-fatty acid.

5.7.2.2   Selectivity of Hydrogenation

The products of hydrogenation are complex. The more double bond the oil contains the more kinds of products it will create. Triene can be catalyzed to diene and then becomes single olefine until the bond becomes totally saturated.

$$\text{Triene} \xrightarrow{K_3} \text{Diene} \xrightarrow{K_2} \text{Single olefine} \xrightarrow{K_1} \text{Saturation}$$

Take α-linolenic acid as an example, seven kinds of product could obtain:

The selectivity of products can be controlled by the ratio of two hydrogenation step reaction rate:

$$S_{32} = \frac{K_3}{K_2}; \quad S_{21} = \frac{K_2}{K_1}; \quad S_{31} = \frac{K_3}{K_1}$$

The higher of S_{32}, the rate of triene to become diene is faster. For example:

$$\text{Linolenic} \xrightarrow{K_3} \text{Linoleic acid} \xrightarrow{K_2} \text{Oleic acid} \xrightarrow{K_1} \text{Stearic acid}$$

$$S_{21} = K_2/K_1 = 0.159/0.013 = 12.2$$

This indicates the hydrogenate rate of linoleic acid is 12.2 times faster than oleic acid. The K values are related to reaction condition and catalyst. The products could be improved by reaction condition and catalyst optimization. For example, the saturate products could be avoided when Cu acts as a catalyst as Cu could not catalyze isolated double bond.

After hydrogenation, the stability is improved while the content of polyunsaturated fatty acid is decreased, and vitamin A and carotenoid are destroyed. Besides, *trans*-fatty acid is generated during hydrogenation. These are not beneficial to the lipid nutrition.

5.7.3   Interesterification of Lipids

The composition of fatty acid in natural oil determined its characteristics such as crystalline and melting point. These characteristics might limit the lipid application. The fatty acid composition can be changed by interesterification to change the physical property of oil to adapt the special needs. For example, the crystal particle of natural lard is coarse, with tough taste that goes against the consistency of the product. Besides, it is not good to use in pastry production. The modified lard can crystallize into fine particles. The melting point and viscosity decrease and the consistency are improved, thus suitable in margarine and candy production. Interesterification is a process that involves the rearranging of fatty acid in triacylglycerols. This process includes intramolecular and intermolecular rearrangement, resulting in random distribution. Interesterification includes chemical interesterification and enzymatic transesterification.

5.7.3.1   Chemical Interesterification

Sodium methoxide is commonly used to accelerate transesterification. This reaction can active at low temperatures (50–70°C) and does not need a long time.

1.   Mechanism of chemical interesterification

S_3, U_3 means tri-saturated and tri-unsaturated glycerides, respectively. The first reaction step is between sodium methoxide and triacylglycerols to produce diacylglycerol.

$$U_3 + NaOCH_3 \longrightarrow U_2ONa + U-CH_3$$

The transition complexation then reacts with other triacylglycerols and the fatty acid is transferred to the diacylglycerols.

$$S_3 + U_2ONa \rightleftharpoons SU_2 + S_2ONa$$

The produced S_2ONa then exchanges fatty acid with other triacylglycerols. Most reactions are conducted until an equilibrium has been reached.

2.   Random interesterification

Random interesterification reactions performing at temperatures above the melting point of the component with the highest melting point in a mixture result in complete randomization of fatty acids among all triacylglycerols according to the laws of probability. For example, interesterification between the equal amount of tri-saturate and glycerol trioleate is exhibited as follows (Fig. 5.21).

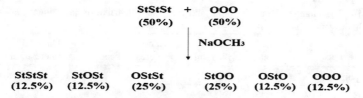

Fig. 5.21 Diagram of random interesterification

Random interesterification results in a significant change in the characteristic of crystallization and viscosity. Lard can be used as shortening in baked food after lard random interesterification.

3.   Directed interesterification

If the reaction is carried out at temperatures below the melting point of component with the highest melting point, as the tri-saturate with high melting point is produced, it crystallizes and falls from the solution. The reaction equilibrium in the remaining liquid phase is pushed toward increased production of the crystallizing tri-saturate. Crystallization continues until all triacylglycerols capable of crystallizing have been eliminated from the reaction phase. After directed interesterification, mixed glyceride can generate high melting point product S_3 and low melting point product U_3 as shown in Fig. 5.22.

The directed interesterification can transfer liquid oil to shortening-like products without hydrogenation or saturated fat addition. The viscosity and melting point of liquid oil (cottonseed oil, peanut oil) are improved as it will contain saturated fatty acid. Table 5.10 shows the change of melting point and fatty acid distribution after interesterification of partially hydrogenated palm oil.

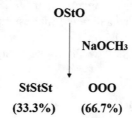

Fig. 5.22 Diagram of directed interesterification

Fig. 5.23 Phospholipid bilayer

Table 5.10 The change of melting point and fatty acid distribution after interesterification in partially hydrogenation palm oil

	Before interesterification	Random interesterification	Directed interesterification
Melting point/°C	41	47	52
Mole percentages (%) of triacylglycerol			
S_3	7	13	32
S_2U^a	49	38	13
SU_2	38	37	31
U_3	6	12	24

[a]S and U stand for saturated fatty acid and unsaturated fatty acid, respectively

5.7.3.2 Enzymatic Interesterification

Since enzymatic reaction is specific and calls for milder processing conditions with less side-products produced, this advantage indicates enzymatic interesterification would be an ideally method of directed interesterification. Besides, the cost of production could be even reduced if enzyme immobilization technology is applied to recycle the enzyme.

The use of lipases as catalyst for lipid interesterification is called enzyme interesterification. Lipase is a kind of hydrolase coming from bacteria, yeast, and fungus. Long-chain triacylglycerol can be hydrolyzed when there is enough water, while ester can be synthesized when there is limited water.

$$\text{Triacylglycerol} \xrightleftharpoons[\text{Lipases, Organic solvent}]{\text{Lipases, H2O}} \text{Fatty acid} + \text{Glycerol}$$

Enzymatic interesterification is random interesterification when the lipases are non-specific, while it can also be a directed interesterification when the lipases are specific. For example, taking Sn-1,3 lipases as the catalyst, the interesterification can only happen in Sn-1,3 and the Sn-2 would not be changed.

$$B\left[\begin{array}{c}A\\ \\A\end{array}\right.+C \longrightarrow B\left[\begin{array}{c}A\\ \\C\end{array}\right. + B\left[\begin{array}{c}C\\ \\C\end{array}\right. + B\left[\begin{array}{c}A\\ \\A\end{array}\right.+C+A$$

The large amount of Sn-POP is included in palm oil. When stearic acid or stearyl glycerin is added and Sn-1,3 lipases are used, Sn-POSt and Sn-StOST are obtained that are the main contents of cocoa butter. This is the main way to synthesize cocoa butter.

$$O-\begin{bmatrix} P \\ P \end{bmatrix} + St \longrightarrow O-\begin{bmatrix} P \\ St \end{bmatrix} + O-\begin{bmatrix} St \\ St \end{bmatrix} + O-\begin{bmatrix} P \\ P \end{bmatrix} + \ldots\ldots$$

Enzymatic interesterification is a promising method to produce some common products such as shortening and margarines, although its cost is relatively expensive. It is impossible to replace the chemical interesterification totally at present. The specificity of enzymatic interesterification permits the production of high-value lipid products, which is not possible by chemical means, such as cocoa butter substitutes, candy oil, and so on. The future direction of enzymatic interesterification is still the development of low-caloric products. Besides, new lipases discovery, especially Sn-2 lipases would be also a promising direction.

5.7.4 Oil Fractionation

Different triacylglycerols are separated by using stepwise crystallization at a certain temperature, which bases on their difference in the melting point of various triacylglycerols in oils and solubility in different solvents, and this processing is called oil fractionation. The fractionation method includes dry fractionation, solvent fractionation, and surfactant fractionation.

The dry fractionation means that triacylglycerol with higher melting point is selectively precipitated and separated by filtration during the slowly cooling of molten oils in the absence of an organic solvent. The crystallization of high-melting point triacylglycerol in the oils at 5.5°C is called 'winterization', which can be used to produce margarine. The wax crystals in oils were precipitated at 10 °C, which is called 'dewaxing'. Rapeseed oil, cottonseed oil, and sunflower oil will not be turbid after dealing with winterization and dewaxing during the refrigeration process.

The solvent fractionation means that cooling and crystallization after the addition of an organic solvent to the oils. The organic solvent can facilitate the formation of stable crystals, which improves the separation efficiency and in favor of the fractionation of fatty acids with long carbon chain. Commonly used organic solvents are n-hexane, acetone, 2-nitropropane, and so on, but this method is not commonly used due to its relative expensiveness.

Surfactant fractionation is to follow the above two methods, during the cooling and crystallization process, followed by the addition of surfactant (such as sodium dodecyl sulfonate) to dabble the crystals and subsequently suspending in the aqueous phase to separate the crystals. After separating the aqueous phase containing the crystals, the crystals are melted by heating and separated from the water layer and then centrifuged to obtain a liquid triacylglycerol. This method shows high efficiency, yielding triacylglycerol with high quality, suitable for large-scale continuous production.

5.8 Compound Lipids and Lipid Derivatives

This section highlights lecithin and cholesterol, which are very important in foods. Fat-soluble vitamins will be described in Chap. 6 of this book.

5.8.1 Phospholipid

Phospholipids are lipids containing phosphoric acid, mainly including phosphoglycerides and sphingomyelins. Glycerol acts as a skeleton of glycerol phospholipids, where hydroxyl groups at C^1 and C^2 positions form an ester with two fatty acids, and the hydroxyl group at C^3 position forms an ester with phosphoric acid that is called phosphatidyl acid. The phosphate group in the phosphatidic acid can be further esterified with other alcohols, followed by forming a variety of phospholipids, such as lecithin (phosphatidylcholine), cephalin (phosphatidylethanolamine), phosphatidylserine, and phosphatidylinositol.

$$
\begin{array}{c}
\overset{\overset{O}{\|}}{CH_2OC-R_1} \\
R_2COO-CH \qquad O^- \\
CH_2-O-\overset{\overset{}{|}}{\underset{\underset{O}{\|}}{P}}-OR_3
\end{array}
$$

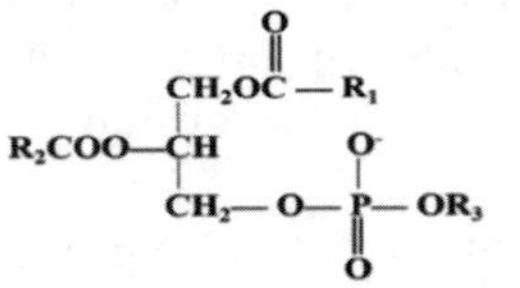

R_1, R_2 is representative for fatty acid. R_1 is usually representative for saturated fatty acids; and R_2, unsaturated fatty acids.
When:

$R_3 = $ —H, is phosphatidyl acid
$R_3 = $ —$CH_2CH_2N^+(CH_3)_3$, is phosphatidyl choline (PC)
$R_3 = $ —$CH_2CH_2NH_2$, is phosphatidyl ethanolamine (PE)
$R_3 = $ —$CH_2CH(NH_2)$ N^+COOH, is phosphatidyl serine (PS)
$R_3 = $ — glycerol, is phosphatidyl glycerol (PG)

$R_3 = $ (inositol ring structure with OH, OH, OH, HO, OH groups) , is phosphatidyl inositol (PI)

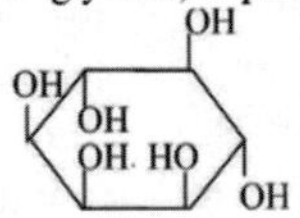

Biological membrane is composed of phospholipids, and PC, PE, and sphingomyelin, and phospholipids are important constituents. Phospholipid is amphiphilic due to a non-polar group of a long hydrocarbon chain and a polar group, phosphatidyl. When they are in polar medium, such as water, the tail of fatty acid hydrocarbon chain (non-polar side) has a strong tendency to evade aqueous phase and gather together, followed by forming a hydrophobic inner layer. While the hydrophilic polar terminus faces the bilateral surface to form a phospholipid bilayer (Fig. 5.23), which is the basic skeleton of the biological membrane. Due to the tight and orderly arrangement of phospholipid bilayer, biological membrane can maintain the relative stability of the membrane microenvironment and prevent various nutrients such as water and inorganic salts from being lost, which creates suitable conditions for the survival and physiological activities of various organisms. The amphiphilic phospholipid is soluble in a non-polar solvent containing a small amount of water, and thus it can be extracted from biological tissues with a chloroform–methanol (2:1) solvent. The phospholipids can be separated from other lipids due to insolubility of acetone. The solubility of different phospholipids in organic solvents is also different, and this difference can be used to further separate PC, PE, and so on. For example, PC is soluble in ethanol, and PE is insoluble. Phospholipids and proteins are present in the bound state in tissues. For example, the seeds of oilseed crop have high levels of phospholipids, which are one of the energy sources when seeds are germinated.

In the food industry, lecithin is an important type of phospholipid and of high content in egg yolk (8% ~ 10%), which is also one of the most widely distributed in the biosphere, such as plant seeds, animal eggs, and nerve tissues. The commercial lecithin obtained from soybean is usually called 'raw lecithin', which is also a lipid mixture containing three phosphatidic acid derivatives. The composition of the raw lecithin obtained from soybean is shown in Table 5.11.

The phosphatidylcholine (PC) purified from lecithin is an important component of biological membrane, which also has many functions in life phenomena. For example, it participates in the metabolism of fat in the body, lowers the blood cholesterol level, prevents atherosclerosis and fatty liver, and strengthens the functions of brain and memorization.

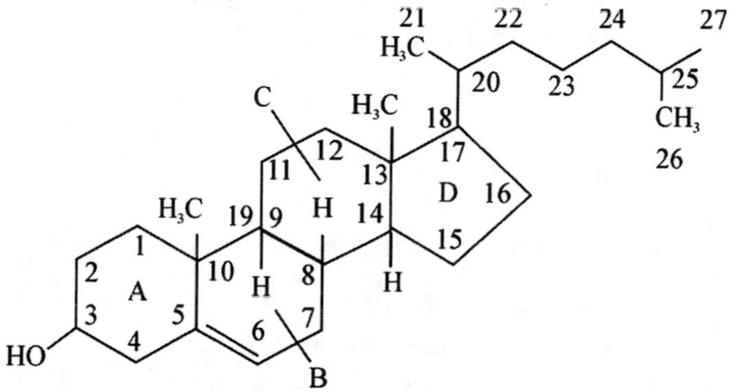

Fig. 5.24 Structure of cholesterol

Table 5.11 Approximate composition of commercial soy lecithin

Composition	Mass fraction (%)	Composition	Mass fraction (%)
PC	20	Other phospholipids	5
PE	15	Triacylglycerol	35
PI	20	Carbohydrates, alcohols, glycol esters	5

Under the effect of lecithin cholesterol acyltransferase, lecithin transfers the unsaturated fatty acids in lecithin to the hydroxyl position of cholesterol (Fig. 5.24). Esterified cholesterol is not easily deposited on the blood vessel wall, so lecithin has the effect of softening blood vessels and preventing coronary heart disease.

Acetylcholine is a compound necessary for the nervous system to transmit information. Human memory loss is related to acetylcholine deficiency. The human brain can directly take lecithin from the blood and quickly convert it to acetylcholine.

Lecithin is an important emulsifier in the food industry because of its amphiphilic nature. Pure lecithin is a W/O emulsifier with an HLB value of 3. Since lecithin contains unsaturated fatty acids, lecithin is easily oxidized and can also be used as an antioxidant in the food industry.

5.8.2 Sterol

There is a large class of substances in the organism that are based on cyclopentane polyhydrophenyl, known as sterols, which are unsaponifiable in lipids. There is a methyl group between the AB rings and between the CD rings, which is called angular methyl group, and cyclopentane polyhydrophenanthrene with angular methyl group is called 'steroid'. Figure 5.24 shows the structure of cholesterol and is the most important animal sterol. Vertebrates can synthesize cholesterol in the body, while most invertebrates lack enzymes that synthesize sterols and must take in sterols from food.

5.8.2.1 Cholesterol

As early as the eighteenth century, cholesterol was found in gallstones, the chemist Bencher named the substance with lipid properties as cholesterol in 1816. Cholesterol is an indispensable substance in animal tissue cells. It is not only involved in the formation of cell membranes but also a raw material for the synthesis of bile acids, vitamin D, and sex hormones. Bile acids can emulsify fat and play an important role in the digestion and absorption of fat. A small amount of cholesterol is essential for human health and low cholesterol is highly correlated with high mortality caused by non-vascular sclerosis in malnourished people; however, excessive cholesterol is deposited as gallstones in the biliary tract, and deposition on the walls of the blood vessels causes atherosclerosis. Cholesterol is widely present in animals, especially in the brain and nerve tissue, and is also high in the kidney, spleen, skin, liver, and bile. Cholesterol needs to be combined with lipoproteins to be transported to various parts of the body. There are two types of lipoproteins that transport cholesterol, low-density lipoprotein (LDL), and high-density lipoprotein (HDL). Low-density lipoprotein cholesterol (LDL-C) has a strong adhesion and can adhere to the blood vessel wall. It is the main culprit in vascular embolism and it is 'bad' cholesterol. High-density lipoprotein cholesterol (HDL-C) is 'benign' cholesterol by transporting

Table 5.12 Cholesterol content in some foods (mg/100 g)

Food	Calf brain	Yolk	Pig kidney	Pork liver	Butter	Lean pork	Lean beef	Fish (flounder)
Content	2000	1010	410	340	240	70	60	50

'bad' cholesterol from the blood vessels back to the liver and avoiding vascular occlusion.

The currently accepted theory leading to the development of arteriosclerosis is 'vascular injury and cholesterol oxidation modification', that is, the cause of atherosclerosis is mainly vascular endothelial cell injury: ① causing the barrier and permeability of the vascular endothelium to change when LDL-C penetrates into the arterial endothelium. Due to the microporous filtration of vascular endothelial cells, many endogenous natural antioxidants are blocked, and LDL-C is no longer protected by antioxidants in plasma or intercellular fluid after leaving the blood. At this time, induced by drugs, hypertension, or diabetes, endothelial cells and smooth muscle cells will produce a large number of oxygen free radicals, which will cause oxidative modification of LDL-C under the endothelium; ② Interfering with the antithrombotic properties of endothelial cells; ③ Affecting the release of vasoactive substances by endothelial cells. All these changes lead to the development of arteriosclerosis.

Cholesterol is insoluble in dilute acid and dilute alkali, cannot be saponified, and is hardly destroyed in food processing. About 2/3 of the cholesterol in adults is synthesized in the liver, and about 1/3 is derived from food. High levels of serum cholesterol are risky factors for cardiovascular disease, so it is necessary to limit the intake of high cholesterol foods in the diet. The cholesterol content of some foods is shown in Table 5.12.

5.8.2.2 Phytosterol

Phytosterols are widely found in rice bran oil, soybean oil, rapeseed oil, and corn oil. Generally, the content of sterol in the seeds of oil corps is higher than 1%, and the sterols contained in each plant are often a mixture of several sterols, where stigmasterol and sitosterol are most abundant and widely distributed in plants, while ergosterol is mainly found in the fungi and yeast, which can be converted into vitamin D under ultraviolet light. Phytosterols are characterized by one more alkyl group at C^{24} compared with cholesterols. Since sterols are unsaponifiable, when the lipid is refined and deacidified by alkali, most of the sterol remains adsorbed by the soap stock. Therefore, the alkali-smelting soap of the lipid and the deodorized distillate of the lipid can be used as a raw material for extracting phytosterols. The pure phytosterols are obtained by solvent extraction, concentration, crystallization, and subsequent purification.

Ergosterol   Sitosterol   Stigmaesterol

As phytosterol has the functions of preventing and treating coronary atheroscle-rosis, preventing thrombosis, treating ulcerative skin diseases, and anti-inflammation, it has been widely used in cosmetics, edible oils, health-care products, and feed addi-tives in recent years. Stigmasterol can be used to make 'pregnancy hormones' and androstenone. The main physiological function of plant sterols is to regulate blood lipids. In the 1950s, it was recognized that the more phytosterols ingested from the diet, the lower the rate of cholesterol absorption. Studies have found that adding a fat-soluble alkanol ester to mayonnaise or margarine can reduce serum total cholesterol levels by 10–15%. In 1995, Finland took the lead in the production of Benecol, a plant steranol ester product. This steranol ester can be hydrolyzed by enzymes in the small intestine and release free sterols with physiological activity, which can significantly reduce the serum cholesterol level of patients with heart disease and even reduce to the normal level in one third of the people. Studies have also shown that sterol esters can reduce low-density lipoprotein cholesterol by 14% but have no effect on high-density lipoprotein cholesterol levels that are beneficial to human health, so it is safe and effective. The main theories about the mechanism of action of phytosterols to lower cholesterol are as follows: ① Phytosterol precipitates cholesterol in the small intestine so that it does not dissolve and is therefore not absorbed. ② Cholesterol can be dissolved in the micro-micelle of bile acids in the small intestine, which is a necessary condition for absorption, and phytosterol can replace cholesterol from it and hinder the absorption of cholesterol. ③ Phytosterol competes with cholesterol in the small intestine chorion for the absorption site, thereby reducing the absorption rate of cholesterol.

5.9   Summary

Lipids are a major component of foods, which contain a wide range of bioactive compounds that provide necessary nutritional ingredients and beneficial functions for human bodies, such as oils, phospholipids, glycolipids, sphingolipids, steroids, waxes, etc. This chapter introduces oil-related knowledge as well as related content of phospholipids, sterols, and oil substitutes. The types, basic compositions, structures, and naming methods of lipids were mainly introduced. The main physical properties of oils and fats, the main chemical reactions, refining, and deep processing techniques were also discussed.

Questions

1. Why does hoarfrost appear on chocolate? How to prevent chocolate from hoarfrost?
2. Why do water and fat in milk not stratify?
3. What is the relationship between fat oxidation and water activity? Is it right that 'the more oxidative the oil, the higher the POV value'?
4. What is the antioxidant principle of phenolic antioxidants? How about the antioxidant dosage? Is it better with the dosage as much as possible?
5. How to make margarine from vegetable oil?
6. Briefly describe the mechanism and influencing factors of lipid oxidation.
7. What are the mechanisms for lipid antioxidants?
8. Briefly describe the homopolycrystalline of oils and fats and their application in the food industry.
9. What are the changes in fats and oils during high temperature heating and the impact on food quality?
10. How is *trans*-fat produced and how safe is it?
11. In daily life, why are fat-containing foods difficult to preserve and prone to deterioration?
12. Why do fried foods easily lead to obesity?
13. Terminology: neutral fat, phospholipids, lipid derivatives, solid fat index, emulsifier, homopolycrystalline, auto-oxidation of fats, hydrogenation of oils, transesterification of fats and oils, fat fractionation, photooxidation of oils, enzymatic oxidation of oils, antioxidants.

Bibliography

1. Ali, A.M.M., Bavisetty, S.C.B., Prodpran, T., Benjakul, S.: Squalene from fish livers extracted by ultrasound-assisted direct in situ saponification: purification and molecular characteristics. J. Am. Oil Chem. Soc. **96**(9), 1059–1071 (2019)
2. Ballesteros, L.F., Ramirez, M.J., Orrego, C.E., Teixeira, J.A., Mussatto, S.I.: Encapsulation of antioxidant phenolic compounds extracted from spent coffee grounds by freeze-drying and spray-drying using different coating materials. Food Chem. **237**, 623–631 (2017)
3. Belitz, H.D., Grosch, W., Schieberle, P.: Food Chemistry. Springer-Verlag, Berlin, Heidelberg (2009)
4. Berton-Carabin, C.C., Ropers, M.-H., Genot, C.: Lipid oxidation in oil-in-water emulsions: involvement of the interfacial layer. Compr. Rev. Food Sci. Food Saf. **13**(5), 945–977 (2014)
5. Damodaran, S., Parkin, K.L., Fennema, O.R.: Fennema's Food Chemistry. CRC Press/Taylor & Francis, Pieter Walstra (2008)
6. de Souza, R.J., Mente, A., Maroleanu, A., Cozma, A.I., Ha, V., Kishibe, T., Uleryk, E., Budylowski, P., Schuenemann, H., Beyene, J., Anand, S.S.: Intake of saturated and *trans* unsaturated fatty acids and risk of all-cause mortality, cardiovascular disease, and type 2 diabetes: systematic review and meta-analysis of observational studies. BMJ-Br. Med. J. **351** (2015)

7. Dominguez, R., Pateiro, M., Gagaoua, M., Barba, F.J., Zhang, W., Lorenzo, J.M.: A comprehensive review on lipid oxidation in meat and meat products. Antioxidants **9**(10) (2019)
8. Forouhi, N.G., Koulman, A., Sharp, S.J., Imamura, F., Kroger, J., Schulze, M.B., Crowe, F.L., Huerta, J.M., Guevara, M., Beulens, J.W.J., van Woudenbergh, G.J., Wang, L., Summerhill, K., Griffin, J.L., Feskens, E.J.M., Amiano, P., Boeing, H., Clavel-Chapelon, F., Dartois, L., Fagherazzi, G., Franks, P.W., Gonzalez, C., Jakobsen, M.U., Kaaks, R., Key, T.J., Khaw, K.-T., Kuhn, T., Mattiello, A., Nilsson, P.M., Overvad, K., Pala, V., Palli, D., Quiros, J.R., Rolandsson, O., Roswall, N., Sacerdote, C., Sanchez, M.-J., Slimani, N., Spijkerman, A.M.W., Tjonneland, A., Tormo, M.-J., Tumino, R., van der A.D.L., van der Schouw, Y.T., Langenberg, C., Riboli, E., Wareham, N.J.: Differences in the prospective association between individual plasma phospholipid saturated fatty acids and incident type 2 diabetes: the EPIC-InterAct case-cohort study. Lancet Diabetes Endocrinol. **2**(10), 810–818 (2014)
9. Geranpour, M., Assadpour, E., Jafari, S.M.: Recent advances in the spray drying encapsulation of essential fatty acids and functional oils. Trends Food Sci. Technol. **102**, 71–90 (2020)
10. Golding, M., Wooster, T.J.: The influence of emulsion structure and stability on lipid digestion. Curr. Opin. Colloid Interface Sci. **15**(1–2), 90–101 (2010)
11. Hallahan, B., Ryan, T., Hibbeln, J.R., Murray, I.T., Glynn, S., Ramsden, C.E., SanGiovanni, J.P., Davis, J.M.: Efficacy of omega-3 highly unsaturated fatty acids in the treatment of depression. Br. J. Psychiatry **209**(3), 192–201 (2016)
12. He, W.-S., Zhu, H., Chen, Z.-Y.: Plant sterols: chemical and enzymatic structural modifications and effects on their cholesterol-lowering activity. J. Agric. Food Chem. **66**(12), 3047–3062 (2018)
13. Kalaycioglu, Z., Erim, F.B.: Total phenolic contents, antioxidant activities, and bioactive ingredients of juices from pomegranate cultivars worldwide. Food Chem. **221**, 496–507 (2017)
14. Kan, J.: Food Chemistry. China Agricultural University Press, Beijing (2016)
15. Luo, J., Jiang, L.-Y., Yang, H., Song, B.-L.: Intracellular cholesterol transport by sterol transfer proteins at membrane contact sites. Trends Biochem. Sci. **44**(3), 273–292 (2019)
16. Mariutti, L.R.B., Bragagnolo, N.: Influence of salt on lipid oxidation in meat and seafood products: a review. Food Res. Int. **94**, 90–100 (2017)
17. McClements, D.J.: Enhanced delivery of lipophilic bioactives using emulsions: a review of major factors affecting vitamin, nutraceutical, and lipid bioaccessibility. Food Funct. **9**(1), 22–41 (2018)
18. Nieva-Echevarria, B., Goicoechea, E., Guillen, M.D.: Food lipid oxidation under gastrointestinal digestion conditions: a review. Crit. Rev. Food Sci. Nutr. **60**(3), 461–478 (2020)
19. Papuc, C., Goran, G.V., Predescu, C.N., Nicorescu, V., Stefan, G.: Plant polyphenols as antioxidant and antibacterial agents for shelf-life extension of meat and meat products: classification, structures, sources, and action mechanisms. Compr. Rev. Food Sci. Food Saf. **16**(6), 1243–1268 (2017)
20. Shahidi, F., Zhong, Y.: Lipid oxidation and improving the oxidative stability. Chem. Soc. Rev. **39**(11), 4067–4079 (2010)
21. Sun, M., Liu, S., Zhang, Y., Liu, M., Yi, X., Hu, J.: Insights into the saponification process of di(2-ethylhexyl) phosphoric acid extractant: thermodynamics and structural aspects. J. Mol. Liq. **280**, 252–258 (2019)
22. Waraho, T., McClements, D.J., Decker, E.A.: Mechanisms of lipid oxidation in food dispersions. Trends Food Sci. Technol. **22**(1), 3–13 (2011)
23. Xie, B.: Food Chemistry. China Science Press, Beijing (2011)

Dr. Hui He works in the Department of Food Chemistry and is the Deputy Director of Key Laboratory of Environment Correlative Dietology at Huazhong Agricultural University (HZAU) in China. She graduated from the Department of Chemistry, Wuhan University, in 1982, and later taught at Wuhan University of Technology and Huazhong Agricultural University. Professor He has long been engaged in the teaching of food chemistry, spectral analysis, and other courses and focuses on the researches of food-based bioactive peptides. Dr. He edited the national planning textbook "Food Analysis" and presided over the National Sharing Course of high-quality resources "Food Chemistry and Analysis". Dr. He has presided over three National Natural Science Foundation projects and one National 863 project, and has published more than 70 academic papers in important academic journals. She and her team mostly study bioactive peptides such as Ganoderma lucidum peptide, corn peptide, soybean peptide, egg white peptide, and chickpea peptide. In recent years, the activities and mechanisms of corn peptides in ethanol metabolism and liver-protecting have been systematically studied. Currently, Dr. He focuses on researches about desalted duck egg white peptides (DPs), the processing byproduct hydrolysates of salted eggs, on calcium bioavailability, and structure-activity relationship.

Dr. Tao Hou works at College of Food Science and Technology in Huazhong Agricultural University (HZAU), China. He obtained his Ph.D. degree from HZAU in 2017, and academically visited Cornell University from 2016 to 2017. Dr. Tao Hou has been engaged in the teaching of food chemistry and food analysis, candy technology and focuses on the researches of food-based bioactive peptides, trace element, and molecular nutrition. He has presided on one National Natural Science Foundation project and one Fundamental Research Funds for the Central Universities. He has published more than 20 academic papers in Comprehensive Reviews in Food Science and Food Safety, Food Chemistry, Journal of Agricultural and Food Chemistry, Journal of Functional Foods, Nutrients, etc. Currently, Dr. Tao Hou focuses on researches about promoting calcium absorption and bone formation peptides, and selenium nutrition, such as desalted duck egg white peptides (DPs) and casein phosphopeptides.

Chapter 6
Vitamins

Yali Yang

Abstract In this chapter, it has been introduced the types of vitamins and their main functions in the body. It is very important for food processors to be familiar with the physical and chemical changes of vitamins during food processing and storage, as well as the impact on food quality. It has been illustrated the content and distribution of important vitamins such as vitamin A (retinol), vitamin D, vitamin E (tocopherol), vitamin B_1 (thiamine), vitamin B_2 (riboflavin), vitamin PP (niacin), vitamin C (ascorbic acid), and their derivatives, etc. in food, as well as their transformations due to the environmental changes in food.

Keywords Fat-soluble vitamins · Water-soluble vitamins · Provitamins · Processing · Bioactive

6.1 Overview

Vitamins are the nutrient required in small amounts for normal body function. With few exceptions, vitamins are not made in the body and must be supplied from an outside source to maintain normal physiological functions. A vitamin must be an organic chemical; this is, every vitamin has at least one carbon atom in its molecular structure and plays an important role in human growth, metabolism and development. Vitamins do not participate in the formation of human cells, nor do they provide energy to the body.

Although the chemical structures and characteristics of vitamins are different, they have something in common as following: ① Vitamins are present in food in the form of provitamins. ② Vitamins are not the composition of tissues and cells, nor do they provide the energy of body. They play an important role in the metabolism regulation. ③ The majority of vitamins are not synthesized to meet the needs of body and need to be obtained from foods. ④ There is a specific set of symptoms or a specific disease associated with a deficiency of each vitamin, and it can be corrected

Y. Yang (✉)
College of Food Engineering and Nutritional Science, Shaanxi Normal University, Xian 710062, China
e-mail: yangyali@snnu.edu.cn

© The Author(s), under exclusive license to Springer Nature Singapore Pte Ltd. 2021
J. Kan and K. Chen (eds.), *Essentials of Food Chemistry*,
https://doi.org/10.1007/978-981-16-0610-6_6

255

by taking the appropriated amount of the vitamin. In addition, vitamins are essential organic compounds in the metabolism. There are a series of biochemical reactions of body which are closely related to the catalysis of enzymes. The coenzyme must be involved in the reactions to retain the active of enzyme, and many vitamins are the components of coenzyme and enzyme. Moreover, the best vitamins are found in human tissues and in the form of "bioactive substances". Therefore, vitamins are the important substance to maintain and regulate the normal metabolism of body.

Vitamins are impossible to be classified by their chemical structure and physiological characteristic due to the complex chemical structure. They are generally classified into two categories based on their solubility characteristics: fat-soluble vitamins and water-soluble vitamins. Fat-soluble vitamins include vitamin A, vitamin D, vitamin E, and vitamin K; the water-soluble vitamins include thiamine (vitamin B_1), riboflavin (vitamin B_2), pantothenic acid (vitamin B_5), vitamin B_6, vitamin B_{12}, niacin (vitamin PP), vitamin C, folic acid and biotin. The water-soluble vitamins are instable and easy to be lost during food processing, while the stability of fat-soluble vitamins is higher. In addition, there are some vitamin analogs which are always called "other organic micronutrient", such as choline, carnitine, pyrroloquinoline, inositol, orotic acid and taurine.

6.2　Fat-Soluble Vitamins

6.2.1　Vitamin A

Vitamin A (all-*trans*-retinol, vitamin A alcohol) is the parent molecule of a family compounds with shared biological activities, and these compounds are essential for visual functions and for the maintenance of healthy epithelial tissues (skin, immune system organs, gastrointestinal tract, reproductive organs, lungs, and others). The nutritional term A includes both preformed vitamin A and provitamin A. Preformed vitamin A (retinol and its esters) occurs naturally in animal foods and is added in the preparation of some fortified foods and nutritional supplements. Provitamin A compounds (β-carotene, α-carotene, and β-cyptoxanthin), which are synthesized by plants as accessory photosynthetic pigments, can be metabolized by humans and animals to retinol. Vitamin A is a kind of bioactive and unsaturated hydrocarbon which includes vitamin A_1 (retinol) and its esters (ester, aldehyde, acid), and vitamin A_2. The structure of retinol includes a methyl-substituted cyclohexenyl (β-ionone) ring and a conjugated tetraene side chain ending in a terminal hydroxyl group (Fig. 6.1a), which can be converted into all the other essential forms of vitamin A. Vitamin A_2 (3, 4-didehydroretinol) is a minor variant found in some freshwater fishes (Fig. 6.1b). Its biological activity is qualitatively like that of retinol, but quantitatively about half as much.

Vitamin A is the component of rhodopsin in visual cells and related with dark vision. Lack of vitamin A in the human body can affect dark adaptation ability, such

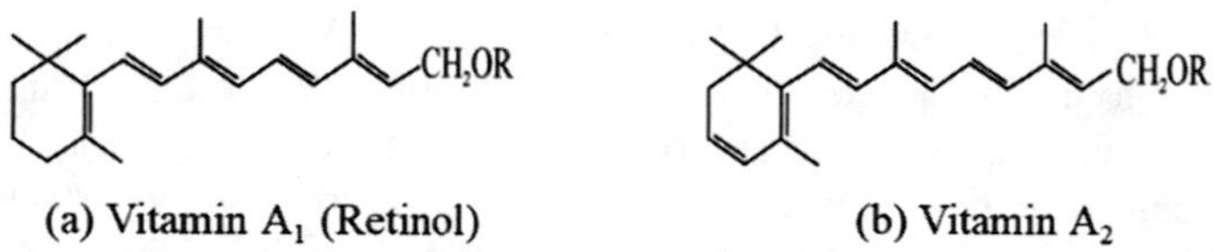

(a) Vitamin A₁ (Retinol)   (b) Vitamin A₂

Fig. 6.1 The chemical structure of vitamin A (R = H or $COCH_3$ Acetate or $CO(CH_2)_{14}CH_3$ Palmitate)

Table 6.1 The content of vitamin A and carotene in some foods

Foods	Vitamin A (mg/100 g)	Carotene (mg/100 g)
Beef	37	0.04
Butter	2363 ~ 3452	0.43 ~ 0.17
Cottage cheese	553 ~ 1078	0.07 ~ 0.11
Egg (cooked)	165 ~ 488	0.01 ~ 0.15
Herring (canned)	178	0.07
Milk	110 ~ 307	0.01 ~ 0.06
Tomato (canned)	0	0.5
Peach	0	0.34
Cabbage	0	0.10
Cauliflower (cooked)	0	2.5
Spinach (cooked)	0	6.0

as child dysplasia, dry eye, night blindness, etc.; however, excessive intake of vitamin A will cause the symptoms such as dry skin, desquamation, hair loss, etc.

The main structural unit of vitamin A is an unsaturated hydrocarbon of 20 carbons. Vitamin A has varieties of *cis* and *trans* stereoisomers because of the chemical structure with conjugated double bond. The retinol of food is mainly all-*trans* structure, and its biological potency is the highest. Dehydro-retinol, vitamin A₂, has a biological potency of 40% of vitamin A₁, while the 13-*cis* isomer is called new vitamin A, and its bio-potency is 75% of all-*trans* retinol. In the natural vitamin A, the content of new vitamin A is about 1/3, and lower in synthetic vitamin A. The content of vitamin A is usually expressed as retinol equivalent (RE), and 1RE = 1 μg of retinol.

Vitamin A is mainly found in animal tissues; vitamin A₁ occurs in animals and marine fish; vitamin A₂ is mainly found in freshwater fish and not present in terrestrial animals. Although vitamin A does not exist in vegetables, the carotenoids of vegetables can be converted to vitamin A by animals. For example, one molecule of β-carotene can be converted into two molecules of vitamin A, so the carotenoid is usually called provitamin A. The optimal ratio of vitamin A to provitamin A is 1:2 in the daily diet. The daily foods which is rich in vitamin A or provitamin A are: cod liver oil, animal liver, dairy products, eggs, carrots, spinach, pea sprouts, red sweet potatoes, green peppers, yellow green fruits and vegetables, etc. (Table 6.1).

Vitamin A is active and easily oxidized in the air or destroyed by ultraviolet radiation due to the unsaturated bonds. To maintain the physiological function, the preparation of vitamin A should be stored in a brown bottle away from light. Both vitamin A_1 and A_2 interacted with antimony trichloride could show dark blue, which can be used as the basis for the quantitative determination of vitamin A. Many plants such as carrot, tomato, green leafy vegetables, corn, etc. contain the carotenoids such as α, β, γ-carotene, cryptoxanthin, lutein and so on. Some carotenoids have the ring structure like vitamin A_1 and can be converted in vivo to vitamin A, so they are known as provitamin A. β-carotene has the highest conversion rate because of the two ring structures of vitamin A_1. One molecule of β-carotene and two molecules of water can produce two molecules of vitamin A_1. The oxidation process with water is catalyzed by β-carotene-15,15'-oxygenase and primarily occurs in the intestinal mucosa of animals.

Vitamin A in food produced by cleavage of β-carotene forms an ester with the fatty acids in the intestinal mucosal cells, and then incorporated into chylomicrons, entering the body through the lymphatic absorption. Vitamin A is mainly stored in the liver of animals and released into the blood when the body needs. In the blood, the retinol (R) is combined with the retinol binding protein (RBP) and serum prealbumin (PA) to produce the R-RBP-PA complex and is transported to all tissues throughout the body.

Vitamin A is relatively stable in the conditions of heating, alkaline and weak acid and unstable in inorganic strong acid. There are many changes of vitamin A and provitamin A in the lack of oxygen. Especially, β-carotene could be converted into a new β-carotene by *cis–trans* isomerization, reducing its nutritional value, and this reaction occurs during the cooking and filling processing of vegetables. The destruction of β-carotene by metallic copper ions is very strong, as is by iron, but to a lesser extent. Figure 6.2 summarizes some pathways of vitamin A destruction.

6.2.2 Vitamin D

Vitamin D is a group of steroid compounds that have the biological activity of cholecalciferol which possess antirachitic activity. The two most prominent members of this group are ergocalciferol (vitamin D_2) and cholecalciferol (vitamin D_3), and their chemical structures are shown in Fig. 6.3. Their chemical structures are very similar, vitamin D_2 just has one more methyl group and double bond than vitamin D_3. The ergosterol in plant food and yeast, is converted into ergocalciferol (vitamin D_2) after ultraviolet irradiation. Cholecalciferol is the form of vitamin D obtained when radiant energy from the sun strikes the skin and converts the precursor 7-dehydrocholesterol into vitamin D_3 in the skin of human and animal.

Vitamin D_3: three double bonds, melting point 84–85 °C, insoluble in H_2O and soluble in benzene, chloroform, ethanol and acetone;

Vitamin D_2: four double bonds, melting point 121 °C, same solubility as Vitamin D_3.

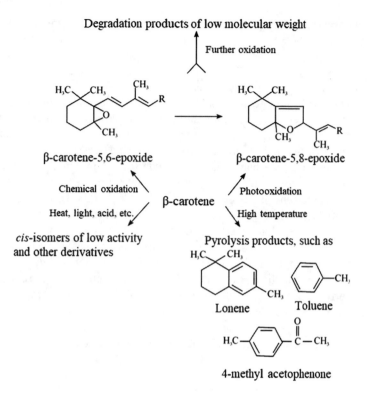

Fig. 6.2 The main pathways and products of degradation of vitamin A

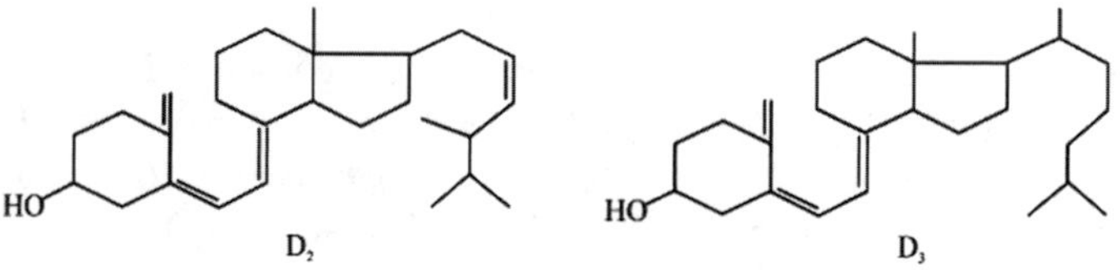

Fig. 6.3 Chemical structure of vitamin D₂ and vitamin D₃

Vitamin D₂ and D₃ are often present in the form of esters in nature. They are white crystals, soluble in fats and organic solvents, and their chemical characteristics are stable. Vitamin D is resistant to high temperature and oxidation in neutral and alkaline solutions. However, it is sensitive to light and is easily destroyed by ultraviolet radiation, so it needs to be stored in a sealed and opaque container. In addition, vitamin D is gradually broken down in the acidic solution. The rancidity of fat in food can also cause the destruction of vitamin D. Moreover, the storage and processing do not affect the physiological activity of vitamin D, but a few of toxic compounds are produced by the excessive ultraviolet radiation, which lead to the inactivity of anti-rickets.

The important physiological function of vitamin D is to regulate the metabolism of calcium and phosphorus in the body. It is also a new neuroendocrine-immunomodulatory hormone. In addition, it could maintain the normal concentration of amino acids in the blood and regulate the metabolism of citric acid, and has the function of preventing infant rickets and adult osteoporosis.

Vitamin D is widely found in animal foods. The cod liver oil has the higher content of vitamin D, while eggs, milk, butter, and cheese have the less content. In general, it is not easy to get enough vitamin D from ordinary food, and sunbathing is an important way for the body to synthesize vitamin D. Therefore, to supplement all vitamin D of body, the face, hands and arms need to be exposed in the sun directly for two–three times weekly, 15 min each time. Sunbathing promotes the body's synthesis of vitamin D as the sunshine contains solar ultraviolet B (UVB). But not everyone has the potential to effectively contact solar UVB. Inadequate sun exposure, obesity, insufficient intakes of vitamin D, gastrointestinal disorders, malabsorption, renal and liver diseases, and other various health conditions are contributing to vitamin D deficiency. At present, the prevalence of vitamin D deficiency is high in all sex and age groups in all around the world. Although vitamin D supplements are considered a simple and effective short-term solution for vitamin D deficiency treatment; however, it seems that at a population level, food fortification, using staple foods, is the best method to increase vitamin D. Flour and dairy products should be mandatory fortified with vitamin D using novel, practical, and cost-effective technologies, while oils and beverages could be optionally fortified with vitamin D.

6.2.3 Vitamin E

Vitamin E is designated as a group of all tocol and tocotrienol derivatives qualitatively exhibiting the biological activity of α-tocopherol (Fig. 6.4). Tocopherol could promote the secretion of sex hormones and improve fertility. All eight naturally occurring tocopherol compounds isolated from plant sources: 4 tocopherols and 4 tocotrienols. The structure of tocotrienols have a structure like that of tocopherols, except that the side chain contains three isolated double bonds at the 3′, 7′, 11′ positions. Higher content of tocopherol is found in daily food.

Vitamin E is widely found in animal and plant foods. There are α-, β- and γ-tocopherol in the cottonseed oil and δ-tocopherol in the soybean oil. α-tocopherol

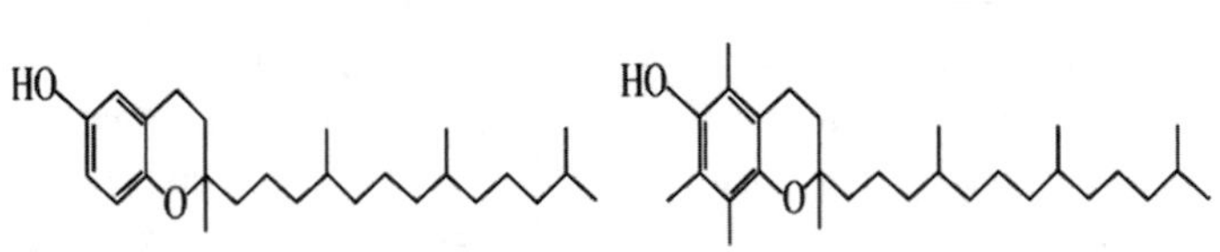

Tocol   α-tocopherol

Fig. 6.4 Structural formula of tocol and α-tocopherol

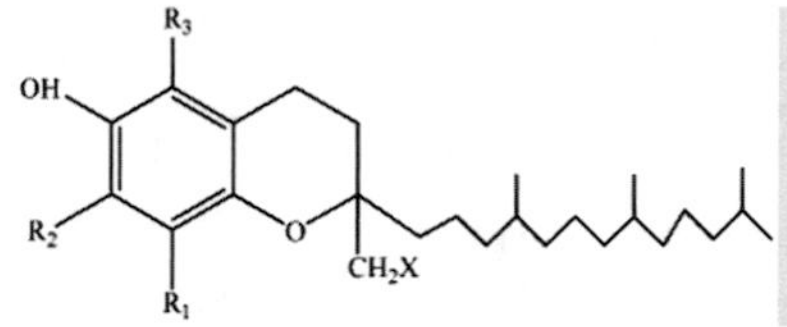

		Positions		
		R_1	R_2	R_3
	α-tocopherol	CH_3	CH_3	CH_3
	β-tocopherol	CH_3	H	CH_3
	γ-tocopherol	H	CH_3	CH_3
	δ-tocopherol	H	H	CH_3
	tocopherol	H	H	H

Fig. 6.5 Structure formula of chemical substitution pattern of tocopherol

is the most widely distributed, most abundant, and most active form of vitamin E in nature. The chemical structures of four common tocopherols found in nature are shown in Fig. 6.5. They all have the same physiological function, but α-tocopherol has the highest physiological activity.

Vitamin E is an important antioxidant which is widely used in foods, especially in animal and vegetable oils and its antioxidant capacity is in order of δ > γ > β > α. Conversely, the order of antioxidant capacity of tocopherol in vivo is α > β > γ > δ. The relative antioxidant activity of tocopherols in vitro is depending on the experimental conditions and the determination method. In addition, the antioxidant activities of tocopherol in vitro depend not only the chemical reaction toward hydroperoxyl and other free radicals, but also on other factors such as the tocopherol concentrations, temperature, light, the reaction substrate and so on, which may act as the pro-oxidants and synergists, in the experimental systems. The antioxidant process of vitamin E is shown in Fig. 6.6.

Vitamin E in food is stable to heat and acid, even when heated to 200 °C, it is not destroyed. The loss of vitamin E is not great during the cooking processing, but its activity obviously reduced when fried. Vitamin E is unstable to alkali and sensitive to oxidants and ultraviolet rays. In addition, the rancidity of oil could accelerate the destruction of vitamin E, and the metal ions such as Fe^{2+} can promote the oxidation of vitamin E. Moreover, vitamin E in dehydrated foods is easily oxidized.

Vitamin E in food is mainly absorbed in the upper part of the small intestine of animals, and is mainly carried by β-lipoprotein in the blood and transported to various tissues. Isotope tracer experiments showed that α-tocopherol could be oxidized into α-tocopherol quinone in tissues. Then, α-tocopherol quinone is reduced to α- tocopherol hydroquinone which can be combined with glucuronic acid in the liver and enter the intestine with bile and is discharged through the feces. The metabolism pathways of other tocopherols are like α-tocopherol. Vitamin E is essential for animal reproduction. In the absence of vitamin E, the testis of male mice is degenerated and cannot form the normal sperm; the female mice of embryo and placenta are atrophied and absorbed, causing miscarriage. The lack of vitamin E of animals may also cause muscle atrophy, anemia, brain softening and other neurodegenerative diseases. If accompanied by insufficient protein at the same time, it can cause acute cirrhosis. Although the metabolic mechanisms of these lesions have not been fully elucidated, the various functions of vitamin E may be related to its antioxidation. In the lack of vitamin E, the symptoms of some diseases in the human body are like the symptoms in animals. As the content of vitamin E in general foods is enough and easy

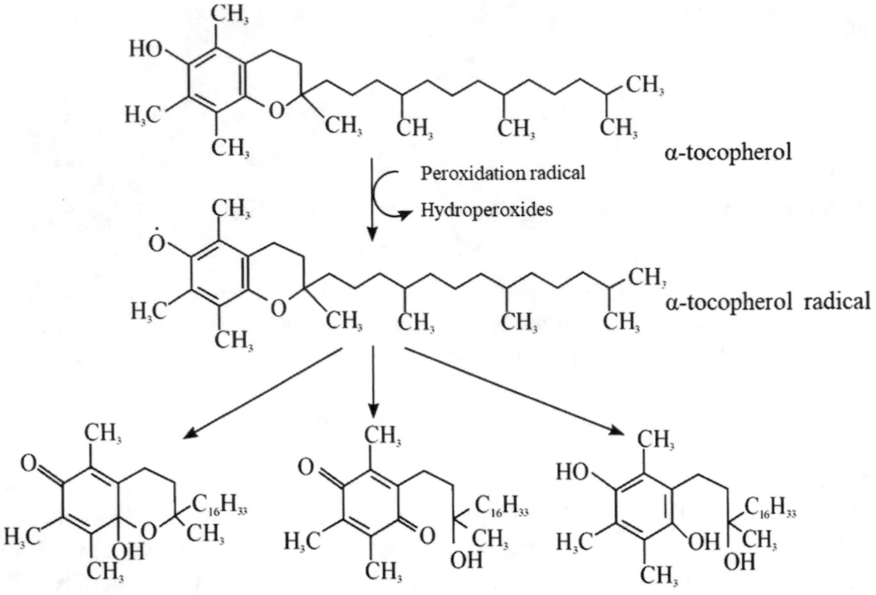

Fig. 6.6 Oxidative degradation pathway of vitamin E

Table 6.2 The content of vitamin E in common foods	Food name	Content (mg/100 g)	Food name	Content (mg/100 g)
	Cottonseed oil	90	Bovine liver	1.4
	Corn oil	87	Carrot	0.45
	Peanut oil	22	Tomato	0.40
	Sweet potato	4.0	Apple	0.31
	Fresh cream	2.4	Chicken	0.25
	Beans	2.1	Banana	0.22

to be absorbed, so it is difficult to cause the deficiency of vitamin E, only when the intestinal absorption of lipids is incomplete.

Vitamin E is widely distributed in nature and mainly exists in plant foods. Vitamin E is abundant in fats and oils, leafy plants and avocados, as is shown in Table 6.2.

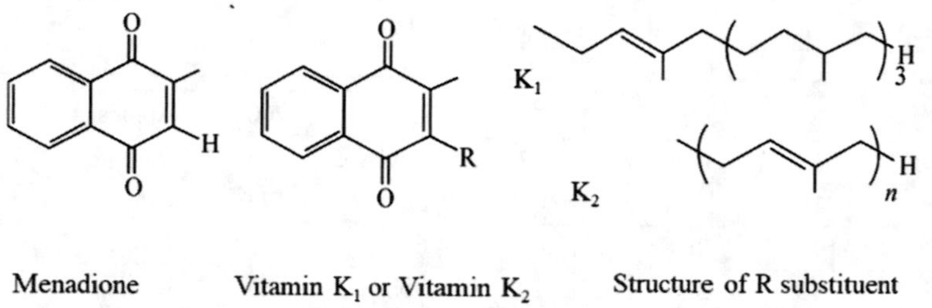

Menadione   Vitamin K₁ or Vitamin K₂   Structure of R substituent

Fig. 6.7 Chemical structure of vitamin K

6.2.4 Vitamin K

Vitamin K is a kind of bioactive compounds of phylloquinone, the common natural vitamin K are vitamin K_1 (phylloquinone) and vitamin K_2 (menaquinones), as well as the synthetic water-soluble vitamins K_3, K_4. However, vitamin K_1 and vitamin K_2 are still the most important of vitamins (Fig. 6.7).

The chemical properties of vitamin K are relatively stable. They are resistant to acid and heat, and have little loss during the cooking, but are sensitive to light and easily decomposed by alkali and ultraviolet rays. Some derivatives, such as methyl naphthalene hydroquinone acetate, have the higher activity of vitamin K and are insensitive to light.

Vitamin K is related to blood coagulation, and its major function is to accelerate blood coagulation. Besides, it can promote the liver to synthesize prothrombin. Therefore, it could be called the coagulation vitamin. Vitamin K also has reducibility. It can eliminate free radicals (the same as β-carotene and vitamin E) in the food system, protecting food ingredients from being oxidized and reducing the formation of nitrosamines in bacon.

There are two sources of vitamin K in humans: on the one hand, it is synthesized by the intestinal bacteria, accounting for 50 to 60%; on the other hand, it is obtained from food, accounting for 40 to 50%. The content of vitamin K of green leafy vegetables is high, while the content of milk and meat, fruits and cereals are low. The content of vitamin K in foods are shown in Table 6.3.

6.3 Water-Soluble Vitamins

6.3.1 Vitamin C

Vitamin C has the physiological function of preventing scurvy and it shows significant acid taste, which is known as ascorbic acid. It is an essential nutrient for the higher primates and a few specialized creatures. Vitamin C is a polyhydroxy carboxylic acid lactone containing 6 carbon atoms and an olefinic diol. It is the antioxidant in vivo,

Table 6.3 The content of vitamin K in foods (μg/100 g)

Animal food	Content	Cereal	Content	Vegetables	Content	Fruit drink	content
Milk	3	Millet	5	Kale	200	Applesauce	2
Cheese	35	Whole wheat	17	Cabbage	125	Banana	2
Butter	30	Flour	4	Lettuce	129	Orange	1
Pork	11	Bread	4	Beans	19	Peach	8
Ham	15	Oat	20	Spinach	89	Raisin	6
Smoked pork	46	Green beans	14	Radish	650	Coffee	38
Bovine liver	92			Potato	3	Coca Cola	2
Pig liver	25			Pumpkin	2	Green tea	712
Chicken liver	7			Tomato	5		

protecting the body from free radicals. It is also a coenzyme. Additionally, it has the positive effect on the absorption of iron and prevention of various diseases and plays an important role in the synthesis of collagen. There are two natural ascorbic acids, type L and type D, and the latter has no biological activity (As shown in Fig. 6.8).

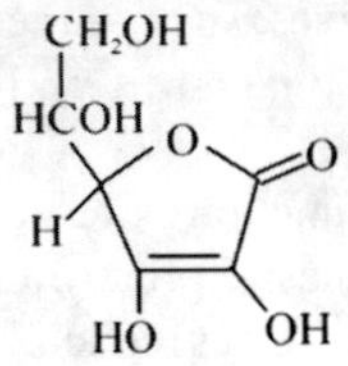

L-ascorbic acid (reduced type)

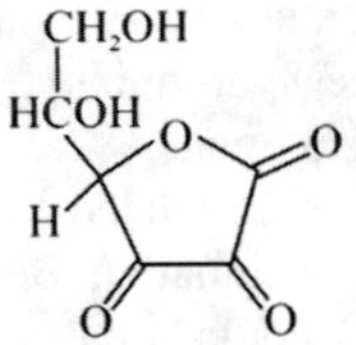

L-dehydroascorbic acid (oxidized type)

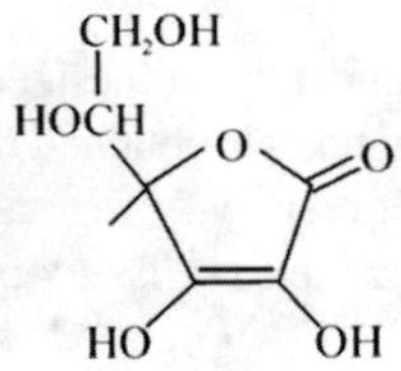

L-isoascorbic acid

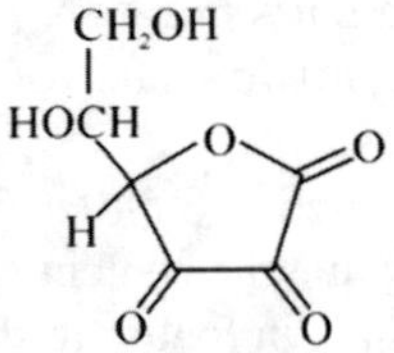

L-isohydroascorbic acid

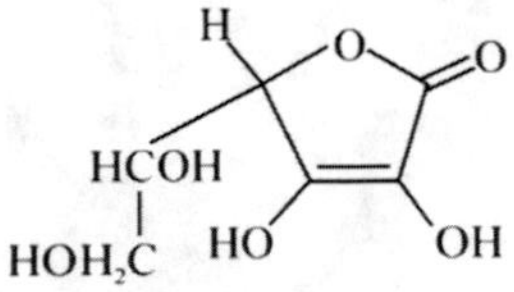

D-ascorbic acid

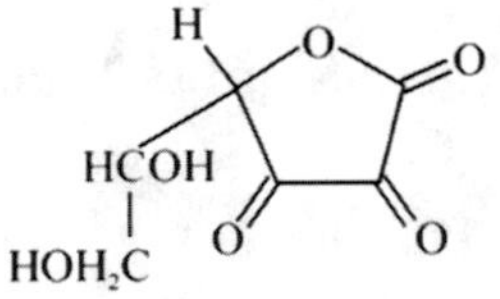

D-dehydroascorbic acid

Fig. 6.8 Structures of vitamin C

Vitamin C is a colorless crystal with the melting point of 190–192 °C; when it dissolves in water, the liquid is acidic. Its chemical properties are active, but stable in acidic environment (pH < 4). When exposed to oxygen, heat, light and alkaline (pH > 7.6), it can be destroyed by oxidation. Besides, the oxidase and a trace of metal ions such as copper and iron could speed up the oxidation of vitamin C. The content of oxidase is higher in vegetables, so vitamin C will loss during the storage. But some fruits contain bioflavonoids, which can help to maintain the stability of vitamin C. Therefore, vitamin C of foods can be destroyed during the storage, cooking and chopping, and the fresh vegetables and fruits are rich in vitamin C.

In the presence of oxygen, ascorbic acid is degraded into univalent anion (HA^-), which can form the ternary complex with metal ion and oxygen. According to the concentration of metal catalyst (Mn^{2+}) and oxygen partial pressure, there are many pathways for the oxidation of univalent anion HA^-. Once the HA^- is formed, it is quickly converted to dehydroascorbic acid (A) by the oxidation pathway of one single electron. When the metal catalyst is Cu^{2+} or Fe^{3+}, the degradation rate usually is larger than that of auto-oxidation, and the catalytic reaction rate of Cu^{2+} is 80 times greater than that of Fe^{3+}. Even if the content of metal ions are a few parts per million, they could still cause serious loss of vitamin C in food.

During the degradation of ascorbic acid, dehydroascorbic acid (A) can still be converted to ascorbic acid by the reduction reaction. Therefore, the activity of vitamin C is lost as the lactone ring-opening is hydrolyzed into 2,3-dione gulonic acid (DKG). And then DKG could be degraded into xylose ketone (X) or 3-deoxypentose Ketone (DP). The xylose ketone can be degraded into the reducing ketone and ethyl glyoxal, while the degradation of DP is to form the furfural (F) and 2-furancarboxylic acid (FA). All these products could be combined with the amino to cause the browning of the food (non-enzymatic browning), which ultimately forms the precursor of the flavor compound. Therefore, ascorbic acid similar with the reducing sugar in carbohydrates, can also react with amino acids or proteins to take place Maillard reaction. Some sugars and sugar alcohols can prevent the oxidative degradation of ascorbic acid, which may be due to that they can be combined with metal ions to reduce the catalytic activity of metal ions and protect vitamin C in foods. Their mechanism still needs to be further studied. The degradation pathway of ascorbic acid is shown in Figs. 6.9 and 6.10.

Vitamin C is one of the essential vitamins in human body and its mainly physiological functions are: ① Maintain the metabolism of cells and protect the activity of enzymes; ② It can detoxify lead compounds, arsenide, benzene and bacterial toxins; ③ the ferric iron is reduced to ferrous iron, which is beneficial to the absorption of iron and participates in the synthesis of ferritin; ④ It is involved in the synthetic process of hydroxyproline in collagen, avoiding the increase of capillary fragility and facilitating the healing of tissue wounds; ⑤ It could promote the synthesis of glucose and myocardial glycogen in the myocardium and dilate coronary arteries; ⑥ It is a good free radical scavenger in the body.

Vitamin C is widely distributed in nature, and the main sources of food are fresh vegetables and fruits, especially the fruits with a strong acidic taste and the fresh leafy vegetables. Vegetables such as pepper, garland chrysanthemum, bitter melon, beans,

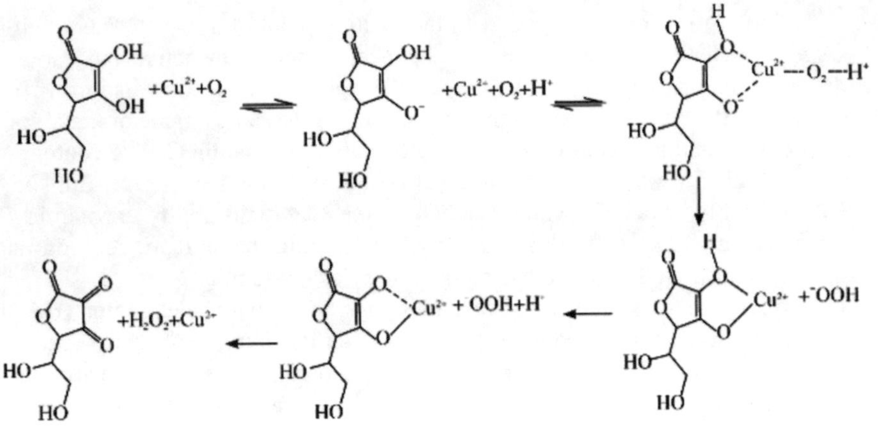

Fig. 6.9 The oxidation of vitamin C catalyzed by Cu^{2+}

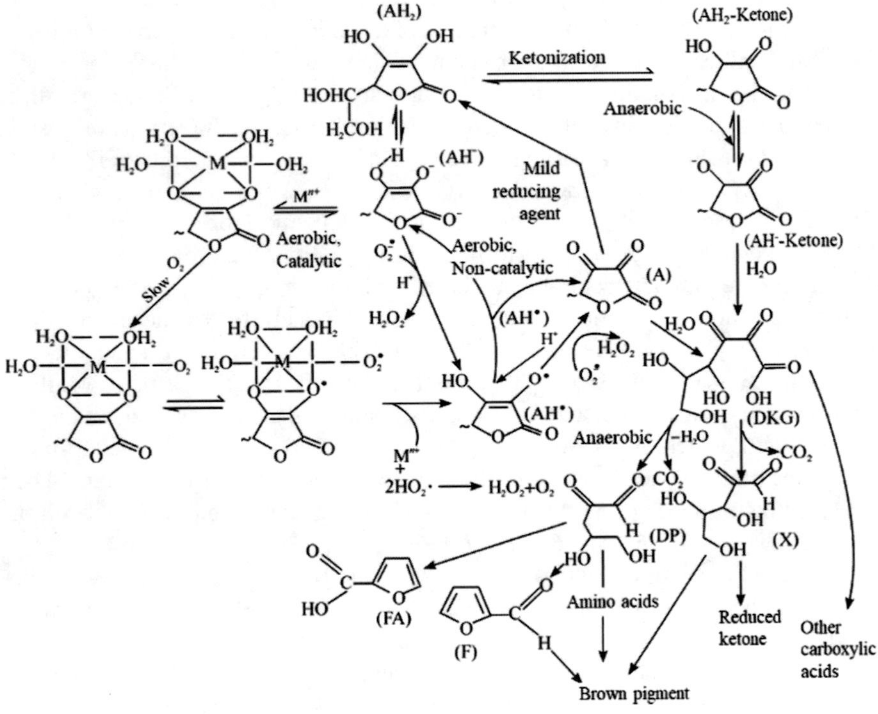

Fig. 6.10 The degradation pathway of ascorbic acid

Table 6.4 Content of vitamin C in plants (mg/100 g edible portion)

Food	VC content	Food	VC content	Food	VC content
Jujube (fresh)	243	Radish tassel (white)	77	Guava (chicken fruit, peach)	68
Chili (red small)	144	Stem with mustard (green head)	76	Rapeseed	65
Jujube (candied)	104	Mustard (large leaf mustard)	72	Cauliflower (cauliflower)	61
Garlic (dehydrated)	79	Green pepper (bell pepper, bell pepper, big pepper)	72	Red cabbage	57

spinach, potatoes, leeks, etc., are rich in vitamin C, while the content of vitamin C of the fruits such as jujube, thorn pear, kiwi, rose hip, strawberry, citrus, lemon, etc., are also very high. Additionally, there are a small amount of vitamin C in the internal organs of animals. The content of vitamin C in some common plants is shown in Table 6.4.

6.3.2 Vitamin B₁

Vitamin B_1, called thiamine or anti-beriberi vitamin, is one of the B group vitamins which are formed by the combination of a pyrimidine ring and a thiazole ring through a methylene group. It is widely found in the tissues of animal and plant (Fig. 6.11). It was discovered for the first time by the Dutch scientist Ikerman in 1896 and extracted and purified from the rice bran by the Polish chemist Fonck in 1910. It is white powder that is easily soluble in water and broken down under the condition of alkali. The physiological function of vitamin B_1 is to improve appetite, and to maintain the normalization of nerve activity, etc. However, vitamin B_1 deficiency could cause beriberi, neurodermatitis and so on. Thiamine have many forms in foods, including free thiamine, pyrophosphate (co-carboxylase), and hydrochloride and nitrate of thiamine. There are two base of nitrogen atoms in the molecule of thiamine, one in the primary amino group and the other in the quaternary amine with strong basicity, so thiamine is combined with inorganic acid or organic acid to forms salts. The natural thiamine has a base of primary alcohol, which could form the phosphate with phosphoric acid. There are many different forms of phosphate depending on the pH of the solution.

Thiamine is very stable in acidic solution and unstable in alkaline solution, which is easy to be oxidized and broken down by heat. Reductive substances such as sulfites and sulfur dioxide can deactivate vitamin B_1, and when exposed to light and heat the potency of vitamin B_1 will reduce. Thiamine is the most unstable vitamin. Other

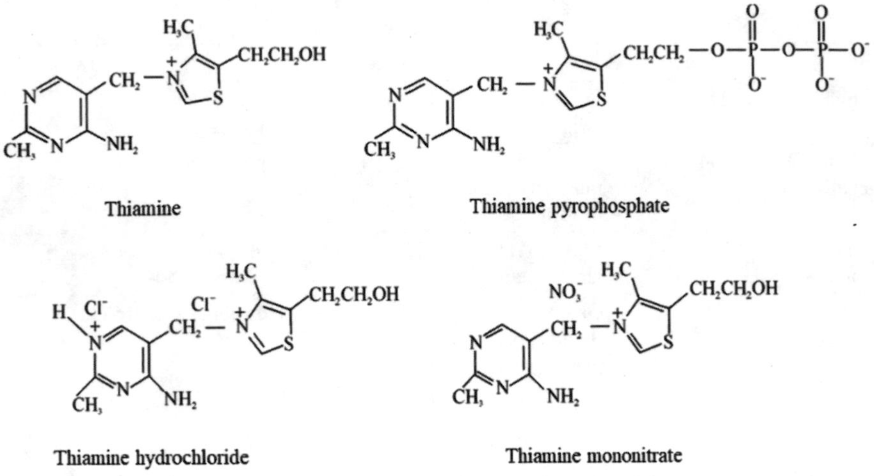

Fig. 6.11 The structure of various forms of thiamine

components in food also affect the degradation of thiamine, such as tannins, sulfur dioxide, sulfite, flavonoids, choline, and so on. Tannins is combined with thiamine to form a complex and induces the activity loss. Sulfur dioxide or sulfite can also destroy it. Flavonoids could change the molecule of thiamine. Choline can cause its molecules to break and accelerate degradation. However, proteins and carbohydrates reduce the thermal degradation of thiamine, mainly for, proteins can form disulfide with the thiol of thiamine, thus preventing its degradation. The degradation process of thiamine is shown in Fig. 6.12.

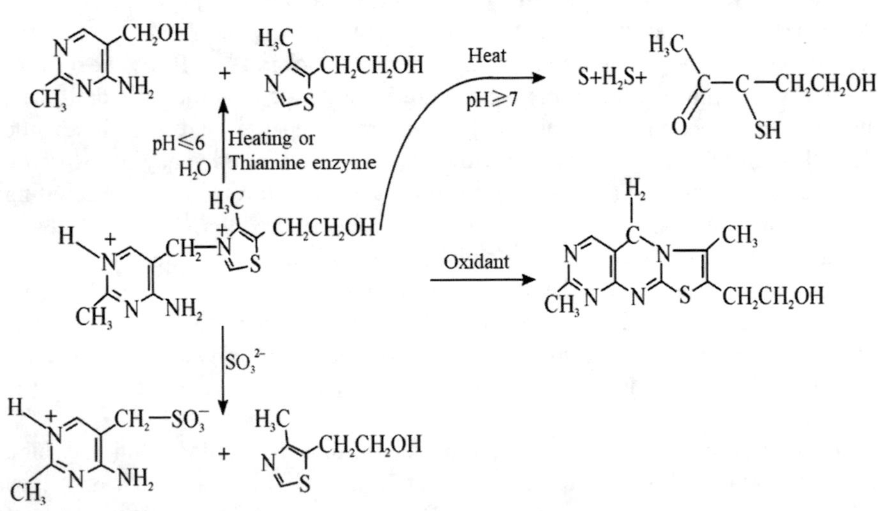

Fig. 6.12 The degradation process of thiamine

Thiamine are relatively stable at room temperature with low water activity, but the storage of long-term, high water activity and high temperature could result in great losses (Table 6.5). As shown in Fig. 6.13, when the temperature is lower than 37 °C and the water activity is from 0.1 to 0.65, there is little or no loss of thiamine in the simulated cereal breakfast; when the temperature is 45 °C and the water activity is higher than 0.4, the degradation rate of thiamine significantly increases, especially when the water activity is between 0.5 and 0.65. However, when the water activity increases from 0.65 to 0.85, the degradation rate of thiamine decreases.

The instability of thiamine in some fish and crustaceans has been thought to be caused by the thiamine enzymes after slaughter, but it is now thought to be partially due to the non-enzymatic catalysis of degradation of thiamine caused by heme containing proteins (myoglobin and hemoglobin). There is the hemoprotein that promotes the degradation of thiamine in the muscle tissues of tuna, pork and beef, which suggests that denatured myoglobin may be involved in the degradation of thiamine during food processing and storage.

Thiamine is phosphorylated to thiamine pyrophosphate in the liver, which form an important coenzyme involved in the metabolism of body. Thiamine is involved in the oxidative decarboxylation of α-ketonic acids in vivo, which is important for

Table 6.5 Retention rate of thiamine in canned foods

Food name	Retention rate after 12 months of storage /%		Food name	Retention rate after 12 months of storage /%	
	38 °C	1.5 °C		38 °C	1.5 °C
Apricot	35	72	Tomato juice	60	100
Green beans	8	76	Pea	68	100
Lima bean	48	92	Orange juice	78	100

Fig. 6.13 Effect of water activity and temperature on the retention rate of thiamine in simulated breakfast (8 months)

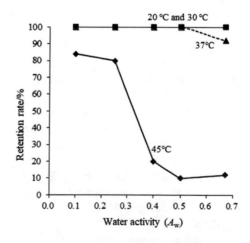

the glucose metabolism. Additionally, thiamine is also involved in the pentose phosphate pathway as a coenzyme of transketolase, which is the only way to make ribose for the synthetic RNA.

There is a small amount of thiamine in the body. Refined cereals are used as the staple food and other foods are scarce over a long period of time, which would cause the thiamine deficiency (beriberi). This disease can damage the vascular system of nerve, leading to multiple peripheral neuritis and cardiac dysfunction. There may be fatigue, irritability, headache, loss of appetite, constipation and reduced work ability in the early stages of the disease. On the contrary, excessive intake of thiamine may be excreted by the kidneys, and until now, there is no record of thiamine poisoning in humans.

Thiamine is widely distributed throughout the animal and plant and is existed in the various foods. The good sources of thiamine are the internal organs of animal (liver, kidney and heart), lean meat, whole grains, beans and nuts, especially in the epidermis of cereals and unrefined cereals (0.3–0.4 mg/100 g), but there is great loss of thiamine in the excessively refined rice and refined flour. At present, cereals are still the main source of thiamine in traditional Chinese diets.

6.3.3 Vitamin B₂ (Riboflavin)

Vitamin B_2 is also called riboflavin, which is discovered from the whey by the British chemist Bruce in 1879. And then, it was extracted from milk by the United States chemist Golderg in 1933, and synthesized by the German chemist Cohen in 1935. Vitamin B_2 is an orange-yellow needle-like crystal with a slightly bitter taste. The aqueous solution has yellow-green fluorescence. It is easily decomposed under alkaline condition or light. As the promethazine derivative containing the side chain of ribitol, it is usually phosphorylated in the natural condition and a coenzyme which plays an important role in the metabolism of human body. One form of vitamin B_2 is flavin mononucleotide (FMN), and the other form is flavin adenosine dinucleotide (FAD). They are ingredient of some enzymes such as reductase of cytochrome C, flavoprotein, etc. The FAD is an electron carrier and plays a role in the oxidation of glucose, fatty acids, amino acids and purine. Therefore, riboflavin is an important vitamin (Fig. 6.14).

Riboflavin is stable when heated in neutral medium or acidic medium, but rapidly degraded in alkaline conditions. The loss of riboflavin is not significant during conventional heat treatment, dehydration and cooking. Riboflavin is a photosensitizer. There are two bioactive products after photodegradation, which are lumiflavin and lumichrome (Fig. 6.15). Lumiflavin is a strong oxidizer that significantly destroys other vitamins, especially ascorbic acid. The nutritional value of bottle milk is seriously reduced and produce an unpleasant taste because of the reaction, that is "sunlight odor", so the package of dairy products should be away from light.

Riboflavin does not accumulate in the body, so it usually necessary to supplement it with food or dietary supplements. The good food sources of riboflavin are mainly

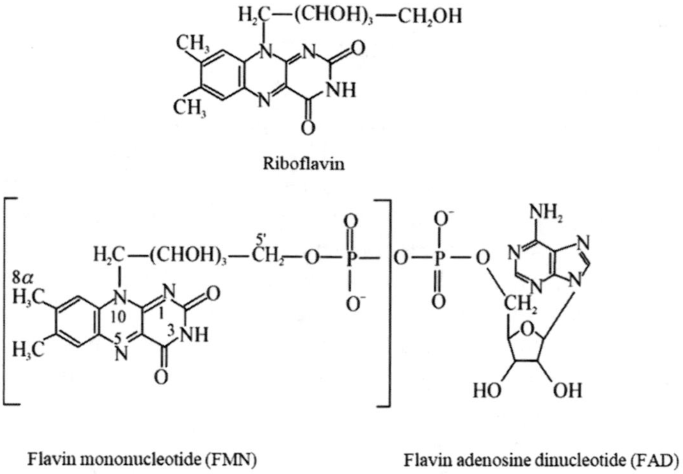

Fig. 6.14 The structure of riboflavin, flavin mononucleotide and flavin adenine dinucleotide

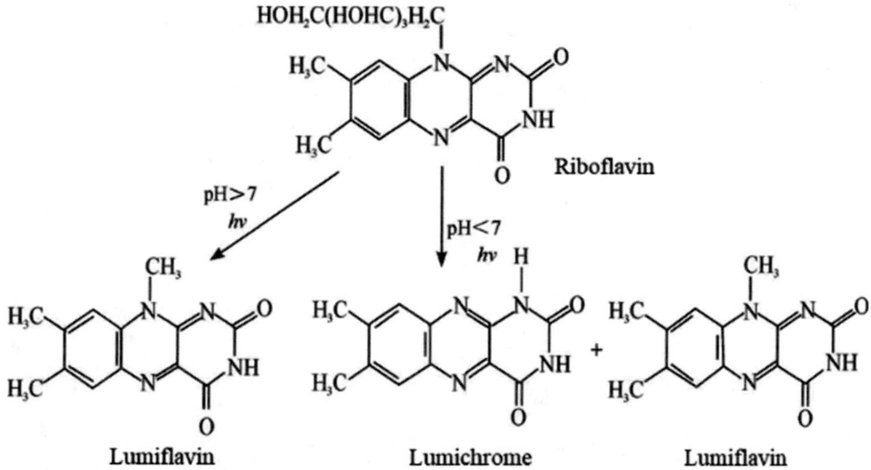

Fig. 6.15 Decomposition of riboflavin in alkaline and acid and light

animal food, especially the internal organs of animal such as liver, kidney, heart, egg yolk, milk and fish. In the plant foods, the content of lutein in the green leafy vegetables such as spinach, leeks, rapeseed and beans are high, and the content of wild vegetables is also high. The content of riboflavin in cereals is related to the extent of processing, the high level of processing of grain has the low content of riboflavin. The dietary composition for Asia residents is based on plant foods, so riboflavin is also easy to be in lack in our diet.

6.3.4 Niacin

Niacin, also known as vitamin PP, nicotinic acid, and the factor of anti-skin disease, is a general term for pyridine 3-carboxylic acid and its derivatives, as well as nicotinamide (Fig. 6.16). Nicotinamide is a component of two important coenzymes, nicotinamide adenine dinucleotide (NAD$^+$) and nicotinamide adenine dinucleotide phosphate (NADP$^+$), which play an important role in glycolysis, fat synthesis and respiration. Niacin is also the factor for preventing ecdysis, which is a serious problem in the area where corn is the staple food. As the content of tryptophan in corn protein is low, and tryptophan could be converted into niacin in vivo. In addition, the lower utilization of niacin in corn and other cereals may be due to that it is combined with sugar to form a complex, but it can be released after the alkali treatment. The main form of vitamin PP in animal tissues is nicotinamide.

Niacin is one of the most stable vitamins. It is not sensitive to heat, light, air and alkali, and there is no loss in food processing. However, the content of niacin of food can also be changed through the conversion reaction in the cooking. For example, the niacin of corn could be released to the free nicotinamide from NAD$^+$ and NADP$^+$ during the boiling.

Niacin is widely found in animals and plant foods. Good sources of niacin are animal liver, kidney, lean meat, whole grains, beans, etc. Milk and eggs are low in niacin, but are rich in tryptophan, which can be converted into niacin in the body. Niacin in some plants is often combined with macromolecules and cannot be absorbed by mammals. For example, about 64 to 73% of the niacin in corn and sorghum is combined niacin, which cannot be absorbed by human body. However, the combined

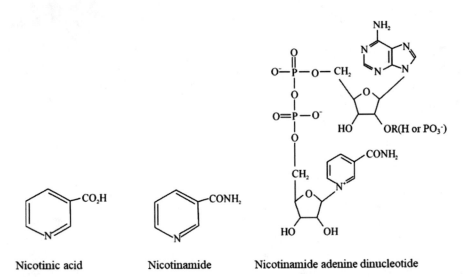

Fig. 6.16 The structure of niacin, nicotinamide and nicotinamide adenine dinucleotide

niacin can be converted to free niacin in the alkaline solution and then can be used by animals and humans.

6.3.5 Vitamin B₆

Vitamin B_6 is also called pyridoxine, including pyridoxal, pyridoxine and pyridoxamine, which exists in organisms as phosphate. Pyridoxal phosphate plays as a coenzyme role in the metabolism of amino acid (such as ammonification, racemization and decarboxylation), which can help the decomposition and utilization of sugars, fats and proteins in the body, as well as the glycogen (Fig. 6.17).

Vitamin B_6 is stable in acid solution, but it can be easily destroyed in alkali solution. Pyridoxine is heat-resistant, while pyridoxal and pyridoxamine are sensitive to high-temperature. Three forms of vitamin B_6 are all sensitive to light, especially ultraviolet light and the alkaline condition. When pyridoxal and pyridoxamine are exposed to the air, they are quickly destroyed by heating and light, resulting in the formation of inactive compounds such as 4-pyridoxic acid.

In addition to the loss of vitamin B_6 caused by heating, pyridoxal can also be reacted with the amino acids of proteins, such as cysteine, to form derivatives containing sulfur, or interacted with amino acids directly to form the Schiff's base, thereby reducing its biological activity during a long-term storage. The stability of vitamin B_6 in some foods is shown in Table 6.6.

Insufficiency of vitamin B_6 intake can lead to vitamin B_6 deficiency disease, mainly characterized by seborrheic dermatitis, stomatitis, dry lips, glossitis, irritability, depression and so on. Vitamin B_6 can be obtained by two routes: food intake and synthesis of intestinal bacteria. Although vitamin B_6 is contained in many foods, the content of vitamin B_6 in foods is not high. Most of the vitamin B_6 in animal food exists in the form of pyridoxal and pyridoxamine. For example, the content of vitamin B_6 in white meat (chicken and fish), liver, eggs and so on is relatively high, but is less in milk and dairy products, while the content of vitamin B_6 in plant foods like legume, cereal, fruit and vegetable is also high, but is less in lemon fruit. Besides, vitamin B_6 in plant foods is mostly combined with protein, so it is not easy to be absorbed.

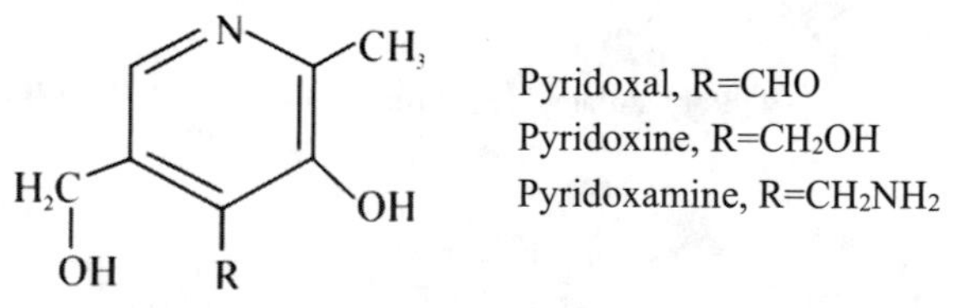

Fig. 6.17 The chemical structure of vitamin B_6

Pyridoxal, R=CHO
Pyridoxine, R=CH₂OH
Pyridoxamine, R=CH₂NH₂

Table 6.6 Stability of vitamin B$_6$ in foods

Food	Process	Retention/%
Bread (with vitamin B$_6$)	Bake	100
Fortified corn flour	50% relative humidity, stored at 38 °C for 12 months	90 ~ 95
Strengthen macaroni	0% relative humidity, stored at 38 °C for 12 months	100
Full fat milk	Evaporation and high temperature sterilization	30
	Evaporate and autoclave, store at room temperature for 6 months	18
Milk replacer (liquid)	Processing and disinfection	33 ~ 55
Milk replacer (solid)	Spray drying	84
Boneless chicken	Filling	57
	Radiation (2.7 Mrad)	68

6.3.6 Folic Acid

Folic acid, also known as vitamin M or vitamin Bc, are consisted of a series of compounds with similar chemical structures and physiological activities. Their molecular structure includes three parts: the pterin, p-aminobenzoic acid and glutamic acid (Fig. 6.18). The tetrahydrofolate is the biologically active form of folic acid in vivo, which is transformed by the synergistic action of folate reductase, vitamin C and coenzyme II. Folic acid plays an important role in the metabolism

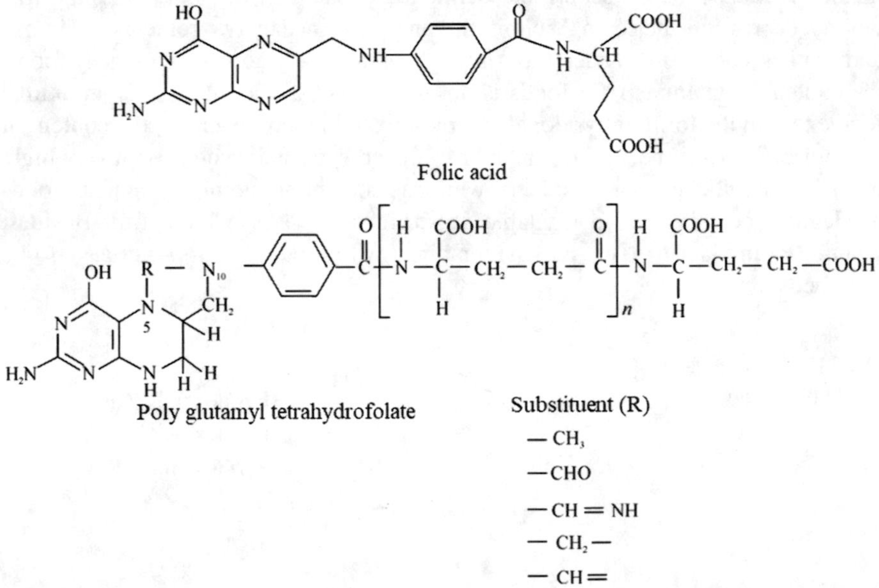

Fig. 6.18 The structure of folic acid

of nucleotides and amino acids. Lack of folic acid can cause various diseases such as anemia and stomatitis. Several studies have demonstrated that folic acid plays an important role in the prevention of some congenital anomalies such as neural tube defects and spina bifida, congenital heart defects, and orofacial clefts, as well as chronic diseases in adults, such as cancer, depression, and age-related hearing loss.

Folic acid is yellow crystal, slightly soluble in water, and its sodium salt is easily soluble in water, insoluble in ethanol, ether and other organic solvents. The aqueous solution of folate acid is unstable under light and heat, and easy to lose its activity. For example, the content of folic acid of vegetables will loss 50 to 70% after a storage for 2 to 3 days; cooking results in loss of 50 to 95% in foods; the content of folic acid in vegetables will also be greatly lost after soaking in salt water, so the body doesn't get much folic acid from food. However, it is stable under the neutral and alkaline conditions and will not be damaged even if heated to 100 °C for 1 h. Folic acid can be interacted with sulfite and nitrite, which can cause the side chain of folic acid to be dissociated and form the reduced pterin-6-carboxaldehyde and aminobenzoyl glutamic acid. At the low temperatures, the reaction of folic acid and nitrite will generate N-10-nitro derivatives, which is a weak carcinogen for mice.

In the oxidation reaction of folic acid, copper ions and iron ions are important catalysts, and the catalyzing reaction of copper ions is greater than that of iron ions. Tetrahydrofolate is oxidized and degraded into two products, pterin compound and ρ-aminobenzoyl glutamate (Fig. 6.19), resulting in loss of its biological activity. If the reducing substance such as vitamin C, thiol and so on, is added in the reaction,

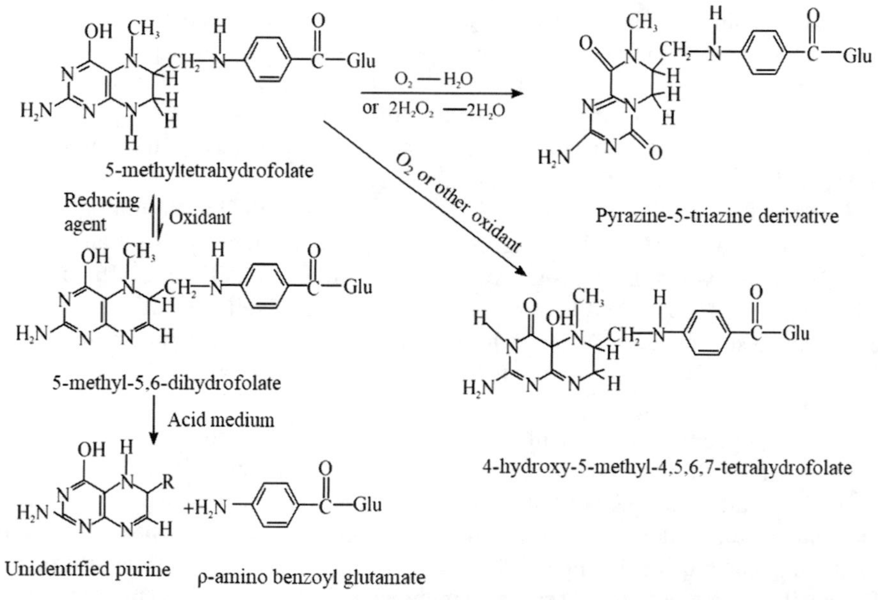

Fig. 6.19 Oxidative degradation of 5-methyltetrahydrofolate

Table 6.7 Effects of processing on the content of folic acid in vegetables

Vegetables (cooked in water for 10 min)	Total folic acid content/(μg/100 g Fresh sample)		
	Fresh	After cooking	The content of folic acid in cooking water
Asparagus	175 ± 25	146 ± 16	39 ± 10
Green vegetables	169 ± 24	65 ± 7	116 ± 35
Brussels sprouts	88 ± 15	16 ± 4	17 ± 4
Cabbage	30 ± 12	16 ± 8	17 ± 4
Cauliflower	56 ± 18	42 ± 7	47 ± 20
Spinach	143 ± 50	31 ± 10	92 ± 12

5-methyl-dihydrofolate can be reduced to 5-methyltetrahydrofolate and increase the stability of folic acid.

Folic acid is widely found in animals and plant foods, and its good sources are liver, kidney, green leafy vegetables, potatoes, beans, wheat germs, nuts and so on. The effects of processing on folic acid in foods is shown in Table 6.7.

6.3.7 Vitamin B_{12}

Vitamin B_{12} is also called cobalamin, which is B group vitamin, composed of a cobalt-containing porphyrin compound and the only vitamin containing metallic element. Vitamin B_{12} is a conjugated complex and its central ring is porphyrin ring. The trivalent cobalt ion of its central ring is bonded to the four nitrogen atoms of the porphyrin ring and the sixth valence position of cobalt ions is occupied by cyanide, so it is also called cyanocobalamine. It is a red crystalline substance that can be used to fortify foods because of its good stability (Fig. 6.20).

The aqueous solution of vitamin B_{12} is unstable under alkaline conditions and is sensitive to ultraviolet light. The range of suitable pH for its stability is 4–6, and within this range, even with high-pressure heating, there is only a small loss of vitamin B_{12}. Vitamin B_{12} is decomposed in strong acid (pH < 2) or alkaline solution. It can be destroyed to some extent when it is heated, but the loss by high-temperature short-time sterilization is small. It is easily to be destroyed by strong light or ultraviolet light, while the loss of vitamin B_{12} during normal cooking is about 30%. Vitamin B_{12} can also be destroyed by ascorbic acid or sulfites, and the low concentrations of reducing agents such as sulfhydryl compounds can prevent the destruction of vitamin B_{12}, but the higher content may damage vitamin B_{12}. Vitamin B_{12} will be slowly destroyed by the combination of thiamine and niacin in solutions. Iron ions and hydrogen sulfide of thiamine can protect vitamin B_{12}. Ferric salts have the stabilizing effect on vitamin B_{12}, while ferrous salts could rapidly damage vitamin B_{12}.

Vitamin B_{12} is the only vitamin that can be absorbed with the help of intestinal secretions (endogenous factors). Its main physiological function is to be involved in

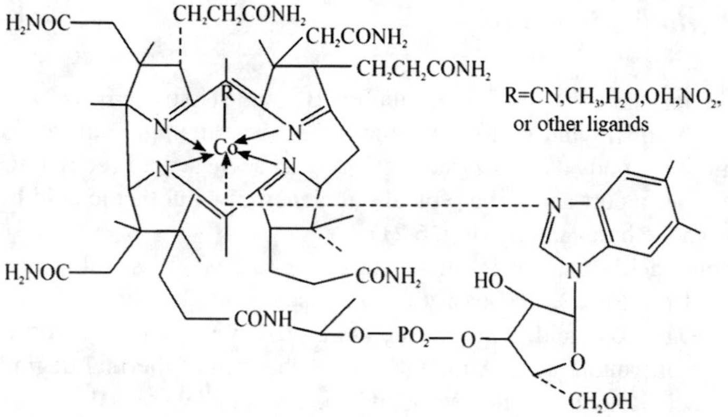

Fig. 6.20 The chemical structure of vitamin B$_{12}$

making the red blood cells of bone marrow, preventing pernicious anemia and the destruction of brain nerves. Vitamin B$_{12}$ is a coenzyme of several mutase, such as methylmalonyl mutase and diol dehydratase.

Vitamin B$_{12}$ is mainly found in animal tissues (Table 6.8), and the major of its dietary sources are animal foods, fungi foods and fermented foods. The loss of vitamin B$_{12}$ is less in other conditions except the cooking in alkaline solution. For example, when the liver is boiled at 100 °C for 5 min, the loss of vitamin B$_{12}$ is 8%; When the meat is baked at 170 °C for 45 min, the loss is 30%. The content of vitamin B$_{12}$ of frozen foods such as fish, fried chicken, turkey and beef, can be retained from 79 to 100% after heating on the traditional oven.

The lack of vitamin B$_{12}$ has been widespread worldwide, especially for the persons who consume small amounts of animal-derived food. The addition of vitamin B$_{12}$ in the flour product as an additive contains intact vitamin B$_{12}$, which can maintain about 50% bioavailability when fed to healthy subjects in small doses, while the bioavailability of foods such as meat and fish is also 50%. Wheat flour can supplement with vitamin B$_{12}$ in regions where intake of the vitamin B$_{12}$ is low.

Table 6.8 The content of vitamin B$_{12}$ in food	Food	The content of Vitamin B$_{12}$ (μg/100gWet weight)
	Organs (liver, kidney, heart), shellfish (clam, oyster)	>10
	Degreased concentrated milk, some fish, crab, egg yolk	3 ~ 10
	Muscle, fish, cheese	1 ~ 3
	Liquid milk, cheddar cheese, farm cheese	<1

6.3.8 *Pantothenic Acid*

Pantothenic acid is called polyacid, vitamin B_5, and its structure is D (+)-N-2,4-dihydroxy-3,3-dimethylbutyryl-oxime-alanine. It plays an important role in the lipid metabolism of the body as coenzyme A (CoA) and acyl carrier protein (ACP). Until now, there is no report about the typical deficiency of pantothenic acid because of its ubiquity in the human body (Fig. 6.21).

Pantothenic acid is stable in air but unstable to heat. It is stable in the range of pH value from 5 to 7. It is easily decomposed in the alkaline solution to form β-alanine and pantoic acid, while in the acidic solution it can be hydrolyzed into the γ-lactone of pantoic acid. Although the mechanism of thermal degradation for pantothenic acid is not fully understood, it is believed that there is the acid-catalyzed hydrolysis of the linkage between β-alanine and 2,4-dihydroxy-3,3-dimethylbutyric acid. However, pantothenic acid does not react with other components of food under other conditions.

Pantothenic acid has the high stability during food processing and storage, especially under the condition of low water activity. In the process of cooking and heat treatment, the loss rate of pantothenic acid is usually from 30 to 80%, and as the temperature rises, the degree of loss increases.

Pantothenic acid is widely distributed in organisms. The food rich in pantothenic acid are mainly meat, unrefined cereal products, malt and wheat bran, animal kidney and heart, green leafy vegetables, brewer yeast, nuts, chicken, unrefined molasses and so on. The distribution of pantothenic acid in foods is shown in Table 6.9.

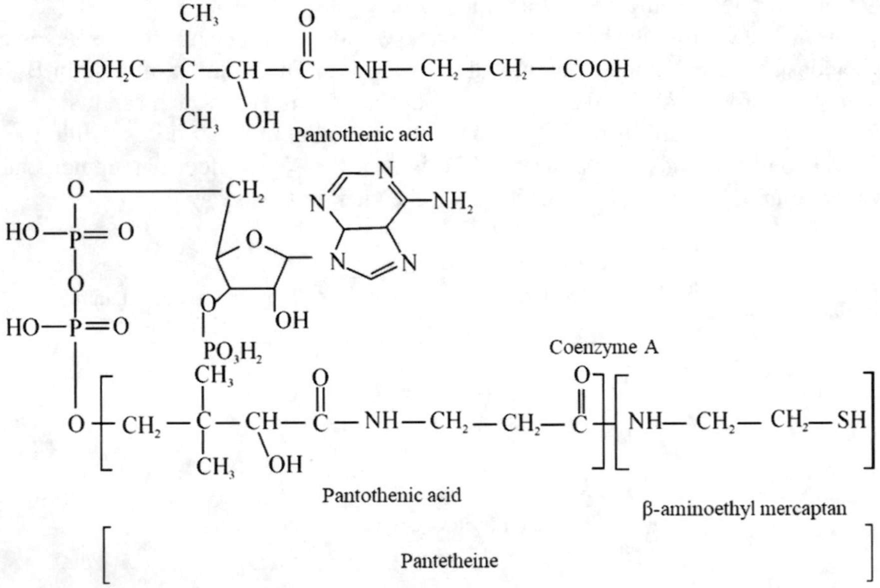

Fig. 6.21 The structure of various pantothenic acids

Food	Pantothenic acid content (mg/g)	Food	Pantothenic acid content (mg/g)
Dry beer yeast	200	Buckwheat	26
Bovine liver	76	Spinach	26
Yolk	63	Roasted peanut	25
Wheat bran	30	Whole milk	24

Table 6.9 The content of pantothenic acid in common foods

6.3.9 Biotin

Biotin (vitamin H) also called vitamin B_7 and Coenzyme R, is an essential substance for the synthesis of vitamin C. It is involved in the normal metabolism of the three major nutrients and is an essential nutrient for maintaining the growth and health of human beings. As there are three asymmetric carbon atoms in the molecular structure of biotin, it has eight stereoisomers. Dextral D-biotin is natural and of all eight stereoisomers, only Dextral D-biotin has the biological activity. Biotin is reacted with lysine residues in proteins to form biocytin, both biotin and biocytin are natural vitamins (Fig. 6.22).

The chemical character of biotin is stable and is not easily damaged by acid, alkali and light. However, biotin may be deactivated due to hydrolysis of the amide bond under the high or low pH value. When biotin is oxidized to sulfoxide or sulfone compounds or reacted with nitrite to form the nitroso compounds, its biological activity is destroyed.

Many animals including human need biotin to keep health and biotin is the coenzyme of a variety of carboxylase, which plays a role as CO_2 carrier in the carboxylase reaction. A mild biotin deficiency in vivo can lead to dry skin, dandruff, brittle hair, etc., while a severe deficiency can cause reversible baldness, depression, muscle pain and atrophy, etc.

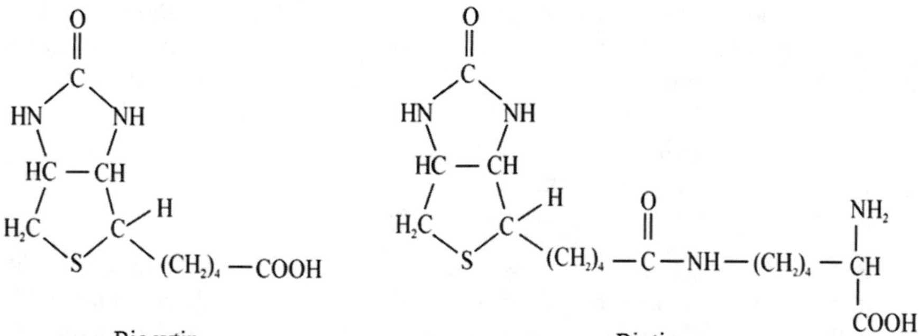

Fig. 6.22 The structure of biotin and biocytin

Table 6.10 The content of biotin in common foods

Food	Biotin content (μg/g)	Food	Biotin content (μg/g)
Apple	0.9	Mushroom	16.0
Soy	3.0	Orange	2.0
Beef	2.6	Peanut	30.0
Bovine liver	96.0	Potato	0.6
Cheese	1.8 ~ 8.0	Spinach	7.0
Lettuce	3.0	Tomato	1.0
Milk	1.0 ~ 4.0	Wheat	5.0

Biotin is widely distributed in plants and animals (Table 6.10), it exists on free form in vegetables, milk, and fruits, and is combined with proteins in animal viscera, seeds and yeast. The content of biotin in beer is higher. The supply of biotin in human body is partly dependent on dietary intake, and most is relying on intestinal bacterial synthesis. Biotin can be inactivated by eating raw egg white, which is caused by an anti-biotin glycoprotein.

6.4 Vitamin-Like Compounds

At present, some organic substances are also found to be essential for human physiological functions, which is called the biological nutrient enhancer or vitamin-like compounds. Although whether they are vitamins or not is controversial, it also fully reflects the development of modern science and technology.

6.4.1 Choline

Choline is a component of lecithin and a precursor of acetylcholine. Although humans can synthesize choline by themselves, it is still necessary to obtain choline from the diet for mammals, especially for growing children or infants who need the higher content of choline, so choline in baby foods is often strengthened in the form of choline chloride or choline tartrate.

The chemical structure of choline: $(CH_3)_3 \; N^+CH_2CH_2OH$.

The important functions of choline in the body are as follows:

(1)   Prevent the fatty liver. Choline is a "lipophilic agent" which can promote the transport of fat in the form of lecithin or enhance the utilization of fatty acids in the liver and prevent the abnormal accumulation of fat in the liver to ensure normal liver function.

(2) Nerve conduction. It can help to cross the gaps in the nerve cells and produce conduction pulses.
(3) Promote metabolism. Choline is widely found in foods. Egg yolks, eggs, and liver are the most abundant in choline and is usually present in the form of lecithin and acetylcholine. Choline is stable to heat and has almost no loss during processing, cooking, and storage.

6.4.2 Inositol

Inositol is a growth factor of animal and microorganism. In theory, it has nine stereochemical structures. Four of them are found in nature, but only myo-inositol has the biologically activity.

At present, the function of inositol is not fully understood and the functions we have already know are:

(1) The lipophilic action, like choline, has affinity for fat. It also can promote the body to produce phospholipids which can help transport liver fat to cells and reduce cholesterol.
(2) Inositol can be combined with choline to prevent fatty arteriosclerosis and protect the heart.
(3) Inositol is a precursor of phosphoinositide which is found in various tissues of the body, especially in the brain.
(4) Recent studies have shown a potential link between free inositol content and improved growth response in animals. Within cells, the main function of inositol appears to be as a messenger in several signaling pathways that regulate various cellular processes, including cell survival, growth and metabolism. Inositol also plays a crucial role in fat transport and deposition, as well as for normal development. With deficient inositol levels these processes are severely disrupted. In some circumstances, inositol could be considered an essential nutrient (Fig. 6.23).

There is abundant inositol in nature, which mainly comes from kidney, brain, liver, heart, yeast, orange fruits, etc. Inositol in the form of phytic acid and phytate is widely present in cereals, which has an impact on the mineral nutrition for cereals.

Fig. 6.23 The chemical structure of inositol (meso with 1,4 as the axis)

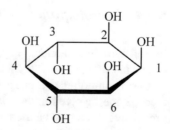

Table 6.11 The content of L-carnitine in some raw materials (mg/kg)

Name	Content	Name	Content	Name	Content
Corn	5 ~ 10	Dish	10	Whey powder	300 ~ 500
Wheat	3 ~ 12	Clover	20	Skimmed milk powder	120 ~ 150
Soy	0 ~ 10	Barley	10 ~ 18	Fish bone meal	85
Sorghum	15	Fish meal	85 ~ 145	Milk	6 ~ 50

Inositol is highly stable and resistant to acids, alkalis and heat, so there is little loss of inositol in food processing.

6.4.3 Carnitine

Carnitine was discovered and extracted in the muscles as early as 1905. Its chemical name is β-hydroxy-γ-Trimethylaminobutyric acid, and its chemical formula is $(CH_3)_3 N^+CH_2CH(OH)CH_2COO^-$. There are two optical isomers, levorotatory (L) and dextrorotatory (D). Only L-carnitine exists in nature and has a nutritional effect on animals.

The main function of L-carnitine is to be involved in the metabolism of fatty acids in the body as a carrier, providing energy and reducing serum cholesterol. It also can promote the absorption of fat-soluble vitamins and Ca and P, regulate the normal metabolism of some branch chain amino acids and the ratio of CoA/free CoA in the mitochondria, stimulate the formation of ketone body in the liver and glyconeogenesis and maintain normal metabolism. Although L-carnitine is abundant in human tissues, the synthesis ability of infants and young children is very low and they are mainly relying on the supply of breast milk and complementary foods. Moreover, humans or other mammals may also have carnitine deficiency in the body due to congenital or metabolic diseases. Therefore, the Ministry of Health of China has approved that L-carnitine as a safe and nutritional fortifier can be used in milk powder, anti-aging food, sports food and diet food. The content of L-carnitine in common foods is shown in Table 6.11.

6.4.4 Pyrroloquinolone Quinone

Pyrroloquinolone quinone (PQQ) is a tricyclic quinone and its chemical structure is shown in Fig. 6.24.

Pyrroloquinoline quinone is widely distributed in organisms. In addition to being a coenzyme involved in the catalytic reaction of certain oxidoreductases in living organisms, it can also be used as a growth stimulation factor to promote the growth and development of microorganisms and plants.

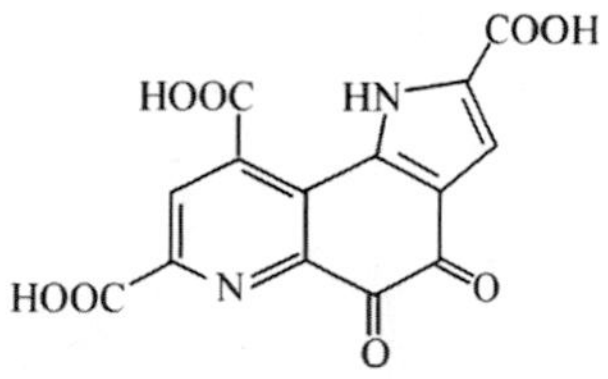

Fig. 6.24 Structure of
pyrroloquinoline quinone

Up to now, scientists have found that the best source of pyrroloquinoline quinone is natto which is a traditional Japanese food fermented with soybeans. Other foods which are rich in pyrroloquinolone quinone include parsley, green tea, green pepper, kiwi and papaya. Due to the ubiquity of pyrroloquinoline and that it can be synthesized by intestinal bacteria, there does not seem to be the deficiency of pyrroloquinoline quinone in humans and caries animals.

6.5 Changing of Vitamins During Food Processing and Storage

All foods are inevitably suffering from vitamin loss to some extent in food processing and storage. Therefore, in addition to maintaining the minimum loss of nutrients and food safety during food processing, the influence of various conditions before processing on the nutrient of food should also be assessed, such as maturity, growth environment, soil condition, fertilizer, water, temperature, light and so on.

6.5.1 The Effect of Raw Materials

(1) Maturity

The content of vitamins in fruits and vegetables changed varies with maturity, growth and climate. For example, the highest content of vitamin C in tomato is in the early stage of maturity, while the highest content in pepper is in the mature stage (Table 6.12).

(2) Changes of the content of vitamins after harvest (slaughter)

The nutritional value of food changes significantly during the time from harvesting or slaughtering to processing. As the derivatives of many vitamin are cofactors of enzymes, they are easily to be degraded by enzymes, especially the endogenous enzymes released from the dead animals or harvested plants. When the cells are damaged, the originally separated oxidase and hydrolase are released from the intact cells, thereby changing the chemical form and activity of vitamins. For example,

Table 6.12 Effect of maturity on the content of ascorbic acid in tomato

Number of weeks after flowering	Average quality/g	Color	Ascorbic acid content/(mg/100 g)	Number of weeks after flowering	Average quality/g	color	Ascorbic acid content/(mg/100 g)
2	33.4	Green	10.7	5	146	Red-yellow	20.7
3	57.2	Green	7.6	6	160	Red	14.6
4	102	Green-yellow	10.9	7	168	Red	10.1

dephosphorylation of vitamin B_6, thiamine or riboflavin coenzyme, deglucosylation of glucoside of vitamin B_6, and deconjugation of polyglutamyl folate can lead to the changes in the distribution and natural state of vitamins in plants or animals after harvesting or slaughter, and the degree of change in the storage and processing is related to temperature and time. In general, the change of vitamins content is small, but it could cause the changes of their bioavailability. The oxidation of lipoxygenase can reduce the content of many vitamins, while ascorbate oxidase specifically causes the loss of ascorbic acid. For example, there is obvious reductive reaction in the vitamins of peas from the harvest to the processing within 1 h. The vitamins of fresh vegetables can be severely damaged if they are not handled properly and stored at room or higher temperatures for 24 h or longer.

6.5.2 Effect of Treatments Before Processing

(1) Cutting and peeling

There is some loss of nutrients in plant tissues after trimming or peeling. The content of ascorbic acid in apple skin is higher than that in flesh. The pineapple heart contains higher content of vitamin C than the edible part. The niacin content of the epidermis of carrot is higher than other parts. There are also differences in the content of nutrient in different parts of plants such as potatoes, onions, beets and so on. Therefore, when these vegetables and fruits are trimmed, such as picking the stems and stalks of spinach, broccoli, mung beans, asparagus and other vegetables, some nutrients are lost. In addition, the use of strong chemicals in the peeling, such as lye treatment, will destroy the vitamins of the outer skin (such as folic acid, ascorbic acid and thiamine). Animal tissues are damaged by cutting or other treatments, which may cause loss of water-soluble vitamins due to leaching in the water and aqueous solutions. The milling and grading of grains including the remove of bran (seed coat) and germ, also can lead to the loss of vitamins as the germ and bran contain high content of vitamins.

(2) Rinsing and blanching

There are some losses of vitamins in rice during the rinsing process. When the rice was rinsed, the loss of B series vitamin was about 60% and the loss of total vitamin was 47%. The more times and force are used during washing, the more content of B vitamins is lost. This is mainly due to that most B series vitamins are present in the rice bran which is on the surface of the rice grains.

Hot blanching is an indispensable process in the processing of fruits and vegetables to deactivate the harmful enzymes, reduce the pollution of microorganism and remove air from tissues. Hot blanching methods include hot water, steam, hot air or microwaves, and hot water blanching can cause great loss of water-soluble vitamins (Fig. 6.25).

Fig. 6.25 Retention rate of ascorbic acid of peas after blanching at different temperatures for 10 min

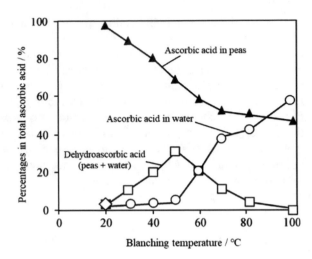

6.5.3 Effects of Processing and Storage

There are lots of treatments before cooking to ensure the safety and nutritional value of foods during transportation and distribution, and the treatment methods vary depending on the type of food and the purpose of processing.

(1) Freezing

Freezing is the most common method of food storage. The freezing includes three stages: pre-freezing, freezing storage and thawing, and the loss of vitamins mainly includes chemical degradation during storage and loss of water-soluble vitamins in the thawing. For example, vegetables will lose 37 to 56% of vitamins after freezing; and meats, 21 to 70% of pantothenic acid. Similarly, the losses of vitamin C of kale, cauliflower and spinach were 49%, 50% and 65%, respectively, when stored at $-18\,^\circ\text{C}$ for 6 ~ 12 months. The degree of loss is related to the type of vegetables. After the fruits and their products are frozen, the loss of vitamin C is complicated and related to many factors, such as variety, ratio of solid to liquid, packaging materials and so on.

(2) Irradiation

Irradiation is usually used for the sterilization and preservation of meat and the preservation of vegetable and fruit. For example, the preservation of onion, potato, apple and strawberry by the γ radiation of ^{60}Co does not only prolong the preservation period, but also improve the quality. The effect of irradiation on B series vitamin is dependent on the temperature, irradiation dose and radiance. Compared with traditional heat-sterilization, irradiation can reduce the loss and degradation of B series vitamin, and it has little effect on vitamin B_2 and niacin.

6.5.4 The Impact of Food Additives

To prevent food spoilage and improve the quality of food during processing, it is usually necessary to add some food additives, which have some effects on vitamins. For example, oxidants usually damage vitamin A, vitamin C and vitamin E, so oxidants such as bleach used in flour can reduce the content of vitamins.

Sulfite (or SO_2) is commonly used to prevent the enzymatic and non-enzymatic browning of fruit and vegetable and protect vitamin C from being oxidized as a reducing agent, but as a nucleophile it can break down vitamin B_1.

To improve the color of meat products, nitrates and nitrites are usually added into meat. However, some vegetables, such as spinach and sugar beets, also contain the high levels of nitrite. The nitrite in foods can not only react quickly with vitamin C, but also destroy carotene, vitamin B_1 and folic acid.

6.6 Summary

Vitamins are a kind of organic compounds that are indispensable micronutrients in the process of life activity. Although vitamins we need are in trace amount, but they have a great effect in our body. Lack of vitamins will lead to the disease of vitamin deficiency and affect the normal metabolism of human body. Most of the vitamins we need are relied on the supplement of food. Vitamins mainly include fat-soluble vitamins and water-soluble vitamins. The fat-soluble vitamins include vitamin A, vitamin D, vitamin E, vitamin K, etc., while water-soluble vitamins include B series vitamins (vitamin B_1, vitamin B_2, vitamin PP, vitamin B_6, vitamin B_{12}, etc.) and vitamin C. They have great differences in stability, also depending on the components of food. For example, some vitamins are very unstable and can easily be lost, especially in food processing and storage, so the processing methods and storage conditions are very important for maintaining the content of vitamins. In addition, the content and variety of vitamins in foods are different, and vitamins usually need to be fortified in some foods to meet the needs of human body. Table 6.13 summarized the physiological functions and main food sources of various vitamins.

Questions

1. What is vitamins? What are their common characteristics?
2. How many types of vitamins are there? What is the difference between water-soluble vitamins and fat-soluble vitamins?
3. What is the effect of heat treatment on vitamins during food processing?
4. What are the factors that affect the degradation of vitamin C? What are the roles of vitamin C in the food industry?
5. During the food storage process, what factors are related to the loss of vitamins?
6. What are the main physiological functions of vitamin C?
7. What are the changes in vitamins in food after harvest or slaughter?

Table 6.13 Classification, physiological functions and sources of vitamins

Classification	Name	Common name	Physiological function	Main source
Water-soluble vitamins	Vitamin B$_1$	Thiamine	Maintain nerve conduction and prevent beriberi	Yeast, cereal, liver, germ
	Vitamin B$_2$	Riboflavin	Promote growth, prevent lip, glossitis, seborrheic dermatitis	Yeast, liver
	Vitamin B$_5$	Niacin	Prevention of dermatosis, dermatitis, glossitis	Yeast, germ, liver, rice bran
	Vitamin B$_6$	Pyridoxine	Related to amino acid metabolism	Yeast, germ, liver, rice bran
		Folic acid	Prevention of pernicious anemia, stomatitis	Liver and kidney, lean meat, nuts
	Vitamin B	Cyanocobalamin	Prevention of pernicious anemia	liver
	Vitamin H	Biotin	Promotes lipid metabolism and prevents skin diseases	Liver, yeast, cheese
		Pantothenic acid	Promote metabolism	Meat, cereals, fresh vegetables
	Vitamin C	Ascorbic acid	Prevent and treat scurvy and promote cell interstitial growth	vegetable and fruit
	Vitamin P	Rutin	Maintain normal vascular permeability	lemon
Fat-soluble vitamins	Vitamin A	Retinol	Prevent epidermal cell keratinization and prevent dry eye disease	Liver, cod liver oil, carotene
	Vitamin D	Calcitonin	Regulate calcium and phosphorus metabolism to prevent rickets	Cod liver oil, milk
	Vitamin E	Tocopherol	Prevention of infertility	Cereal germ and its oil
	Vitamin K	Hemostatic vitamin	Promote blood coagulation	Liver, spinach, green tea

8.   Why are coarse grains more nutritious than refined grains?
9.   What are the main reasons for the lack of vitamins in the human body?
10. What is the effect of sulfur dioxide used in food processing on the nutrition of processed foods?
11. Which vitamins are lacking in children can cause slow growth, angular cheilitis, cheilitis and glossitis?
12. What are the factors that affect the bioavailability of vitamins?

Bibliography

1. Belitz, H.D., Grosch, W., Schieberle, P.: Food Chemistry. Springer, Berlin, Heidelberg (2009)
2. Bolland, M.J., Grey, A., Avenell, A.: Effects of vitamin D supplementation on musculoskeletal health: a systematic review, meta-analysis, and trial sequential analysis. Lancet Diab. Endocrinol. **6**(11), 847–858 (2018)
3. Cai, D.L.: Practical Dietitian Handbook. China People's Health Publishing House, Beijing (2009)
4. Cashman, K.D.: Vitamin D deficiency: defining, prevalence, causes, and strategies of addressing. Calcif. Tissue Int. **106**(1), 14–29 (2020)
5. Cha, X.L.: Biochemistry. China People's Health Publishing House, Beijing (2010)
6. Fennema, O.R.: Food Chemistry. Marcel Dekker, New York (1996)
7. Galli, F., Azzi, A., Birringer, M., Cook-Mills, J.M., Eggersdorfer, M., Frank, J., Cruciani, G., Lorkowski, S., Ozer, N.K.: Vitamin E: emerging aspects and new directions. Free Radical Biol. Med. **102**, 16–36 (2017)
8. Ge, K.Y.: Public Dietitian Basics. China Labor and Social Security Publishing House, Beijing (2013)
9. Gil, A., Plaza-Diaz, J., Dolores Mesa, M., Vitamin, D.: Classic and novel actions. Ann. Nutr. Metab. **72**(2), 87–95 (2018)
10. Green, R., Allen, L.H., Bjorke-Monsen, A.-L., Brito, A., Gueant, J.-L., Miller, J.W., Molloy, A.M., Nexo, E., Stabler, S., Toh, B.-H., Ueland, P.M., Yajnik, C.: Vitamin B-12 deficiency. Nat. Rev. Dis. Primers **3**, 17041 (2017)
11. Holick, M.F.: The vitamin D deficiency pandemic: approaches for diagnosis, treatment and prevention. Rev. Endocr. Metab. Disord. **18**(2), 153–165 (2017)
12. Jiang, Q.: Natural forms of vitamin E: metabolism, antioxidant, and anti-inflammatory activities and their role in disease prevention and therapy. Free Radical Biol. Med. **72**, 76–90 (2014)
13. Kan, J.: Food Chemistry. China Agricultural University Press, Beijing (2016)
14. Lee, S.A., Bedford, M.R.: Inositol—an effective growth promotor? Worlds Poult. Sci. J. **72**(4), 743–759 (2016)
15. Li, F.L., Li, F.Y., Zhang, Z.: Food Nutrition. Beijing, China Chemical Industry (2009)
16. Lu, D., Yang, Y., Sun, C., Wang, S.: Determination of tocopherols and tocotrienols in cereals and nuts by dispersive solid-phase microextraction-gas chromatography-mass spectrometry. Anal. Meth. **11**(42), 5439–5446 (2019)
17. Niki, E.: Role of vitamin E as a lipid-soluble peroxyl radical scavenger: *in vitro* and *in vivo* evidence. Free Radical Biol. Med. **66**, 3–12 (2014)
18. Sassi, F., Tamone, C., D'Amelio, P.: Vitamin D: Nutrient, hormone, and immunomodulator. Nutrients **10**(11), 1656–1670 (2018)
19. Sijilmassi, O.: Folic acid deficiency and vision: a review. Graefes Arch. Clin. Exp. Ophthalmol. **257**(8), 1573–1580 (2019)

20. Slagman, A., Harriss, L., Campbell, S., Muller, R., McDermott, R.: Folic acid deficiency declined substantially after introduction of the mandatory fortification programme in Queensland, Australia: a secondary health data analysis. Public Health Nutr. **22**(18), 3426–3434 (2019)
21. Stabler, S.P.: Vitamin B-12 deficiency. N. Engl. J. Med. **368**(2), 149–160 (2013)
22. Wu, K.: Nutrition and Food Hygiene. China People's Health Publishing House, Beijing (2003)
23. Yang, Y.X.: Food Nutrition Book. China Science Press, Beijing (2009)
24. Yang, C.S., Luo, P., Zeng, Z., Wang, H., Malafa, M., Suh, N.: Vitamin E and cancer prevention: studies with different forms of tocopherols and tocotrienols. Mol. Carcinog. **59**(4), 365–389 (2020)

Dr. Yali Yang has been working in the Department of Food Engineering and Nutritional Science at Shaanxi Normal University, China, since 2016. She has got her Ph.D. degree at Universitat Politècnica de Catalunya, Spain, from 2012 to 2015. She teaches undergraduate and graduate courses in food flavor chemistry and food microorganism. She has obtained two Chinese patents and published over 15 academic papers. She has obtained two government grants including the National Natural Science Foundation of China and Key Research and Development Programme of Shaanxi Province. Her research includes extensive work on the nutritional and functional properties of fruits and vegetables and the flavor contributions of peptides in the Maillard reaction

Chapter 7
Minerals

Aidong Sun and Hui Li

Abstract In this chapter, various minerals including macro-element such as calcium, phosphorus, potassium, sodium, chlorine and micro element such as iron, iodine, copper, zinc, and selenium have been introduced according to their distribution and forms, resources, and bioavailability in human, as well as their bioactive functions in human body. Some toxic minerals have also been illustrated about their toxicity and contamination pathways in food industry. It should be acquired that the mineral changes during food processing, storage, and their impact for the human utilization.

Keywords Macro-element · Micro-element · Function · Bioactivity · Toxic mineral

7.1 Overview

There are mineral elements of varying amounts in food, many of which are essential for human nutrition. These mineral elements exist either in the form of inorganic or organic salts or in combination with organic substances, such as phosphorus in phosphoproteins and other metal elements in enzymes. Among these mineral elements, it has been found that about 25 kinds of mineral elements are necessary for constituting human tissues, maintaining physiological functions, and biochemical metabolism. In addition to carbon, hydrogen, oxygen, and nitrogen that are present in organic compounds, all the others are referred to inorganic salts or minerals. Meanwhile, these mineral elements cannot be synthesized in the body but must be provided by food.

A. Sun (✉)
College of Biological Science and Technology, Beijing Forestry University, Beijing 100083, China
e-mail: adsun68@163.com

H. Li
Institute of Quality Standards and Testing Technology for Agro-Products, Chinese Academy of Agricultural Sciences, Beijing 100081, China

© The Author(s), under exclusive license to Springer Nature Singapore Pte Ltd. 2021   291
J. Kan and K. Chen (eds.), *Essentials of Food Chemistry*,
https://doi.org/10.1007/978-981-16-0610-6_7

Table 7.1 Main elements in the human body

Element	Content (g/kg)
Calcium	10–20
Phosphorus	6–12
Potassium	2–2.5
Sodium	1–1.5
Chlorine	1–1.2
Magnesium	0.4–0.5

According to the different levels and required quantity of these mineral elements in the human body, they are habitually divided into two categories. One type is macro-element, such as six elements including calcium, phosphorus, sodium, potassium, chlorine, and magnesium; their amounts are about 60–80% of the total ash in the human body, with the body content > 0.01% (Table 7.1), and the amounts of body needs are more than 50 mg/d. The other type is microelements or trace elements, which contain only trace or ultra-trace, including 14 elements such as Fe, I, Cu, Zn, Se, Mo, Co, Cr, Mn, F, Ni, Si, Sn, and V; the first eight elements are currently considered to be essential trace elements in the human body; the latter ones are possibly essential for the human body. Trace elements are divided into three types: (1) essential elements including Fe, Cu, I, Co, Mn, and Zn; (2) non-nutritive but non-toxic elements including Al, B, Ni, Sn, and Cr; (3) Non-nutritional and toxic elements including Hg, Pb, As, Cd, and Sb.

The main functions of minerals in the human body:

(1) Be an important part of the body. The minerals in the body are mainly found in the bones and maintain the rigidity of the bones. For example, 99% of the calcium and a large amount of phosphorus and magnesium are present in the bones and teeth. In addition, phosphorus and sulfur are constituent elements of proteins, while potassium and sodium are commonly contained in cells.

(2) Maintain the osmotic pressure of the cells and the acid-base balance of the body. Minerals and protein maintain the balance of osmotic pressure of the cells, and play an important role in the retention and movement of body fluids. In addition, buffer systems composed of carbonates and phosphates together with proteins constitute the acid-base buffer system of the body, maintaining the acid-base balance of the body.

(3) Maintain excitability of nerves and muscles. Certain ratio of plasmas, such as K, Na, Ca, and Mg, plays an important role in maintaining the excitability of nerves and muscle tissues, and the permeability of cell membranes as well.

(4) Perform special physiological effects, such as the importance of iron for hemoglobin, cytochrome enzymes, and the importance of iodine for synthesis of thyroxine.

(5) The effects on the sensory quality of food. Minerals also play an important role in improving the sensory quality of foods, such as the effect of phosphates on

water retention and sequestration of meat products, and the effect of calcium on the formation of some gels and the hardening of food textures.

Some macroelements in food, especially monovalent, including cations such as sodium and potassium, and anions such as chlorine and sulfate, generally exist in a soluble state, and most of them are in a free state. Some polyvalent ions are often in a steady state of free, dissolved, rather than ionized, colloidal forms, for example, such homeostasis exists in meat and milk; metal elements are often present in a chelated state, such as cobalt in vitamin B12.

The change of mineral amounts in foods primarily depends on environmental factors, such as the composition of the soil on which plant grow, or the composition of animal feed. The loss of minerals in foods caused by chemical reactions are not as severe as those caused by physical removal or unavailable forms for organisms. Minerals are initially lost through the leaching of water-soluble substances and the elimination of non-edible parts of plant. For example, minerals in rice are mainly lost during grain milling, and the higher the processing accuracy, the more seriously the minerals are lost. Therefore, it is necessary to supplement some trace minerals in the diet. The interaction between minerals and other ingredients in foods is also important. For example, polyvalent anions such as oxalic acid and phytic acid present in some foods can form extremely insoluble salts with divalent metal ions, as a result, they cannot be absorbed and utilized by intestine. Therefore, it is very necessary to determine the bioavailability of minerals.

7.2 Basic Properties of Mineral Absorption and Utilization in Foods

To make full and reasonable use of minerals, firstly, it is necessary to understand their properties, states and changes in processing or storage of foods. The following is a brief introduction to the relevant physical and chemical properties.

7.2.1 Solubility

Water is present in all biological systems, and most mineral elements are transported and metabolized in aqueous solution. Therefore, the bioavailability and activity of minerals are largely directly related to their solubility in water. Magnesium, calcium, and strontium are homologous elements and exist only in the +2-oxidation state. Although the halides of this group are soluble, their important salts including hydroxides, carbonates, phosphates, sulfates, oxalate, and phytate are all extremely difficult to dissolve. When food is decomposed by certain bacteria, the magnesium there can form a very insoluble complex $NH_4MgPO_4 \cdot 6H_2O$, which is commonly known as struvite. Copper is present in the oxidation state of $+1$ or $+2$ and forms complex ions, whose halides and sulfates are soluble. Minerals of various valences may form different types of compounds in the water with organic substances in life body, such as proteins, amino acids, organic acids, nucleic acids, nucleotides, peptides, and sugars, which is beneficial to the stability of minerals and transport of minerals between organs. In addition, the chemical form of the element also affects the utilization and function of the element. For example, the ferric ion is difficult to be absorbed by the human body, but the ferrous ion is more easily absorbed and utilized (see Table 7.2); trivalent chromium ions are essential nutrients for the body, while hexavalent chromium ions are toxic (Table 7.3).

7.2.2 Acidity and Alkaline

Any mineral has cations and anions. However, considering nutrition of minerals, only the anions of fluoride, iodide, and phosphate are important. Fluoride components in water are more common than in foods, and their intake is highly dependent on geographic location. Iodine exists in the forms of iodide (I^-) or iodate (IO_3^-). Phosphates exist in many different forms, such as phosphate (PO_4^{3-}), hydrogen phosphate (HPO_4^{2-}), dihydrogen phosphate ($H_2PO_4^-$), or phosphoric acid (H_3PO_4), with the ionization constants $k_1 = 7.5 \times 10^{-3}$, $k_2 = 6.2 \times 10^{-8}$ and $k_3 = 1.0 \times 10^{-12}$, respectively. The complex biological processes involved in various trace elements can be explained by Lewis's acid-base theory. Since the same element of different valence states can participate in different biochemical processes by forming a variety of complexes, it shows different nutritional values.

7.2.3 Redox Property

Iodides and iodates are relatively strong oxidants compared to other important inorganic anions in foods such as phosphates, sulfates, and iodates. Cations are more numerous than anions and more complex in structure, and their general chemical properties can be determined by their located family in the periodic table of elements. Some metal ions are important as a nutritional factor, while others are very harmful toxic pollutants and even cause carcinogenic effects. Carbonates and phosphates are more difficult to dissolve. Other metal elements have multiple oxidation states, such as tin and aluminum (+2 and +4), mercury (+1 and +2), iron (+2 and +3), chromium (+3 and +6), manganese (+2, +3, +4, +6, and +7), so many of these metal elements can form zwitterions, both as an oxidant and as a reducing agent. For instance, molybdenum and iron can catalyze the oxidation of ascorbic acid and unsaturated lipids. These valence changes of trace elements and the equilibrium reaction of mutual conversion will affect the environmental characteristics in tissues and organs, such as pH, ligand composition, and electrical effects, thus affect their physiological functions.

Table 7.2 Key characteristics of iron compounds commonly used for food fortification purpose: solubility, bioavailability, and cost [1]

Compound	Iron content (%)	Relative bioavailability[a]	Relative cost[b] (per mg iron)
Water soluble			
Ferrous sulfate. $7H_2O$	20	100	1.0
Ferrous sulfate, dried	33	100	1.0
Ferrous gluconate	12	89	6.7
Ferrous lactate	19	67	7.5
Ferrous bisglycinate	20	>100[c]	17.6
Ferric ammonium citrate	17	51	4.4
Sodium iron EDTA	13	>100[c]	16.7
Poorly water soluble, soluble in dilute acid			
Ferrous fumarate	33	100	2.2
Ferrous succinate	33	92	9.7
Ferric saccharate	10	74	8.1
Water insoluble, poorly soluble in dilute acid			
Ferric orthophosphate	29	25–32	4.0
Ferric pyrophosphate	25	21–74	4.7
Elemental iron		–	–
H-reduced	96	13–148[d]	0.5
Atomized	96	(24)	0.4
CO-reduced	97	(12–32)	<1.0
Electrolytic	97	75	0.8
Carbonyl	99	5–20	2.2
Encapsulated forms			
Ferrous sulfate	16	100	10.8
Ferrous fumarate	16	100	17.4

EDTA, ethylenediamineteraacetate; H-reduced, hydrogen reduced; CO-reduced, carbon monoxide reduced

[a]relative to hydrated ferrous sulfate ($FeSO_4.7H_2O$), in adult humans. Values in parenthesis are derived from studies in rats

[b]relative to dried ferrous sulfate. Per mg of iron, the cost of hydrated and dry ferrous sulfate is similar

[c]absorption is two-three times better than that from ferrous sulfate if the phytate content of food vehicle is high

[d]the high value refers to a very small particle size which has only been used in experimental studies

Table 7.3 Iron salts for food fortification and their bioavailability

Compounds	Fe content (mg/kg)	Relative bioavailability	
		Human	Mice
Ferrous sulfate	200	100	100
Ferrous lactate	190	106	–
Ferric pyrophosphate	250	–	45
Ferric sodium pyrophosphate	150	15	14
Ferrous ammonium citrate	165-185	–	107
Ferrum	960–980	13–90	8–76

7.2.4 Concentration of Trace Elements

The concentration and state of trace elements will affect various biochemical reactions. Many unexplained diseases (such as cancer and endemic diseases) are related to trace elements. However, in fact, the determination of essential trace elements is not easy, because of the differences on the valence, concentration of mineral elements, and even the order and state of the arrangement, and the subsequent different life activities of living things as well.

7.2.5 Chelate Effect

Many metal ions can act as ligands or chelating agents for organic molecules, such as iron in heme, copper in cytochromes, magnesium in chlorophyll, and cobalt in vitamin B12. Chromium with a biologically active structure is called glucose tolerance factor (GTF), which is an organic complex form of trivalent chromium. It is 50 times more potent than inorganic Cr^{3+} ions in glucose tolerance bioassays. The GTF contains niacin, cysteine, glycine, and glutamic acid in addition to about 65% chromium, but the precise structure is unclear currently. Cr^{6+} has no biological activity. The chelation effect of metal ions and the stability of the chelate are affected by its own structure and environmental factors. Generally, five-membered and six-membered cyclic chelates are more stable than other larger or smaller rings. Lewis alkaline of metal ions also affects its stability: generally, the stronger basicity, and higher stability. Charged ligands facilitate the formation of stable chelates. Different electron donors have different coordination bond strengths: $H_2O>ROH>R_2H$ for oxygen, $H_3N>RNH_2>R_3N$ for nitrogen, and $R_2S>RSH>H_2S$ for sulfur. In addition, the conjugated structure and effect of steric hindrance in molecules facilitate the stabilization of the chelate.

7.2.6 Utilization of Minerals in Foods

The determination of the total amount of a mineral element in a food or diet provides only a limited nutritional value, while the determination of the amount used by the human body has greater practical significance. The utilization of iron and iron salts in food depends not only on their form of existence, but also on the various conditions that affect their absorption or utilization. Methods for determining the bioavailability of minerals include chemical equilibrium, bioassays, tests in vitro, and isotope tracers. These methods have been widely used to determine the digestibility of minerals in livestock feed.

The radioisotope tracer technology is an ideal method for detecting the use of minerals by the human body. It refers to add radioactive iron to the medium in which the plant grows, or to inject radioactive tracer substances (^{55}Fe and ^{59}Fe) before the animal is slaughtered. The radioactive tracer substance is made into a labeled food by biosynthesis, and then the labeled food is consumed, consequently the absorption of radioactive tracer substances is determined, which is called the internal standard method. The external standard method can also be used to study the absorption of iron and zinc in foods, which refers to add radioactive elements to foods.

The bioavailability of minerals is related to many factors. There are:

(1) The solubility and states of minerals in water. The better the water solubility of minerals, the better the absorption and utilization of the body. In addition, the states of minerals also affect the utilization of the elements.

(2) The interaction between minerals. The absorption of minerals by the body sometimes antagonizes, which may be related to their competitive carriers. For example, the absorption of excessive iron will affect the absorption of zinc and manganese.

(3) Chelate effects. Metal ions can interact with different ligands to form corresponding complexes or chelates. Chelates in food systems not only increase or decrease the bioavailability of minerals, but also play other roles, such as preventing the pro-oxidant effect of iron and copper ions. The ability of minerals to form chelates is related to their own properties.

(4) The effects of intakes of other nutrients. The intake of protein, vitamins, and fats will affect the absorption and the utilization of minerals by the body. For example, the intake level of vitamin C is related to the absorption of iron. The effect of vitamin D on the absorption of calcium is more obvious. The insufficient intake of proteins causes the decreased absorption level of calcium, while excessive intake of fat affects the absorption of calcium. Excessive phytate, oxalate, and phosphate in food can also reduce the bioavailability of minerals.

(5) The physiological state of the human body. The human body can mediate the absorption of minerals to maintain the relative stability of the body environment. For example, when a certain mineral is lacking in food, its absorption rate will increase; when the food is adequately supplied, the absorption rate will decrease. In addition, the state of the body, such as diseases, age, and individual differences will cause changes in the utilization of minerals in the body.

For example, iron-deficient or iron-deficient anemia patients have an increased absorption rate of iron; women's absorption of iron is higher than that of men; and as children age increases, the absorption of iron is reduced. The utilization of zinc is also affected by a variety of dietary and individual factors.

(6)   The nutritional compositions of foods. The nutritional compositions of foods also affect the absorption of minerals in the body. For example, the absorption rate of minerals in meat products is higher, while the absorption rate of minerals in grains is lower (Fig. 7.1). Heme iron from consumption of meat, poultry, and fish is highly bioavailable (15–35%) and dietary factors have little effect on its absorption, whereas nonheme iron obtained from cereals, pulses, legumes, fruits, and vegetables is much lower absorbed (2–20%) and strongly influenced by the presence of other food components [9]. Vitamins enhance the absorption of iron. In the case of very low calcium levels, phosphate and sugar reduce the absorption of iron; all of proteins, amino acids, and carbohydrates affect the utilization of iron. Moreover, the conditions of the intake of calcium salt, the type of the calcium salt, other ingredients in the diet (such as phosphoric acid, phytic acid, tannin, dietary fiber) may affect the intake of calcium in the body. Vitamin D, lactose, lactic acid, and oligosaccharides are beneficial for the absorption of calcium; the calcium absorption rate of calcium from L-calcium lactate is much higher than from calcium phosphate, and higher than calcium carbonate and citric acid/calcium malate.

In addition, the valence of iron affects the absorption, and the ferrous salt is easier to be utilized than the ferric salt; the size of the iron particles and the type of foods also affect the absorption of iron.

A rational and effective food fortification plan requires complete information on the sources of foods and the availability of minerals in the diet, which is also important for the assessment of the nutritional properties of alternative foods and similar foods. Further research regarding human nutrition is necessary to determine the bioavailability of various essential trace elements, and to understand various factors affecting the utilization of minerals in modern diets.

Fig. 7.1 Absorption of iron in various foods by adults

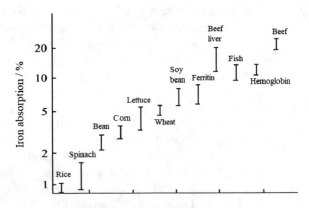

7.2.7 Safety of Minerals in Foods

From the point of nutrition, some minerals not just have no nutritional value, but even cause harm to human health, such as mercury and cadmium. At the same time, all excessive minerals, even if they are essential trace elements, are toxic to humans. The relationship between the safe dose of several minerals and the toxic dose is shown in Table 7.4. As shown in Table 7.4, the range of minerals overcommitted overlaps between different degrees of reaction, and the range of doses required to produce each degree of reaction is often quite large. For many mineral elements, a small amount of intake is toxic while a large quantity of intake is not, the reason of which is unclear at present. This indicates that the safety of some minerals is uncertain when the intake of them exceeds the required amounts.

Table 7.4 The relationship between the excessive intake of minerals [a] and the severity of toxic reactions

Minerals	Symptoms[b]	Toxic levels		
		None	Medium	Severe
Copper	Acute	–	2–16	125–50,000
	Chronic	–	–	0.5–4
Cobalt	Chronic	–	150	35–600
Fluorine	Acute	–	3–19	80–3,000
	Chronic	–	3–19	2–180
Iron	Acute	–	–	12–1,500
	Chronic	7	–	6–15
Iodine	Chronic	8–35,000	1–15,000	1–180,000
Tin	Acute	–	130	23–700
Zinc	Acute	13–23	–	8–530
	Chronic	2–10	10	–

[a]excessive oral intake means the actual intake of the reaction divided by the required intake or the equivalent average intake
[b]acute, within 24 h; chronic, from one day to several generations
Source Food and Drug Administration (1975). Toxicity of the Essential Minerals. DHEW, Washington, D.C

7.3   Common Macro Minerals

7.3.1   Calcium

Calcium (Ca), with atomic number 20, and relative atomic weight 40.078, belongs to the periodic system IIA, is a member of alkaline earth metal. It has a melting point of 839 °C, with the boiling point of 1484 °C, and the relative density of 1.54.

Calcium is the most important and most abundant mineral element in the human body. In general, the body of an adult contains 1200–1500 g of calcium. About 99% of calcium form hydroxyapatite crystals $Ca_3(PO_4)_2$. $(OH)_2$ and calcium phosphate together with phosphorus, concentrating in bones and teeth. The remaining 1% of calcium either chelates with citric acid or binds to proteins, and most are presenting in soft tissue, cell fluid and blood in an ionic state, which is called miscible calcium pool. Among them, liver contains 100–360 mg/kg, muscle contains 140–700 mg/kg, blood contains 60.5 mg/dm^3, and bone contains 170000 mg/kg, with daily intake 600–1400 mg.

The physiological function of calcium is to form bones and teeth, maintain nerve and muscle activity, and promote the activity of certain enzymes in the body. In addition, calcium is involved in the blood coagulation process, hormone secretion, maintenance of body fluid acid-base balance, and intracellular gelatin stability.

The physiological function of calcium is to constitute bones and teeth, maintain the activity of nerves and muscles, and promote the activity of certain enzymes in the body. In addition, calcium is involved in the processes of blood coagulation and hormone secretion, maintenance of acid-base balance of body fluid, and intracellular stability of gelatin.

Some ingredients in foods will affect the absorption and utilization of calcium (Table 7.5). For example, oxalate and phytate in the diet can combine with calcium to form a salt which is difficult to be absorbed; dietary fiber interferes with the absorption of calcium, possibly caused by the combination of aldonic acid residues and calcium; vitamin D can promote the absorption of calcium; lactose can chelate with calcium to form a soluble complex with low molecular weight, which is beneficial to the absorption of calcium; dietary protein is beneficial to the absorption of calcium. Therefore, the absorption of calcium in the diet is very incomplete and can only be absorbed by 20–30%.

Table 7.5 Effects of dietary ingredients on the absorption and utilization of calcium

Promotion	Reduction	No effects
Lactose	Phytate	Phosphorus
Some amino acids	Dietary fiber	Protein
Vitamin D	Oxalate	Vitamin C
	Fat (in case of dyspepsia)	Citric acid
	Alcohol	Pectin

Table 7.6 Amount of calcium in common foods (mg/100 g edible part)

Food	Amount	Food	Amount	Food	Amount
Human milk	34	Kelp (dried)	1177	Broad bean	93
Cow milk	120	Hair weeds	767	Yuba	280
Cheese	590	Tremella	380	Peanut kernel	67
Yolk	134	Agaric	357	Almond (raw)	140
Standard flour	24	Seaweed	343	Watermelon seed (dried)	237
Standard rice	10	Soy	367	Pumpkin seeds (fried)	235
Dried shrimp	2000	Tofu skin	284	Walnut	119
Pork (lean)	11	Tofu	240–277	Bok-choy	93–163
Beef (lean)	6	Green beans	240	Cabbage	61
Lamb (lean)	13	Cowpea	100	Rape	140
Chicken (lean)	11	Pea	84	Chives	105

Food sources of calcium should involve two considerations, one is the amount of calcium in the food, and the other is the absorption rate of calcium. Milk and dairy products are the best sources of calcium. They are not only rich in calcium, but also a high absorption rate of calcium, therefore they are the best sources of calcium for infants and young children. Small dried shrimps, kelp, and hairy vegetables are also rich in calcium; addition of bone meal and eggshell powder is also an effective way to supplement dietary calcium; there are also high amount of calcium in beans, soy products, and oilseeds (Table 7.6).

7.3.2 Phosphorus

Phosphorus (P), with atomic number 15, and relative atomic weight 30.973762, is one of the most abundant elements in the human body, second only to calcium. Both phosphorus and calcium are important constituents of bone and teeth, with a calcium/phosphorus ratio of 2:1 approximately. Normal adult body contains 1% of phosphorus, with 80% of it, about 600–900 g locating in bone, and the remaining 20% is distributed in soft tissues such as nervous tissue. The human body contains 35–45 mg/100 ml of phosphorus in whole blood, 3–8.5 mg/kg in liver, 3000–8500 mg/kg in muscle, 67000–71000 mg/kg in bone, and the daily intake of phosphorus is 900–19000 mg.

Phosphorus is an essential component of bone tissue. It is about 650 g in adult body, about 1% of body weight, with the ratio to calcium 1:2. Nearly 85% (700 g) of phosphorus in adult body is distributed in bones. Phosphorus exists as a soluble phosphate ion in soft tissues and as a bonded form of ester or glycoside in fat, protein, carbohydrates, and nucleic acids. Within enzymes, it exists as a regulator of enzyme activity. Phosphorus also plays an important role in many different biochemical reactions in the body. Most of the energy required for metabolism is derived from the phosphate bond of adenosine triphosphate, creatine phosphate, and similar compounds.

Phosphorus physiologically functions as follows: firstly, it functions as the important components of bone, teeth, and soft tissues; secondly, phosphorus is also the regulator of energy release, because the energy in the metabolism of the body is mostly stored in the form of ADP + phosphoric acid + energy = ATP and phosphoinositide; thirdly, phosphorus is the important component of life substances such as phosphoric acid, phosphoprotein, and nucleic acid; fourthly, it is an important component of enzymes, for example, the activity of thiamine pyrophosphate, pyridoxal phosphate, coenzyme I, coenzyme II and other coenzymes or prosthetic groups all require the participation of phosphorus. In addition, phosphate is also involved in the regulation of acid-base balance.

Phosphorus is widely found in animal and plant tissues, and binds to proteins or fats to form nucleoproteins, phosphoproteins and phospholipids, and small amounts of other organic and inorganic phosphorus compounds as well. Most forms of phosphorus can be utilized by the body except that phytic acid cannot be fully absorbed. The main form of phosphorus in cereal seeds is phytic acid, therefore, the utilization rate is very low. However, fermentation of flour with yeast or immerse of grain in hot water can greatly reduce the amount of phytate phosphorus, thereby increase the absorption rate. Intake of cereals for long terms can make the body adapted to phytic acid, and improve the absorption rate of phytate phosphorus to a certain extent.

Phosphorus is widely distributed in foods, especially in cereals and protein-rich foods such as lean meat, eggs, fish (eggs), viscera, kelp, peanuts, beans, nuts, and coarse grains. Therefore, the general diet can meet the requirement of the human body.

7.3.3 Magnesium

Magnesium (Mg), with atomic number 12, and relative atomic weight 24.3050, has a melting point 649 °C, with the boiling point 1090 °C.

Magnesium accounts for 0.05% of body weight, of which about 60% is present in bones and teeth in the form of phosphate, 38% is combined with protein to form complexes in soft tissues, and 2% is present in plasma and serum. Among them, liver contains magnesium 590 mg/kg; muscle contains 900 mg/kg; blood contains 37.8 mg/L; bone contains 700–1800 mg/kg; and daily intake is 250–380 mg.

Physiological functions of magnesium are as follows: firstly, magnesium is one of the most abundant cations in the human body, which is the main component of bones, teeth and cytoplasm, regulating and inhibiting muscle contraction and nerve impulses and maintaining acid-base balance in the body, normal function and structure of the heart muscle; secondly, magnesium is also an activator of many enzymes that activates many enzyme systems (such as alkaline phosphatase, enolase, and leucine aminopeptidase), and also is a necessary cofactor for oxidative phosphorylation.

Magnesium is widely distributed in various foods. Fresh green vegetables, seafood, and beans are good sources for magnesium. Coffee (instant), cocoa powder, cereals, peanuts, walnuts, whole wheat flour, millet, and bananas also contains much magnesium, but milk contains little magnesium. Therefore, the lack of dietary magnesium generally does not occur. However, chronic diarrhea will cause excessive discharge of magnesium, which may cause symptoms of magnesium deficiency with symptoms including depression, dizziness, and muscle weakness.

7.3.4 Potassium

Potassium (P), with atomic number 19, and relative atomic weight 39.0983, belongs to the periodic family IA and is a member of alkali metal. The melting point of it is 63.25 °C, the boiling point is 760 °C, and the density is 0.86 g/cm^3 (20 °C). Normal human body contains about 175 g of potassium, 98% of which is stored in the cell fluid, which is the most important cation in the cell. Among them, liver contains potassium 16000 mg/kg; muscle contains 16000 mg/kg; blood contains 1620 mg/L; bone contains 2100 mg/kg; and the daily intake is 1400–1700 mg.

Potassium's physiological functions: maintains normal metabolism of carbohydrates and proteins; maintains normal osmotic pressure in cells; maintains neuromuscular stress and normal function; maintains normal function of myocardium; maintains the normal acid-base balance and ionic equilibrium of cells; lowers blood pressure.

Potassium is widely distributed in foods. Meat, poultry, fish, various fruits and vegetables are all good sources of potassium, such as dehydrated fruits, tomato, spinach, radish, watercress, papaya, red pepper, lamb, beef, potato, banana, syrup, rice bran, seaweed, soy flour, spices, sunflower seeds, and wheat bran and beef. However, when sodium is restricted, potassium of these foods is also limited. The people who urgently need to supplement with potassium are those who drink a lot of coffee, those who often drink alcohol and who like to eat sweets, those who have hypoglycemia, and those who are on diet for a long time.

7.3.5 Sodium

Sodium (Na), with atomic number 11, and relative atomic weight of 22.990, has the melting point of 98 °C and the boiling point of 883 °C. Sodium is present as a salt in the body fluid. Among them, liver contains sodium 2000–4000 mg/kg; muscle contains 2600–7800 mg/kg; blood contains 1970 mg/L; bone contains 2100 mg/kg; and the daily intake is 2000–15000 mg.

The physiological functions of sodium: (1) is the main positive ion in the extracellular fluid, participating in the metabolism of water to ensure the balance of water in the body; (2) cooperate with potassium to maintain the acid-base balance of the body fluid; (3) are components of gastric juice together with chlorine, functioning for digestion, and constitute pancreatic juice, bile, sweat, and tears; (4) regulate cell excitability and maintain normal myocardial movement; (5) composed of salt with chloride ions which is an indispensable condiment.

In addition to cooking, processing, and flavoring salt (sodium chloride), sodium is present in all foods at varying levels. In general, protein-containing foods contain more sodium than vegetables and grains, while there is little or no sodium in fruits. The main food sources of sodium are smoked pork, red sausage, gluten, corn flakes, cucumber, ham, green olives, luncheon meat, oatmeal, potato chips, sausage, seaweed, shrimp, soy sauce, ketchup, etc. Therefore, people rarely have problems of sodium deficiency. However, in the case of taking vegetarian diet without salt or persistent sweating, diarrhea, or vomiting, sodium deficiency will occur, which may cause symptoms of slow growth, loss of appetite, weight loss, muscle spasm, nausea, diarrhea, and headache.

7.3.6   Chlorine

Chlorine (Cl), with atomic number 17, and relative atomic weight 35.4527, belongs to the periodic family VIIA and is a member of halogen. It has the melting point of -100.98 °C, the boiling point of -34.6 °C, and the density of 3.214 g/L. Chlorine accounts for about 0.15% of the body weight, distributed in the forms of chloride in the tissues of whole body and in form of salts in body fluid, and most abundant in cerebrospinal fluid and gastrointestinal secretions. Among them, liver contains chlorine of 3000–7200 mg/kg; muscle contains 2000–5200 mg/kg; blood contains 2890 mg/L; bone contains 900 mg/kg; and the daily intake is 3000–6500 mg.

The physiological function of chlorine is to compose the digestive tract secretions such as gastric acid and intestinal fluid, which is related to digestion. Cl^- and Na^+ are important ions that maintain the osmotic pressure of cells and the acid-base balance of the body fluid, and participate in the metabolism of water.

Salt and salty foods are sources of chlorine. It is worth noting that it is usually balanced between the loss of chlorine and the loss of sodium in the body. When the intake of sodium chloride is limited, the amount of chlorine in the urine decreases, and that in tissues also decreases. When sweating and diarrhea occur, the loss of sodium increases, which also causes loss of chlorine.

7.4 Common Trace Elements

7.4.1 Iron

Iron (Fe), with atomic number 26, and relative atomic weight 55.847, belongs to the periodic family VIII. Its melting point is 1535 °C and the boiling point is 2750 °C. Iron is the most abundant trace element in the body. A healthy adult contains 3–5 g iron, about 0.004% of the body weight in the body. About 60–70% of iron is present in hemoglobin; about 30% is present in the forms of ferritin and hemosiderin in liver, spleen, and bone marrow; about 3% is present in myoglobin; < 1% is present in various enzyme systems (cytochrome, cytochrome oxidase, peroxidase, and catalase), and a small part is present in blood transferrin. Among them, liver contains iron 250–1400 mg/kg, muscle contains 180 mg/kg; blood contains 447 mg/L; bone contains 3–380 mg/kg; and the daily intake is 6–40 mg.

The physiological roles of iron: (1) to combine with protein to form hemoglobin and myoglobin, participating in the transport of oxygen, promoting hematopoiesis, and to maintain the normal growth and development of the body; (2) is the component of many important enzyme systems of the body, such as cytochrome, catalase, and peroxidase, participating in the respiration of tissues and promoting the biological redox reaction; (3) is an alkaline element and one of the basic substances to maintain the acid-base balance in the body; (4) to increase the resistance of the body to diseases.

The resources of iron in foods are hemoglobin iron and non-hemoglobin iron. The former has an absorption rate of 23% and the latter is 3–8%. High amount of iron is found in animal foods such as liver, animal blood, meat, and fish (Fig. 7.2), and the iron they contain is hemoglobin iron (also known as ferrous iron) and can be directly absorbed by the intestinal tract. The iron contained in plant foods including cereals, fruits, vegetables, legumes, and animal foods including milk and eggs is non-hemoglobin iron, which exists as a complex. The organic part of the complex is

Fig. 7.2 Content of Fe in foods. Data are in mg/100 g edible portion (average values). The figure was created according to data from: Belitz et al., Food Chemistry (book), 4th edition, 2009

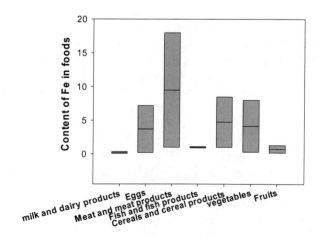

Table 7.7 Amount of iron in common foods (mg/100 g edible part)

Food	Amount	Food	Amount	Food	Amount
Rice	2.3	Black fungus (dried)	97.4	Celery	0.8
Standard flour	3.5	Pork (lean)	3.0	Rape	7.0
Millet	5.1	Pork liver	22.6	Chinese cabbage	4.4
Corn (fresh)	1.1	Chicken liver	8.2	Spinach	2.5
Bean	8.2	Eggs	2.0	Dried dates	1.6
Red bean	7.4	Shrimp	11.0	Raisin	0.4
Green bean	6.5	Kelp (dried)	4.7	Walnut kernel	3.5
Sesame paste	58.0	Hairtail	1.2	Longan	44.0

protein, amino acid, or organic acid. Non-hemoglobin iron must be separated from the organic part by gastric acid and become ferrous ions before absorbed by the intestine. Therefore, iron in animal foods is easier to be absorbed than that in plant foods. To prevent the deficiency of iron, animal food should be preferable.

However, milk has little iron (Fig. 7.2). Some countries limit the daily intake of fresh milk for children, weekly milk-free days is arranged to prevent iron deficiency anemia. Although iron is abundant in bean foods, it is not easy to be absorbed; more iron is contained in egg yolk, but the absorption rate of it is not high because of yolk phosphoprotein. The absorption rate of iron is only 2% in spinach as it contains high amount of oxalic acid. Therefore, the long-term deficiency of utilized iron in the diet can often lead to iron deficiency anemia, especially for infants, pregnant women, and nursing mothers. The amount of iron in vegetables is not high, and the absorption rate of iron in them such as in rape, leeks, spinach and chives is not high (Table 7.7).

7.4.2   Zinc

Zinc (Zn), with atomic number 30, and relative atomic weight 65.39, has a melting point 419.73 °C with the boiling point 907 °C. The amount of zinc in human body is about half of iron (1.4–2.3 g), which is widely distributed in human nerves, immunity, blood, bones, and digestive system, with a big part present in bones and skin. The concentration of zinc on the erythrocyte membrane is very high, mainly present in the components of metalloenzyme, carbonic anhydrase, and alkaline phosphatase; zinc in plasma is mainly combined with protein, and the amount of free zinc is little. The amount of zinc in hair can reflect the long-term supply of dietary zinc and the nutritional status of zinc in the body. Among them, liver contains zinc 240 mg/kg, and muscle, 240 mg/kg; blood, 7 mg/L; bone, 75–170 mg/kg; and the daily intake is 5–40 mg.

Physiological functions of zinc: it is a component or an activator of many enzymes in the body (such as alcohol dehydrogenase, glutamate dehydrogenase); it is closely related to the synthesis with nucleic acids and proteins, the metabolism of carbohydrates and vitamin A, and the activity of pancreas, gonads and pituitary gland. Zinc maintains the health of the digestive system and skin, and maintains normal night vision.

It is generally believed that high-protein foods contain high levels of zinc. Seafood is a good source of zinc, followed by dairy and egg products, while zinc in vegetables and fruits is not high. Fermented foods contain increased zinc, such as gluten and malt. However, phytic acid in cereals affects the absorption of zinc. Refined rice and flour have less zinc, that is, the more refined the diet, the less zinc it provides. Therefore, children who take cereals as their main food, or those who only take vegetables with no meat, are prone to have zinc deficiency.

Zinc is widely available, but its content and the absorption rate in animal foods differs a lot from that in plant foods significantly. In 100 g of edible part, the amount of zinc in oysters can be up to 100 mg or more, compared with 2–5 mg in the livestock and poultry meat, liver and eggs, 1.5 mg in fish and other seafood, and 0.3–0.5 mg in the livestock and poultry products, 1.5–2.0 mg in beans and cereals, and < 1.0 mg in vegetables and fruits.

7.4.3 Iodine

Iodine (I), with atomic number 53, and relative atomic weight 126.90447, belongs to the periodic family VIIA and a member of halogen. The melting point of it is 113.5 °C, the boiling point is 184.35 °C, and the density is 4.93 g/cm^3.

The human body contains about 25 mg of iodine, of which about 15 mg is present in the thyroid gland, and others are distributed in muscles, skin, bones, and other endocrine and central nervous systems. The liver contains iodine 0.7 mg/kg, and the muscle, 0.05–0.5 mg/kg; the blood, 0.057 mg/L; the bone, 0.27 mg/kg; and the daily intake is 0.1–0.2 mg.

The physiological role of iodine: iodine is mainly involved in the synthesis of thyroxine (triiodothyronine, T3, and tetraiodothyronine, T4); to promote biooxidation, coordinate oxidative phosphorylation, and regulate energy transformation; to promote protein synthesis, regulate protein synthesis and decomposition; to promote sugar and fat metabolism; to regulate water and salt metabolism in tissues; to promote vitamin absorption and utilization; to activate more than 100 cytochrome enzymes and succinate oxidase, etc., which constitute an enzyme system that promotes both biological oxidation and metabolism; to promote nervous system development, tissue development and differentiation.

The iodine required by the body can be obtained from water, food, and salt. The iodine content is mainly determined by the biogeochemical status of each region. Under normal circumstances, inland mountains area is far from the ocean, where the soil and air contain less iodine, and the water and food contain less iodine as well. Therefore, it may become a risky area with thyroid disease.

The food containing high iodine content is seafood, such as kelp (dry) with 24000 μg per 100 g, laver (dry) with 1800 μg per 100 g, mussel (dry) with 1000 μg per 100 g, sea cucumber (dry) with 600 μg per 100 g. For areas where seafood cannot be eaten frequently, the need for iodine in the body can also be obtained by adding potassium iodide to the diet. For example, if you eat 15 g of salt per day, you can get 150 μg of potassium iodide, which is equivalent to 115 μg of iodine and this can meet the needs for the body.

7.4.4 Selenium

Selenium (Se), with atomic number 34 and relative atomic mass 78.96, has a melting point of 221 °C and boiling point of 685 °C. The amount of selenium in the human body is 14–21 mg, which is widely distributed in all tissues and organs. The most is present in nails, followed by the liver and kidney; the amount of selenium in muscle and blood is about 1/2 of that in liver or 1/4 of that in kidney. Liver contains selenium 0.35–2.4 mg/kg, and the muscle, 0.42–1.9 mg/kg; the blood, 0.171 mg/L; the bone, 1–9 mg/kg; and the daily intake is 0.006–0.2 mg.

The physiological functions of selenium: (1) selenium is a component of glutathione peroxidase, which can remove peroxides from the body and protect cells and tissues from damage; (2) selenium has a good function of scavenging free radicals in the body. It can improve the immune ability of the body, and show the anti-aging and anti-carcinogenic properties; (3) to maintain the normal structure and function of the cardiovascular system, preventing cardiovascular diseases; (4) selenium has the antagonism to some toxic heavy metal elements such as cadmium and lead.

Selenium deficiency is an important cause for Keshan disease, which can also induce liver necrosis and cardiovascular disease. Animal foods including liver, kidney, meat, and seafood are good sources of selenium. However, the amount of selenium in foods is greatly affected by the amount of selenium in the local water and soil.

7.4.5 Copper

Copper (Cu), with atomic number of 29, and relative atomic weight of 63.546, has a melting point of 1084.6 °C, and a boiling point of 2567 °C. The body contains about 50–100 mg of copper, which is widely distributed in various organs and tissues, and the highest amount is in liver, kidney, heart, hair, and brain. The amount of copper in liver accounts for about 15% of the total copper in the body, and the brain, about 10%. The concentration in the muscle is low, but the total amount in muscle accounts for about 40% of the total copper in the whole body. Liver and spleen are the storage organs of copper, and the amount of copper in liver and spleen in the body of infants is relatively higher than that of adults. The level of copper in the serum is 10–24 μmol/L, which is very close to that in red blood cells.

The physiological functions of copper: (1) It is involved in the formation of various enzymes in the body. More than ten kinds of enzymes are known to contain copper, and all are oxidases, such as ceruloplasmin, cytochrome oxidase, superoxide dismutase, tyrosine acidase, dopa-β-hydroxylase, and lysyl oxidase. (2) It can promote the absorption of iron in the gastrointestinal tract, and deliver iron to bone marrow to hematopoietic and promote maturation of red blood cells. (3) There is a copper-containing enzyme in the elastic tissue and connective tissue of the body, which can catalyze the maturation of collagen, maintain the elasticity of blood vessel and the toughness of bone, and maintain the elasticity and moisturization in skin, and the pigments and normal structures in hair. (4) It participates in important life activities such as growth hormone, pituitary hormones, and sex hormones to maintain the health of the central nervous system. (5) It can regulate heartbeat, and copper deficiency can induce coronary heart disease.

The amount of copper is high in liver and kidney of animals, fish, shrimp, and clam, but is low in beans, fruits, and milk.

7.4.6 Chromium

Chromium (Cr), is an element with atomic number 24, and relative atomic weight 51.9961. Its melting point is 1857 °C, and the boiling point is 2672 °C, with the density of 7.22 g/cm^3 for single crystal and 7.14 g/cm^3 for polycrystalline. The total amount of chromium in the human body is about 6–7 mg, which is widely distributed in organs, tissues, and body fluids. Among them, liver contains chromium 0.02–3.3 mg/kg, and muscle, 0.024–0.84 mg/kg; blood, 0.006–0.11 mg/L; bone, 0.1–0.33 mg/kg; and the daily intake is 0.01–1.2 mg.

The physiological functions of chromium: (1) it is a component of glucose tolerance factor (GTF), which plays an important role in regulating glucose metabolism and maintaining normal glucose tolerance in the body; (2) chromium affects the metabolism of lipid in the body and lowers the amount of blood cholesterol and triglycerides, thus preventing cardiovascular diseases; (3) chromium is a stabilizer for nucleic acids (DNA and RNA), which prevents the mutation of certain genetic substances in cells and prevents cancer. Therefore, chromium deficiency will be mainly manifested in impaired glucose tolerance, and may be accompanied by hyperglycemia and urine sugar; chromium deficiency can also lead to the disorders of lipid metabolism, being prone to induce coronary sclerosis and cardiovascular diseases.

The main food sources of chromium are coarse grains, meat, brewer's yeast, cheese, black pepper, and cocoa powder. The more refined is the food processing, the less is the chromium content. The refined white sugar and flour contain almost no chromium.

7.4.7 Molybdenum

Molybdenum (Mo), is an element with atomic number 42, and relative atomic weight of 95.94. Its melting point is 2617 °C, and the boiling point is 4612 °C. The total amount of molybdenum is less than 9 mg in the human body, and it accumulates in organs such as liver and heart. Among them, liver contains molybdenum 1.3–5.8 mg/kg; and muscle, 0.018 mg/kg; blood, 0.001 mg/L; bone, <0.7 mg/kg; and the daily intake is 0.05–0.35 mg.

The physiological function of molybdenum is to act as a component of human xanthine oxidase or dehydrogenase, aldehyde oxidase and sulfite oxidase. It can participate in the transmission of intracellular electrons, and prevent tumor occurrence, showing anti-cancer and anti-cancer effect. In recent years, molybdenum has been found to be one of the seven trace elements (Fe, Cu, Zn, Mn, Mo, I, Se) essential for the brain. Deficiency of molybdenum can cause neurological abnormalities, mental retardation, and abnormal bone growth.

When the body is deficient in molybdenum, myocardial hypoxia will occur and cause palpitations and shortness of breath. The excretion of uric acid is reduced, and consequently kidney stones and urinary calculi are formed. Caries can be caused by molybdenum deficiency, and supplement of molybdenum can enhance the anti-caries effect of fluoride.

Molybdenum is abundant in meat, coarse grains, beans, wheat, and is rich in leafy vegetables. In general, the more refined is the food processing, the less is the molybdenum content.

7.4.8 Cobalt

Cobalt (Cobalt, Co), with the atomic number of 27, and the relative atomic weight of 58.9332, belongs to the periodic system VIII. Its melting point is 1495 °C, and boiling point is 2870 °C, with the relative density of 8.9.

The amount of cobalt in the human body is generally 1.1–1.5 mg, which is widely distributed in various parts of the body, and the amount in liver, kidney, and bone is high. The amount of cobalt in red blood cells is 0.059–0.13 mg/kg, 0.005–0.40 mg/kg in serum, and about 0.238 mg/kg in the whole blood. The amount of cobalt in the blood of persons over 50 years old is lower than that aged between 20 and 50 years old. At all stages of human growth, the cobalt amount in blood of men is always higher than that of women. The amount of cobalt in the blood of normal people is the highest in August and the lowest in January, which is due to the highest cobalt intake from vegetables and dairy products in May to July, and relatively the least of the intake in January. Liver contains cobalt 0.06–1.1 mg/kg; and the muscle, 0.28–0.65 mg/kg; the blood, 0.0002–0.04 mg/L; the bone, 0.01–0.04 mg/kg; and the daily intake is 0.005–1.8 mg.

The physiological functions of cobalt are as follows. It is mainly stored in liver in the form of vitamin B_{12} and acts as coenzyme of vitamin B_{12} to exert its biological effects. It plays an important role in the metabolism of proteins, fat and carbohydrate, and the synthesis of hemoglobin, and it can also dilate blood vessels and lower blood pressure. It can prevent fat from accumulating in liver cells and prevent fatty liver, and activate many enzymes, such as increasing the activity of amylase, pancreatic amylase and lipase. It stimulates the hematopoietic system of bone marrow, promoting the synthesis of hemoglobin, and increases the number of red blood cells. It can promote the absorption of zinc in intestine. Therefore, cobalt deficiency can cause nutritional anemia.

Cobalt is high in animal organs (kidney, liver, pancreas); oysters and lean meat also contain a certain amount of cobalt; all fermented soy products such as stinky tofu, soybean meal, and soy sauce contain a small amount of vitamin B_{12}, which can be used as a food source of cobalt; however, dairy products and cereals usually contain less cobalt.

7.5 Changes in Minerals During Food Processing and Storage

The amount of minerals in food is largely affected by various environmental factors, such as the content of minerals in the soil, regional distribution, seasons, water sources, fertilizer application, pesticides, pesticides and fungicides, and the nature of the diet. In addition, minerals can be directly or indirectly transported into the food during processing, such as the increase of iron, zinc, and calcium after the addition of water or passing through the equipment. Therefore, the amount of minerals in food varies largely (Table 7.8).

In food processing, the minerals present in the food, no matter they are natural or artificially added, affect the nutrition and sensory quality of the food to some extent. For example, the discoloration of fruit and vegetable products is mostly caused by the formation of a complex of polyphenols (anthocyanins) with metals. Oxidation loss of ascorbic acid is caused by some metal-containing enzymes, and iron-containing lipoxygenase can cause undesirable flavor in foods. The use of a chelating agent can eliminate or mitigate the adverse effects of the above metals in food.

During food processing, the loss of minerals is different from vitamins, as in most cases it is not caused by chemical reactions, but by the loss of them or form chemicals difficult for the absorption and utilization by the body with other substances.

The impact of food processing and cooking on minerals is a common cause of mineral loss in foods, such as tanking, blanching, leaching, steaming, boiling, and milling. According to previous reports, the canned spinach lost 81.7% of manganese, 70.8% of cobalt, and 40.1% of zinc. Canned tomato lost 83.8% of the zinc, and canned carrots, beets and green beans lost 70%, 66.7%, and 88.9% of cobalt, respectively. The processing of fruits and vegetables usually goes through blanching which also

Table 7.8 Concentration of trace elements in drinking water and foods

Elements	Concentration
Arsenic	0–100 μg/L,daily intake 137–330 μg
Barium	<1 mg/L
Beryllium	<1 μg/L
Cadmium	<10 μg/L
Chromium	<100 μg/L
Cobalt	0.5 mg/kg in spinach
Copper	Be found in animal and plant foods, 1–280 μg/L
Lead	20–600 μg/L
Manganese	0.5–1.5 mg/L
Magnesium	6–120 mg/L
Mercury	<1 μg/L, 10 μg by daily intake of food
Molybdenum	<100 μg/L, 100–1000 μg/kg in food
Nickel	1–100 μg/L, 300–600 μg by daily intake of food
Selenium	<10 μg/L, 100–300 μg per kg of grains, meat and seafood
Silver	Trace
Tin	1–2 μg/L, 1–30 mg by daily intake of food
Vanadium	2–300 μg/L
Zinc	3–2000 μg/L

cause the loss of minerals due to the use of water. Table 7.9 shows the effect of blanching on minerals of spinach, showing that the loss of minerals is related to their solubility. Sometimes the amount of minerals in processing increases, such as the case with calcium in Table 7.9. However, the loss of minerals in cooked peas is slightly different from that in spinach mentioned above, that is, the loss of calcium in peas is the same as that of other minerals, and the loss of trace elements is like the above (Table 7.10). The increase in trace elements and minerals during processing may

Table 7.9 Effects of blanching on minerals of spinach

Minerals	Content g/100 g		Loss (%)
	Raw	Blanching	
Potassium	6.9	3.0	56
Sodium	0.5	0.3	43
Calcium	2.2	2.3	0
Magnesium	0.3	0.2	36
Phosphorus	0.6	0.4	36
Nitrite	2.5	0.8	70

Table 7.10 Amount of minerals in raw and boiled peas

Minerals	Content mg/100 g		Loss (%)
	Raw	Boiled	
Calcium	135	69	49
Copper	0.80	0.33	59
Iron	5.3	2.6	51
Magnesium	163	57	65
Manganese	1.0	0.4	60
Phosphorus	453	156	65
Potassium	821	298	64
Zinc	2.2	1.1	50

be caused by the addition of processing water, contacting with metal containers and packaging materials. It may also be related to tin-plating of cans. For example, nickel in milk is mainly caused by containers of stainless steel for processing (Table 7.11).

In addition, milling also influences the mineral in cereals. Since the minerals in cereals are mainly distributed in the aleurone layer and the embryo tissue, the milling process can cause loss of minerals. The amount of loss increases with the degree of milling, but the loss of various minerals is different. For example, after the wheat is milled, the loss of iron is very serious, and copper, manganese, zinc and cobalt are also lost a lot. When the rice is finely ground, zinc and chromium are largely lost, and manganese, cobalt and copper are also affected. However, it is different

Table 7.11 Distribution of trace metal elements in canned vegetables

Vegetable	Can[a]	Form[b]	Metal element content g/kg		
			Aluminum	Tin	Iron
Green bean	La	L	0.10	5	2.8
		S	0.7	10	4.8
Kidney bean	La	L	0.07	5	9.8
		S	0.15	10	26
Small green peas	La	L	0.04	10	10
		S	0.55	20	12
Dried celery heart	La	L	0.13	10	4.0
		S	1.50	20	3.4
Sweet corn	La	L	0.04	10	1.0
		S	0.30	20	6.4
Mushroom	P	L	0.01	15	5.1
		S	0.04	55	16

[a]La = painted cans, P = metal cans
[b]L = liquid, S = solid

in the processing of soybeans, as the processing of soybeans is mainly a process of degreasing, separation and concentration, with which the amount of proteins is increased. As many minerals are combined with proteins components, therefore, after the processing of soybeans, there is basically no loss of minerals except for silicon.

Another way to lose minerals in food is the interactions of minerals with other ingredients in the food, which leads to a decline in bioavailability. Some polyvalent anions, such as oxalic acid and phytic acid that are widely present in plant foods, can form salts with divalent metal cations such as iron and calcium, and these salts are very insoluble and can pass through the digestive tract without absorption by the body. Therefore, they have a great influence on the biological potency of minerals.

In brief, research on the impact of food processing on minerals is still relatively rare currently. Inconsistent sampling techniques and analytical methods, and different types and sources of foods are used in the studies, so that limited data cannot be directly used for comparison, and the effect of processing on minerals cannot be fully explained. However, the lack of minerals in the human body can cause different degrees of harm to the body, so it is necessary to fortify minerals in food.

7.6 Summary

There are mineral elements with different amounts in foods. They exist in the form of either inorganic or organic salts. Among them, about 25 mineral elements are necessary for constituting tissues of human body, maintaining physiological functions, and biochemical metabolism. Elements cannot be synthesized in the body and need to be provided by foods. The changes in the amounts of minerals in foods depend primarily on environmental factors. The loss of minerals induced by chemical reactions in foods is not as severe as those caused by physical removal or formation of unavailable forms. Developing a sound and effective program of food fortification requires complete information on the sources of foods and the availability of minerals in the diet. From a nutritional point of view, some minerals not only have no nutritional value, but also are hazard to human health, such as mercury and cadmium. Meanwhile, all minerals with a certain excessive amount, even if the essential trace elements, are toxic to the human body. Table 7.12 briefly summarizes the physiological functions of various mineral elements and the main food sources. Table 7.13 also briefly summarizes the recommended daily intake of various mineral elements.

Questions

1. The basic properties of the absorption and utilization of minerals in food and their roles in the body.
2. The basic physical and chemical properties of common minerals (macro- and micro-elements).
3. Why is the absorption and utilization of calcium in cereals and legumes very low, and how to improve its absorption and utilization?

Table 7.12 Functions and main food sources of minerals

Minerals	Functions	Symptoms of deficiency	Food sources
Calcium	1% of calcium reduces the permeability of capillary and cell membrane, preventing exudation, inflammation, and edema; Ca^{2+} is an important activator of many enzymatic reactions; it interacts with phosphorus to construct healthy bones and teeth; it's the triggering agent necessary for the basic reactions between coagulation protein, actin, and ATP; it is an essential factor for blood coagulation and participates in the process of blood coagulation	Rickets for infants and young children, osteomalacia and osteoporosis for adults	Milk and dairy products are best, and beans and vegetables are also rich in calcium. The amount of calcium in shrimp skin, clam, egg yolk, crisp fish, bone meal, kelp, sesame, and soy product are also quite high
Phosphorus	It is an important component of cells in tissues, and participates in metabolism as a component of nucleic acids, phospholipids, and coenzymes; participates in storage materials of energy such as ATP and C-P, playing a very important role in production and transmission of energy; VB family (B1, B2, Vpp, etc.) plays a coenzyme role only by the activity of phosphorylation; buffer system of phosphate maintains osmotic pressure of body fluid and acid-base balance; interact with calcium to construct healthy bones and teeth	Bone fragility; osteoporosis; gum sputum; rickets, retardation of growth, weakness, fatigue; anorexia; hand, foot, facial muscle spasm on hand, foot, and face	Phosphorus is widely distributed in foods, especially in cereals and protein-rich foods such as lean meat, eggs, fish (eggs), viscera, kelp, peanuts, beans, nuts, and coarse grains

(continued)

Table 7.12 (continued)

Minerals	Functions	Symptoms of deficiency	Food sources
Sulfur	It can protect the health of hair and nail, reduce pain of the joint, alleviate side-effect of alcohol, environmental pollutes and cyanide to the human body; help cells to resist bacterial infection; help the body to form insulin, control the levels of blood sugar; help the liver to secrete bile	Related to the deficiency of proteins	Eggs, beans, meat, fish, milk, cabbage, wheat germ, and other protein foods are rich in sulfur
Potassium	The relationship of potassium between intracellular and extracellular can affect the polarization of cell membrane and important cellular processes; it works with sodium to maintain acid-base balance in human body fluids; it participates in the metabolism of intracellular sugar and proteins; it contributes to allergies treatment; in the case of high blood pressure by the intake of high sodium, potassium has a blood pressure lowering effect; it can assist with oxygen transport for the brain, improving the brain function	Irregularity and acceleration of the heartbeat, abnormal electrocardiogram, muscle weakness, and irritability which eventually lead to cardiac arrest	Meat, poultry, fish, various fruits, and vegetables are all good sources of potassium

(continued)

Table 7.12 (continued)

Minerals	Functions	Symptoms of deficiency	Food sources
Sodium	It is the main positive ion in the extracellular fluid, which is involved in the metabolism of water to ensure the balance of water in the body; it can maintain the acid-base balance of body fluids with potassium; it is the component of pancreatic juice, bile, sweat, and tears; it regulates cell excitability and maintenance of normal myocardial movement; it composes salt with chloride ions to be an indispensable condiment	Slow growth, loss of appetite, weight loss due to dehydration, decreased maternal milk during lactation, muscle spasm, nausea, diarrhea, and headache	Protein foods contain more sodium than vegetables and grains. There is little or no sodium in fruits
Chlorine	Cl^- is an important ion that maintains the osmotic pressure of the cell and the acid-base balance of the body fluid, and participates in the metabolism of water; it is the main component of water; it is the main component of the digestive tract secretions such as gastric acid and intestinal juice	The loss of chlorine is always accompanied with the loss of sodium	Salt and salty foods are sources of chlorine
Magnesium	It is the main component of bones, teeth, and cytoplasm; it can regulate and inhibit the contraction of muscles and nerve impulses; maintain the acid-base balance of the body; it is an activator of many enzymes, and can activate many enzyme systems; it is also an essential cofactor for oxidative phosphorylation	Irritability, tension, and stress. Severe deficiency of magnesium can cause and convulsions	Fresh green leafy vegetables, seafood, and beans are good sources of magnesium; cocoa, walnuts, and bananas also contain much magnesium

(continued)

Table 7.12 (continued)

Minerals	Functions	Symptoms of deficiency	Food sources
Iron	It combines with protein to form hemoglobin and myoglobin, participates in the transport of oxygen, promotes hematopoiesis, and maintains the normal growth and development of the body; it is the component of many important enzyme in the body, such as cytochrome, catalase and peroxidase, participating in the respiration of tissues and promoting the biological redox reaction; it is an alkaline element and one of the basic substances to maintain the acid-base balance in the body; it increases the resistance of the body to diseases	Loss of appetite, irritability, paleness, pale hair, dizziness, decreased immune function, thin nails, and nail depression	Liver and whole blood of animals, meat, fish and certain vegetables (cabbage, rape, leeks, etc.) are rich in iron
Zinc	it is a component or an activator of many enzymes in the body (such as alcohol dehydrogenase, glutamate dehydrogenase); it is closely related to the synthesis with nucleic acids and proteins, the metabolism of carbohydrates and vitamin A, and the activity of pancreas, gonads and pituitary gland. Zinc maintains the health of the digestive system and skin, and maintains normal night vision	Stagnation of growth and development, loss of appetite, insensitivity of taste, inhibition of sexual maturity, and poor healing of wounds	Animal foods generally contain higher levels of zinc, much in oysters, pancreas, liver fish, and meat. Milk contains less zinc, sugar, and fruits contain lower

(continued)

Table 7.12 (continued)

Minerals	Functions	Symptoms of deficiency	Food sources
Copper	It is involved in the formation of various enzymes in the body. More than ten kinds of enzymes are known to contain copper, and all are oxidases, such as ceruloplasmin, cytochrome oxidase, superoxide dismutase, tyrosine acidase, dopa-β-hydroxylase, lysyl oxidase. Copper also participates in many functions in the form of the above enzymes in the body. Copper can promote the absorption of iron in the gastrointestinal tract, and deliver iron to bone marrow to hematopoietic and promote maturation of red blood cells. Copper constitutes an enzyme in the elastic tissue and connective tissue of the body, which can catalyze the maturation of collagen, maintain the elasticity of blood vessel and the toughness of bone, and maintain the elasticity and moisturization of skin, and the normal structure of hair pigments. Copper participates in important life activities such as growth hormone, pituitary hormones, and sex hormones to maintain the health of the central nervous system. Copper can regulate heartbeat, and copper deficiency can induce coronary heart disease	Anemia, osteoporosis, discoloration of skin and hair, decreased muscle tone, and psychomotor disorders	The amount is high in animal liver, kidney, fish, shrimp and cockroach

(continued)

Table 7.12 (continued)

Minerals	Functions	Symptoms of deficiency	Food sources
Manganese	It can promote the growth and development of bones, protect the integrity of mitochondria in cells, maintain normal function of brain, maintain normal metabolism of glucose and fat, and improve the hematopoietic function of the body	Declined reproductive ability; abnormal formation of bone and cartilage and impaired glucose tolerance; causes neurasthenic syndrome, affecting mental development; leads to the reduction of the synthesis and secretion of insulin, affecting glucose metabolism	Manganese is the most abundant in tea, and some in brown rice, malt, lettuce, dried beans, peanuts, potatoes, soybeans, and sunflower seed
Iodine	It is mainly involved in the synthesis of thyroxine including triiodothyronine (T3) and tetraiodothyronine (T4); it can activate more than 100 cytochrome enzymes	Goiter; iodine deficiency in the fetus and newborn can cause severe dysplasia such as stagnation, mental retardation, and poor physical strength	Iodized salt, sugar cane, honey, seafood, vegetables, milk and dairy products, eggs, whole wheat, etc.
Chromium	It is a component of glucose tolerance factor (GTF), which plays an important role in regulating glucose metabolism and maintaining normal glucose tolerance in the body; it affects the metabolism of lipid in the body and lowers the amount of blood cholesterol and triglycerides, thus preventing cardiovascular diseases; it is a stabilizer for nucleic acids (DNA and RNA), which prevents the mutation of certain genetic substances in cells and prevents cancer	Glucose tolerance is impaired and may be accompanied by hyperglycemia and urine sugar. Deregulation of lipid metabolism, and induction of coronary artery disease which leads to cardiovascular disease	Whole grains, meat, brewer's yeast, cheese, black pepper and cocoa powder contain chromium, while refined white sugar, and flour contain almost no chromium

(continued)

Table 7.12 (continued)

Minerals	Functions	Symptoms of deficiency	Food sources
Selenium	It is a component of glutathione peroxidase, which can remove peroxides from the body and protect cells and tissues from damage; it has a good function of scavenging free radicals in the body. It can improve the immunity of the body, showing anti-aging, and anti-carcinogenic properties; it maintains the normal structure and function of the cardiovascular system, and prevent cardiovascular diseases; it has the antagonism to some toxic heavy metal elements such as cadmium and lead	Selenium deficiency is an important cause of Keshan disease; it can induce liver necrosis and cardiovascular disease	Liver, kidney, seafood and meat are good sources of food for selenium
Molybdenum	It is a component of human xanthine oxidase or dehydrogenase, aldehyde oxidase, and sulfite oxidase. It can participate in the transmission of intracellular electrons, showing anti-cancer effect	Myocardial hypoxia is induced which can cause palpitations and shortness of breath; the excretion of uric acid is reduced, and kidney stones and urinary calculi are formed; dental caries are induced	Rich in meat, coarse grains, dried beans, wheat, and leafy vegetables
Cobalt	It is mainly stored in liver in the form of vitamin B_{12} to exert its biological effects. It plays an important role in the metabolism of proteins, fat and carbohydrate, and the synthesis of hemoglobin, and it can also dilate blood vessels and lower blood pressure	Nutritional anemia	The amount in animal internal organs (kidney, liver, pancreas) is high, and oysters, lean meat, and fermented soy products also contain a certain amount of cobalt
Fluorine	It is a component of human bones and teeth, preventing dental caries and osteoporosis in the elderly	The enamel is destroyed and dental caries appear; bones are fragile	Saltwater fish and tea are rich sources of fluoride, but the main source is drinking water

Table 7.13 Recommended daily intake of various minerals. Ref: DRI reports from www.nap.edu

Life Stage Group	Calcium (mg/d)	Chromium (μg/d)	Copper (μg/d)	Fluoride (mg/d)	Iodine (μg/d)	Iron (mg/d)	Magnesium (mg/d)	Manganese (mg/d)	Molybdenum (μg/d)	Phosphorus (mg/d)	Selenium (μg/d)	Zinc (mg/d)	Potassium (g/d)	Sodium (g/d)	Chloride (g/d)
Infants															
0–6 mo	200	0.2	200	0.01	110	0.27	30	0.003	2	100	15	2	0.4	0.12	0.18
6–12 mo	260	5.5	220	0.5	130	11	75	0.6	3	275	20	3	0.7	0.37	0.57
Children															
1–3 y	700	11	340	0.7	90	7	80	1.2	17	460	20	3	3.0	1.0	1.5
4–8 y	1000	15	440	1	90	10	130	1.5	22	500	30	5	3.8	1.2	1.9
Males															
9–13 y	1300	25	700	2	120	8	240	1.9	34	1250	40	8	4.5	1.5	2.3
14–18 y	1300	35	890	3	150	11	410	2.2	43	1250	55	11	4.7	1.5	2.3
19–30 y	1000	35	900	4	150	8	400	2.3	45	700	55	11	4.7	1.5	2.3
31–50 y	1000	35	900	4	150	8	420	2.3	45	700	55	11	4.7	1.5	2.3
51–70 y	1000	30	900	4	150	8	420	2.3	45	700	55	11	4.7	1.3	2.0
>70 y	1200	30	900	4	150	8	420	2.3	45	700	55	11	4.7	1.2	1.8
Females															
9–13 y	1300	21	700	2	120	8	240	1.6	34	1250	40	8	4.5	1.5	2.3

(continued)

Table 7.13 (continued)

Life Stage Group	Calcium (mg/d)	Chromium (µg/d)	Copper (µg/d)	Fluoride (mg/d)	Iodine (µg/d)	Iron (mg/d)	Magnesium (mg/d)	Manganese (mg/d)	Molybdenum (µg/d)	Phosphorus (mg/d)	Selenium (µg/d)	Zinc (mg/d)	Potassium (g/d)	Sodium (g/d)	Chloride (g/d)
14–18 y	1300	24	890	3	150	15	360	1.6	43	1250	55	9	4.7	1.5	2.3
19–30 y	1000	25	900	3	150	18	310	1.8	45	700	55	8	4.7	1.5	2.3
31–50 y	1000	25	900	3	150	18	320	1.8	45	700	55	8	4.7	1.5	2.3
51–70 y	1200	20	900	3	150	8	320	1.8	45	700	55	8	4.7	1.3	2.0
>70 y	1200	20	900	3	150	8	320	1.8	45	700	55	8	4.7	1.2	1.8
Pregnancy															
14–18 y	1300	29	1000	3	220	27	400	2.0	50	1250	60	12	4.7	1.5	2.3
19–30 y	1000	30	1000	3	220	27	350	2.0	50	700	60	11	4.7	1.5	2.3
31–50 y	1000	30	1000	3	220	27	360	2.0	50	700	60	11	4.7	1.5	2.3
Lactation															
14–18 y	1300	44	1300	3	290	10	360	2.6	50	1250	70	13	5.1	1.5	2.3
19–30 y	1000	45	1300	3	290	9	310	2.6	50	700	70	12	5.1	1.5	2.3
31–50 y	1000	45	1300	3	290	9	320	2.6	50	700	70	12	5.1	1.5	2.3

4. Explain the changes of minerals during food processing and storage and the impact on the utilization rate by the body.
5. Describe the existing forms of iron in food and the factors affecting the absorption rate.

Bibliography

1. Allen, L., De Benoist, B., Dary, O., Hurrell, R.: Guidelines on food fortification with micronutrients. World Health Organization, Geneva (2006)
2. Avery, J.C., Hoffmann, P.R.: Selenium, selenoproteins, and immunity. Nutrients **10**(9), 1203–1222 (2018)
3. Bailey, R.L., Dodd, K.W., Goldman, J.A., Gahche, J.J., Dwyer, J.T., Moshfegh, A.J., Sempos, C.T., Picciano, M.F.: Estimation of total usual calcium and vitamin D intakes in the United States. J. Nutr. **140**(4), 817–822 (2010)
4. Belitz, H.D., Grosch, W., Schieberle, P.: Food Chemistry. Springer-Verlag Berlin, Heidelberg (2009)
5. Fonseca-Nunes, A., Jakszyn, P., Agudo, A.: Iron and cancer risk-A systematic review and meta-analysis of the epidemiological evidence. Cancer Epidemiol. Biomark. Prev. **23**(1), 12–31 (2014)
6. Gammoh, N.Z., Rink, L.: Zinc in infection and inflammation. Nutrients **9**(6), 624–649 (2017)
7. Gharibzahedi, S.M.T., Jafari, S.M.: The importance of minerals in human nutrition: Bioavailability, food fortification, processing effects and nanoencapsulation. Trends Food Sci. Technol. **62**, 119–132 (2017)
8. Groeber, U., Schmidt, J., Kisters, K.: Magnesium in prevention and therapy. Nutrients **7**(9), 8199–8226 (2015)
9. Hurrell, R., Egli, I.: Iron bioavailability and dietary reference values. Am. J. Clin. Nutr. **91**(5), 1461S–1467S (2010)
10. Kahwati, L.C., Weber, R.P., Pan, H., Gourlay, M., LeBlanc, E., Coker-Schwimmer, M., Viswanathan, M.: Vitamin D, calcium, or combined supplementation for the primary prevention of fractures in community-dwelling adults evidence report and systematic review for the US preventive services task force. Jama J. Am. Med. Assoc. **319**(15), 1600–1612 (2018)
11. Kan, J.: Food Chemistry. China Agricultural University Press, Beijing (2016)
12. Kumssa, D.B., Joy, E.J.M., Ander, E.L., Watts, M.J., Young, S.D., Walker, S., Broadley, M.R.: Dietary calcium and zinc deficiency risks are decreasing but remain prevalent. Sci. Rep. **5**, 10974 (2015)
13. Kuria, A., Fang, X., Li, M., Han, H., He, J., Aaseth, J.O., Cao, Y.: Does dietary intake of selenium protect against cancer? A systematic review and meta-analysis of population-based prospective studies. Crit. Rev. Food Sci. Nutr. **60**(4), 684–694 (2020)
14. Lopez, A., Cacoub, P., Macdougall, I.C., Peyrin-Biroulet, L.: Iron deficiency anaemia. Lancet **387**(10021), 907–916 (2016)
15. Potter, N.N., Hotchkiss, J.H.: Food Chemistry. China Light Industry Press, Beijing (2001)
16. Priemel, M., von Domarus, C., Klatte, T.O., Kessler, S., Schlie, J., Meier, S., Proksch, N., Pastor, F., Netter, C., Streichert, T., Pueschel, K., Amling, M.: Bone mineralization defects and vitamin D deficiency: Histomorphometric analysis of iliac crest bone biopsies and circulating 25-hydroxyvitamin D in 675 patients. J. Bone Miner. Res. **25**(2), 305–312 (2010)
17. Reid, I.R., Bolland, M.J., Grey, A.: Effects of vitamin D supplements on bone mineral density: a systematic review and meta-analysis. Lancet **383**(9912), 146–155 (2014)

18. Roohani, N., Hurrell, R., Kelishadi, R., Schulin, R.: Zinc and its importance for human health: An integrative review. J. Res. Med. Sci. **18**(2), 144–157 (2013)
19. Torti, S.V., Manz, D.H., Paul, B.T., Blanchette-Farra, N., Torti, F.M.: Iron and cancer. In: Stover, P.J., Balling, R. (eds.) Annual Review of Nutrition, vol. 38, pp. 97–125 (2018)
20. Veronese, N., Demurtas, J., Pesolillo, G., Celotto, S., Barnini, T., Calusi, G., Caruso, M.G., Notarnicola, M., Reddavide, R., Stubbs, B., Solmi, M., Maggi, S., Vaona, A., Firth, J., Smith, L., Koyanagi, A., Dominguez, L., Barbagallo, M.: Magnesium and health outcomes: an umbrella review of systematic reviews and meta-analyses of observational and intervention studies. Eur. J. Nutr. **59**(1), 263–272 (2020)

Dr. Aidong Sun professor and doctoral supervisor at College of Biological Sciences and Biotechnology, Beijing Forestry University, China, is the Dean of the Department of Food Science and Technology. Dr. Aidong Sun is a member of the Teaching Steering Committee of the Ministry of Education for Food Science and Engineering and a member of the Professional Technical Committee of the Fruit and Vegetable Processing Technology, one branch of the Chinese Institute of Food Science and Technology. She is also a member of the Snack Food Branch, a member of the Non-thermal Processing Branch, a member of the Forest Food Expert Committee, a member of the Beijing Food Institute, and a member of the Advisory Committee of the National Innovation Alliance of the Fruit Freeze-Dried Industry. She is also a member of the editorial board for the Journal of Beijing Forestry University and the Journal of Chinese Fruits and Vegetables. Dr. Aidong Sun is mainly engaged in research on the screening and efficacy evaluation of active substances in economic forest products, non-thermal food processing technology, authenticity identification based on DNA barcode technology, and the formation of aromatic substances and aroma regulation during food processing. In the past 5 years, she has presided over three general projects of the National Natural Science Foundation of China and two public welfare projects from the State Forestry Administration of the People's Republic of China. She has published more than 150 research papers and 4 monographs.

Dr. Hui Li born in 1986, Assisstant Professor at the Institute of Quality Standards and Testing Technology for Agro-products, Chinese Academy of Agricultural Sciences, is mainly focused on the research of food microbiology and biological testing. Dr. Hui Li has studied/worked in China Agricultural University, University of Alberta in Canada, and Technical University of Munich in Germany, and has presided one project from the National Natural Science Foundation of China and one project from central public welfare research institutes, and one sub-project from national key research and development plan. Dr. Hui Li has published 17 scientific papers in the field of food science such as Comprehensive Reviews in Food Science & Food Safety, Food Research International, Frontiers in Microbiology, Food Microbiology, Innovative Food Science & Emerging Technology, etc. She is also a member of review editor for the journals Frontiers in Microbiology and Frontiers in Nutrition.

Chapter 8
Enzyme

Hui Shi

Abstract Enzymes are the biocatalysts that have the characteristics of strong specificity, high catalytic efficiency, and mild action conditions compared with other
non-biocatalysts. It takes part in various reactions and is mainly found in plants,
animals, and microorganisms. Its catalytic rate is affected by many factors. Kinetics
of enzyme-catalyzed reactions refers to study enzyme-catalyzed reaction rate and the
factors affect this rate. Browning usually happens when plant tissue is damaged or
in an abnormal environment. It is produced by enzymatic or non-enzymatic oxidation of phenolic compounds. Browning in fruits and vegetables, such as bananas,
apples, and potatoes, is not expected to appear. However, browning is sometimes
desirable, as it can enhance the sensory properties of some products such as tea and
cocoa beans. So, controlling enzymatic browning is necessary to obtain a consumer-
acceptable product. Presently, microbial enzymes are used in starch processing, dairy
processing, fruit processing, alcoholic brewing, meat processing, and bakery food
manufacturing as a green alternative to traditional chemical methods. In this chapter,
the source of different microbial enzymes and their applications in food sector will be
highlighted. The chapter will also include enzyme catalytic kinetics and enzymatic
browning.

Keywords Enzymes · Catalytic kinetics · Browning · Application in food
processing

8.1 Overview

8.1.1 Nature of Enzymes

Enzymes were defined as a protein with catalytic activity by Dixon and Webb
in 1979. The catalytic activity was due to their power of specific activation and
conversion of substrates to products. However, according to recent studies, not all

H. Shi (✉)
College of Food Science, Southwest University, Chongqing 400715, China
e-mail: shi_hui_1986@163.com

biomolecules with catalytic capacity are proteins. For example, some RNA molecules (like ribozyme) also have catalytic capacity. In fact, most of the enzymes that are currently used in the food industry are proteins. Therefore, the nature of enzymes that were mentioned in this book is protein.

According to the characteristics of enzyme molecules, enzymes can be divided into monomeric enzyme, oligomeric enzyme, and multienzyme complex. Enzymes that belong to monomeric enzyme are usually hydrolases, such as lysozyme and trypsin. Oligomeric enzyme is composed of several subunits and the number of subunits of most oligomeric enzyme is even. Subunits of oligomeric enzyme are not covalently bonded, so subunits are easily separated from each other. Multienzyme complex is a compound that has several enzymes chimeric with each other by non-covalent bonds, generally with molecular weight of more than several million. Multienzyme complex is beneficial to a series of reactions, for example, fatty acid synthase complex is used for the synthesis of fatty acids.

Some enzymes are simple proteins such as urease and protease. But some enzymes are conjugated protein which is composed of protein part (apoenzyme) and nonprotein part (co-enzyme or prosthetic group). Co-enzymes loosely combine with enzyme proteins, which can be separated by dialysis, but prosthetic group binds to protein more firmly. It is necessary that co-enzyme or prosthetic group on catalysis of enzyme. Only holoenzyme has catalytic activity, so when the enzyme protein is separated from the co-enzyme or prosthetic group, neither of them can play a catalytic role. Moreover, compared to co-enzyme or prosthetic group, enzyme protein plays different roles in catalytic reaction (the specificity and high efficiency of enzyme reaction depend on enzyme protein; the function of co-enzyme or prosthetic group is to transfer hydrogen, electrons, or some chemical groups).

Co-enzyme or prosthetic group includes metal ions (Fe^{2+}, Cu^{2+}, Zn^{2+}, Mg^{2+}, Ca^{2+}, Na^+, K^+, and so on) and small molecule organic compounds like nicotinamide adenine dinucleotide (NAD), nicotinamide adenine dinucleotide phosphate (NADP), flavin adenine dinucleotide (FAD), and flavin mononucleotide (FMN). There are many kinds of enzymes in organism, but co-enzyme or prosthetic group is few. Usually, a kind of enzyme protein can only combine with one co-enzyme or prosthetic group, but a kind of co-enzyme or prosthetic group can couple with many kinds of enzyme protein to form many kinds of enzyme.

8.1.2 Enzyme Specificity

Enzymes are the biocatalysts that have the characteristics of strong specificity, high catalytic efficiency, and mild action conditions compared with other non-biocatalysts. Enzyme specificity is one of the most important characteristics of enzyme and the biggest difference between enzymes and non-biocatalysts. Enzyme specificity is a selectivity of enzyme to substrate. This means that an enzyme can only act on a kind of substrate or a certain chemical bond, even a substance. Enzyme specificity ensures the metabolic activities of organism proceed methodically in a certain direction and manner, and maintains the normal activity of the organism. According to the degree of specificity of enzyme to substrate, enzyme specificity can be divided into the following types.

8.1.2.1 Bond Specificity

It is "bond specificity" that some enzymes only act on a certain bond of substrate and have not strict requirement to the groups at both ends of the bond. For example, some glycosidases and proteases only require the substrate to have glycosidic or peptide bonds but no strict requirement to the types of sugar or amino acid residues that make up glycosidic or peptide bonds.

8.1.2.2 Group Specificity

Group specificity means that enzymes not only specifically require to the chemical bond of the substrate acted by enzymes, but also have certain requirements for the groups at both ends of the chemical bond. For example, trypsin can only hydrolyze the peptide bonds of arginine and lysine on the side of carboxyl (this property was used in protein sequence analysis). Phosphomonoesterase can hydrolyze many phosphoric acid monoester compounds like glucose 6-phosphate and various nucleotides, but cannot hydrolyze the phosphodiester compound.

8.1.2.3 Absolute Specificity

Absolute specificity is also called structural specificity. Some enzymes are strict to their substrates and only catalyze the reaction of one substrate. For instance, urease only catalyzes the hydrolysis of urea and has no effect on the derivatives of urea. Maltin only acts on maltose and cannot act on other disaccharides.

8.1.2.4 Stereospecificity

Enzyme can only act on a certain kind of substrate which has stereoisomer. The specificity is called stereospecificity and divided into optical isomerism and geometric heterogeneity. This property of the enzyme can be used to separate chiral compounds, so the stereo specificity of the enzyme is very significant in food analysis and processing.

8.1.3 Nomenclature and Classification

The nomenclature of enzymes mainly includes common nomenclature and international systematic nomenclature. Common nomenclature is named according to the following three principles: (1) all enzymes are classified into six major classes based on the nature of the chemical reaction catalyzed (oxidoreductases, transferases, hydrolases, lyases, isomerases, and ligases); (2) the nature of enzyme substrate and action, such as amylase, lipase, and protease; (3) enzyme is named according to the source of them and combined the above two conditions, such as pepsin, trypsin, etc. Common nomenclature is relatively simple, and the application history is long. Although it is not systematic, it is still used by people.

International Commission on Enzymes of the International Union of Biochemistry recommended a new systematic nomenclature scheme and classification method, and stipulated the systematic nomenclature of enzyme in 1961. The principle of international systematic nomenclature specifies the name of each enzyme and clarifies the substrate nature of enzyme and catalytic reaction, which based on the whole reaction catalyzed by enzymes. Enzyme that can catalyze the reaction of two substrates should include the name of the two substrates in their systematic name, and be separated by ";". The substrate can be omitted if it is water.

According to the types of enzymatic reactions, enzymes are divided into oxidoreductases, transferases, hydrolases, lyases, isomerases, and ligases in systematic classification method of enzyme, which are expressed as 1, 2, 3, 4, 5, and 6. Then, these classes are divided into several subcategories according to the characteristics of the functional groups or bonds in the substrate, and each subcategory is coded into numbers 1, 2, 3, 4, etc. Each subcategory can also be subdivided into sub-subcategories, still using 1, 2, 3, 4, etc. Therefore, the classification number of each enzyme is an EC number composed of four Arabic numerals. For example, a-amylase (common nomenclature) is named as a-1,4-glucose-4-glucose hydrolase and its international committee on enzymology number is EC3.2.1.1. EC represents international commission on enzymology, the first number represents which of the six classes the enzyme belonging to, and the second number is the subcategory of the major category of enzymes, and the third number is the sub-subclass of the subclass to which the enzyme belongs, which is used to supplement the deficiency of the second number classification. The first three numbers indicate the way enzyme is working, the fourth number indicates number of enzymes in the sub-subcategory. Although this name system is strict, it has not been widely used because it is very complex.

Hydrolases are frequently used in food processing, followed by oxidoreductases and isomerases. These enzymes can also be classified into exogenous enzymes and endogenous enzymes. For example, lipoxygenase, chlorophyllase, and polyphenol oxidase which are key enzymes which lead to the change of fruits and vegetables; pectinase, cellulase, and pentosanase that affect texture; amylase that hydrolyzes starch in animals, higher plants, and microorganisms; and cathepsin in animal tissue cells which are common endogenous enzymes. To change the characteristics of food, exogenous enzymes can be added to food materials during processing, so that certain components in the food materials have desired changes. For example, α-amylase, glucoamylase, and glucose isomerase are used in starch to produce fructose syrup; lipolytic enzymes are used to modify triglycerides; proteases are used to produce high-quality cheese, beer, and soy sauce; and papain is used to prepare the meat tenderizer. Table 8.1 lists the system classification of important enzymes in food processing.

8.1.4 Catalysis Theory of Enzymes

Emil Fischer, a German chemist, proposed "lock and key theory" first in 1894. This theory indicates that enzymes have strict specificity for the substrate acted by them. The structure of the substrate must be consistent with the structure of the active site of the enzyme, like the lock and the key, then be tightly combined to form an intermediate product. However, the theory is difficult to explain the obvious change of certain groups on the enzyme molecule when the substrate is combined with the enzyme, and the phenomenon that the enzyme can often catalyze the reaction in two inverse directions.

Therefore, biochemists Michaelic and Menten proposed the enzyme intermediates theory in 1913. The key of this theory is that reaction involving enzymes and substrate generates unstable intermediates, thus allowing it to proceed rapidly along a pathway with lower activation energy. In fact, the intermediate products theory has been confirmed by many experiments, and intermediate products really exist. If E, S, ES, and P represent enzyme, substrate, intermediate, and final product of the reaction, respectively, the reaction process can be expressed as follows:

$$E + S \rightarrow ES \rightarrow E + P$$

At the same time, D. E. Koshland proposed the "induction fit theory" in 1958. It indicates that the active part of the enzyme is not a rigid structure and has some flexibility. When the substrate meets the enzyme, the conformation of the enzyme protein can be induced and change, resulting in the enzyme and the substrate forming an intermediate product, and catalyze the reaction of the substrate.

The central theory of enzyme activity thinks that enzyme is a biological macro-molecule, and the volume of enzyme molecule is larger than substrate. When the enzyme is combined with the substrate, the site that catalyzes a chemical reaction of a substrate is called active center of enzyme. The groups that constitute active site can be divided into binding group (a group that binds to a substrate) and catalytic group (a group that does not bind to a substrate but participates in a catalytic reaction), but some groups have both effects. The active center of the enzyme not only determines the specificity of the enzyme, but also plays a decisive role in the catalytic properties of the enzyme.

Table 8.1 Classification of important enzymes in food processing

Class and subclass	Enzymes	Enzymes
1. Oxidoreductase		
1.1 Donor is CH–OH		
1.1.1 Receptor is NAD$^+$ or NADP$^+$	Alcohol dehydrogenase	1.1.1.1
	Butanediol dehydrogenase	1.1.1.4
	L-iditol-2-dehydrogenase L-lactose dehydrogenase	1.1.1.14
	Malate dehydrogenase	1.1.1.37
	Galactose-1-dehydrogenase	1.1.1.48
	Glucose-6-phosphate-1-dehydrogenase	1.1.1.49
1.1.3 Receptor is oxygen	Glucose oxidase	1.1.3.4
	Xanthine oxidase	1.1.3.22
1.2 Donor is aldehyde		
1.2.1 Receptor is NAD$^+$ or NADP$^+$	Aldehyde dehydrogenase	1.2.1.3
1.8 Donor is sulfur compounds		
1.8.5 Receptor is quinone or quinone compounds	Glutathione dehydrogenase (ascorbic acid)	1.8.5.1
1.10 Donor is dienol or diphenol		
1.10.3 Receptor is oxygen	Ascorbate oxidase	1.10.3.3
1.11 Receptor is hydroperoxide	Catalase	1.11.1.6
	Peroxidase	1.11.1.7
1.13 Act on a single donor		
1.13.11 Bind to oxygen	Lipoxygenase	1.13.11.12
1.14 Act on a pair of donors		
1.14.18 Bind to an oxygen atom	Monohydric phenol monooxygenase (poly-phenol oxidase)	1.14.18.1
2 Transferase		
2.7 Transfer phosphoric acid		
2.7.1 Receptor is OH	Hexokinase	2.7.1.1
	Glycerol kinase	2.7.1.30
	Pyruvate kinase	2.7.1.40
2.7.3 Receptor is N-base	Creatine kinase	2.7.3.2
3 Hydrolase		
3.1 Cut the ester bond		
	Carboxylesterase	3.1.1.1
	Triacylglycerol esterase	3.1.1.3
3.1.1 Carboxylate hydrolase	Phosphatase A$_2$	3.1.1.4
	Acetylcholinesterase	3.1.1.7
	Pectin methylesterase	3.1.1.11

(continued)

Table 8.1 (continued)

Class and subclass	Enzymes	Enzymes
	Phosphatase A_1	3.1.1.32
3.1.3 Phosphomonoester hydrolase	Alkaline phosphatase	3.1.3.1
3.1.4 Phosphodiester hydrolase	Phospholipase C	3.1.4.3
	Phospholipase D	3.1.4.4
3.2 Hydrolysis the O-glycosyl compounds		
3.2.1 Glycosidase	α–amylase	3.2.1.1
	β–amylase	3.2.1.2
	Glucoamylase	3.2.1.3
	Cellulase	3.2.1.4
	Polygalacturonase	3.2.1.15
	Lysozyme	3.2.1.17
	α–D–glycosidase (Maltase)	3.2.1.20
	β–D–glycosidase	3.2.1.21
	α–D–galactosidase	3.2.1.22
	β–D–galactosidase (Lactase)	3.2.1.23
	β–fructofuranosidase (Invertase or sucrase)	3.2.1.26
	1,3–β–D–xylanase	3.2.1.32
	α–L–rhamnosidase	3.2.1.40
	Pullulanase	3.2.1.41
	Exo-polygalacturonase	3.2.1.67
3.2.3 Hydrolysis S-glycosyl compounds	Glucosidase (Myrosinase)	3.2.3.1
3.2 Peptidase		
3.4.21 Serine peptide bond endonuclease	Subtilisin	3.4.21.62
3.4.23 Aspartate peptide bond endonuclease		3.4.23.4
3.4.24 Metal peptide bond endonuclease	Chymosin	3.4.24.27
3.5 Act on C-N bonds except peptide bonds		
3.5.2 In cyclolactam	Thermolysin	3.5.2.10
4 Lyase		
4.2 C–O–Lyase		
4.2.2 Act on polysaccharide	Creatininase	4.2.2.2
	Exo-polygalacturonic acid lyase	4.2.2.9

(continued)

Table 8.1 (continued)

Class and subclass	Enzymes	Enzymes
	Pectate lyase	4.2.2.10
5 Isomerase		
5.3 Intramolecular oxidoreductase		
5.3.1 Interconversion between aldose and ketose	Xylose isomerase	5.3.1.5
	Glucose–6–phosphate isomerase	5.3.1.9

8.1.5 Enzyme Activity

The quantitative problem of enzyme is involved in both theoretical research and practical production. The enzyme content of enzyme preparation is expressed by its ability to catalyze a certain reaction because the enzyme preparation often contains many impurities.

Enzyme activity is the ability that enzyme catalyzes a certain chemical reaction, and it can be expressed by the reaction rate. Therefore, the enzymatic reaction rate can be obtained by measuring the amount of substrate reduction or product production at per unit time. The amount of enzyme is indicated by the ability of a certain amount of enzymic preparations that catalyze a reaction. Therefore, enzyme units are defined by the enzyme activity.

Enzyme Commission regulate that under certain conditions (25 °C, pH and substrate concentration are the optimum reaction conditions for the enzyme), the amount of enzyme that converts 1 µmol of substrate into product in 1 min is 1 active unit or enzyme unit, called the International Unit of Enzyme (IU) in 1964. Although the regulation facilitates the comparison of the research results, it is often inconvenient in practical applications, so it is not often used except scientific research.

In 1972, the International Enzyme Commission recommended a new international unit of enzyme activity, namely, Katal (Kat) unit. Under the optimum reaction conditions (temperature 25 °C), the amount of enzyme required to catalyze the conversion of 1 mol substrate per second to product is 1 Kat unit (Kat = 1 mol $\cdot$ s^{-1}). The conversion relationship between Kat units and IU units is given as follows:

$$1 \text{ Kat} = 60 \times 10^6 \text{ IU}$$

In the production of enzyme preparation, manufacturers sometimes develop their own enzyme activity units according to different products and specify the corresponding substrates or products. For example, one protease unit is prescribed as the amount of enzyme that hydrolyzes casein to produce 1 µg tyrosine within 1 min; one α- amylase unit is prescribed as the amount of enzyme that liquefies 1 g starch to produce 1 g liquefied starch within 1 h, etc. In the determination of enzyme activity, the temperature of the reaction, pH, substrate concentrations, and action time are uniformly regulated, so that the products of the same type of enzyme preparation are compared with each other.

The unit of enzyme activity is merely a basis for comparison, and does not represent directly the absolute quantity of enzymes. In practical applications, besides use per gram (or per milliliter) enzyme preparation containing how many units of enzyme activity to represent the size of the enzyme activity, per mg enzyme protein containing how many units of enzyme activity also represent the size of the enzyme activity, which is called specific activity, and the specific activity of enzyme is also expressed as enzyme activity unit number of per (milli) liter enzyme solution or per gram enzyme preparation.

8.2 Catalytic Kinetics

8.2.1 The Rate of Enzyme Catalysis

Kinetics of enzyme-catalyzed reaction are used to study the rate of enzyme catalysis and various factors affecting this catalysis rate. The rate of various chemical reactions can vary greatly, and the same reaction can also have significantly different reaction rates due to different conditions. Therefore, we usually need to change the reaction conditions to control the reaction rate. In addition, there are many chemical reactions accompanied by side reactions. It is necessary to reduce the rate of side reactions so that the rate of the main reaction increases. Through the study of chemical kinetics, the mechanism of chemical reaction can be clarified theoretically and the specific process and way of chemical reaction can be understood. In practical applications, the rate of chemical reaction can be used to estimate the required time of the reaction achieving to certain degree. According to the factors affecting rate of chemical reaction, the measures can be further explored to control the progress of reaction.

The reaction rate is expressed as change in concentration of reactants or products in per unit time. As the reaction is progressing, the reactants are gradually consumed and the chance of molecular collision is gradually reduced, so the reaction rate will gradually slow down. Because the reaction rate is different at each instant, the instantaneous rate is usually used to express reaction rate. dc is assumed as the small change in reactant concentration in the instantaneous dt, then

$$v = -\frac{dc}{dt}$$

where "−" indicates the decrease of reactant concentration. The reaction rate can also be expressed as the increase of product concentration in per unit time:

$$v = +\frac{dc}{dt}$$

where "+" indicates that the product increases with increase in time, reactants or products concentration can be used to express reaction rate. In fact, the concentration of reactants or products at different times can be measured quantitatively by chemical or physical methods.

The distinguishing feature of enzymes from other proteins is that enzymes bind their substrates stereospecifically in the active site and convert the substrates to products. Kinetics of enzyme-catalyzed reactions refers to study of enzyme-catalyzed reaction rate and the factors which affect this rate. Reaction rate is expressed as a change in the concentration of the reactants or products at per unit time.

8.2.2 Factors Affecting the Rate of Enzyme Catalysis

8.2.2.1 Effect of Substrate Concentration

If the other conditions are constant, the reaction rate depends on enzyme concentration and substrate concentration. If enzyme concentration is constant, the effect of substrate concentration on enzyme-catalyzed reaction is shown in Fig. 8.1, as rectangular hyperbola. The reaction velocity increases sharply with low substrate concentration increasing. Reaction velocity and substrate concentration are in a direct ratio. With the substrate concentration further increasing, the incremental range of the reaction velocity gradually decreases. But if the concentration of substrate is sequentially increasing, the reaction velocity will not increase. At this time, the active center of the enzyme has been saturated by substrate. The influence of substrate concentration on the enzyme reaction velocity can be explained by the theory proposed by Michaelis and Menten in 1913.

The Michaelis–Menten theory assumes the formation of an enzyme–substrate intermediate, and that the rate with which the intermediate is transformed into product in the reaction depends on the rate with which the intermediate is transformed into reaction product and enzyme. The relationship is given as follows:

Fig. 8.1 Effect of substrate concentration on enzyme reaction rate

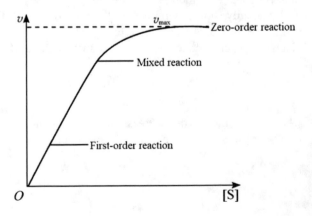

$$E + S \underset{K_{-1}}{\overset{K_1}{\rightleftharpoons}} ES \overset{K_2}{\longrightarrow} E + P$$

enzyme   substrate   enzyme-substrate   enzyme   product
intermediate

K_1, K_{-1} and K_2 are the velocity constants. Derived by mathematics:

$$v = \frac{K_2[E_t][S]}{[S] + \frac{K_{-1}+K_2}{K_1}}$$

where v is velocity of the formation of products and E_t is the total concentration of enzyme.

If $K_m = \frac{K_{-1}+K_2}{K_1}$, $V_{max} = K_2[E_t]$:

$$v = \frac{v_{max}[S]}{K_m + [S]}$$

It is famous Michaelis–Menten equation. K_m is an important parameter of enzyme, named Michaelis constant. The physical meaning of Michaelis constant (Km) is substrate concentration (mol/L) when the enzyme-catalyzed rate is half maximum rate. It is related to the nature of enzyme, the substrate type of enzyme, pH, and temperature in enzyme catalysis. But it has no relationship with the concentration of enzyme. For most enzymes, K_m can indicate the affinity between enzyme and substrate. As Km increases, the affinity decreases. The Lineweaver–Burk double reciprocal plot method is usually used to determine K_m. It takes the reciprocal form of the Michaelis–Menten equation:

$$\frac{1}{v} = \frac{K_m}{V_{max}} \cdot \frac{1}{[S]} + \frac{1}{V_{max}}$$

According to the formula, the straight line of Fig. 8.2 can be obtained, wherein the intercept on the horizontal axis is $-1/K_m$, and the slope of the line is k_m/V_{max}. By measuring the intercept of a straight line on two coordinate axes or the incept of a straight line on any coordinate axis and combining with the slope value, k_m and V_{max} can be easily obtained.

8.2.2.2 Effect of Enzymes Concentration

Under excess of substrate and other fixed conditions, and other factors that are un-conducive to the action of the enzyme, the speed is proportional to the concentration of the enzyme reaction in the reaction system. Enzyme firstly forms an intermediate with the substrate, and this step is the rate-limiting step of the entire catalytic reaction. When the substrate concentration exceeds the enzyme concentration, the generation rate of intermediate is determined by enzyme concentration. Therefore,

Fig. 8.2 Double reciprocal
mapping

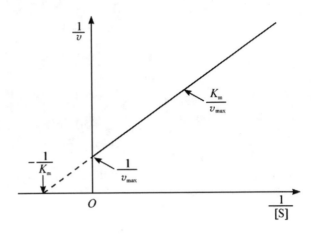

Fig. 8.3 Relationship
between reaction rate and
enzyme concentration

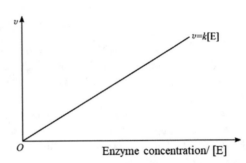

if the concentration of enzyme increases at this time, the reaction rate can increase
(the enzyme reaction rate is linear with the enzyme concentration (Fig. 8.3)).

8.2.2.3 Effect of Temperature

The effect of temperature on enzyme-catalyzed reaction rates consists of two aspects
(Fig. 8.4). (1) The reaction velocity increases as the temperature increases. (2) The

Fig. 8.4 Effect of
temperature on enzyme
reaction rate

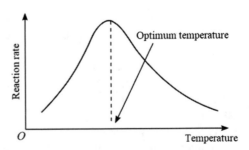

reaction velocity reduces at high temperature because enzyme denatures. Therefore, as the temperature increases, the reaction rate increases at low temperatures. As the temperature is continuously increasing, denaturation of enzymes is predominant. Although the rise of temperature facilitates the enzyme reaction rate, the reaction rate decreases. Under certain conditions, each enzyme exhibits maximum vigor at an appropriate temperature. The temperature is called the optimum temperature of enzyme. In general, optimum temperatures of most enzymes obtained from animal are usually in the range of 37–50 °C. And optimum temperatures of enzymes obtained from plants are more than the range of 50–60 °C. The optimal temperature of enzyme is not a fixed constant and its value is affected by factors such as type of substrates and reaction time. For example, the optimal temperature is different because of different enzyme reaction time. The optimal temperature decreases with the reaction time prolonging.

8.2.2.4 Effect of pH

pH changing within a certain range does not influence the catalysis, but it significantly influences the catalytic reaction rate of enzyme. Each enzyme shows its activity only during a range of pH and has the highest activity at a certain pH called the optimum pH (Table 8.2). Generally, the enzyme activity on both sides of the optimum pH suddenly drops (Fig. 8.5). However, not all enzymes have such characteristics. Figure 8.6 lists the pH activity curves of several enzymes.

Therefore, it is necessary to understand the optimum pH range in research and application of enzymes. The pH of reaction solution must be controlled with a buffering, so that the enzyme has the highest activity. The optimal pH of some enzymes is extreme pH. For example, the optimal pH of pepsin is 1.5–3 and the optimal pH of arginase is 10.6. Due to the numerous and complicated ingredients in

Table 8.2 Optimum pH of different enzymes

Enzyme	Source	Substrate	pH
Pepsin	Stomach	Protein	2.0
Chymotrypsin	Pancreas	Protein	7.8
Papain	Tropical plants	Protein	7–8
Lipase	Microorganisms	Olive oil	5–8
α-Glucosidase(maltase)	Microorganisms	Maltose	6.6
β-Amylase	Malt	Starch	5.2
β-Fruclofuranosidase(invertase)	Tomato	Saccharose	4.5
Pectin lyase	Microorganisms	Pectic acid	9.0–9.2
Xanthine oxidase	Milk	Xanthine	8.3
Lipoxygenase, type I[a]	Soybean	Linoleic acid	9.0
Lipoxygenase, type I[b]	Soybean	Linoleic acid	6.5

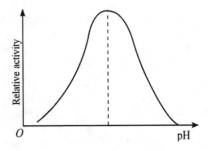

Fig. 8.5 Clock curve of the effect of pH on enzyme activity

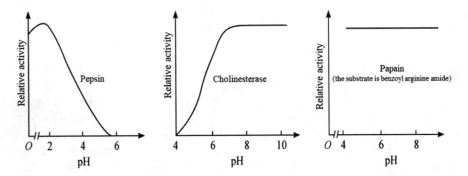

Fig. 8.6 pH activity curve of some enzymes

food, it is important to control pH in food processing. If certain enzyme is necessary, the pH can be adjusted to its optimal pH to maximize its activity. On the contrary, if the function of certain enzyme needs to be avoided, the pH can be adjusted reasonably to inhibit its activity. For example, phenolase can produce enzymatic browning and its optimal pH is 6.5. If pH is reduced to 3.0, browning can be prevented. Therefore, acidulants are usually added in fruit processing, such as citric acid, malic acid, and phosphoric acid.

The reasons for the activity of the enzyme to changes in pH are three aspects: (1) peracid or over-base will affect the conformation of the enzyme protein and denature the enzyme; (2) when the change of pH is not very intense, the enzyme is not denatured, but its viability is affected. pH can affect the dissociation of the enzyme and the substrate, which affect the affinity of the enzyme and the substrate. The pH also affects the dissociation of the intermediate, which is disadvantageous for the generation of catalytic product; and (3) pH affects the dissociation of the relevant groups that maintain the spatial structure of the enzyme, thereby affecting the conformation of the active site and the activity of enzyme.

8.2.2.5 Effect of Water Activity

Enzyme activities usually occur in aqueous media in vitro although in vivo enzyme reactions can occur not only in the cytoplasm but in cell membranes, in lipid depots, and in the electron transport system, where transfer of electrons is known to occur in a lipid matrix. Three methods can be used to study the effect of water activity on enzyme activity:

(1) Carefully dry (without heating) a biomaterial (or model system) containing enzyme activity, then equilibrate the dried sample to various water activities and determine the enzyme activity in the sample. For example, when Aw is lower than 0.35 (water content <1%), phospholipase cannot hydrolyze lecithin. When Aw exceeds 0.35, the enzyme activity non-linearly increases. When Aw is 0.9, it still does not reach the highest activity. Only when Aw is higher than 0.8 (water content <2%), β-amylase shows the activity of hydrolyzing starch. If Aw rises to 0.95, the enzyme activity will increase by 15 times. From these examples, it can be concluded that water content in food material must be less than 1–2% to inhibit enzyme activity.

(2) The method of replacing the water with an organic solvent to determine the concentration of water required for the action of the enzyme. For example, water is replaced by glycerin which is miscible with water. When water content reduces to 75%, the activity of lipoxygenase and peroxidase begins to decrease. The activity of lipoxygenase and peroxidase reduces to 0 when water content reduces to 20% and 10%, respectively. Viscosity and the special effects of glycerin may affect these results.

(3) Most of the water can be substituted with organic solvents in the ester-transfer reaction of glycerol tributyrate in various alcohols catalyzed by lipase. The initial reaction rate is 0.8 μmol, 3.5 μmol, 5 μmol, and 4 μmol transesterification/(h·100 mg lipids) when "dry" lipase particles (0.48% water content) are suspended in the dry n-butanol with water content of 0.3%, 0.6%, 0.9%, and 1.1% (mass fraction), respectively. Therefore, porcine pancreatic lipase has the highest initial rate of catalyzing transesterification at 0.9% water content.

The effect of organic solvents on the enzymatic reaction of the enzyme has two main aspects: the stability of the enzyme and the direction of the reaction (if the reaction is reversible). The effects are different in organic solvents that are immiscible and miscible with water. In organic solvents that are immiscible with water, the specificity of the enzyme shifts from catalytic hydrolysis to catalytic synthesis. For example, when "dry" (1% water content) enzyme particles are suspended in an organic solvent that is not miscible with water, the transesterification rate of lipase-catalyzed lipid increases by more than 6 times, while the ester hydrolysis rate decreases by 16 times. The stability and catalytic activity of enzymes in water and water-soluble solvent systems are different from those in water and water-insoluble solvent systems. For example, K_m increased, V_{max} and enzyme stability decreased at the reaction of casein hydrolysis catalyzed by protease in 5% ethanol–95% buffer or a 5% propionitrile–95% buffer compared with buffer system. It is well known that alcohols and amines compete with water in hydrolase-catalyzed reactions.

8.2.2.6   Effect of Activator

Any substance that can improve the enzyme activity is called enzyme activator. Enzyme activator is mostly metal ions or other inorganic ions. For example, Cl^- is activator of salivary amylase; Mg^{2+} is activator of RNase; Mg^{2+}, Mn^{2+}, and Co^{2+} are activator of decarboxylase; Mn^{2+} is activator of aldolase; etc.

Certain reducing agents, such as cysteine or glutathione, can also activate certain enzymes and reduce the disulfide bonds of enzyme to sulfhydryl groups, thereby enhancing enzyme activity, such as papain.

8.2.3   Enzyme Inhibition and Inhibitors

Inhibitor can reversibly or irreversibly bind to certain enzymes and inhibits the catalysis of the enzyme, such as drugs, antibiotics, poisons, antimetabolites, etc. The inhibition of enzymes can be divided into two categories which are reversible inhibition and irreversible inhibition.

8.2.3.1   Irreversible Inhibition

The inhibitor and the essential active group of the enzyme are bound by a very firm covalent bond to cause loss of enzyme activity, and the activity of enzyme cannot be recovered by physical means such as dialysis or ultrafiltration to remove inhibitor, which is called irreversible inhibition. For example, diisopropyl fluorophosphate (DIFP) can combine with hydroxyl of active site of cholinesterase serine residue bound to inactivate the enzyme. The inactivation of cholinesterase leads to accumulation of acetylcholine, which causes the excitotoxic state of vagus nerve.

8.2.3.2   Reversible Inhibition

Reversible inhibition reduces enzymatic activity by the reversible binding of a non-covalent bond to an enzyme and (or) an intermediate (ES). The inhibitor can be removed by dialysis or ultrafiltration to restore the activity of enzyme. Reversible inhibition can be divided into three types.

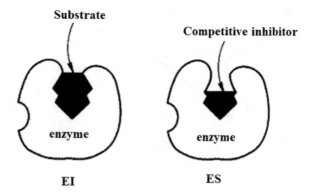

Fig. 8.7 An intermediate which is produced by an enzyme binds to a substrate or a competitive inhibitor

(1) Competitive inhibition

Some compounds can reversibly bind to the active center of the enzyme and compete with the substrate at the same active center, especially those compounds that are structurally like the natural substrate. When the inhibitor binds to enzyme, it hinders the binding of the substrate to enzyme, reducing the chance of action of enzyme, thus decreasing the activity of enzyme. The effect, competitive inhibition, is the most common reversible inhibition (Fig. 8.7). It can be expressed by the following balance:

$$E + S \underset{K_2}{\overset{K_1}{\rightleftharpoons}} ES \xrightarrow{K_3} E + P$$
$$+$$
$$I$$
$$K_{i1} \big\| K_{i2}$$
$$EI$$

where I is inhibitor, Ki is inhibitor constant, and EI is enzyme-inhibitor complex. The enzyme-inhibitor complex cannot react with the substrate to produce EIS, because the formation of EI is reversible and the substrate and inhibitor constantly compete at the active center of enzyme. A typical example of competitive inhibition is catalysis of succinate dehydrogenase which catalyzes the following reaction when appropriate hydrogen acceptor (A) exists:

$$\text{succinate} + \text{acceptor} \rightleftharpoons \text{fumaric acid} + \text{reduced acceptor}$$

Many compounds with structure like succinic acid can bind to succinic acid dehydrogenase, but do not dehydrogenate. These compounds block active center of enzyme, thereby inhibiting normal reaction progress. The compounds that inhibit succinate dehydrogenase include oxalic acid, malonic acid, and glutaric acid, with malonic acid as the strongest inhibitor. The inhibition of competitive agents depends

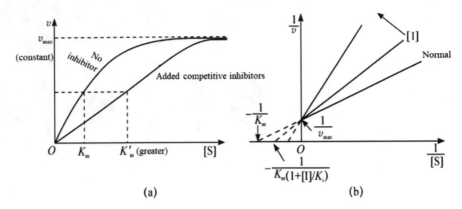

Fig. 8.8 Competitive inhibition curve

on their affinity to enzymes and the relative ratio of substrate concentration. If substrate concentration increases, this inhibition can be weakened.

The velocity equation of competitive inhibition is derived by deducing Michaelis–Menten equation as follows:

$$v = \frac{v\text{max}[S]}{K_m\left(1 + \frac{[I]}{K_i}\right) + [S]}$$

It is plotted by Michaelis–Menten equation and Lineweaver–Burk (Fig. 8.8). It was shown in Fig. 8.8 that V_{max} does not change and Km becomes large after the addition of the competitive inhibitor. And Km increases as [I] increases. The double reciprocal plot intersects the vertical axis, which is characteristic of competitive inhibition.

(2)   Non-competitive inhibition

Some inhibitors and substrates can bind to different parts of the enzyme at the same time, forming enzyme–substrate–inhibitor ternary complex (ESI) which cannot be further decomposed (Fig. 8.9). The enzyme activity is reduced, and this inhibition is called non-competitive inhibition. For instance, leucine is a non-competitive inhibitor of arginase. Heavy metals such as Ag^+, Hg^{2+}, and Pb^{2+} and organic mercury compounds can form a complex with -SH in the enzyme to inhibit enzyme activity. Certain enzymes that require metal ions to maintain activity can also be inhibited by non-competitive inhibitors. Non-competitive inhibition can be expressed by the following balance:

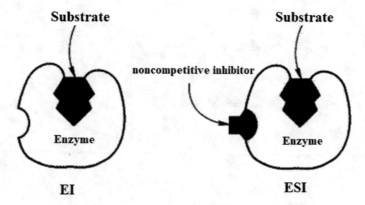

Fig. 8.9 An intermediate which is produced by an enzyme binds to a substrate or a non-competitive inhibitor

$$E + S \xrightleftharpoons{K_m} ES \longrightarrow E + P$$

The velocity equation of non-competitive inhibition is derived by Michaelis–Menten equation as follows:

$$v = \frac{v_{max}[S]}{(K_m + [S])\left(1 + \frac{[I]}{K_i}\right)}$$

It is plotted by Michaelis–Menten equation and Lineweaver–Burk (Fig. 8.10). It was shown in Fig. 8.10 that Km does not change, Vmax becomes small, and Vmax decreases as the [I] increases after the addition of the non-competitive inhibitor. The double reciprocal plot intersects the horizontal axis, which is characteristic of non-competitive inhibition.

(3) Uncompetitive inhibition

Enzymes can only bind to inhibitors after they bind to substrates, which is called uncompetitive inhibition. It is often seen in multi-substrate reactions but is relatively rare in single-substrate reactions. For example, the impact of alkaline phosphatase by L-phenylalanine, L-arginine, and other amino acids is uncompetitive inhibition, and the inhibitory effect of hydrazine compounds on pepsin and cyanide on aromatic sulfate esterase also belong to anti-competitive inhibition. The velocity equation of competitive inhibition is derived by Michaelis–Menten equation as follows:

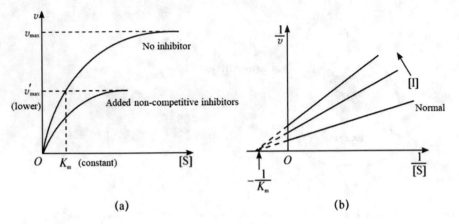

Fig. 8.10 Non-competitive inhibition curve

$$v = \frac{v_{\max}[S]}{K_m + [S]\left(1 + \frac{[I]}{K_i}\right)}$$

It is plotted by Michaelis–Menten equation and Lineweaver–Burk (Fig. 8.11). It was shown in Fig. 8.11 that both Km and Vmax become small and decrease with the increase of [I] after the addition of the uncompetitive inhibitor. The double reciprocal plot is a set of parallel lines, which is characteristic of uncompetitive inhibition.

The changes of the Michaelis–Menten equation and Km and Vmax with inhibitors or not are summarized in Table 8.3.

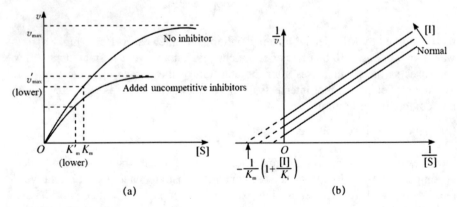

Fig. 8.11 Uncompetitive inhibition curve

Table 8.3 Comparison of competitive inhibition, non-competitive inhibition, and normal enzyme reaction

	Michaelis–Menten equation	V_{max}	K_m
No inhibition	$v = \frac{v_{max}[S]}{K_m + [S]}$	–	–
Competitive inhibition	$v = \frac{v_{max}[S]}{K_m(1 + [I]/K_i) + [S]}$	Invariant	Increase
Non-competitive inhibition	$v = \frac{v_{max}[S]}{(K_m + [S])(1 + [I]/K_i)}$	Reduce	Invariant
Uncompetitive inhibition	$v = \frac{\frac{v_{max}}{1 + [I]/K_i}[S]}{\frac{K_m}{1 + [I]/K_i} + [S]}$	Reduce	Invariant

8.3 Enzyme Browning

Browning can be classified as enzymatic browning (biochemical browning) and non-enzyme browning according to their mechanism. Enzymatic browning occurs in fresh fruits and vegetables. Post-harvest fruits and vegetables still have active metabolic activities in their tissues. Under normal conditions, redox reactions in the tissues of full fruits and vegetables are coupled. When fruits and vegetables are mechanically damaged (cutting, crushing, insect biting) or under exceptional environment (freezing and heating), it will affect the balance of oxidation reduction, leading to the accumulation of oxidation products and causing discoloration. The function of discoloration, enzymatic browning, requires to contact with oxygen and be catalyzed by enzyme. In most cases, enzymatic browning is a change that is unexpected to occur in fruits and vegetables. For example, bananas, apples, pears, eggplants, and potatoes are easily browned after being peeled. These changes should be avoided. However, suitable enzymatic browning in foods (tea and cocoa beans) is necessary to obtain good flavor and color.

8.3.1 Mechanism of Enzymatic Browning

Enzymatic browning is a reaction process in which phenolase catalyzes the formation of quinones and their polymers by phenolic compounds. Phenolic substances in plant tissues act as respiratory transport substances in intact cells and maintain a dynamic balance between phenol and hydrazine. Cell tissues are destroyed and oxygen invades promptly, causing the imbalance between the formation and reduction reaction of quinones. Hence, the massive accumulation of quinones further oxidizes to form the brown pigment that is known as melanin or melanoid.

Enzymatic browning is initiated by the enzymatic oxidation of phenolic compounds by polyphenol oxidase (PPO, EC 1.10.3.1) under aerobic conditions. Quinones produced by oxidation rapidly polymerize into brown pigment and causes browning of the tissue. PPO can be found in most fruits and vegetables and is the main enzyme for enzymatic browning. In most cases, PPO not only damages the perception of fruits and vegetables, but also leads to a negative effect on flavor and quality.

The systematic name of phenolic enzymes is catechol (Oxygen-oxidoreductase). Oxygen-oxidoreductase is a terminal oxidase which must use oxygen as hydrogen acceptor and Cu as a prosthetic group. The phenolase may use a monohydric phenol or a dihydric phenol as a substrate. It is considered that the phenolase is an enzyme that acts on both monohydric phenols and dihydric phenols. However, others hold that the phenolase is a complex enzyme of two phenolic. One is phenolhydroxylase (cresylase) and the other is polyphenoloxidase (catecholase). The suitable pH of phenolase is close to 7, which has heat resistance. According to different sources, phenolase is inactivated at 100 °C for 2–8 min.

The function of phenolase is illustrated by the browning of the potato after being cut (Fig. 8.12). The substrate for phenolic action is the richest phenolic compound tyrosine in potatoes.

This is also the mechanism of melanin formation in the animal's skin and hair. In fruits, catechol is widely distributed that is easily oxidized to quinones under the influence of catecholase:

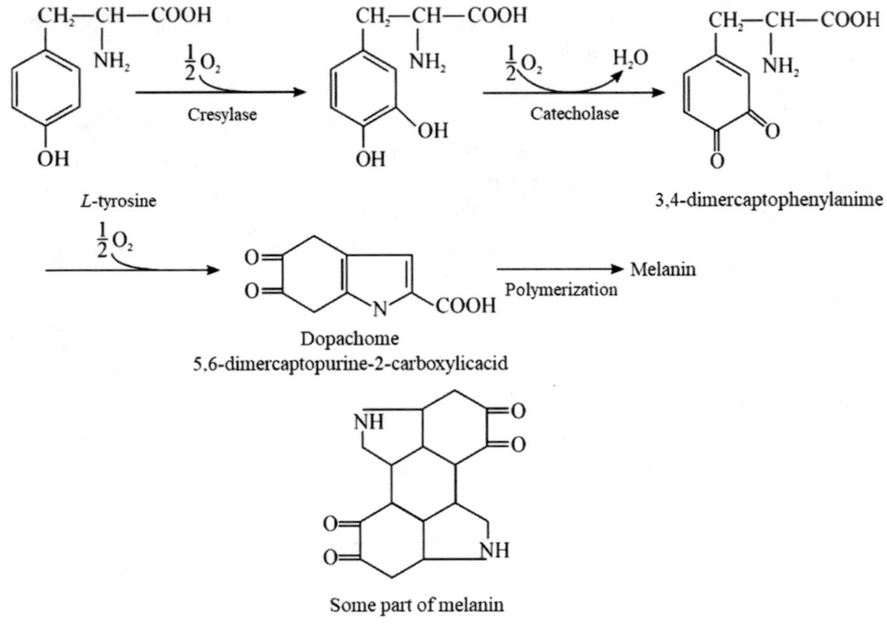

Fig. 8.12 Mechanism of enzymatic browning

The formation of quinone is catalyzed by oxygen and enzymes. Quinone produces further hydroquinone which is non-enzymatic auto-reaction. The color of product polymerized by hydroquinone depends on the degree of polymerization and finally becomes melanin substance.

The most abundant phenolic substrates in fruits and vegetables are o-diphenols and monohydric phenols. In general, phenolase acts faster on o-hydroxyphenolic structures than monohydric phenols. Para-diphenols can also be utilized, but meta-diphenols cannot be used as substrates and has negative effect on phenolic enzymes. However, substituted derivatives of o-diphenol are also not catalyzed by phenolases as guaiacol and ferulic acid.

Chlorogenic acid is a key substance in the browning of many fruits, especially peaches and apples.

As mentioned above, the major substrate for potato browning is tyrosine. In bananas, the main browning substrate is also a nitrogen-containing phenolic derivative which is 3,4-dihydroxyphenol ethylamine.

Amino acids and nitrogen compounds react with o-diphenol to produce a deep colored complex. The mechanism is probably that the phenol is firstly enzymatically oxidized to form the corresponding quinone which generates a non-enzymatic condensation reaction with the amino group. This is the reason why pink color usually forms in processing of white onions, garlic, and allium porrum.

Other phenolic derivatives with intricate structures, such as anthocyanins, flavonoids, and tannin, can be used as substrates of phenolase. All of them have the structures of o-diphenol or monophenol.

During the fermentation of black tea, the activity of polyphenol oxidase in fresh tea leaves aggrandizes and the catechins are catalyzed to form colored substances as theaflavins and thearubigins which are the main components of the color of black tea. The processing of black tea is a beneficial application of enzymatic browning of polyphenol oxidase in food processing.

Caffeic acid residue   Quinine residue

Chlorogenic acid

In addition to the common enzymatic browning of foods caused by polyphenol oxidase, ascorbate oxidase and peroxidase existing extensively in fruits and vegetables can also cause enzymatic browning.

8.3.2 Control of Enzymatic Browning

A small number of enzymatic browning that occur in food processing are expected, such as processing of black tea, cocoa, and certain dried fruits (raisins, prunes). However, most enzymatic browning adversely affects the color of foods and needs to be controlled.

Enzymatic browning requires three indispensable conditions: phenolic substrate, polyphenol oxidase, and oxygen. The possibility of removing the phenolic substrate is infinitesimal to control enzymatic browning. It has been conceived to change the structure of the phenolic substrate without success. Therefore, it mainly inhibits enzymatic browning from both phenolase and oxygen. The primary ways are: (1) passivation of phenolase activity (scalding, inhibitor, etc.); (2) changing conditions of phenolic enzyme action (pH value, water activity, etc.); (3) isolation of oxygen (cut off oxygen contact); and (4) adding antioxidants (ascorbic acid, SO_2, etc.). The primary methods to control enzymatic browning are followed.

8.3.2.1 Heat Treatment

Heat treatment is the most common method to stabilize color of foods because of its capacity to destroy microorganisms and inactivate enzymes. The key of heat treatment is passivating enzymes in the shortest time, but excessive heating will affect the quality of the food. On the contrary, the heat treatment time is not enough, leading to the cell structure destroyed but the enzyme non-passivated, and it will enhance the contact between the enzyme and the substrate to promote browning. If white onions and leeks are not blanched enough, they will turn pink even more intensely than unblanched ones.

The sensitivity of polyphenol oxidases to thermal treatment is diverse because of disparate sources. According to research, most of the polyphenol oxidases are inactivated by heating at 70–95 °C for approximately 7 s.

Boil and steam are still the most popular blanching methods at present. The application of microwave provides a novel means for thermal inactivation of enzyme activity, rapidly and equably heating tissues. Furthermore, it is extremely beneficial to the maintenance of texture and flavor and an ideal method to inhibit enzymatic browning by heat treatment.

8.3.2.2 pH Adjustment

The adhibition of acid to control enzymatic browning is an extensive way. Used acids usually are citric acid, malic acid, phosphoric acid, ascorbic acid, etc. In general, they decrease the pH to control the phenolase activity, because the optimal pH of phenolase is between 6 and 7, the phenolase is inactive when the pH is lower than 3.0.

Citric acid is wide-applied edible acid. It reduces the pH and chelates phenolic Cu prosthetic group. As a browning inhibitor, the effect of using citric acid alone is not obvious. It is usually combined with ascorbic acid or sulfite. The fresh-cut fruits are usually immersed in a dilute solution of acid. For alkaline peeled fruits, these acids also neutralize residual alkali.

Malic acid is the principal organic acid in apple juice, and the inhibition of phenolase in apple juice is more remarkable than citric acid.

Ascorbic acid is a telling phenolase inhibitor, which is odorless at extremely high concentrations and is not corroding metals. In addition, as a vitamin, its nutritional value is distinguished. Others argue that ascorbic acid can inactivate phenolase. The anti-browning effect of ascorbic acid in fruit juice may also be used as a substrate of ascorbate oxidase. The research indicated that 660 mg ascorbic acid is added into a kilogram of fruit products, which can effectively control browning and cut down the oxygen content in bottom clearance of canned apples.

8.3.2.3 Sulfur Dioxide and Sulfite Treatment

Sulfur dioxide (SO_2) and commonly used sulfites such as sodium sulfite (Na_2SO_3), sodium hydrogen sulfite ($NaHSO_3$), sodium metabisulfite ($Na_2S_2O_5$), sodium dithionite, and sodium hyposulfite ($Na_2S_2O_4$) are extensively utilized in phenolase inhibitors of food industry. It has been applied in the processing of mushrooms, potatoes, peaches, and apples.

Fruits and vegetables are treated by SO_2. SO_2 infiltrates tissue fleetly, nevertheless, the merit of sulfite solution conveniently uses. No matter what form it is taking, only free SO_2 can act on phenolic enzymes. The inhibition of SO_2 and sulfite solutions to phenolase is best under slight acid (pH $= 6$).

Ideally, 10 mg/kg SO_2 can almost completely inhibit phenolase. In practice, due to volatilization loss and reaction with other substances, the actual dosage is larger and often amounts to 300–600 mg/kg. The control mechanism of SO_2 on enzymatic browning is still inconclusive. Some scholars believe that SO_2 inhibits enzyme activity. It is assumed that SO_2 reduces quinone to phenol. Others hold that the interaction between SO_2 and quinone prevents the polymerization of quinone.

The advantages of the sulfur dioxide method are convenience, reliability, and beneficial to the conserve of vitamin C. The residual SO_2 can be removed by vacuuming, boiling, or using H_2O_2. The shortcoming is food discoloration, corrosion of the inner wall of can, undesirable smell, and the destruction of vitamin B.

8.3.2.4 Oxygen Isolation

Specific measures include: (1) immersing the peeled fruits and vegetables in water, sugar water, or salt water; (2) coating ascorbic acid to form isolated layer on the surface; and (3) applying vacuum infiltration method to drive out the air. The fruits such as apple and pear which have more gas in the interspace of flesh are suitable to use in this method. Generally, the vacuum was maintained for 5–15 min at 1.028×10^5 Pa. The vacuum was suddenly broken, and the soup could be forcibly penetrated into tissues, thus driving out the gas in intercellular space.

Sodium chloride is also known to prevent enzymatic browning and is usually used with citric acid and ascorbic acid. When it is used alone, the concentration of sodium chloride is more than 20% and can inhibit the activity of polyphenol oxidase. Beyond that, vacuum or nitrogen-filled packaging can also effectively prevent or retard the enzymatic browning caused by polyphenol oxidase.

8.3.2.5 Addition of Phenolase Substrate Analogs

The phenolase substrate analogs can effectively control the enzymatic browning of apple juice. Cinnamic acid, para-coumaric acid, and ferulic acid are examples of phenolase substrate analogs. Among the above three homologs, cinnamic acid has the highest efficiency. When the concentration of cinnamic acid is more than 0.5 mmol/L, the browning of apple juice in the atmosphere can be effectively controlled for 7 h. These three acids are aromatic organic acids and naturally exist in fruits and vegetables, which is safe without toxic side effects.

CH=CH—COOH CH=CH—COOH CH=CH—COOH

OCH₃

OH OH

Cinnamic acid *p*-cumaric acid Ferulic acid

8.3.2.6 Modification of Substrate

Methyltransferase is used to proceed methylation of ortho-dihydroxy compound and then form a methyl substituted derivative, which can effectively prevent browning. For example, under the action of methyltransferase, catechins, caffeic acid, and chlorogenic acid can be methylated into calcitriol, ferulic acid, and 3-ferulic acid sodium gallate, respectively.

8.4 Applications in Food Processing

8.4.1 Enzymes Commonly Used in Food Processing

The purpose of adding enzymes in food processing is usually to: (1) improve the quality of food; (2) manufacture synthetic foods; (3) increase the speed and yield of extracting food composition; (4) improve the flavor of food; (5) stabilize the quality of food; and (6) increase the utilization rate of by-products. Compared to standard biochemical reagent, the enzymes used in the food processing industry are quite rough. Most enzyme preparations still contain many impurities and other enzymes. Enzyme preparations used in food processing are extracted from edible and non-toxic animal or plant, also from non-pathogenic, non-toxic microorganisms. The benefits of using microorganisms to prepare enzymes include different aspects as: (1) microorganisms are widely used and can be used to produce any kind of enzyme; (2) microorganisms can be altered by mutation or genetic engineering to efficiently produce enzymes; (3) it is very easy to isolate and extract enzymes because most microbial enzymes are extracellular enzymes; (4) source of the culture medium is easily available; and (5) the growth rate of microorganisms and the yield of enzymes are very high.

Enzymes are widely used in food processing, as they are more selective, efficient, and significantly less hazardous. As shown in Table 8.4, the total number of enzymes used in food processing is still small relative to the type and number of enzymes found. The most used enzyme in food processing is hydrolase, mainly carbohydrate hydrolase, followed by protease and lipase; small amounts of oxidoreductase are

Table 8.4 Application of enzymes in food processing

Type of enzyme	Food	Enzyme function
Amylase	Baked goods	Increase the sugar content of the yeast fermentation process
	Brewing	Turbidity caused by conversion of starch to maltose to remove starch during fermentation
	Chocolate	Convert starch into a flowing form
	Candy	Recover sugar from candy crumbs
	Fruit juice	Remove starch to increase foaming
	Jelly	Remove starch and increase gloss
	Pectin	As an adjuvant in the preparation of pectin from apple peel
	Syrup and sugar	Convert starch to low molecular weight dextrin
	Vegetables	Hydrolysis of starch during softening of peas
Invertase	Artificial honey	Convert sucrose to glucose and fructose
	Candy	Production of invert sugar for confectionery
Lactase	Ice cream	Prevent the crystallization of lactose
	Feed	Convert lactose to galactose and glucose
	Milk	Remove lactose from milk to stabilize protein
Cellulase	Brewing	Hydrolyze complex carbohydrates in cell walls
	Coffee	Hydrolyze cellulose during drying of coffee beans
	Fruit	Remove the granules from the pears and accelerate the peeling of apricots and tomatoes
Hemicellulase	Coffee	Reduce the viscosity of espresso
Pectinase	Chocolate—cocoa	Increase the hydrolysis activity of cocoa beans during fermentation
	Coffee	Increase the hydrolysis of gelatinous coatings during the fermentation of cocoa beans
	Fruit	Soften
	Fruit juice	Increase the output of the juice, prevent flocculation, and improve the concentration process
	Olive	Increase oil extraction
	Alcohol	Clarify

(continued)

Table 8.4 (continued)

Type of enzyme	Food	Enzyme function
	Orange juice	Destroy and separate pectin substances from juice
	Fruit	Excessive softening
Naringinase	Tangerine	Hydrolysis of naringin and other glycosides to decimate the pectin and juice made from citrus
Pentosanase	Milling	Increase the recovery of starch during flour production
Stasusase	Bean products	Reduce the formation of gas in the intestines caused by oligosaccharides such as raffinose and stachyose
Tannase	Brewing	Removal of polyphenolic compounds
	Tea	Prevent turbidity of cold tea extract
Protease	Baked goods	Soften the dough, reduce mixing time, increase dough extensibility, and improve bread texture
	Beer brewing	Produce flavor and nutrients, increase filtration and clarity during the brewing process
	Cereals	Modify proteins to increase the drying rate of food
	Cottage cheese	Increase casein aggregation and increase flavor during ripening
	Chocolate—cocoa	Cocoa beans for fermentation
	Eggs and egg products	Improve drying properties
	Feed	Handling the waste to convert it into feed
	Fish and meat	Tenderize meat, recycle proteins from bones, fish, and other wastes
	Milk	Preparation of soy milk
	Protein hydrolysate	Used in the production of soy sauce, soup, seasonings, and meat
	Alcohol	Clarify
Lipase	Cottage cheese	Accelerate ripening
	Grease	Convert fat to glycerol and fatty acids
	Milk	Make milk chocolate special flavor
Phosphatase	Baby food	Increased effectiveness of phosphate
	Beer fermentation	Hydrolysis of phosphate compounds
	Milk	Check the effect of pasteurization
Ribonuclease	Flavor enhancer	Add 5'-nucleotides
Peroxidase	Vegetables	Check the blanching effect

(continued)

Table 8.4 (continued)

Type of enzyme	Food	Enzyme function
	Measure of glucose	Determination of grapes in combination with glucose oxidase
Lipoxygenase	Bread	Improve bread texture, flavor, and bleach
Diacetaldehyde reductase	Beer	Reduce the concentration of diacetaldehyde in beer
Polyphenol oxidase	Tea, coffee, tobacco	Brown during ripening and fermentation

also used in food processing. Currently, only a few isomerases are used in food processing.

The enzyme can also control the storage and quality of food materials. Some crops are harvested when they are not fully mature, and it takes a period of ripening to reach edible quality. In fact, enzymes control changes in the maturation process, such as the disappearance of chlorophyll, the production of carotene, the conversion of starch, the softening of tissues, the production of aroma, etc. If the role of the enzyme in the maturation process is mastered, we can improve the storage and quality of food.

8.4.1.1 Hydrolase

Amylase

Amylases are crucial enzymes which hydrolyze internal glycosidic linkages in starch and produce dextrin and oligosaccharides as primary products. Amylases are classified into α-amylase, β-amylase, and glucoamylase based on their three-dimensional structures, reaction mechanisms, and amino acid sequences.

α-Amylase can be isolated from animals, plants, and microorganisms. Also, the content of α-amylase is particularly high in germinating seeds, human saliva, and animal pancreas. At present, high-purity α-amylase can be prepared by microorganisms such as *Bacillus subtilis*, *Aspergillus oryzae*, and *Aspergillus Niger* in industry. α-Amylase is an incision enzyme that catalyzes the hydrolysis of internal α-1,4-glycosidic bonds in starch, glycogen, and cyclodextrin molecules to yield products like glucose and maltose, retaining α-configuration of anomeric carbon. The viscosity of the amylose is quickly lowered, the color development of the iodine liquid disappears rapidly, and the reducing power is increased by the formation of the reductive group. α-Amylase cannot cleave α-1,6-glycosidic bonds, but it is possible to continue to hydrolyze the α-1,4-glycosidic bonds over this bond and unable to hydrolyze the α-1,4-glycosidic bonds in maltose. Hence, starch is hydrolyzed to form a mixture of maltose, glucose, and dextrin. The relative molecular mass of α-amylase is approximately 50000. It contains a well-knit calcium which maintain the optimal conformation of the enzyme protein, resulting in the highest stability and maximum activity of enzyme. The optimum temperature of α-amylase from different sources is different, ranging from 55 to 70 °C. However, the optimum temperature of α-amylase in a few bacteria is as high as 92 °C, such as *Bacillus licheniformis*. When starch mass fraction is 30–40%, they still have short-term catalytic capacity under 110 °C. When α-amylase is used for the catalytic reaction at a higher temperature, it is preferred to add a certain amount of calcium to maintain the stability and activity of enzyme. The optimum pH of α-amylase from different sources is different, ranging from 4.5 to 7.0.

Primary sources of β-amylase are the seeds of higher plants and sweet potatoes. Not found in mammals, but in recent years it has been found in a small number of microorganisms. β-Amylase is an exo-hydrolase enzyme that only hydrolyzes the α-1,4-glycosidic bond and cannot hydrolyze α-1,6-glycosidic bond of starch. It hydrolyzes the starch from the nonreducing end of a polysaccharide chain. The maltose units are cutted successively and the cut α-maltose is converted into β-maltose. Since β-maltose has sweetness, β-amylase is also known as saccharifying enzyme. The 1,3-glycosidic bonds in amylose and the α-1,6-glycosidic bonds in amylopectin are not hydrolyzed by β-amylase, then the reaction stops, and the remaining compound is called limit dextrin.

Glucoamylase, also known as α-1,4-glucosidase, is mainly derived from *Rhizopus*, *Aspergillus*, etc. The optimal pH of glucoamylase ranges from 4.0 to 5.0. The optimal temperature of glucoamylase ranges from 50 to 60 °C. Glucoamylase is an exo-hydrolase enzyme that not only hydrolyzes α-1,4-glycosidic bonds, but is also hydrolyzing α-1,6-glycosidic bonds and α-1,3-glycosidic bonds. However, the hydrolysis rate of two bonds is very slow. For example, the hydrolysis rate of α-1,6-glycoside bond is only 4–10% of that of α-1,4-glycoside bond. When the glucoamylase hydrolyzes starch, one glucose unit is sequentially cut from the nonreducing end of a polysaccharide chain, and the cut α-glucose is converted into β-glucose. Therefore, glucoamylase yields glucose acting on amylose or amylopectin. Since the hydrolysis of the α-1,6-glycoside bond is slow when glucose amylase acts on amylopectin alone, it takes a long time for amylopectin to be completely hydrolyzed. So, in the industry, it is often used to hydrolyze starch with α-amylase, speeding up hydrolysis and improving efficiency.

In addition, amylase also includes pullulanases and isoamylases. They can hydrolyze 1,6-α-D-glycosidic bonds in amylopectin and glycogen to form a segment of straight chain. If mixed with β-amylase, it can produce maltose-rich starch syrup.

Figure 8.13 illustrates the mode of action of the above several amylases.

α-D-Galactosidase

α-D-galactosidase can hydrolyze nonreducing terminal monosaccharides of disaccharides, oligosaccharides, and polysaccharides. Stachyose in legumes can produce gas in the stomach and intestines because there are some anaerobic microbes in the intestine that can hydrolyze oligosaccharides or monosaccharides to generate CO_2, CH_4, and H_2. However, when the stachyose is hydrolyzed by α-D-galactosidase, the flatulence in the stomach is eliminated.

Fig. 8.13 Schematic
diagram of the action of
several amylases

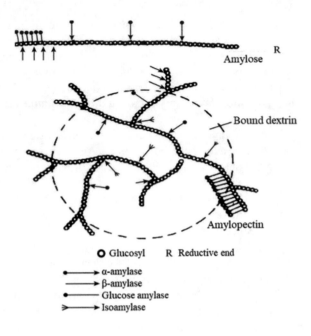

β-D-Galactosidase

β-D-galactosidase is also called lactase which can catalyze the hydrolysis of lactose.
It is widely distributed and found in higher animals, plants, bacteria, and yeast.
Some people lack lactase and can't utilize lactose, leading to they can't digest milk.
They should drink milk with β-D-galactosidase. Galactose inhibits the hydrolysis of
lactose by lactase, but glucose does not. In addition, the solubility of lactose is very
low, which is not conducive to the production of skimmed milk powder or ice cream.
The use of β-D-galactosidase can hydrolyze lactose to improve the quality of food
processing.

β-D-Fructosidases

β-D-fructosidases is an enzyme preparation isolated from a special yeast strain and is
commonly used in confectionery industry to hydrolyze sucrose and produce inverted
sugar. Inverted sugar is more soluble and sweeter than sucrose.

α-L-Rhamnosidases

Some orange juice, plum juice, and grapefruit juice contain hesperidin, which has a bitter taste. Treated with a mixture of α-L-rhamnosidases and β-D-glucosidase, hesperidin can produce a non-bitter compound—Naringin ligand, 4,5,7-trihydroxyflavanone (naringenin).

Glycosidase Mixture

Glycosidase mixture is a pentosanase preparation. It is a mixture of glycosidase (including extracellular and intracellular cellulase, α-, β-mannosidase, pectinase, etc.). The baking quality of rye flour and the shelf life of rye bread are improved by the partial hydrolysis of pentosans by this enzyme.

To solubilize the main component of the plant, it is achieved by impregnating the glycosidase mixture in a milder and shorter period. For example, it can degrade the puree product and the leaf of the vegetables to solubilization. This enzyme can also be used to increase the mechanical breakage of cell walls, thus preventing too much gelatinized starch in cell from rinsing, causing the vegetable purees not too sticky.

Glycosidase extracted from *Aspergillus niger* is a mixed enzyme preparation (a mixture of cellulase, amylase, and protease). It can be used for shelling shrimp. Because glycosidase can make the shell of shrimp loose, the shell can be eluted by using water vapor.

Pectic Enzyme

Pectic enzyme is a heterogeneous group of related enzymes that hydrolyze the pectic substances, which is present mostly in higher plants and microorganisms. The classification of pectic enzyme is based on they attacking on the different substrates, and pectic enzyme is broadly classified into three types: pectinesterase, polygalacturonase and pectinlyase.

(1) Pectinesterase

Pectinesterase catalyzes the removal of methyl ester groups from pectin to produce polygalacturonic acid chains and methanol, also known as pectase or pectin methoxylase or pectin demethoxylase. It is specific to galacturonate and requires the presence of a free carboxyl group near the esterified group of the galacturonate in which it acts, and does not decompose other methyl esters. Pectin esterase is found in bacteria, fungi, and higher plants and is abundant in citrus and tomato, which is usually present with polygalacturonases. In the processing of fruits and vegetables, if pectin esterase is activated, a large amount of methyl ester group will be removed from pectin, thereby affecting the texture of fruits and vegetables, and also the methanol is toxic to the human body. Especially in the wine brewing, the content of methanol in wine may exceed the standard due to the action of pectin esterase. Therefore, the fruit should be pre-heat treatment and passivate the activity of the pectin esterase to control the content of methanol in the wine.

(2)   Polygalacturonase

Polygalacturonase is an enzyme that degrades pectic acid. It can be divided into two types according to the way of action on substrate: one is a glycosidic bond called polygalacturonidase to random hydrolysis of pectic acid, which is found in higher plants, molds, bacteria, and some yeasts; another is polygalacturonic acid exo-nuclease, which cuts glycosidic bonds from the end of the pectate chain. It is not only found mostly in higher plants and molds, but also found in bacteria and insect intestines. Sodium chloride is a necessary cofactor of polygalacturonase to achieve the highest activity. In addition, some polygalacturonases require copper ions as auxiliary factors.

(3)   Pectinlyase

Pectinlyase which is also called pectin transductase belonging to lyases catalyzes the trans-elimination of hydrogen at the C_4 and C_5 of pectin galacturonic acid residue, splitting the glycosidic bonds to produce a galacturonic acid containing unsaturated bond. Pectinlyase is a generic term for endo-polygalacturon lyase, exo-polygalacturon lyase, and endo-polygalacturonan lyase. Most pectinlyases are obtained from mildew, not found in plants.

Figure 8.14 summarizes the enzymic mode of action of the above three pectic enzymes on the pectin.

Pectic substances form the major components of the middle lamella, a thin layer of adhesive extracellular material found between the primary walls of adjacent young plant cells. Their degree of polymerization and esterification change the texture of fruits and vegetables during post-harvest, post-harvest preservation, or processing. Endogenous pectinase is the main enzyme that catalyzes the hydrolysis of endogenous pectin in fruit ripening to cause tissue softening. Post-harvest ripening of fresh fruits is artificially promoted. Transgenic technology was used to cultivate tomatoes with low endogenous pectinase activity, thus extending the storage period after harvest and reducing the damage in transportation. The addition of pectic enzymes which results in a rapid decline of viscosity and the flocculation of the micelles increases juice yield and clarify juices. However, it is necessary to reduce the effect of pectinase to maintain the stability of the insoluble particles in the production of turbid juices. For example, the cut fruit pieces are first hot in canned fruit processing, which is a measure of blunting enzyme. It is called soft root that mold and bacterial contamination can cause rapid softening and decay of fruits and vegetables.

Mold and bacterial contamination can cause rapid softening and decomposition of fruit and vegetable tissues, known as soft root, which reflects the adverse aspects of microbial-derived pectinase. Pectinase is extracted from microorganisms and used in food production, which reflects beneficial aspects of microbial-derived pectinase.

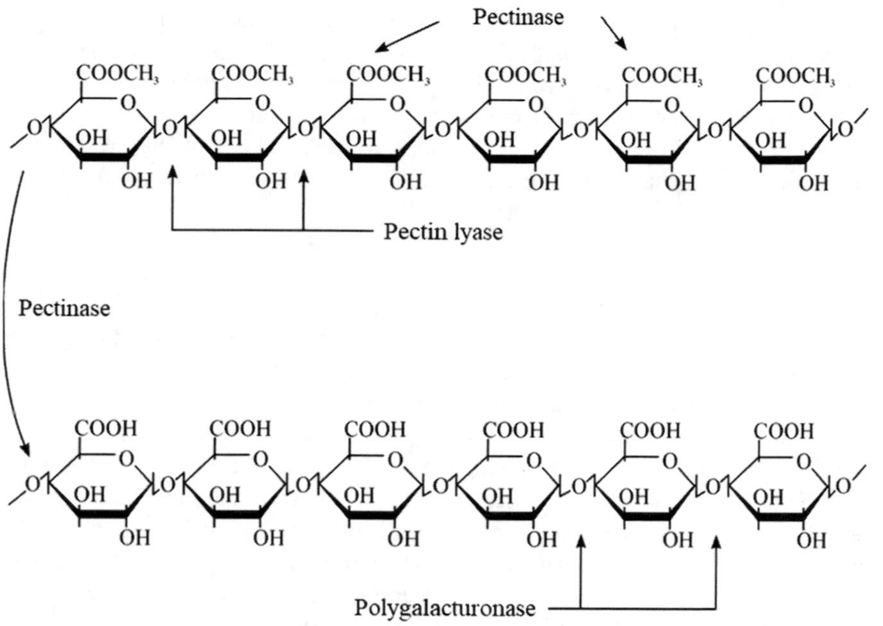

Fig. 8.14 Mechanism of pectinase action

Cellulase

Cellulase is an enzyme that hydrolyzes cellulose. It remains inconclusive that whether cellulase plays an important role in the softening process of vegetables. Microbial cellulase is of great significance in transforming insoluble cellulose into glucose and destroying the cell wall in the production of fruit and vegetable juice to increase juice yield.

Based on the different of cellulose and intermediates of degradation, cellulose is classified into three types: endoglucanase, cellobiohydrolase, and β-glucosidase.

Lipases

Lipases hydrolyze ester bonds of triglycerides presented at the oil/water interface to generate fatty acids and glycerol. Lipase is specific for ester bond of the hydrolyzed triglyceride. The lipase firstly hydrolyzes the ester bond at the first and third positions to produce a monoglyceride. The second ester bond is transferred to the first or the third position after non-enzymatic isomerization, and then completely hydrolyzed to glycerol and fatty acid by lipase. Lipase only acts on fat molecules at the glycerol–water interface, so the addition of an emulsifier to fat increases the oil–water interface to greatly increase the catalytic capacity of lipase.

Primary sources of lipase are the tissues of animals, plants, and microorganisms that contain fat. Lipase can catalyze fats into fatty acids leading to food spoilage, while the activity of lipase is required to produce flavor in another cases, such as moderate hydrolysis of milk fat in cheese production produces a good flavor.

The type of lipases includes phosphatase, sterolase, and carboxylesterase which can hydrolyze phosphates, sterol esters, and triglycerides, respectively.

Protease

Proteases are a class of hydrolases that represent one of the most important groups of food industry and biological systems. They are commonly present in a wide diversity including animals, plants, or microbes. Due to their enormous diversity, the classification of proteases is mainly based on the type of reaction catalyzed, their structure, and their active site. These enzymes are classified as exo- or endopeptidases depending on mode of action of protease. Endopeptidases randomly cleave peptide bonds distal to the terminal groups in the middle of the chain, making them smaller peptide fragments and small amounts of free amino acids. Exopeptidases cleave peptide bonds near the terminal groups to free the amino acids, which can be classified into two categories: aminopeptidases hydrolyze peptide bond from amino end of peptide chain and carboxypeptidases hydrolyze peptide bonds from the carboxy end of peptide chain. Proteases can also be named according to the range of pH in which they act: acidic, neutral, and alkaline. According to the chemical nature of the active center of protease, the protease also can be divided into serine protease (the active center contains a serine residue), thiol protease (the active center contains a sulfhydryl group), metalloprotease (the active center contains a metal ion), and acid protease (the active center contains two carboxyl groups). In addition, proteases can be classified into animal proteases, plant proteases, and microbial proteases depending on their sources.

Acid proteases include pepsin, chymosin, and many microbial and fungal proteases. The chymosin is used as a coagulant in cheese processing. The mechanism of action of chymosin with condensation action is that the chymosin can hydrolyze the peptide bond between the k-casein Phe_{105}-Met_{106}, the casein micelles lose stability, and then aggregate into a clot. And it contributes to the formation of flavor substances, while other proteases can also precipitate cheese, but the yield and hardness will be reduced. However, since chymosin is extracted from the stomach of calf and is very expensive, it has been replaced by a substitute in recent years. The addition of an acid protease to the flour changes the rheological properties of the dough and firmness of products in the baked food. Microbial proteases can be used to make fermented foods such as soy sauce.

Serine proteases mainly include chymotrypsin, trypsin, and elastase, it can be used to soften and tender connective tissues in meat, through effect on myosin–actin complex.

The thiol protease has a thiol group at its active center, which is mostly isolated from plants and widely used in food processing, such as papain, bromelain, and ficin.

The cooling turbidity of beer is related to the sedimentation of proteins; thus, thiol protease can be used as clarifying agents for beer. Also, thiol protease is used as a tenderizer for meat that can be injected into the meat to partially hydrolyze elastin and collagen. Protease can also be used to produce hydrolysates of complete or partial hydrolysates of proteins, such as liquefaction of fish proteins yielding products with good flavor.

Flavor Enzyme

Flavor enzymes act on flavor precursors to produce flavor in fruits and vegetables. For example:

(1) Fruits such as bananas, apples, or pears have no flavor during growth and harvest. Until the early stage of ripening, a small amount of ethylene is produced to stimulate the synthesis of flavor substances. For example, the precursors of banana flavor are non-polar amino acids and fatty acids, which are converted to aromatic esters, alcohols, and acids by a series of flavor enzymes during maturation to form the characteristic flavor of banana.

(2) The flavors of cabbage and onion are produced by the direct action of specific enzymes on specific flavor precursors. The flavor of these plants is mainly derived from the mustard oil (Isothiocyanate) produced by thioglycosides enzyme acting on thioglycosides.

$$R-C\begin{array}{c} {}^{\nearrow S-C_6H_{11}O_5} \\ {}_{\searrow N-O-SO_2O^-K^+} \end{array} + H_2O \xrightarrow[\text{enzyme}]{\text{Thioglycoside}} R-N=C=S+C_6H_{12}O_6+KHSO_4$$

Thioglycoside   Isothiocyanate

Onion flavor comes from sulfur-alkyl-L-cysteine sulfoxide dissociation enzyme acting on sulfur-substituent L-cysteine sulfoxide to produce volatile sulfur-containing compounds.

$$2R-\overset{\uparrow}{S}-CH_2-\underset{\underset{NH_2}{|}}{CH}-COOH \xrightarrow[\substack{\text{Sulfur-alkyl-}L\text{-cysteine sulfoxide} \\ \text{dissociation enzyme}}]{H_2O \quad 2NH_3} 2CH_3-\overset{O}{\overset{||}{C}}-COOH+R-\overset{\overset{O}{\uparrow}}{S}-S-R$$

Sulfur-substituent L-cysteine sulfoxide   Thiosulfonate (allicin)

(3) The flavor of black tea is produced by the oxidation of enzymes. First, catecholoxidase oxidizes flavonol, and oxidized flavonol reoxidizes amino acids, carotene, and unsaturated fatty acids to produce a unique aroma component in black tea. Lipoxygenase is widely found in plants, which can catalyze oxidation of unsaturated fatty acids to form hydroperoxides and lyses with enzymes

or non-enzymes to produce aldehydes or ketones and other components with flavor characteristics.

(4) Flavor enzymes can be used to restore flavor during food processing. Because most of the volatile flavor compounds are volatilized during the heat treatment of food, the food will lose its flavor. The processed food can still retain the special flavor if the original flavor precursors are transformed into flavor substances by adding external enzymes.

Vitamin B_1 Hydrolase

Vitamin B_1 hydrolase can hydrolyze vitamin B_1 to 2-methyl-6-amino-5-hydroxymethylpyrl midine and 4-methyl-5-hydroxyethyl thiazole, which is mainly found in aquatic animals such as fish and shellfish. Owing to eating raw fish and caviar (unheated but made by fermentation), Asians often suffer from deficiency of vitamin B_1. Raw herring cured in salt can destroy 50–60% of added vitamin B1 within 6 h. In addition to aquatic animals, vitamin B_1 hydrolase is also found in ferns, beans, and mustard. Some microorganisms in the mouth can secrete vitamin B_1 hydrolase, which can also cause deficiency of vitamin B_1.

Phytase

Phytase, is widely used in plants as a reservoir of phosphoric acid. Because minerals are incorporated in protein–phytic acid–mineral complexes, the nutrient potency of minerals in plant foods is reduced. For example, the connection between phytases and Ca^{2+} will obviously affect the texture of some vegetables, because Ca^{2+} cannot be involved in the cross-linking of pectin molecules after being linked to phytases. The presence of phytate causes an unfavorable softening on the vegetables.

Phytase can hydrolyze phosphate residues from phytic acid, thus destroying the strong affinity of phytic acid for mineral elements. Thus, phytase increases the nutritional titer of mineral elements and alters texture of plant-based foods by releasing Ca^{2+} to participate in cross-linking or other reactions. Although its concentration is low, the control of endogenous phytase by temperature and water activity can reduce the concentration of phytic acid in plant foods. It is possible to effectively control phytic acid in food by exogenous phytase.

Pigment Degradation Enzyme

Pigment degrading enzyme, including chlorophyllase and anthocyanase, is present in plants. If they are activated in the harvested plants, they can catalyze chlorophyll and anthocyanin.

8.4.1.2 Oxidoreductase

Glucose Oxidase

Glucose oxidase catalyzes the oxidation of β-D-glucose to produce D-gluconolactone and hydrogen peroxide (H_2O_2) utilizing molecular oxygen. The enzyme is isolated from fungi like *Aspergillus niger* and *Penicillium*.

$$C_6H_{12}O_6 + O_2 \xrightarrow{\text{Enzyme}} C_6H_{10}O_6 + H_2O_2$$

For example, glucose oxidase can be used in egg processing to remove glucose, thereby preventing food from discoloring due to Maillard reaction. In addition, it also gives the fried potato chips a golden yellow, which is caused by the presence of excessive glucose. Glucose oxidase also removes oxygen from the closed packaging system to inhibit oxidation of fat and degradation of natural pigments. For example, dipping crab and shrimp in a mixture of glucose oxidase and catalase prevent their pink color from changing to yellow, because glucose oxidase catalyzes glucose absorbing oxygen to produce gluconic acid and catalase catalyzes hydrogen peroxide to produce water and oxygen. The reaction is given as follows:

$$\text{Glucose} + 1/2\ O_2 \xrightarrow[\text{Catalase}]{\text{Glucose oxidase}} \text{Gluconic acid}$$

Catalase

Catalase plays a crucial role in adaptive response to H_2O_2. The enzyme is isolated from liver or microorganism.

$$2H_2O_2 \longrightarrow 2H_2O + O_2$$

Hydrogen peroxide is a by-product of foods treated with glucose oxidase and compound that can be added to foods by special processes in canning. For instance, milk can be pasteurized with H_2O_2 and is relatively stable, and excess H_2O_2 can be eliminated by catalase.

Aldehyde Dehydrogenase

When soybean is processed, the unsaturated fatty acid is enzymatically oxidized to form a volatile degradation compound (n-hexanal, etc.) with a bean flavor. If the acetaldehyde dehydrogenase is added, the compound can be converted into a carboxylic acid to eliminate the bean flavor.

$$n\text{-hexanal} + NAD^+ \xrightarrow{\text{Acetaldehyde dehydrogenase}} \text{Caproic acid} + H^+ + NADH$$

Among various acetaldehyde dehydrogenases, acetaldehyde dehydrogenase extracted from bovine liver mitochondria has a high affinity with n-hexanal and is recommended for soymilk production.

Peroxidase

Peroxidase is widely present in higher plants and milk, which contains hemoglobin as a prosthetic group that catalyzes the following reaction:

$$ROOH + AH_2 \longrightarrow ROH + A + H_2O$$

ROOH is hydrogen peroxide or organic peroxide, and AH_2 is an electron donor. When ROOH is restored, AH_2 is oxidized. AH_2 may be ascorbate, phenol, amine, or other reductive organic matter, which is oxidized to produce color. Based on this, the colorimetric method can be used to determine the activity of peroxidase. Because peroxidase widely present in plant tissues has high heat resistance, it can be evaluated by sensitive and easy colorimetric method. The activity of the peroxidase disappearing means that other enzymes are destroyed after food is heat-treated, so it can be used as an effective indicator such as blanching or disinfection.

Peroxidase is also very important in the aspects of nutrition, color, and flavor because it can oxidize vitamin C and destroy its physiological function, and can catalyze dissociation of unsaturated fatty acid peroxide to produce carbonyl compounds with unpleasant odors and free radicals destroying many components of food. If food has no unsaturated fatty acids, peroxidase can catalyze carotenoid bleaching and anthocyanin decolorizing.

Ascorbic Acid Oxidase

Ascorbic acid oxidase is a copper-containing enzyme that oxidizes ascorbic acid.

$$L\text{-Ascorbic acid} + 1/2O_2 \rightarrow \text{Dehydroascorbic acid} + H_2O$$

It is present in melons, seeds, grains, fruits, and vegetables. The enzyme has a great influence on the processing of citrus due to oxidation for ascorbic acid. Although oxidase and reductase may be in equilibrium in tissue-completed citrus, the reductase is very unstable and is greatly damaged in the processing of juice. During the processing of citrus, low temperature, rapid juice extraction, air pumping, and pasteurization should be carried out to inactivate the activity of enzyme and reduce the damage of vitamin C.

Lipoxygenase

Lipoxygenase is widely present in plants, such as soybeans, mung beans, wheat, oats, barley, and corn. In addition, it is also present in the leaves of potato tubers, cauliflower, alfalfa, and apples.

It is highly specific to the substrate and can only catalyze the oxidation of polyunsaturated fatty acids containing cis-1,4-pentadiene structures and oxidation of glycerides. Linoleic acid, linolenic acid, and arachidonic acid all contain this structure, so they are easily affected by lipoxygenase, especially linolenic acid is a good substrate of lipoxygenase.

Lipoxygenase is important in food processing that it affects the color, flavor, texture, and nutritional value of food. For example, the beany flavor in soybeans and soybean products is caused by the continuous dissociation of hydroperoxide generated by the oxidation of linolenic acid catalyzed by lipoxygenase. In peas froze without blanching, the accumulation of carbonyl compounds is also caused by lipoxygenase and the plant tissues that are not thoroughly blanched still contain this enzyme, which produce peculiar smell. Therefore, to reduce the activity of lipoxygenase in stored vegetables, it must be blanched before freezing or drying. During the processing of macaroni, the lipoxygenase in it can produce an undesirable bleaching effect on the pigment. It can catalyze β-carotene, luteol, chlorophyll, and vitamins to destroy them. The lipoxygenase in wheat also has a great influence on rheological properties of flour because the oxygen in air is mixed into dough during kneading, so that the lipoxygenase catalyzes oxidation of the sulfhydryl groups in protein into disulfide bonds to form network and this structure improves the elasticity of dough. In addition, soybean flour is usually added to flour, which not only increases the protein content of flour, but also uses lipoxygenase in soybean flour to enhance bleaching effect and improve rheological properties of dough.

8.4.2 Application of Enzymes in Food Processing

Enzyme has gained popularity in starch processing, dairy processing, fruit processing, wine making, meat, eggs and fish processing, baked goods manufacturing, food preservation, and sweetener manufacturing during the past few decades.

8.4.2.1 Application of Enzyme in Starch Processing

The enzymes used for starch processing include α-amylase, β-amylase, glucoamylase, glucose isomerase, debranching enzyme, and cyclodextrin glucosyltransferase. The process of starch involves two steps: liquefaction and saccharification. Liquefaction of starch is its hydrolysis into short-chain dextrin by α-Amylase resulting in reduction of the viscosity of the starch suspension. Saccharification is the production of glucose and fructose syrup by further hydrolysis. Starch syrups made from different enzymes are also diverse, such as high maltose syrup, sucrose, glucose, fructose, fructose syrup, conjugated sugars, and cyclodextrins.

In addition, enzymes have other applications in the production of starchy foods. For example, α-amylase is used to hydrolyze starch in the brewing industry. In bread manufacturing, it provides yeast with fermented sugar to improve the texture of bread; it is used in beer production to remove starch turbidity and improve clarity; etc.

8.4.2.2 Application of Enzyme in Dairy Processing

The most important enzymes used in dairy products are chymosin, lactase, catalase, lysozyme, lipase, etc. Chymosin is used to make cheese; lactase is used to break down lactose in milk; catalase is used to sterilize milk; lysozyme is added to milk powder to prevent intestinal infection in infant; and lipase increases the aroma of cheese and butter.

The first step in cheese production is to ferment milk into yogurt with lactic acid bacteria; the second step is to hydrolyze soluble k-casein to insoluble Para-kappa-casein and glycopeptide using rennet; under acidic conditions, Ca^{2+} makes casein coagulate, and then diced, heated, pressed, and matured to make cheese. For example, as rennet can hydrolyze κ-casein:

$$N...Pr•His•Leu•Ser•Phe•Met•Ala•Ile •Pr•Pr...C$$

$$\downarrow$$

$$(Insoluble)Para-\kappa- \; casein\rightarrow | \leftarrow Glycopeptide$$

In the past, chymosin was taken from the calf's stomach. The source of chymosin is insufficient and expensive. Hence, 85% of animal rennin has been replaced by microbial enzymes at present. Microbial chymosin is an acid protease that has a strong curd effect and a weaker ability to hydrolyze casein, but it will also hydrolyze casein to form bitter peptides. Genetic engineering has now been used to transfer bovine prochymosin gene to *E. coli*, which has been successfully expressed. Therefore, the current fermentation method has been able to produce chymosin.

Milk contains certain amount of lactose. Due to the lack of lactase in body, some people usually experience symptoms such as abdominal pain and diarrhea after drinking milk. At the same time, because lactose is difficult to dissolve in water, it is precipitated as sand-like crystals in condensed milk and ice cream and affects quality. So, it is necessary to use lactase to remove lactose in milk to produce lactose-free milk.

In addition, the by-product of cheese production—whey—contains a lot of lactose which is difficult to be digested and has always been discharged as wastewater. Now, lactase can be used to decompose lactose in whey, so that they can be used as feed and medium for yeast production.

8.4.2.3 Application of Enzymes in Fruit Processing

Enzymes used in fruit processing include pectinase, naringinase, cellulase, hemicellulase, hesperidinase, glucose oxidase, and catalase.

Pectin, belongs to fruit, forms a gel in acidic and high-concentration sugar solutions. This property is the material basis for making jelly and jam. Paradoxically, it became difficult to clarify fruit juice because of the presence of pectin. In the industry, the broken fruit is treated with pectinase produced by *Aspergillus niger*, *Aspergillus oryzae*, or *Rhizopus*, to eliminate pectin and suspended solids.

In the manufacture of canned orange, the orange is treated with a mixture of cellulase, hemicellulase, and pectinase produced by *Aspergillus niger* to remove capsule dressing.

The orange juice is treated with naringinase to remove the bitter taste of naringin. The addition of *aspergillus* hesperidin to the orange juice can decompose the insoluble hesperidin into water-soluble hesperetin, preventing the formation of white precipitates, clarifying the orange juice and removing the bitterness. Treating orange juice with glucose oxidase and catalase can remove oxygen in orange juice, so that orange juice maintains its original color and flavor during storage.

8.4.2.4 Application of Enzymes in Wine Brewing

Beer is made from barley malt. During barley germination, starch present in the barley is broken down by respiration. So, the brewery often uses barley, rice, and corn as auxiliary materials to replace part of the barley malt. However, this method causes deficiencies of amylase, protease, and β-glucanase, resulting in insufficient saccharification of the starch and degradation of the protein and β-glucan, thereby affecting the flavor and yield of the beer. In industrial production, adding microbial amylase, neutral protease, and β-glucanase enzyme preparations can make up for the deficiency of enzyme activity in the raw material, increasing the degree of fermentation and shortening the saccharification time. Moreover, the addition of papain, bromelain, or fungal acid protease can prevent beer turbidity and prolong the shelf life before beer pasteurization.

Glycosylase, used in the manufacture of liquor, rice wine, and alcohol, can increase alcohol yield and simplify equipment. Complex enzyme preparations, including pectinase, protease, cellulase and hemicellulase, are used in the production of fruit wine, which not only can improve the yield of fruit juice and fruit wine, but also conducive to filtration and clarification.

8.4.2.5 Application of Enzymes in the Processing of Meat, Eggs, and Fish

The muscles of old cows and sows are difficult to soften due to the mechanical strength of collagen in their connective tissue. Treatment with an enzyme preparation such as papain or bromelain can hydrolyze collagen to make the muscles tender. There are two ways to tenderize meat in the industry: one is to inject the enzyme into the animal before slaughter and the other is to apply the enzyme preparation to the surface of muscle or to soak muscle with an enzyme solution.

Proteases are used to hydrolyze discarded animal blood, miscellaneous fish, and protein in minced meat, and then extract the soluble protein for food or feed. This is an effective measure to develop protein resources. Among them, the use of trash fish is of greatest concern.

Co-treatment with glucose oxidase and catalase to remove glucose in poultry eggs can eliminate the occurrence of "browning" during drying of poultry egg products.

8.4.2.6   Application of Enzymes in the Manufacture of Baked Goods

Due to the low enzyme activity and fermenting power of long-lasting flour, bread made of long-lasting flour is small and poor in color. Adding α-amylase and protease of mold to dough of long-lasting flour can improve the quality of bread.

Let us now turn to actual application of enzymes in the baking industry. The addition of β-amylase prevents the aging of the cake; the addition of sucrase prevents the sucrose in the pastry crystallization from the syrup; and the addition of protease can make the noodles have good taste and extensibility.

8.4.2.7   Application of Enzymes in the Manufacture of Food Additives

Enzymes are widely used in the production of food additives. Mainly contain emulsifiers, thickeners, sour agents (lactic acid, malic acid, etc.), Umami agents (monosodium glutamate, flavoring nucleotides), sweeteners (aspartame), oligosaccharides (palatinose, malto-oligosaccharides), food fortifiers (lysine, aspartic acid, phenylalanine acid, alanine), and so on.

For example, aspartic acid phenylalanine methyl ester (H-Asp-Phe-OMe) is especially suitable for diabetics, which is a low-heat new dipeptide sweetener with a sweetness 200 times that of sucrose. It is based on benzyloxycarbonyl-L-aspartic acid (Cbz-Asp) and L-phenylalanine methyl ester (Phe-Ome), and is catalyzed by immobilized heat-resistant neutral protease in an organic solvent. Finally, aspartic acid phenylalanine methyl ester is prepared by catalytic hydrogenolysis with palladium on carbon (Pd-C). Its synthesis process is

$$\text{Cbz-Asp} + \text{Phe-OMe} \xrightarrow[\text{Neutral protease, Organic solvent}]{\text{Heat-resistant immobilization}} \text{Cbz-Asp-Phe-OMe} \xrightarrow{\text{Pd-C}} \text{H-Asp-Phe-OMe}$$

For another example, the orange mandarin contains 10–20% hesperidin. After extraction and separation, hesperidin is hydrolyzed with hesperidin of *Aspergillus niger* to remove rhamnose in the molecule, and it is used in alkaline solution for hydrolysis and reduction. Then hesperetin-β-glucoside dihydrochalcone which is

70–100 times sweeter than sucrose is obtained. It is safe, low-calorie sweetener, but its solubility is very low (only 0.1%) and has no practical value. If this substance is mixed with starch solution, cyclodextrin glucosyl transferase is used to catalyze the coupling reaction to attach its C4 glucose molecule on two glucose molecules (g-g) and produce hesperidin dihydrochalcone-7-maltoglucoside, leading to sweetness without change but solubility increases 10 times.

Hesperetin dihydrochalcone-7-maltose

8.4.2.8 Application of Enzymes in Food Preservation

Enzymatic food preservation technology utilizes the catalytic action of enzymes to prevent or eliminate the adverse effects of external factors on food, to maintain the original good quality and characteristics of the food. The enzymes used for food preservation mainly include glucose oxidase and lysozyme.

Glucose Oxidase

Glucose oxidase is an ideal oxygen scavenging agent. The principle of preservation is: catalyzing the reaction of glucose with oxygen to produce gluconic acid and hydrogen peroxide, effectively reducing or eliminating oxygen in the sealed container and preventing food oxidation.

The specific method: the glucose oxidase is made into an "oxygen preservation bag." Briefly, the glucose oxidase and glucose are mixed together, packaged in impermeable but breathable film bag and placed in airtight container containing food that needs to be kept fresh. When oxygen in airtight container penetrates film into the bag, it reacts with glucose under catalysis of glucose oxidase to remove oxygen and achieve purpose of preventing oxidation. Glucose oxidase can also be directly added to canned juice and fruit wine to prevent oxidation of canned food.

In addition, appropriate amount of glucose oxidase is added into the protein liquid or whole egg liquid to remove small amount of glucose in egg products under aerobic conditions, thereby effectively preventing the browning of egg products and improving the quality of products.

Lysozyme

Lysozyme catalyzes the hydrolysis of peptide polysaccharides in the bacterial cell wall. The principle of preservation is: the enzyme acts exclusively on the β-1,4 glycosidic bond between N-acetylmuramic acid and N-acetylglucosamine in the peptide polysaccharide molecule, destroying the cell wall of the bacteria and inhibiting their growth. It effectively prevents and eliminates the contamination of bacteria in food, and achieves the purpose of preservative preservation.

Lysozyme can be produced from egg white or microbial fermentation. Different sources of lysozyme have different characteristics. Lysozyme from egg white has strong bacteriolytic properties for many gram-positive bacteria other than *Staphylococcus aureus*, and has no effect or weak effect on gram-negative bacteria.

Lysozyme, generally egg white lysozyme, is used to keep food fresh. Nowadays, lysozyme is widely used in preservation of cheese, aquatic products, low-alcohol wine, dairy products, sausages, butter, wet noodles, and other foods.

8.4.2.9   Application of Enzymes in Food Detoxification

Enzymes can degrade the toxic components of food into non-toxic compounds to detoxification. For example, toxic components of broad beans may lead to hemolytic anemia. The addition of β-glucosidase can degrade the toxins, producing phenolic bases which are unstable. The phenolic bases can be rapidly oxidized and decomposed when they are heated.

Convicine   Vicine   Isouramil   Divicine

Many toxins and anti-nutritional factors in food can also be removed by the action of enzymes (Table 8.5).

8.4.3   Application of Enzymes in Food Analysis

Compared with chemical analysis, enzymatic analysis is fast, specific, high sensitivity, and precision. The analytical error of low-level compounds is small. The enzymatic analysis eliminates the trouble of separating the analyte and other components. For example, to determine the content of glucose in the plant, it is only needed to remove the insoluble matter that interferes the absorbance. In addition, the conditions

Table 8.5 Enzymatic removal of toxins and anti-nutritional factors in food

Substance	Food source	Toxicity	Type of enzyme
Lactose	Milk	Gastric discomfort	β-Galactosidase
Oligogalactose	Beans	Flatulence	α-Galactosidase
Nucleic acid	Single cell protein	Gout	Ribonuclease
Lignan glycosides	Safflower seed	Catharsis	β-Glucosidase
Phytic acid	Bean, wheat	Mineral deficiency	Phytase
Trypsin inhibitor	Soy	Reduce protein digestion and utilization	Urease
Ramie	Castor bean	Respiratory system	Protease
Cyanide	Fruit nuts	Death	Rhodanese, cyanophenylalanine synthase, nitrilase
Tomato	Green fruit	Alkaloid	Enzyme system of mature fruit
Nitrite	Various foods	Carcinogen	Nitrite reductase
Caffeine	Coffee	Excited	Microbial purine demethylase
Saponin	Alfalfa	Bovine bulge disease	β-Glucosidase
Chlorinated pesticide	Vegetable food	Carcinogen	Glutathione-S-transferase
Organic phosphate	Various foods	Neurotoxin	Esterase

required for enzymatic analysis are mild and the reaction time is very short. Therefore, changes in compounds caused by non-enzymes can be avoided. Immobilized enzymes are widely used in food analysis because they can be reused. Currently, immobilized enzyme columns, enzyme electrodes, enzyme-containing sheets, and enzyme-linked immunoassays (ELISA) are used.

The food enzymatic analysis comprises determining enzyme activity in food and food component, and the measured food component can be a substrate of enzyme, an inhibitor or activator of enzyme.

8.4.3.1 The Tested Compound Is the Substrate of Enzyme

When the tested compound is a substrate of enzyme, end point method and kinetic method can be used for the determination according to the substrate consumption.

Fig. 8.15 Effect of substrate concentration on absorbance

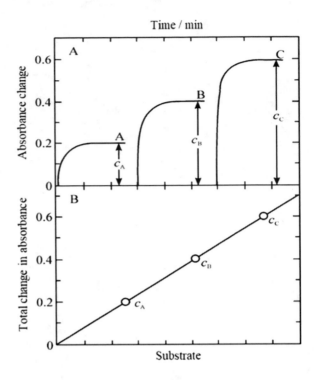

Table 8.6 Enzyme indicator for evaluating food quality

Substrate	Enzyme	K_m	Enzyme concentration (μkat/L)
Glucose	Hexokinase	1.0×10^{-4} (30 °C)	1.67
Glycerin	Glycerol kinase	5.0×10^{-5} (25 °C)	0.83
Uric acid	Uric acid oxidase	1.7×10^{-5} (20 °C)	0.28
Fumaric acid	Fumarase	1.7×10^{-6} (21 °C)	0.03

End Point Method

End point method is reliable only when the reaction is complete. It is based on the total change in absorbance or fluorescence intensity of the enzyme reaction system before and after the reaction to determine the amount of product. Compared with kinetic method, the concentration of the substrate to be analyzed in the food must not be lower than the Michaelis constant K_m of the enzyme catalytic reaction (Fig. 8.15).

The advantage of end point method is that it does not require precise control of pH and temperature of enzyme reaction. It is only needed to ensure enough enzyme to complete the reaction within 2–10 min. As shown in Table 8.6, it lists the concentration of different enzymes used in the end point assay. In certain cases, enzyme-catalyzed reaction has reached equilibrium before it is fully carried out. At this time, it is possible to break the equilibrium by increasing concentration of reactants or

Table 8.7 Determination of enzyme activity in food

Compound category	Representative compound
Alcohol	Ethanol, glycerin
Aldehyde	Acetaldehyde, glycolaldehyde
Acid and its salt	Acetate, lactate, formate, malate, succinate, citrate, isocitrate, pyruvate
Monosaccharides and similar compounds	Glucose, fructose, galactose, pentose, sorbitol, inositol
Disaccharides and oligosaccharides	Sucrose, lactose, raffinose, maltose
Polysaccharide	Starch, cellulose, hemicellulose
L-amino acid	Glutamic acid, arginine
Lipids	Cholesterol

removing certain product of reaction and make it proceed in the direction that is beneficial to product. For example, in reaction catalyzed by lactate dehydrogenase, lactic acid and NAD^+ are the substrates to produce pyruvate, which can be removed by adding oxamate to make reaction proceed toward product. In addition, a standard curve can be prepared under the same reaction conditions, and the concentration of the tested compound can be obtained.

Kinetic Method

Kinetic method is based on measuring the rate at which enzyme catalyzes the reaction and calculate concentration of substrate. The reaction rate is independent of the substrate, and it is apparent that the concentration of the analyte as the enzyme substrate cannot be calculated by measuring the rate of the enzyme-catalyzed reaction when [S] > 100 km. Therefore, the concentration of the tested compound must be less than 100 km. According to the equation, the concentration of a certain analyte as a substrate can be calculated.

The specific steps: according to the Lineweaver–Burk equation, standard curve is drawing with the double reciprocal plot method, and determine concentration of analyte from the standard curve. This method is not susceptible to interference and can be automatically analyzed, but the reaction conditions must be strictly controlled: the conditions for determination of analyte and preparation of standard curve should be identical. In addition, enzyme used in the analysis is required to have a high K_m to determine a higher substrate concentration.

Table 8.7 lists some compounds that can be analyzed by enzymatic methods.

8.4.3.2 The Tested Compound Is an Activator or Inhibitor of Enzyme

Several enzymes have absolute requirements for cofactors. The cofactor may be an organic compound as pyridoxal phosphate NAD^+; it may also be a metal ion like Zn^{2+} or Mg^{2+}.

The initial reaction rate is linear with the concentration of activator when the activator is firmly bound to the enzyme; on the contrary, the initial reaction rate is hyperbolic with the concentration of activator when the activator is loosely bound to the enzyme, like the determination of the substrate concentration.

When the tested compound is an inhibitor of an enzyme, the reaction rate is lowered. Due to the reversible and irreversible binding of inhibitor and enzyme, the determination is very complicated.

8.4.3.3 Enzyme as an Indicator of Food Quality

Enzyme activity levels can be used as an indicator of food materials and product quality. For example, the determination of the activity of remaining peroxidase and the activity of residual alkaline phosphatase can well reflect the degree of heat treatment. Because the two enzymes have higher stability than other enzymes and their activities are easy to be measured, they are very useful enzyme indicators for food quality. Table 8.8 lists enzyme indicators for evaluating food quality.

During freezing and thawing of food, some enzymes are released while the integrity of cells is destroyed. Testing the activity of these enzymes can determine whether the food materials are resistant to freezing and thawing. For example, the increase in activity levels of malic enzyme and glutamate oxaloacetate aminotransferase were used as indicators of tolerance of oysters and meat to freezing and thawing. Different food materials have different indicators of enzymes that tolerate freezing and thawing.

Although the conventional method of detecting the degree of bacterial contamination is plate count, measuring activity level of indicator enzyme can also obtain the same result and save time. The enzymes, which have a low level of viability in food materials and a significant increase in viability levels after being contaminated by bacteria, can be used as indicator enzymes for the degree of contamination of food materials. For example, acid phosphatase and catalase can be used as indicator enzymes for some food materials. The determination of reductase activity with methylene blue can indicate the degree of milk contamination by bacteria. The increase of lysolecithinase activity in fish is related to microbial contamination, so the activity of this enzyme is an indicator of fish freshness. Wheat with high moisture content is prone to germinate during storage, resulting in declining in quality of bakery products. Wheat germination can be determined by measuring the activity of amylase in flour or the activity of peroxidase in wheat (Table 8.9).

During the ripening process of fruits, the activity levels of many enzymes change significantly and can be used to indicate the maturity of fruit, such as sucrose synthase in potatoes and pectinase in pears.

Table 8.8 Enzyme indicator for evaluating food quality

Purpose	Enzyme	Raw materials
Moderate heat treatment	Catalase	Fruits and vegetables
	Alkaline phosphatase	Milk, dairy, ham
	β-acetylglucosaminidase	egg
Freezing and thawing	Malic enzyme	Oyster
	Glutamate oxaloacetate transaminase	Meat
Bacterial contamination	Acid phosphatase	Meat, egg
	Catalase	Milk, green beans
	Glutamate decarboxylase	Milk
	Reductase	Milk
Insect pollution	Uricase	Preserving cereals, fruits
Freshness	Hemolytic lecithinase	Fish
	Xanthine oxidase	Fish
Maturity	Sucrose synthase	Potato
	Pectinase	Pear
Germination	Amylase	Flour
	Peroxidase	Wheat
Color	Polyphenol oxidase	Peach, avocado, wheat, Coffee
	Succinate dehydrogenase	Meat
Flavor	Alliinase	Onion, garlic
	Glutamic acid transpeptidase	Onion
Nutritional value	Protease	Digestive capacity
	Protease	Protein inhibitor
	L-amino acid decarboxylase	Essential amino acid
	Lysine decarboxylase	Lysine

Table 8.9 Determination of enzyme activity in food

Enzyme	Food type
Diphenol oxidase	Cereals, flour, milk, vegetables
Xanthine-oxygen-oxidoreductase	Milk
Lipoxygenase	Soybeans, flour
Peroxidase	Cereals, flour, milk, vegetables
Catalase	Milk, dairy products
Lipase	Milk, dairy products, cereal flours
Phosphatase	Milk, dairy products
Amylase	Honey, flour, malt, milk, bread, starch
Urease	Soy flour, soy products
Creatase	Meat extract, broth

Color and flavor are two key indicators that determine food quality. In terms of the color of food, the level of activity of polyphenol oxidase can be used as an indicator of enzymatic browning of fruits, e.g., peaches and avocados, or as an indicator of desired browning ability of tea, coffee, cocoa, and wheat. The flavor of foods is the result of the action of flavor enzymes. For example, glutaminyl transpeptidase firstly catalyzes the production of glutaminyl peptides which are produced by the action of alliinase to produce the unique flavor of onions and garlic.

The nutritional value of protein is often evaluated by animal feeding experiments or in vitro enzymatic methods. The main factors that affect the nutritional value of protein are: ① protein digestibility; ② influence of protease inhibitors in food on protein digestibility and pancreatic protease secretion; and ③ types and content ratio of essential amino acids in protein. Enzymatic method can measure the above three aspects and save time and money.

In addition, the quality of the food can be known by measuring the enzyme activity. For example, we can determine the quality of malt by measuring α- and β-amylase levels, and the changes in key enzyme activity levels during barley maturation, germination, and fermentation.

8.4.4 Application of Enzymes in Food Waste Treatment

When processing or eating some products, people often abandon processing by-products. In fact, the by-products can be processed into useful things to avoid polluting the environment such as a beverage can be prepared by decomposing citrus pomace with cellulase. Among them, 50% of crude fiber is degraded into short-chain oligosaccharides, which also has certain health value. Cellulase or microorganisms can convert cellulose in agricultural by-products and processed by-products into glucose, alcohol, and single-cell protein.

Aquatic product processing waste accounts for a large proportion of aquatic products. Taking silver carp as an example, about 40% scraps are produced during processing, including 30% fish heads, 8% viscera, 1% swim bladder, and 1% scales. The protein in fish scales is hydrolyzed by protease and the enzymatic hydrolysate can be used for production of condiments and as an additive of functional food. Fish head can be made into flavored food after cooking, enzymatic hydrolysis, and filtration and can be added to soy sauce and chicken essence to make composite condiments. Fish bone power is mainly used as a feed additive and can also be added to surimi products such as fish sausages, which reduce costs and strengthen nutrition. The compound amino acid calcium made by crushing and hydrolyzing fish bones and fresh head is a healthy product that enhances nutrition and supplement calcium. The number of mussel meat after pearl harvesting is quite large. It is rich in protein and essential amino acids and has been used as fresh food, dried, or even feed for a long time. For example, the protein in fish scales is hydrolyzed by protease, and the obtained enzymatic hydrolysate can be used to produce seasonings and functional foods.

During the slaughtering and processing of livestock, many meat by-products, e.g., bone, fat, and oil residue are produced. These by-products are often regarded as low-value products, some of which are sold cheaply as animal feed and some cannot be used, causing storage and environmental problems. Enzymatic biotechnology is used to hydrolyze the animal protein, and these by-products are processed into meat extracts which can be further added to the meat products to improve the quality of the meat products, or added to a variety of foods to improve the flavor of foods.

The use of high-efficiency biocatalysis of enzymes promotes the conversion of non-digestible and non-processable components, increasing the variety of food colors and value of products. This is one of the hotspots of application of enzyme preparations in food.

8.5 Summary

The chemical nature of enzyme is almost protein except for RNA with catalytic activity. Some enzymes are simple proteins and others are binding proteins. After the enzyme protein is combined with a co-enzyme or prosthetic group to form a holoenzyme, it exhibits a catalytic effect. Enzymes are highly specific, such as bond specificity, group specificity, absolute specificity, and stereochemical specificity. The rate of enzyme-catalyzed reaction is affected by substrate concentration, enzyme concentration, temperature, pH, water activity, and inhibitor. According to the mode of action of inhibitors and enzymes, it can be divided into reversible and irreversible inhibition; based on the relationship between reversible inhibitors and substrates, reversible reaction is divided into competitive inhibition, non-competitive inhibition, and uncompetitive inhibition.

Enzymatic browning occurs when plant tissue is damaged or in an abnormal environment, leading to phenolic enzymes catalyzing phenolic substances. To control enzymatic browning, we can control enzyme activity and oxygen concentration. The main pathways are the activity of inactivating enzymes, changing the working conditions of enzymes, isolating oxygen, and using antioxidants.

The purposes of adding enzymes in food processing are usually to: (1) improve food quality; (2) manufacture synthetic foods; (3) increase the speed and yield of extracted food ingredients; (4) improve flavor; (5) stabilize food quality; and (6) increase the utilization rate of by-products. Enzyme is mainly used in starch processing, dairy processing, fruit processing, wine brewing, meat processing, baked food manufacturing, food preservation, and sweetener manufacturing. The enzymes commonly used in the food industry are amylase, lipase, and polyphenol oxidase. If the substance used in food analysis is a substrate of enzyme reaction, the content of the substance can be analyzed by an enzymatic method. The use of high-efficiency biocatalysis of enzymes promotes the conversion of non-digestible and non-processable components, increasing the variety of food colors and value of products. This is one of the hotspots of application of enzyme preparations in food.

Questions:

1. What are endogenous enzymes and exogenous enzymes?
2. Please explain the mechanism of enzymatic browning and its control measures.
3. Please explain the change rule of K_m and V_{max} under competitive, non-competitive, and anti-competitive inhibition from the perspective of enzyme catalytic reaction kinetics.
4. What is the importance of enzymology to food science?
5. What are the commonly used enzymes in baked foods? What are their functions?
6. Why the longer people chew rice and steamed bread, the sweeter they are?
7. Why adding enzymes in washing powder can improve the efficiency of washing clothes?
8. Please explain the definition of enzyme, enzyme activity, enzyme specificity, enzyme active center, prosthetic group, immobilized enzyme, enzyme inhibitor, enzyme activator, and multi-enzyme system.

Bibliography

1. Baskar, G., Aiswarya, R., Renganathan, S.: Applications of asparaginase in food processing. In: Parameswaran, B., Varjani, S., Raveendran, S. (eds.) Green Bio-processes. Energy, Environment, and Sustainability, pp. 83–98. Springer, Singapore (2019)
2. Belitz, H.D., Grosch, W., Schieberle, P.: Food Chemistry. Springer, Berlin, Heidelberg (2009)
3. Binod, P., Papamichael, E., Varjani, S., Sindhu, R.: Introduction to green bioprocesses: industrial enzymes for food applications. In: Parameswaran, B., Varjani, S., Raveendran, S. (eds.) Green Bio-processes. Energy, Environment, and Sustainability, pp. 1–8. Springer, Singapore (2019)
4. Colgrave, M.L., Byrne, K., Howitt, C.A.: Food for thought: selecting the right enzyme for the digestion of gluten. Food Chem. **234**, 389–397 (2017)
5. Damodaran, S., Parkin, K.L., Fennema, O.R.: Fennema's Food Chemistry. CRC Press/Taylor & Francis, Pieter Walstra (2008)
6. Dewdney, P.A.: Enzymes in food processing. Nutr. Food Sci. **73**(4), 20–22 (1973)
7. Fernandes, P.: Enzymes in food processing: a condensed overview on strategies for better biocatalysts. Enzyme Res. **2010**, 862537 (2010)
8. Fraatz, M.A., Rühl, M., Zorn, H.: Food and feed enzymes. In: Zorn, H., Czermak, P. (eds.) Biotechnology of Food and Feed Additives. Advances in Biochemical Engineering/Biotechnology, pp. 229–256. Springer, Berlin, Heidelberg (2013)
9. Jermen, M., Fassil, A.: The role of microbial aspartic protease enzyme in food and beverage industries. J. Food Qual. **2018**, 7957269 (2018)
10. Kan, J.: Food Chemistry. China Agricultural University Press, Beijing (2016)
11. Oey, I.: Effects of high pressure on enzymes. In: Balasubramaniam, V., Barbosa-Cánovas, G., Lelieveld, H. (eds.) High Pressure Processing of Food, pp. 391–431. Springer, New York (2016)
12. Reyes de Corcuera, J.I., Powers, J.R.: Application of enzymes in food analysis. In: Nielsen, S. (ed.) Food Analysis. Springer, Cham, pp. 469–486 (2017)
13. Sindhu, R., Binod, P., Beevi, U.S., Amith, A., Kuruvilla, M.A., Aravind, M., Rebello, S., Pandey, A.: Applications of microbial enzymes in food industry. Food Technol. Biotechnol. **56**(1), 16–30 (2018)
14. Surowsky, B., Fischer, A., Schlueter, O., Knorr, D.: Cold plasma effects on enzyme activity in a model food system. Innov. Food Sci. Emerg. Technol. **19**, 146–152 (2013)

15. Terefe, N.S., Buckow, R., Versteeg, C.: Quality-related enzymes in plant-based products: effects of novel food-processing technologies part 3: ultrasonic processing. Crit. Rev. Food Sci. Nutr. **55**(2), 147–158 (2015)
16. Yu, F., Yu, S., Yu, L., Li, Y., Wu, Y., Zhang, H., Qu, L., Harrington, P.B.: Determination of residual enrofloxacin in food samples by a sensitive method of chemiluminescence enzyme immunoassay. Food Chem. **149**(8), 71–75 (2014)
17. Zeeb, B., Mcclements, D.J., Weiss, J.: Enzyme-based strategies for structuring foods for improved functionality. Annu. Rev. Food Sci. Technol. **8**(1), 21–34 (2017)
18. Zhang, Y., He, S., Simpson, B.K.: Enzymes in food bioprocessing—novel food enzymes, applications, and related techniques. Curr. Opin. Food Sci. **19**, 30–35 (2018)

Dr. Hui Shi born in 1986, associate professor at College of Food Science in Southwest University, is mainly focused on the research of food microbiology and biological testing Dr Hui Shi graduated from China Agricultural University and worked in University of Pennsylvania, and has presided two projects from National Natural Science Foundation of China and one subproject from National Science and Technology Major Project. Dr. Hui Shi has published 16 scientific papers in the field of food science such as Food Control, Applied Microbiology and Biotechnology, Journal of Applied Microbiology, Current Opinion in Food Science, etc.

Chapter 9
Pigments

Kewei Chen

Abstract This chapter gives an overall introduction of food pigments including their roles in the food industry, coloring mechanisms behind their chemical nature and their classification. Based on the chemical structures, some important and common natural pigments are illustrated according to their chemical molecules, changes and treatments during food processing, and related protective techniques including heme, chlorophylls, carotenoids and flavonoids with anthocyanins, catechins and tannins emphasized. In this chapter, plenty of detailed information about surrounding environmental effects on the changes of pigment molecules and the related color is provided to help readers understand the reaction mechanisms and to associate them with food processing techniques such as acid/alkaline treatment, modified atmospheric package and heat treatment.

Keywords Pigment · Food color · Heme · Chlorophyll · Carotenoid · Flavonoid · Anthocyanin · Catechin · Tannin

9.1 Overview

9.1.1 Definition of Pigments and Their Role in the Food Industry

The color of a substance is due to its capacity of selectively absorbing some visible light, thus reflecting the rest of the unabsorbed visible light that visualizes in human eyes. So, the substances in food that show this kind of characteristics leading to the presence of colorful food are collectively referred to as food pigments, including natural pigments inherent in food materials, colored substances formed in food processing, and added food colorants. Food colorants are natural or synthetic chemicals that have been subjected to stringent safety assessment tests and approved for use in the food coloring process.

K. Chen (✉)
College of Food Science, Southwest University, Chongqing 400715, China
e-mail: chenkewei@swu.edu.cn

© The Author(s), under exclusive license to Springer Nature Singapore Pte Ltd. 2021   383
J. Kan and K. Chen (eds.), *Essentials of Food Chemistry*,
https://doi.org/10.1007/978-981-16-0610-6_9

Table 9.1 Food color and associated sensory evaluation for consumers

Color	Sensory evaluation	Color	Sensory evaluation
Red	Strong flavor, ripe, delicious	Gray	Unpalatable, dirty
Yellow	Fragrant, bland, ripe, delicious	Purple	Strong, sweet, warm
Orange	Sweet, nourishing, strong, delicious	Light Brown	Unpalatable, hard, warm
Green	Fresh, refreshing, cool, acid	Dark Orange	Old, hard, warm
Blue	Fresh, refreshing, cool, acid	Cream	Sweet, nourishing, refreshing, delicious
Coffee	Unique flavor, rich texture	Dark Yellow	Not fresh, unpalatable
White	Nutritious, refreshing, hygienic, soft	Light Yellow-Green	Refreshing, cool
Pink	Sweet, soft	Yellow-Green	Refreshing, cool

Color is one of the main sensory indicators for food. Before acquirement of other food information, consumers often judge the food by the color to have a primary picture of food quality, freshness or maturity. For example, the color of a fruit is related to maturity, and the color of fresh meat is inseparable from its freshness. Therefore, how to improve the color characteristics of food is very crucial, which should be taken into consideration by food producers and processors. Food color that meets consumers' psychological requirements can give consumers beautiful enjoyment and increase their appetite and desire to buy.

The color of food can stimulate sensory organs of consumers and make them associate with the taste (Table 9.1). For example, red gives a feeling of ripeness and good taste that people generally like, and therefore plenty of candy, pastries and beverages are in red.

Besides, color can affect people's perception of food flavor. For example, red beverages are considered to have the flavors of strawberry, black strawberry, or cherry; yellow beverages are related to lemon flavor, and green beverages are referred to as lime flavor. Therefore, in the beverage industry, different flavors of beverages are often given different colors that meet the psychological requirements of consumers.

What's more, brightly colored foods can increase appetite. The color from red to orange is most appetizing, and light green and turquoise can also increase appetite, while yellow green depresses, as it is related with the long-term cognition for food, e.g., red apples, orange tangerines, yellow cakes and verdant vegetables. The color of some spoiled foods can make people feel bored, so some colors that are not too bright generally give a bad impression. Even if the same color is used in different foods, it will produce different feelings. For example, purple grape juice is widely accepted, but no one likes purple milk.

The color of foods is mainly decided by their inherent pigments. For example, the color of meat is mainly determined by myoglobin and its derivatives; the color of green leafy vegetables is mainly composed of chlorophyll and its derivatives. In food storage, the color change is often encountered, which sometimes is desirable. For instance, the color becomes more attractive during fruit ripening, and the bread turns out to be a brownish yellow color during the baking process. However, more often it needs to be avoided. For example, the browning on the cut surface after the apple is sliced; green vegetables become brownish green after cooking, and raw meat loses fresh red and gets brown during storage. Most of these changes in food color are due to chemical changes in food pigments. Therefore, understanding different food pigments is of great significance for controlling food color.

In food processing, the control of food color is usually carried out by two methods: color preservation and dyeing. Color protection means less loss of pigment, which calls for raw materials with appropriate maturity, proper processing technology, no excess acidic, alkaline or heat treatment, no access to metal equipment, less exposure to oxygen, etc. Dyeing is another common method to obtain and maintain the desired color for food. Since a coloring agent can produce plenty of color combinations, and its stability is better than inherent pigment, it is very convenient to apply it in food processing. However, the use of synthetic dyeing additives always arouses safety concerns, and it is necessary to comply with food safety regulations and food additive standards to prevent the abuse of colorants.

9.1.2 Mechanism of Food Coloration

Different substances can absorb light with different wavelengths. If the wavelength of light absorbed by a substance is outside the visible light region, the substance appears colorless; if in the visible light region (400–800 nm), the substance will exhibit a certain color. Color is related to the wavelength of the reflected light that is not absorbed. The color seen by the human eye is a composite color composed of visible light with different wavelengths reflected by the object. For example, if an object absorbs only invisible light and reflects all visible light, then it appears colorless; conversely, it appears black or nearly black. When the object selectively absorbs part of visible light, its color is represented by a composite color composed of unabsorbed visible light (also known as the complementary color of the absorbed light wave). Figure 9.1 gives a comparison of different light wavelengths, related colors and complementary colors.

Food pigments are generally organic compounds that often have *chromophores*, which have an absorption peak in the ultraviolet and visible light regions (200–800 nm), in their molecular structure. Always, a chromophore contains a conjugation system with multiple $-C=C-$ bonds, and/or $-C=O$, $-N=N-$, $-N=O$, $-C=S$ combinations. When a molecule contains a chromophore with an absorption wavelength from 200 to 400 nm, the substance is colorless, whereas if a molecule contains two or more

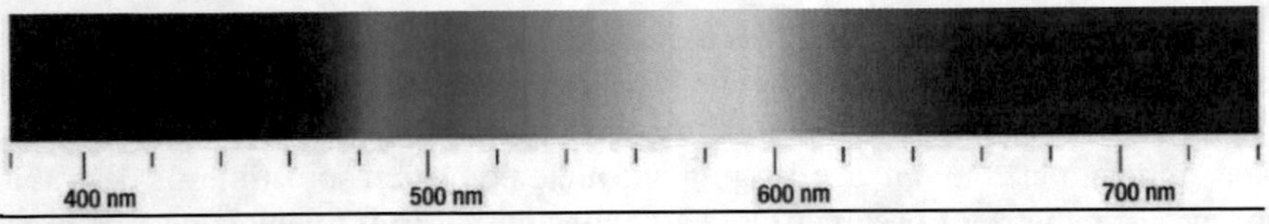

Wavelength (λ nm)	Color	Complementary Color
400	Purple	Yellow-Green
425	Blue-Cyan	Yellow
450	Cyan	Orange-Yellow
490	Cyan-Green	Red
510	Green	Purple
530	Yellow-Green	Purple
550	Yellow	Blue-Cyan
590	Orange-Yellow	Cyan
640	Red	Cyan-Green
730	Red-Purple	Green

Fig. 9.1 Light wavelengths, related colors and complementary color

chromophores, the absorbed light moves from a short wavelength to a long wavelength, causing the substance to develop color. The larger the conjugated system, the longer the wavelength absorbed by the structure, as can be seen by the example given in Table 9.2.

Otherwise, there are also some chemical groups, such as –OH, –OR, –NH$_2$, –SR$_2$, –SR, –CI and –Br, whose absorption bands are in the ultraviolet region and they do not produce color by themselves, but when combined with conjugated systems or attached with chromophores, the absorption wavelength of the entire molecule can be shifted to the long-wave direction, thus producing a color. This kind of group is called **_auxochrome_** and it helps to modify the color presentation of a compound. For example, according to a different arrangement of groups such as –OH and –OCH$_3$ on 2-phenylbenzofuran parent ring, with modified positions and

Table 9.2 Relationship between the maximum absorption wavelength (λ) by conjugated polyene compounds and the number of double bonds

Compounds	Conjugated double bonds (number)	Absorption wavelength (nm)	Color
Butadiene	2	217	Colorless
Hexatriene	3	258	Colorless
Dimethyloctene	4	296	Light Yellow
Vitamin A	5	335	Light Yellow
Dihydro- β-Carotene	8	415	Orange
Lycopene	11	470	Red
Dehydrolycopene	15	504	Purple

group numbers, various anthocyanins form. The structure of food colorants contains chromophores and auxochromes. Understanding their structure and properties is of great significance to the research, development and use of colorants.

9.1.3 Classification of Food Pigments

Food pigments can be divided into natural pigments and synthetic dyes. Natural pigments contain plant pigments such as chlorophyll, carotenoids and anthocyanins, animal pigments such as heme, carotenoids in yolk and shrimp shells, and microbial pigments such as monascus pigments.

Natural pigments can be classified into azoles (or porphyrins), isoprenes, polyphenols, ketones and quinones depending on their chemical nature. For instance, porphyrin pigments include chlorophyll and heme and isoprene pigments contain carotenoids. Anthocyanins and flavonoids belong to polyphenols and ketones, respectively; and shellac colors and cochineal are quinone pigments.

Synthetic pigments can be classified into azo-based pigments and non-azo-based pigments depending on whether the molecule contains a –N=N-chromophore structure. For example, carmine and tartrazine are azo pigments, while etythrosine and brilliant blue are non-azo pigments.

Further, food pigments can be classified into water-soluble and liposoluble decided by their solubility properties. The former are always synthetic pigments for the application in the food industry, while the latter are always natural pigments. In this chapter, natural pigments will be introduced according to their structures and changes during food processing and storage.

9.2 Tetrapyrrole Pigments

Tetrapyrrole pigments are characteristic of their planar macrocycle where four cyclic or linear pyrrole groups are connected. In the food industry, tetrapyrrole pigments are represented by chlorophyll and heme, with the former found widely in green fruits, vegetables and algae and the latter in charge of the color changes of meat and related products.

9.2.1 Chlorophyll

9.2.1.1 Structures and Properties

Chlorophyll is the main pigment in a photosynthetic unit such as green plants, algae and photosynthetic bacteria that helps to capture light for photosynthetic reactions.

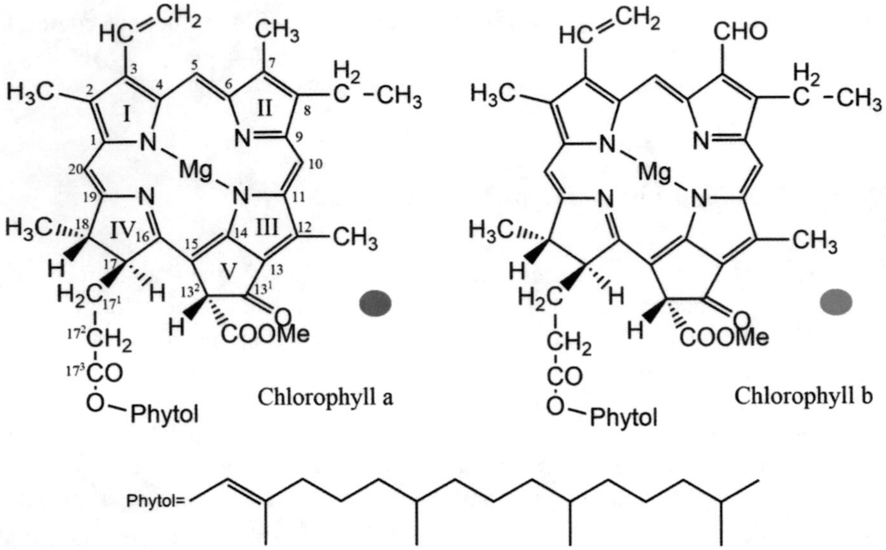

Fig. 9.2 Chemical structures of chlorophyll a and chlorophyll b

Higher plants, only contain chlorophyll a and chlorophyll b, and additionally, marine algae have chlorophyll c and chlorophyll d. Photosynthetic bacteria contain bacteria chlorophyll. This chapter mainly introduces chlorophyll a and chlorophyll b present in higher plants, and their structures are shown in Fig. 9.2.

Chlorophyll is a magnesium-containing tetrapyrrole derivative, which is linked by a four-pyrrole ring and four methylal groups (–CH=) into a macrocycle, also called porphyrin. The magnesium atom is in the center of the porphyrin ring that is easily substituted by two hydrogen atoms and tends to be positively charged, while the surrounding nitrogen atoms tend to have a negative concentration, so porphyrins are polar and can bind to proteins. Additionally, an isocyclic ring V is formed on the porphyrin ring which is quite unstable and involves many reactions. A phytol chain is esterified with propionic acid moiety at C_{17}, and this phytol chain is also called the "tail" of chlorophyll, which is a diterpene composed of 4 isoprene units so that the phytol chain decides the lipophilic nature of chlorophyll compounds. The structural difference between chlorophyll a and chlorophyll b is only that the substituents at the C_3 position are different: chlorophyll a contains a methyl group, chlorophyll b contains a formyl group (Fig. 9.2), and the ratio between chlorophyll a and chlorophyll b in higher plants is around 3 to 1.

Pure chlorophyll a is a black-green powder that is soluble in ethanol solution with a blue-green solution and has deep red fluorescence. Chlorophyll b is a green powder whose ethanol solution is yellowish-green and shows fluorescence. Both are insoluble in water and soluble in organic solvents, and chlorophyll pigments are often extracted from plant homogenates by organic solvents such as acetone, ethanol and ethyl acetate. Chlorophyll in plants is generally present in the chloroplasts, which are complexed with carotenoids, lipids and lipoproteins and distributed on disk-shaped lamellar membranes.

In food processing and storage, *chlorophyll* pigments are chemically transformed into several important derivatives, including phytol-free derivatives (*chlorophyllide*), magnesium-free derivatives (*pheophytin* and *pheophorbide*), pyro-chlorophyll derivatives (*pyropheophytin*, and *pyropheophorbide*), and oxidized derivatives (*C13²-hydroxy chlorophyll* derivatives and *C15¹-hydroxy lactone chlorophyll* derivatives). Some derivatives have been modified with several reactions and Fig. 9.3 illustrates the main reactions that happened to chlorophyll pigments during food processing.

Take chlorophyll a, for instance. The phytol ester bond can be hydrolyzed by the action of chlorophyllase to form phytol and chlorophyllide a that is hydrophilic. Chlorophyll a and chlorophyllide a transform to pheophytin a and pheophorbide a, respectively, when magnesium ions are detached under heating or acidic conditions. In addition, the isocyclic ring V in the chlorophyll molecule is associated with various chemical reactions under heating and aerobic conditions including the following: (1) the methyl ester group at the $C13^2$ position is susceptible to conformational changes in isolated solutions, forming chlorophyll epimer; (2) the methyl ester group is removed under heating to form pyrochlorophyll derivatives; and (3) oxidation reaction. The hydrogen atom at position $C13^2$ is easily oxidized to the hydroxy group under aerobic conditions or with oxidase to form $C13^2$-hydroxy chlorophyll derivatives; besides, at the same time, ring V forms a lactone structure to form $C15^1$-hydroxyl lactone chlorophyll derivatives. In some cases, ring V forms a cyclic anhydride structure, that is, purpurin-18 a derivative. If the cyclic tetrapyrrole structure of chlorophyll a is opened, the chlorophyll structure becomes linear and yields colorless chlorophyll catabolites.

The various reactions that occur to chlorophyll pigments mean changes in molecular structure and related color. Taking chlorophyll acetone solution as an example, chlorophyll a is blue-green, and if magnesium is detached from the tetrapyrrole ring, the color changes to brown immediately. The pure loss of phytol chain or methyl ester group and oxidation of $C13^2$-H to $C13^2$-OH does not affect the chlorophyll molecular tetrapyrrole structure or chlorophyll color. The spectral absorption peak of $C15^1$-hydroxy lactone chlorophyll a is shifted by 10 nm to the short-wave direction, and the color slightly changes, while the purpurin-18 a chlorophyll derivative is purple in color. In the processing of green vegetables and fruits, chlorophyll reactions are always complexed with many factors, resulting in a de-green phenomenon for food manufacturing.

9.2.1.2   Changes in Chlorophyll in Food Processing and Storage

1. Enzymes

There are two types of enzymatic changes that cause chlorophyll degradation. The enzyme that directly uses chlorophyll as a substrate is only chlorophyllase, which is

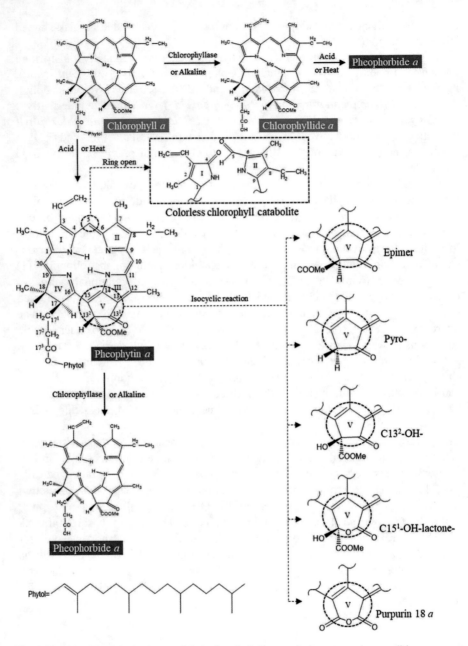

Fig. 9.3 Chlorophyll derivatives and their chemical changes during processing conditions

an esterase that catalyzes the hydrolysis of phytate ester bonds of chlorophyll and pheophytin to produce chlorophyllide and pheophorbide, respectively. The optimum temperature of chlorophyllase is in the range of 60–80 °C, and its activity begins to decrease above 80 °C and can be fully depressed by 100 °C. Indirect enzymes include lipase, protease, pectinesterase, lipoxygenase, peroxidase, etc. Lipase and protease act to destroy the chlorophyll-lipoprotein complex, so that chlorophyll pigments are more susceptible to reactions. The role of pectinesterase is to hydrolyze pectin to pectic acid, thereby lowering the pH of the system and transforming chlorophyll into pheophytin. Lipoxygenase and peroxidase can catalyze their respective substrates and the yielding intermediates cause oxidative degradation of chlorophyll pigments.

2. Heat and acid

The green color of green vegetables after cooking or blanching seems to be strength ened and it may be due to the gas existing in the intercellular space that is heated out, or that the distribution of different components in the chloroplast changes during the thermal process. But during heat treatment, protein denaturation in the chlorophyll-lipoprotein complex causes chlorophyll to be separated. Free chlorophyll is very unstable and sensitive to light, heat and enzymes. At the same time, the tissue cells are destroyed during the heating process, resulting in increased permeability of hydrogen ions across the cell membrane, hydrolysis of triglyceride into fatty acids and carbon dioxide produced by decarboxylation, all of which induce a pH decrease in the surrounding environment.

pH is an important factor in determining the rate of magnesium removal in the chlorophyll molecule. Chlorophyll is relatively stable to heat at pH 9, and easily magnesium-dechelated at pH 3. The decrease in pH induces the formation of pheophytin in plant cells and further produces pyropheophytin, causing the green color of the food to change significantly to brown, and this transformation is irreversible in an aqueous solution. Chlorophyll a is faster in this reaction than chlorophyll b, as the positive charge in the porphyrin ring of chlorophyll b is relatively more, which increases the difficulty of magnesium detachment and makes it more stable than chlorophyll a.

Treating tobacco leaves with NaCl, $MgCl_2$ or $CaCl_2$ and heating them to 90 °C could slow down the magnesium removal reaction by 47%, 70% and 77%, respectively. The role of salt may be as an electrostatic shielding agent. Cations neutralize the negative charge of fatty acids and proteins on the chloroplast membrane, thereby reducing the rate of protons permeating the membrane. The use of cationic surfactants also has a similar effect. It adsorbs to the chloroplast or cell membrane, and restricts the diffusion of protons into the chloroplast, thereby slowing down the effect of magnesium removal.

The previous view believed that the thermal stability of chlorophyllide was higher than that of chlorophyll, but later studies proved the opposite. It is now believed that phytol has a steric hindrance on the substitution of protons for magnesium atom, so chlorophyllide is easier to have magnesium detachment than chlorophyll. In addition, chlorophyllide is water-soluble and will be more likely to encounter protons to cause the magnesium reaction. When enzyme, acid and heat conditions are combined, the

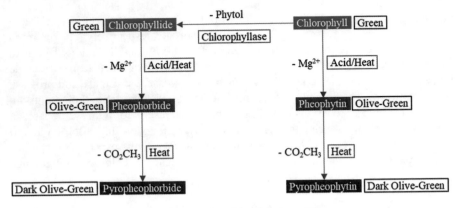

Fig. 9.4 Chlorophyll reactions under enzyme, acid and heat conditions

sequence of chlorophyll changes which can be seen in Fig. 9.4. These reactions are quite common in the curing process of green vegetables, as large sums of lactic acid are accumulated during fermentation.

3. Light

In fresh plants, chlorophyll and protein are combined in the form of protein complexes and they participate in photosynthesis reaction without photodegradation; but when plants are senescent during storage and processing, pigments are exposed to photodegradation due to the cell membrane damage. Under aerobic conditions, chlorophyll or porphyrins can produce singlet oxygen and hydroxyl radicals when exposed to light, which can react with chlorophyll tetrapyrroles to form peroxides and more radicals, and ultimately lead to the decomposition of the porphyrin ring and complete loss of color. The photolytic process of chlorophyll pigments begins with the ring-opening of the methylene group and the main product is glycerin, along with a small amount of lactic acid, citric acid, succinic acid and malonic acid, as shown in Fig. 9.5.

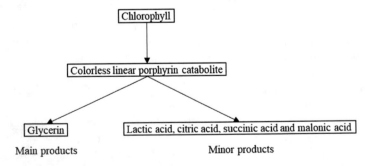

Fig. 9.5 Chlorophyll reactions under light condition

9.2.1.3 Protective Measurements of Chlorophyll Pigments

1. Acid control

Increasing the pH of canned vegetables is an effective method for color preservation. The application of calcium oxide and sodium dihydrogen phosphate to maintain the pH of the blanching solution close to 7.0, or the use of magnesium carbonate, sodium carbonate and sodium phosphate, is an effective method for chlorophyll preservation. However, they are partly limited due to the softening of vegetable tissues and produce an alkali odor.

The traditional Blair method is to use calcium hydroxide or magnesium hydroxide in the blanching solution to increase the pH and maintain the brittleness of vegetables, while this method can only delay chlorophyll loss up to around 2 months as the inside of canned vegetables cannot be treated effectively. Alternatively, coating the inner wall of the tank with ethyl cellulose containing 5% magnesium hydroxide can slowly release the magnesium hydroxide into the food to maintain the pH value of 8.0 for a long time so that the green color can be kept for a relatively long time. A disadvantage of this method is that it will cause partial hydrolysis of glutamine and asparagines to produce an ammonia odor, and cause hydrolysis of lipids. In green peas, this preservative method may also cause the formation of struvite.

2. HTST

High-temperature short-time sterilization (HTST) not only preserves vitamins and flavors, but also significantly reduces the green damage in commercial sterilization. However, after about 2 months of storage, the chlorophyll pigments will also degrade due to the pH decline. It is more effective when HTST is combined with pH adjustment; however, the color protection effect that has been achieved will also fade away due to the drop in pH during storage.

3. Green reoccurrence

Adding zinc ions to the bleaching solution of vegetables is also an effective method for protecting green. The principle is that pheophytin can chelate zinc ions to form zinc metallo complexes (mainly zinc pheophytin and zinc pyropheophytin). This method can produce satisfactory results for the processing of canned vegetables. This method uses a zinc concentration of about a few ten thousandths, and the pH value is controlled at about 6.0, and the heat treatment is performed at a temperature slightly higher than 60 °C. To improve the permeability of zinc in the cell membrane, an appropriate amount of surface-active anionic compounds can also be added to the treatment solution. This method can produce satisfactory results when used in the processing of canned vegetables. Zinc ions can be substituted by copper ions.

Another chlorophyll additive in the food industry is called copper chlorophyllin or zinc chlorophyllin which is widely allowed in Asia and European countries for the improvement of processed vegetable color.

4. Others

The modified atmosphere preservation technology enables the green color to be protected along with the preservation of fresh vegetables. When the water activity is very low, even if acid is present, the chance of H^+ transfer and access to chlorophyll is relatively reduced, so it is not easy to replace Mg^{2+} of chlorophyll pigments. At the same time, due to low water activity, microbial growth and enzyme activity are also inhibited. Therefore, dehydrated vegetables can remain green in color for a long period. When storing green plant foods, separation from light and oxygen can prevent oxidative fading of chlorophylls. Therefore, the proper option of packaging materials and chlorophyll preservative methods, combined with the appropriate use of antioxidants, can maintain the green color of green vegetables and fruits for a long time.

9.2.2 Heme

9.2.2.1 Structures and Properties

Hemes are the main red pigments in animal muscles and blood, mainly present in the form of myoglobin in the muscle and hemoglobin in the blood. The protein part is called globin which consists of 153 amino acid residues. Fe^{2+} in the center of the porphyrin ring has six coordination sites, four of which are coordinated with the nitrogen atom from the tetrapyrrole rings. Another one is coordinately bonded with the histidine residue from the globulin, leaving the sixth coordination site that could be coordinated to small molecules such as O_2 and CO. The structure of heme is illustrated in Fig. 9.6.

Myoglobin is a globular protein composed of one molecule of heme and one globin which has a single polypeptide, and its relative molecular mass is around 16,700. The main role of myoglobin is to receive and store oxygen transported by hemoglobin in muscle cells, helpful for tissue metabolism. The hemoglobin molecule is a tetramer with two alpha-peptide chains (141 amino acid residues for each chain) and two beta-peptide chains (146 amino acid residues for each chain), and each peptide chain

Fig. 9.6 Structure of heme

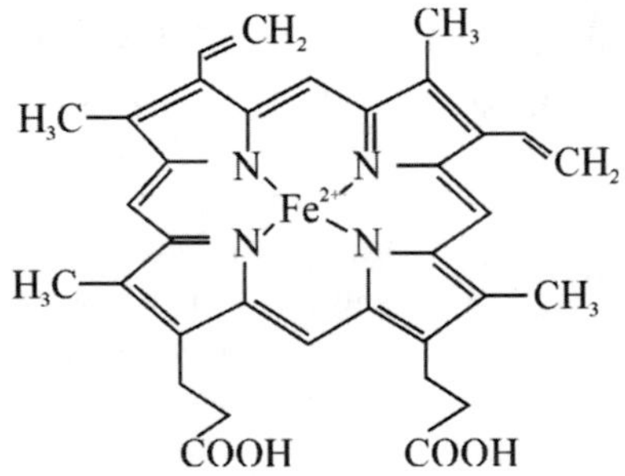

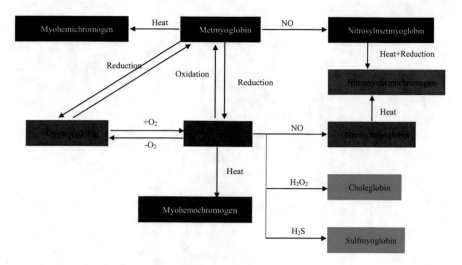

Fig. 9.7 Myoglobin derivatives and related changes during food processing

binds one molecule of heme, with its relative molecular mass around 64,500. The main function is to bind and transport oxygen in the blood.

Nearly 90% of the pigment in the muscle is myoglobin and others are cytochrome enzymes, flavin and vitamin B_{12}. The content of myoglobin in the muscle varies greatly depending on the species, age, sex and location of the animal. For example, the muscle of the veal shows shallow color compared with an adult cow as the lower content of myoglobin. The hemoglobin in shrimps, crabs and insects is hemocyanin containing copper.

In meat processing and storage, myoglobin is converted into a variety of derivatives, the type of which depends mainly on the chemical properties of myoglobin, the valence of iron, the type of ligand of myoglobin and the state of globulin. The heme iron in the porphyrin ring can be in two forms, ferrous iron (Fe^{2+}) and ferric iron (Fe^{3+}). The iron ion of myoglobin is Fe^{2+}, with one coordination site free, and when it combines with oxygen, forms oxymyoglobin. The main derivatives of myoglobin and related reactions are shown in Fig. 9.7 and Table 9.3.

9.2.2.2   Changes in Muscle Color During Storage and Meat Processing

After animal slaughter and bloodletting, due to the stop of oxygen supply to the muscle tissue by hemoglobin, the *myoglobin* in the fresh meat maintains its reduced state, and the color of the muscle is purple-red (the color of myoglobin). When the carcass is divided, the myoglobin in the reduced state changes in two different ways as the muscles are in contact with the air. A part of myoglobin reacts with oxygen to form bright-red *oxymyoglobin*, with typical fresh flesh color; at the same time, another part of myoglobin oxidizes with oxygen to form a brown *metmyoglobin*,

Table 9.3 Main pigments found in fresh, cured and cooked meat

Compounds	Formation pathway	Iron	Heme ring	Globulin	Color
1. Myoglobin	Reduction of metmyoglobin or deoxygenation of oxymyoglobin	Fe^{2+}	Intact	Natural	Purplish red
2. Oxymyoglobin	Oxygenation of myoglobin	Fe^{2+}	Intact	Natural	Bright red
3. Metmyoglobin	Oxidation of myoglobin and oxymyoglobin	Fe^{3+}	Intact	Natural	Brown
4. Nitrosylmyoglobin	Combination of myoglobin to NO	Fe^{2+}	Intact	Natural	Bright red (pink)
5. Nitrosylmetmyoglobin	Combination of metmyoglobin to NO	Fe^{3+}	Intact	Natural	Dark red
6. Metmyoglobin nitrite	Combination of metmyoglobin to excess nitrite	Fe^{3+}	Intact	Natural	Reddish-brown
7. Myohemochromogen	Myoglobin and oxymyoglobin are treated by heat, denaturing reagents or irradiation	Fe^{2+}	Intact (often combined with non-globulin-type denatured proteins)	Denatured (always isolated)	Dark red
8. Myohemichromogen	Myoglobin, oxymyoglobin, metmyoglobin and myohemochromogen are treated with heating and denaturation reagents	Fe^{3+}	Intact (ften combined with non-globulin-type denatured proteins)	Natural (always isolated)	Brown (sometimes gray)
9. Nitrosohemochromogen	Nitrosylmyoglobin are treated with heating and denaturation reagents	Fe^{2+}	Intact, but one double bond has been saturated	Denatured	Bright red (pink)

(continued)

Table 9.3 (continued)

10. Sulfmyoglobin	Myoglobin interacts with H2S and O2	Fe^{3+}	Intact, but one double bond has been saturated	Natural	Green
11. Sulfmetmyoglobin	Oxidation of sulfmyoglobin	Fe^{3+}	Intact, but one double bond has been saturated	Natural	Red
12. Choleglobin	Myoglobin or oxymyoglobin is treated by hydrogen peroxide; oxymyoglobin interacts with ascorbate or other reducing agents	Fe^2 or Fe^{3+}		Natural	Green
13. Nitrohemin	Co-heating of nitrosylmetmyoglobin and excess nitrite	Fe^{3+}	Intact, but the porphyrin ring is open	None	Green
14. Verdohemin	Excessive denaturing reagent	Fe^{3+}	The porphyrin ring is destroyed	None	Green
15. Bilin	High-dose denaturing reagents	None		None	Yellow or colorless

which is gradually dominated as the flesh color turns brownish-red when the meat is placed in the air for an extended period (Fig. 9.8).

The conversion between myoglobin, oxymyoglobin and metmyoglobin is dynamic and strongly influenced by oxygen pressure. Oxygen pressure is favorable for the formation of oxymyoglobin, and low oxygen pressure is beneficial to the formation of metmyoglobin (Fig. 9.9). In fact, the surface of the freshly cut meat is bright red when it meets enough oxygen. In this case, although there is a certain amount of metmyoglobin production on the surface of the meat, the amount is small. With the storage of meat, the production of metmyoglobin is gradually increased, mainly due to two aspects. On the one hand, there is a small number

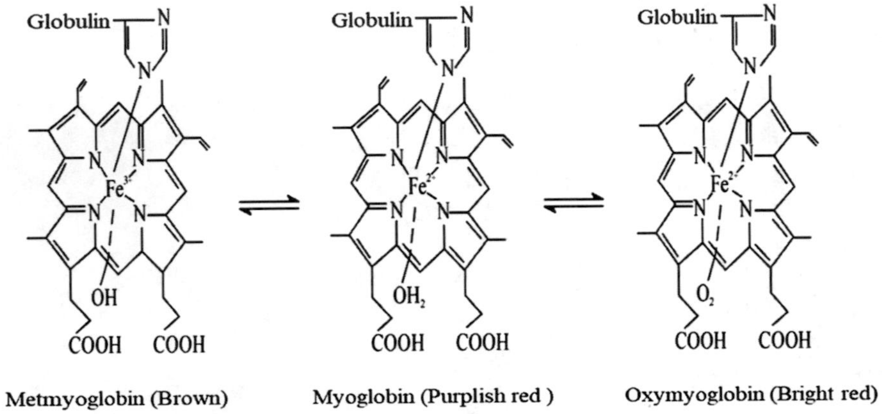

Metmyoglobin (Brown)   **Myoglobin (Purplish red)**   **Oxymyoglobin (Bright red)**

Fig. 9.8 Myoglobin changes in cut meat

Fig. 9.9 The effect of oxygen partial pressure on the conversion among myoglobin, oxymyoglobin and metmyoglobin

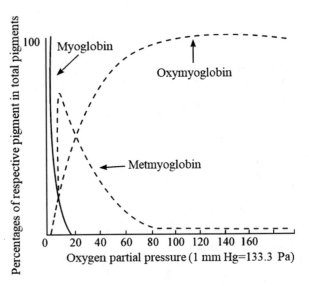

of aerobic microorganisms growing on the surface of the meat, which reduces the oxygen pressure; on the other hand, it is ascribed to the depletion of intrinsic reducing substances such as glutathione and thiol compounds inside the meat which help to reduce metmyoglobin to myoglobin when they are enough.

The reaction from Fe^{2+} in heme to Fe^{3+} is the result of auto-oxidation and the oxidation rate is lower in hemoglobin than in heme, and in oxymyoglobin than in myoglobin, while the auto-oxidation reaction is favored in low pH and with the help of other metal ions such as Cu^{2+}.

When the meat is stored, the myoglobin will be converted into a green substance under certain conditions. This is due to the growth and reproduction of contaminating bacteria that produce hydrogen peroxide or hydrogen sulfide, which react with Fe^{3+} or Fe^{2+} in the heme of myoglobin to produce *choleglobin* and *sulfmyoglobin*, respectively, causing the green color.

$$MbO_2(Myoglobin) + H_2O_2 \rightarrow Choleglobin (Green)$$

$$MbO_2(Myoglobin) + H_2S + O_2 \rightarrow Sulfmyoglobin (Green).$$

Myoglobin and metmyoglobin are denatured during the heating process. In this case, myoglobin and metmyoglobin yield *myohemochromogen* and *myohemichromogen*, respectively, showing a brown color. That is, during heating, due to the increase of meat temperature and the decrease of oxygen partial pressure, the production of myohemochromogen and myohemichromogen is promoted, and the color of meat changes, especially the production of myohemichromogen, which makes the flesh color turn into brown.

In the processing of meat and salted products such as ham and sausage, nitrate or nitrite is used as a coloring agent. As a result, myoglobin and metmyoglobin are converted into *nitrosylmyoglobin* and *nitrosylmetmyoglobin*, and finally form *nitromyohemochromogen*, making the color of the cured meat products vivid and attractive, and showing greater tolerance to heat and oxidation. The sixth ligand of the central iron ion of the three pigments is nitric oxide (NO). The reaction pathway is shown in Fig. 9.10.

In addition to the function of a color former, nitrite and nitrate also have the function of preservatives, which is of great significance for the safe storage of meat products. However, the amount of nitrate and nitrite coloring agent must be strictly controlled, because excessive use will not only produce green compounds, but also form carcinogens with certain ammonia species, as shown in Fig. 9.11.

Nitrate and nitrite can form carcinogens with certain ammonia substances. Therefore, in recent years, there have been many studies and reports on the substitutes of nitrate and nitrite, such as the application of monascus colors in fermented sausage, ham, pork luncheon meat and other foods, which can partially replace nitrite. However, currently, no substance has been found that can completely replace nitrite.

Some substances do not have a coloring function, but the use of a coloring agent can significantly improve the color development of nitrates and nitrites, thereby

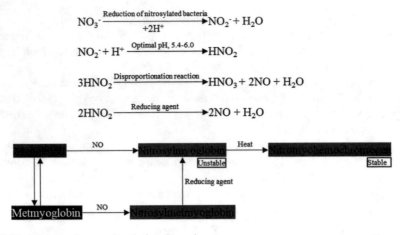

Fig. 9.10 Muscle color reaction during the curing process

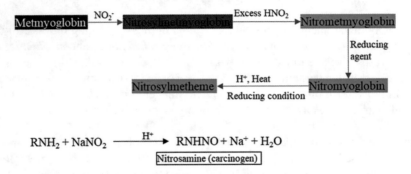

Fig. 9.11 The formation reaction of green substances and carcinogens by excessive use of nitrite

reducing their uses and improving the safety of the meat. Such substances are called coloring auxiliaries that are commonly used in meat products, such as lactic acid, L-ascorbic acid, sodium ascorbic acid and nicotinamide. Lactic acid can promote the formation of nitrous acid; L-ascorbic acid and sodium L-ascorbate can promote the conversion of nitrous acid to nitric oxide. The combination of nicotinamide and myoglobin can form stable nicotinamide myoglobin and prevent myoglobin from getting discolored. By the way, nicotinamide is also a nutrition supplement, with the addition from 0.01 to 0.03%.

Although the color of cured meat products is quite stable under various conditions, visible light can cause them to re-convert to myoglobin and myohemochromogen, while myoglobin and myohemochromogen continue to be oxidized and then converted into metmyoglobin and myohemichromogen. Therefore, bacon products become brown under light. Under the premise of use limit, enough nitrite and antioxidants such as ascorbic acid will help prevent the light browning of cured meat products, as they can re-transform photolytic products into nitrosylmyoglobin and

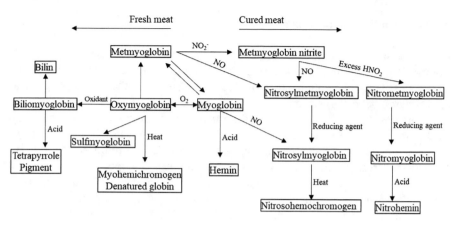

Fig. 9.12 The reaction of myoglobin in fresh meat and cured meat

nitromyohemochromogen. A series of changes in myoglobin in cured meat is shown in Fig. 9.12.

9.2.2.3 Color Protection of Meat and Meat Products

The stability of meat pigments is related to factors such as light, temperature, relative humidity, water activity, pH and microbial growth.

Put the fresh meat in a transparent bag with low air permeability, seal it after vacuuming, and if necessary, add a small amount of oxygen scavenger into the bag to keep the inside oxygen-free. This can make the myoglobin in the meat stay reduced. That is, the iron ions in hemoglobin are Fe^{2+} and have no oxygen combined with them, and the color of the meat can remain unchanged for a long time. Once the bag is opened, a large amount of oxygen is brought into contact with the surface of the meat, and the flesh color is quickly turned to the bright red color of the oxymyoglobin. This is the packaging method for fresh meat in supermarkets.

Modified atmosphere packaging is another effective method for meat or meat products. The color of the meat can be better protected with 100% CO_2 gas. However, when CO_2 pressure is not so high, the meat is prone to brown, mainly due to the conversion of myoglobin to metmyoglobin. If an oxygen scavenger is used in combination, the color protection effect can be improved, but the growth of anaerobic microorganisms must be controlled at the same time.

The color protection method of cured meat products is mainly to protect from light and oxygen. When choosing a packaging method, care must be taken to avoid microbial growth and product water loss. Because choosing the right packaging method can not only ensure the safety of such products and reduce weight loss, but also is one of the important color protection measures.

9.2.2.4   Application of Heme

In the food industry, heme is generally used as a food additive or iron supplement. Compared with iron in plants and other inorganic iron supplements, heme has the advantages of high absorption rate, no toxic side effects, etc., and it has a good clinical effect on iron deficiency anemia.

At present, heme is mainly used in the pharmaceutical industry. Heme can be used as the raw material of protoporphyrin drugs to produce protoporphyrin drugs for the treatment of various liver diseases. For example, disodium protoporphyrin has curative effect on various liver diseases. In addition, hematoporphyrin derivatives prepared with heme as a raw material are used for the treatment of human tumors as they can concentrate in tumors and have an enhanced response to ultraviolet lasers. So, when the patient is treated with red lasers, hematoporphyrins produce free radicals to kill tumor cells.

9.3   Carotenoids

Carotenoids, also known as polyene pigments, are widely distributed pigments in natural food materials. Red, yellow and orange fruits and vegetables are carotenoid-rich foods, and carotenoids are also found in animal materials such as egg yolk and shrimp shells. In general, chlorophyll-rich plant tissues are also rich in carotenoids because chloroplasts and chromoplasts are organelles with abundant carotenoid content.

The structure of carotenoids can be classified into two categories: *carotenes* contain only parent hydrocarbon chains; and *xanthophylls* contain oxygen as a functional group, such as hydroxyl, epoxy, aldehyde and ketone groups, on the parent hydrocarbon chains.

9.3.1   Carotene

9.3.1.1   Structure and Basic Properties

Carotenes include four compounds, namely *α-carotene*, *β-carotene*, *γ-carotene* and *lycopene*, all of which are C_{40} tetraterpenoid pigments and biosynthesized by the linkage of two C_{20} geranylgeranyl diphosphate molecules.

It can be seen from Fig. 9.13 that they are closely related compounds with similar chemical properties, but their nutritional properties are different. For example, α-carotene, β-carotene and γ-carotene are provitamin A, which means they can be converted into vitamin A in the body. One molecule of β-carotene can be converted into two molecules of vitamin A, while each molecule of α-carotene or γ-carotene can only be converted into one molecule of vitamin A. Lycopene is not a provitamin

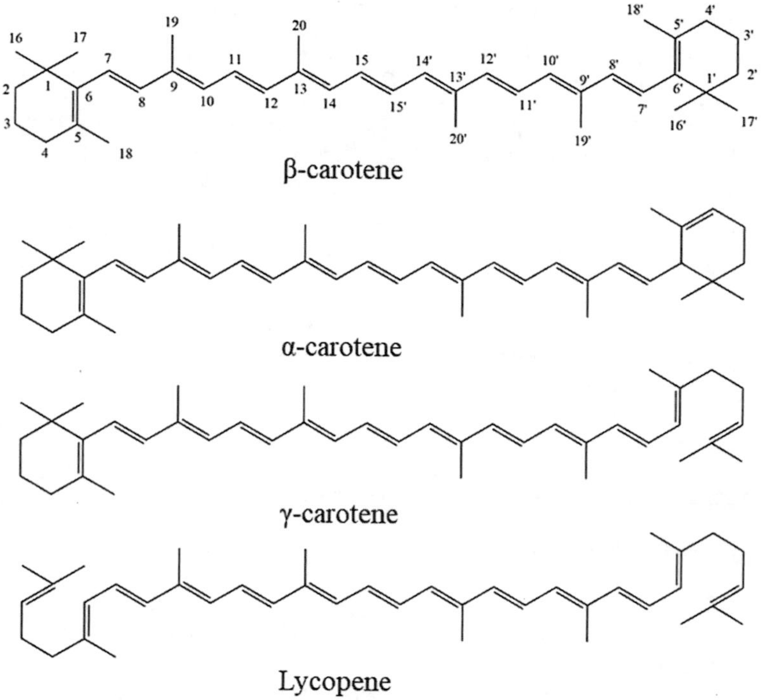

Fig. 9.13 Structures of carotene

A. Geometric isomers of carotene refer to geometric isomerization of one or more of their double bonds and the hydrides of carotenes refer to their hydrogenation products, such as phytoene.

α-Carotene and β-carotene are widely present in foods and biological materials, especially in carrots, sweet potatoes, egg yolks and milk. Lycopene is the main pigment of tomato and is also widely found in fruits such as watermelon, pumpkin, citrus, apricot and peach. In plant tissues, they are mainly found in chromoplasts; in animals, they are mainly distributed in specific tissues rich in lipids, such as egg yolk.

Carotene is a typical fat-soluble pigment, easily soluble in organic solvents such as petroleum ether and ether, and insoluble in ethanol and water. Since carotenes have many conjugated double bonds in their structures, they are easily oxidized and the resulting products are very complicated. Carotenes are susceptible to oxidation when plant tissues are damaged; lipoxygenase, polyphenol oxidase and peroxidase always accelerate the indirect oxidation of carotenes, as they first catalyze the oxidation of their respective substrate to form an intermediate which in turn oxidizes carotene. For example, lipoxygenase catalyzes the oxidation of unsaturated fatty acids to form hydroperoxides, which in turn react with carotenes. Therefore, in food processing,

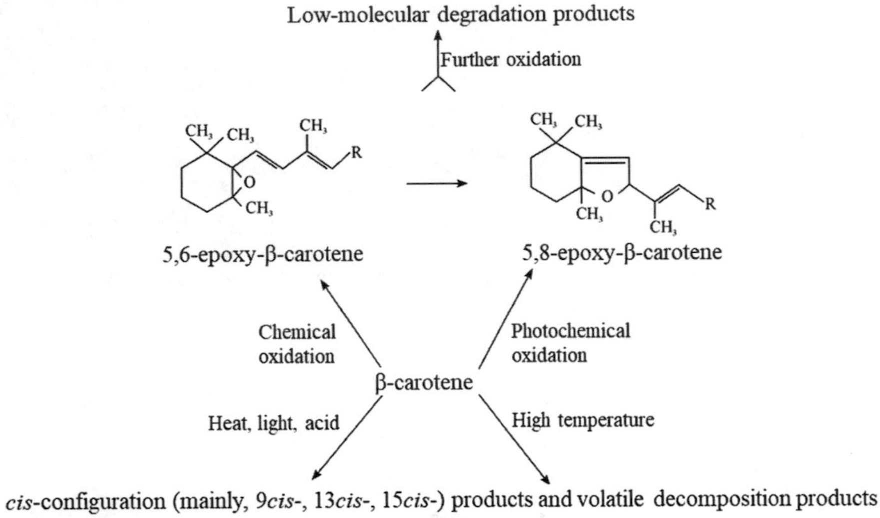
Low-molecular degradation products
Further oxidation
CH3 CH3
CH3
R
O
CH3
5,6-epoxy-β-carotene
CH3 CH3
O
CH3
R
CH3
5,8-epoxy-β-carotene
Chemical oxidation
Photochemical oxidation
β-carotene
Heat, light, acid
High temperature
cis-configuration (mainly, 9cis-, 13cis-, 15cis-) products and volatile decomposition products

Fig. 9.14 β-carotene degradation reaction

appropriate treatment for inactivation enzyme such as blanching is helpful for the protection of carotene pigments.

In general, the conjugated double bonds of carotene are mostly in the all-trans configuration, and only a very small number of cis isomers are present. Carotenoids are highly susceptible to isomerization under heat treatment, organic solvents, acid and light conditions. Since carotene has many double bonds, there are many types of isomers, such as β-carotene, which have 272 possible isomers. Figures 9.14 summarizes the degradation reactions and possible isomerization reactions of β-carotene.

Since carotene is easily oxidized, it undoubtedly has a good antioxidant effect, which can scavenge singlet oxygen, hydroxyl free radicals, superoxide free radicals and peroxyl free radicals. When carotene exerts its antioxidant effect, it may be degraded or recovered after the antioxidant reaction.

9.3.1.2 Carotene Changes in Food Processing and Storage

In most fruit and vegetable processing, the properties of carotenoids are relatively stable. For example, freezing has very little effect on carotene pigments. However, under thermal processing conditions, when the plant tissue is heated, carotene is transferred from the colored tissue and dissolved in the lipid, so that its existing form and distribution in the plant tissue are changed, and it may be degraded under aerobic, acidic and heat conditions, as shown in Fig. 9.12. As provitamin A, the isomerization and degradation reactions of carotene in food during processing and storage sometimes are disastrous, which will definitely reduce the activity of provitamin A.

9.3.2 Xanthophyll

Xanthophyll pigments are widely found in biological materials, and they are more kinds of xanthophyll than carotene due to the addition of different oxygen groups. Some xanthophyll structures are shown in Fig. 9.15.

As the addition of oxygen group increases the hydrophilicity of xanthophyll pigments, they dissolve well in methanol or ethanol, but are less soluble in ether and petroleum ether. When extracting total carotenoids from plants, a compound solvent capable of extracting both carotene and xanthophyll should be taken into consideration, such as a proper ratio of hexane to acetone.

Xanthophyll is often yellow and orange, and a few are red, such as capsanthin. Like carotenes, xanthophylls are also prone to cis/trans isomerization under the action of heat, acid and light, and they are susceptible to degradation by oxidation and photooxidation, which sometimes change the color of the food and related flavor aspect. Some of the xanthophylls are also provitamin A, such as cryptoxanthin. Most of the xanthophyll are the same as carotene and have antioxidant effects.

9.3.3 Carotenoid Changes in Food Processing and Storage

Carotenoids in vegetables and fruits can be found with free form or in esterification with fatty acids, while esterification does not modify their chromophore properties of carotenoids but changes the chemical and biological properties. Some carotenoids are also combined with proteins. For example, scutellin (3,3'-dihydroxy-4,4'-diketone-β-carotene) combines with protein in a fresh lobster shell to form the typical blue color; when the lobster is cooked, the binding of the protein to scutellin is destroyed, and scutellin is oxidized to astaxanthin (3,3',4,4'-tetraketone-β-carotene) with red color.

Carotenoid change in the procedure of fruit and vegetable processing depends on certain conditions. For example, freezing has little effect on carotenoid pigments. However, under heat, aerobic, light and acidic processing conditions, parts of carotenoids are transformed from trans to cis configuration, and further degraded due to oxidation and photooxidation reactions, etc. Some non-thermal processing techniques, such as high electric field pulse, high pressure and modified CO_2 package, may help to preserve carotenoids. Since carotenoids are easily oxidized, they have strong antioxidant properties to remove singlet oxygen, hydroxyl radical, superoxide radical and peroxyl radical. Thus, carotenoids are effective natural antioxidants in the food industry (Fig. 9.16).

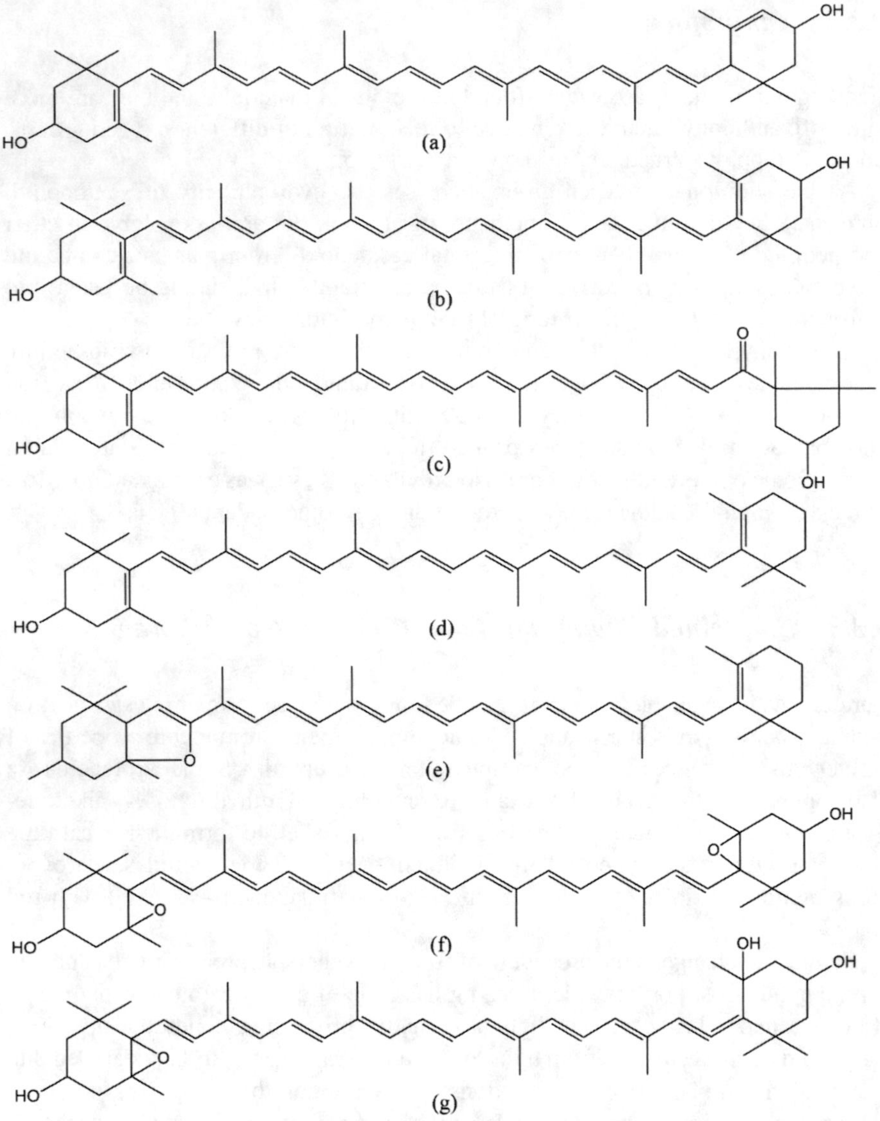

Fig. 9.15 Structures of some xanthophyll pigments that have been found in various plant materials as follows: **a** *lutein*, in marigold flower, orange, pumpkin and green leafy vegetable; **b** *zeaxanthin*, in wolfberry, maize, orange and red pepper; **c** *capsanthin*, in red pepper; **d** *cryptoxanthin*, in buah merah, papaya, mango, orange, maize, and persimmon; **e** *citroxanthin*, in orange; **f** *violaxanthin*, in mango, orange, jio and lamb's quarters; **g** *neoxanthin*, in spider wisp and chanca piedra

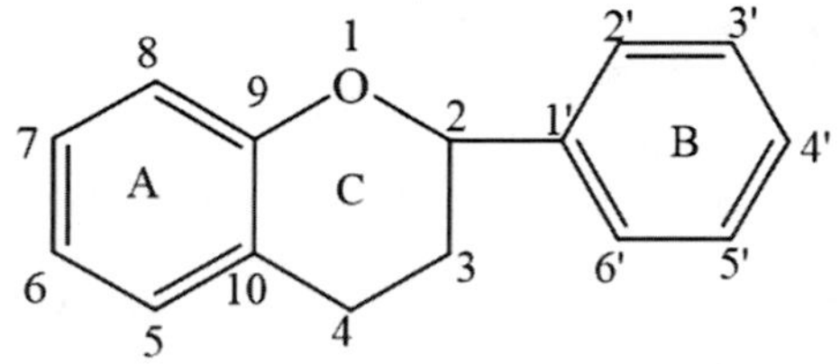

Fig. 9.16 The basic structure (C6–C3–C6) of flavonoid compounds

9.4 Polyphenolic Pigments

Polyphenols are a very broad class of compounds in nature with a basic core structure of **α-phenyl benzopyran**. Since two or more hydroxyl groups are attached to the benzene ring, they are collectively referred to as polyphenol pigments. Polyphenolic pigments are the main water-soluble pigments found in plants, and are mainly associated with flavonoid compounds.

9.4.1 Anthocyanins

In 1835, Marquart first extracted a blue pigment called anthocyan from the cornflower of chrysanthemum. The word cyan is taken from the Greeks Anthos (flowers) and Kyanos (blue). **Anthocyanins** are glycosides of **anthocyanidin**, a type of water-soluble pigment widely found in plants, so anthocyanidin is also called aglycone. Its colors include blue, purple, violet, magenta, red and orange, and is the substance that makes up the visual colors of many flowers and fruits.

9.4.1.1 Structure and Physical Properties

Anthocyanins have a carbon skeleton structure typical of flavonoids, and various anthocyanins and anthocyanidins are formed due to the difference in the number and type of substituents (Fig. 9.17). There are more than 20 anthocyanidins known, among which 6 are the most common anthocyanidins distributed in plant material, namely **cyanidin, pelargonidin, delphinidin, peonidin, petunidin** and **malvidin**. The sugars which are acylated with anthocyanidins are mainly glucose, galactose, xylose, arabinose and their disaccharides or trisaccharides. The aglycone sites of naturally occurring anthocyanins are mostly at the C_3 and C_5 positions of the 2-phenylbenzopyran cation, and a few at the C_7 position, or at the $C_{3'}$, $C_{4'}$ and $C_{5'}$ positions. These glycosyl groups are sometimes acylated by aliphatic or aromatic organic acids, and the main organic acids involved in the above reaction include caffeic acid, *p*-coumaric acid, sinapic acid, *p*-hydroxybenzoic acid, ferulic acid, malonic acid, malic acid, succinic acid or acetic acid.

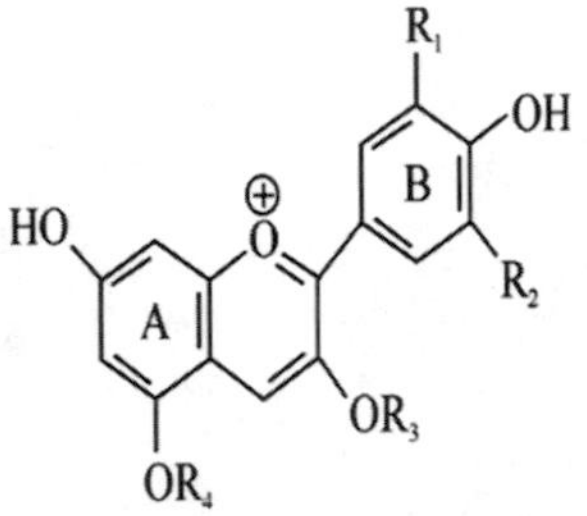

Fig. 9.17 The structure of
anthocyanin

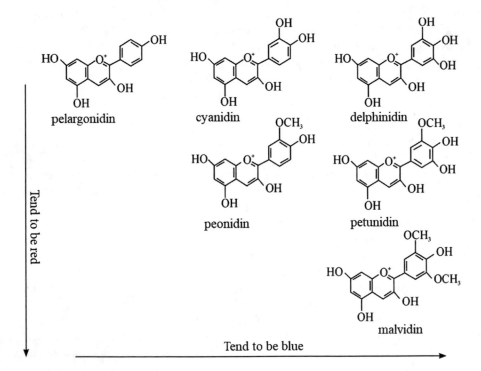

R_1, R_2= -H, -OH, or -OCH$_3$; R_3, R_4=Glycosyl, or –H

The difference in the color of various anthocyanins is mainly caused by the type and number of substituents. The substituents on the anthocyanin molecule are hydroxyl group, methoxy group or sugar group. As an auxochrome, the strength of the color-donating effect of the substituent depends on their electron-donating ability, and the stronger the electron-donating ability, the stronger the color-assisting effect. The electron-donating ability of the methoxy group, similar to the sugar group, is stronger than that of the hydroxyl group; the sugar group may exhibit a space hindrance effect due to the relatively large molecule. Figure 9.18 indicates that with

pelargonidin

cyanidin

delphinidin

peonidin

petunidin

malvidin

Tend to be red

Tend to be blue

Fig. 9.18 The 6 types of common anthocyanidin in foods and their color changes

the increase of the –OH number, the light absorption wavelength shifts to the red light direction (red shift), and the blue color of anthocyanidin is strengthened; as the –CH$_3$ number increases, the light absorption wavelength shifts to the blue light direction (blue shift), and the red color is strengthened; due to red shift and blue shift, the color of anthocyanins deepens.

Both anthocyanins and anthocyanidins are water-soluble pigments, while anthocyanins have additional hydrophilic glycosyl groups, leading to their greater solubility in water.

More than 250 anthocyanins have been found in plants, and the variety of anthocyanins contained in plants varies with different growth stages and maturity periods, ranging from about 20 to 600 mg/100 g fresh weight.

9.4.1.2   Changes in Anthocyanins in the Food Industry

Anthocyanins and anthocyanidins are not stable, and they often change color due to chemical reactions during food processing and storage. Factors that affect their stability include pH, oxygen concentration, oxidants, nucleophiles, enzymes, metal ions and temperature.

The relationship between the structure of different anthocyanins and anthocyanidins and their stability have certain regularity. The stability of anthocyanins and anthocyanidins with more hydroxyl groups is not as high as those with more methoxy groups. Anthocyanins are not as stable as anthocyanidins, and the stability of different glycosyl groups is also different. For example, the color of plants with a high content of cyanidin, pelargonidin and delphinidin is not as stable as that of plants with a high content of petunidin and malvidin. Cranberry contains galactosyl anthocyanins, which are more stable during storage than arabinosyl anthocyanins.

1. Effects of pH

The solution pH significantly influences the color of anthocyanins due to the ionic nature of anthocyanin structures. In an acidic environment, some anthocyanins appear red, purple in natural pH while blue in alkaline conditions. When anthocyanins are native red, which are dominated by the structure of flavylium cations, they are quite stable in acidic solution, as the flavylium cation increases their solubility in water. When the solution pH increases, colorless structures such as alcohol pseudo base and chalcone are yielded, along with the formation of anionic quinonoidal structures which are blue due to hydration reactions of flavylium ion. Quinonoidal compounds are not stable at acidic conditions. If pH is between 4 and 5, anthocyanin solutions always show little color as a result of a few flavylium cations and quinonoidal anions (Figs. 9.19 and 9.20).

2. Temperatures

Temperature strongly affects the stability of anthocyanins and anthocyanins, and the extent of this effect is also affected by environmental oxygen content, anthocyanin species and pH conditions. In general, the thermostability of anthocyanins

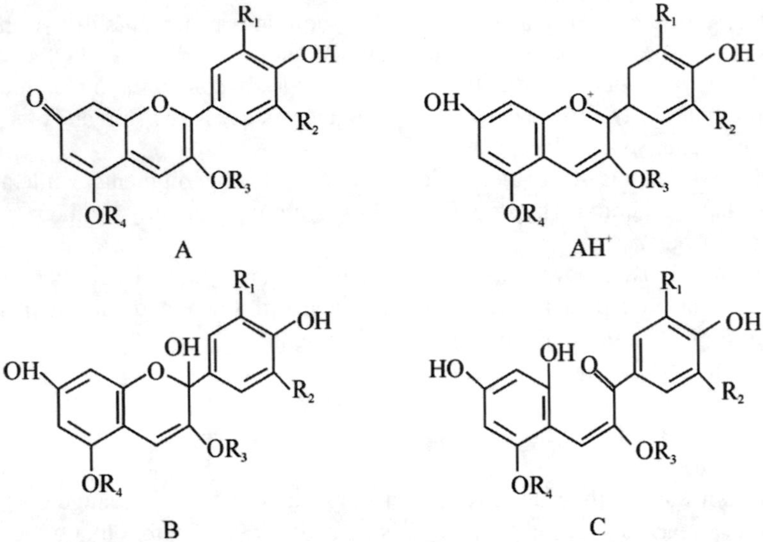

Fig. 9.19 The four forms of anthocyanins in aqueous solution and their colors. A is a quinone structure (blue), AH⁺ is a 2-phenylbenzopyran cation (red), B is an alcohol pseudo base structure (colorless) and C is a chalcone structure (colorless)

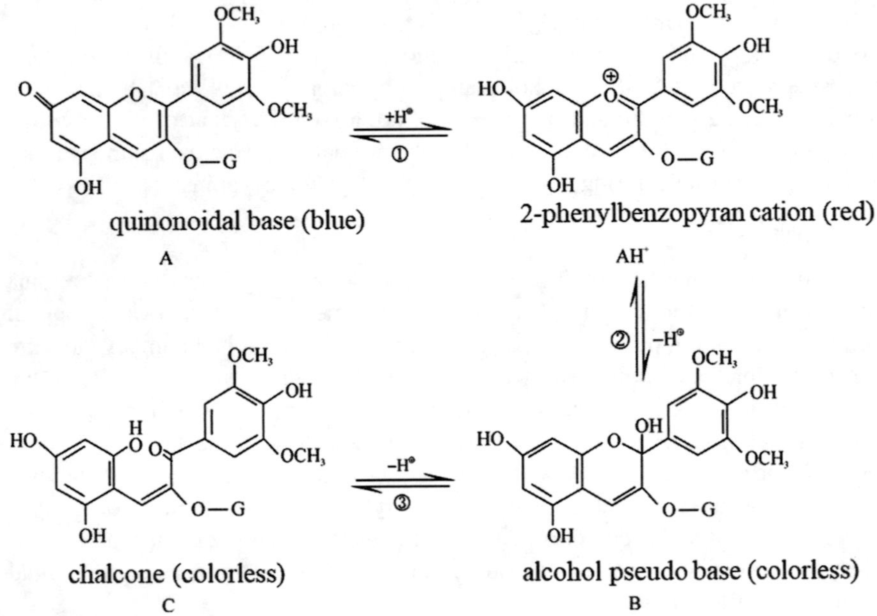

Fig. 9.20 Four kinds of structures with malvidin-3-glucoside appearing in the pH range from 0 to 6

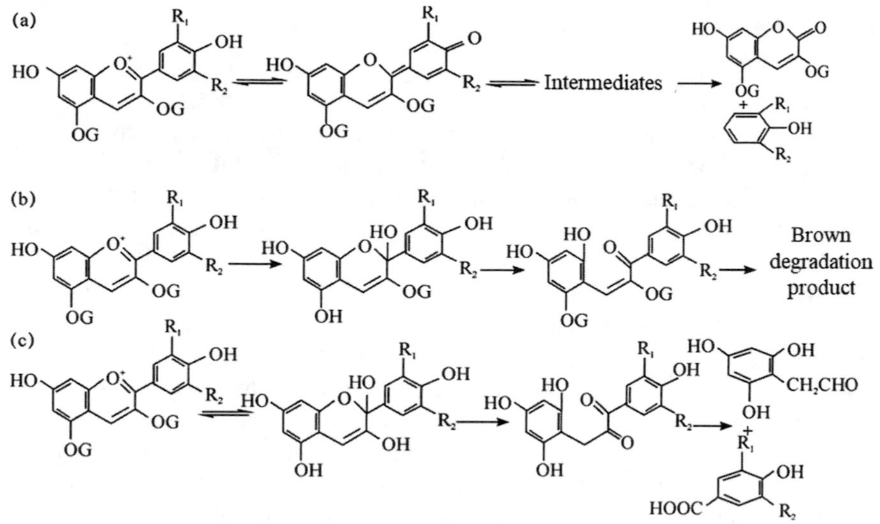

R$_1$, R$_2$ = -OH, -H, -OCH$_3$, -OG; G = Glycosyl

Fig. 9.21 Degradation mechanism of 3,5-diglucose-anthocyanins

and anthocyanidins containing more hydroxyl groups is not as stable as that with more methoxy or glycoside groups.

The conversion balance among the four structural forms of anthocyanins in an aqueous solution is also affected by temperature. When heating, the balance shifts toward the direction of the formation of chalcone structure, and the result is that the content of coloring substances (AH$^+$ and A, Fig. 9.19) decreases. When cooled and acidified, the pseudo base anthocyanins are quickly converted to cation ones (AH$^+$), while the chalcone type changes little.

The exact mechanism of thermal degradation of anthocyanins has not been fully elucidated, and three degradation pathways have been proposed (Fig. 9.21). 3,5-Diglucose-coumarin glycosides are common degradation products of 3,5-diglucose-anthocyanins. Route A (Fig. 9.21a) shows that this product is the first conversion of a 2-phenylbenzopyran cation to the quinone structure, and then the intermediate decomposes to produce coumarin derivatives and phenol compounds. In route B (Fig. 9.21b), the 2-phenylbenzopyran cation is first converted to a pseudo-basic structure, and then it decomposes into a brown degradation product through the chalcone structure. The first few steps of route C (Fig. 9.21c) are similar or identical to those of route B, but the degradation products of chalcone are formed by the insertion of water. These research results indicate that the thermal degradation of anthocyanins is affected by the types of anthocyanins and the degradation temperature. The higher the heating temperature, the faster the color change of anthocyanins, 110 °C is the most tolerable temperature for anthocyanins, and the decomposition rate of anthocyanins is low below 60 °C.

3. Oxygen, water activity and the effects of ascorbic acid

In the presence of oxygen, anthocyanins degrade to produce colorless or brownish substances due to their highly unsaturated structure which makes them sensitive to oxygen. If the grape juice is hot-filled and filled up totally, it takes relatively long time for the color of grape juice to change from purple to brown due to limited accesses of oxygen; if it is replaced by nitrogen filling or vacuum filling, the color change will be more slower. This indicates that oxygen has a destructive effect on anthocyanins or anthocyanidins.

There is not much research data on the influence mechanism of water activity on the stability of anthocyanins, but studies have confirmed that the stability of anthocyanins is relatively highest in the range of water activity from 0.63 to 0.79.

In juices containing ascorbic acid and anthocyanins, the levels of these two substances are reduced simultaneously. This is because ascorbic acid can produce H_2O_2 in oxidation, and H_2O_2 can do a nucleophilic attack on the C_2 of α-phenylbenzopyran cation, thereby splitting the pyran ring to produce a colorless ester and coumarin derivative. Further degradation or polymerization eventually produces a brown precipitate in the juice. Therefore, conditions for promoting or inhibiting oxidative degradation of ascorbic acid also work for anthocyanin degradation. For example, an increase of Cu^{2+} concentration will accelerate the degradation of ascorbic acid and anthocyanins, while ascorbic acid and anthocyanin are simultaneously protected in the presence of antioxidants such as quercetin in foods.

4. The influence of light

Light has a dual effect on anthocyanins. One is beneficial to the biosynthesis of anthocyanins, and the other is the degradation of anthocyanins. Under light conditions, acylated and methylated diglycosides are more stable than non-acylated diglycosides, and diglycosides are more stable than monoglycosides. After anthocyanins condense by themselves or with other organic substances, depending on the environmental conditions, the stability of anthocyanins may be increased or decreased. Polyhydroxyflavonoids, isoflavones, and auron sulfonates are resistant to the photodegradation of anthocyanins, as negatively charged sulfonic acid groups and positively charged 2-phenylbenzopyran cations attract each other, making these molecules form a complex with anthocyanins (Fig. 9.22).

Other radiation energy can also cause anthocyanin degradation. For example, when using ionizing radiation to preserve fruits and vegetables, there is photodegradation of anthocyanins.

5. Effect of sulfur dioxide

SO_2 is a preservative commonly used in the food industry. The decolorization of anthocyanin by sulfur dioxide is either reversible or irreversible. When the use of SO_2 is between 500 and 2000 $\mu g/g$, the color can be partially recovered by elution with a large amount of water in the subsequent processing. The irreversible bleaching effect suggests that the bleaching mechanism is that SO_2 forms bisulfite under the

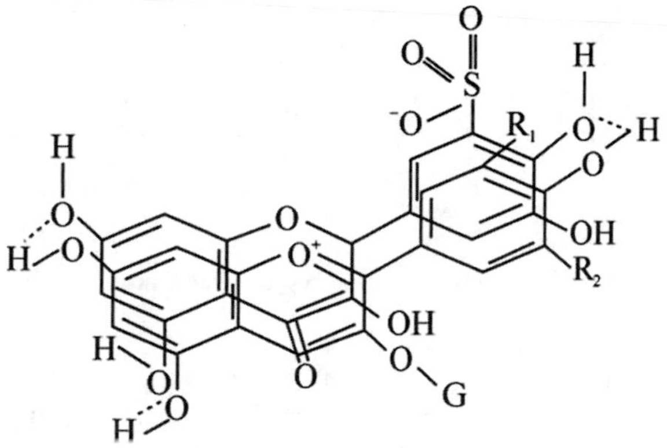

Fig. 9.22 Anthocyanin-polyhydroxyflavone sulfonyl complex

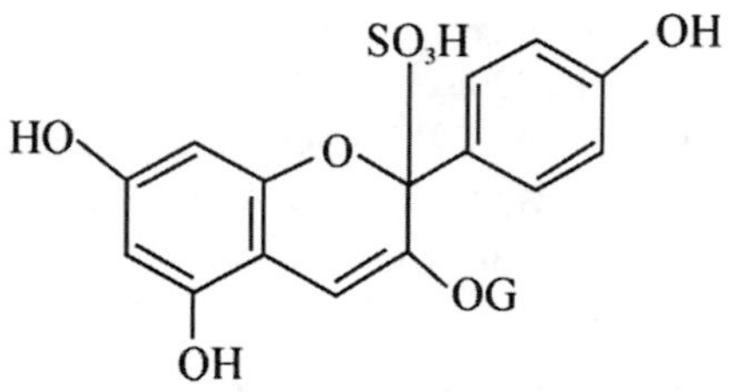

Fig. 9.23 Anthocyanin sulfite complex

acid condition in juice, and nucleophilic attacks occur on C2 or C4 of anthocyanin to produce colorless anthocyanin sulfite (Fig. 9.23).

6. Effects of sugar and sugar degradation products

When the sugar concentration is high, the color of the anthocyanin is well protected due to the low water activity. However, when the sugar concentration is low, i.e., in juice, the degradation or discoloration of anthocyanins is accelerated. These sugars first degrade (non-enzymatic browning) into furfural or hydroxymethyl furfural, and then react with anthocyanins to form a brown substance, which will be promoted by increasing temperature and oxygen concentration. This effect is more obvious when fructose, arabinose, lactose and sorbose are used than glucose, sucrose or maltose, which causes more sever problems in fruit juice production.

7. Metal ion

Anthocyanins can complex with metal ions such as Al^{3+}, Fe^{2+}, Fe^{3+}, Sn^{2+} and Ca^{2+}, thereby stabilizing the color of anthocyanins. However, this reaction can only occur when the anthocyanin contains vicinal hydroxyl groups on the B ring, and the product may be dark red, blue, green and brown (Fig. 9.24). Such metal ion complexes are common in plants. For example, the color of fresh flowers is brighter than anthocyanins as part of the anthocyanins in fresh flowers forms complexes with metal ions. The color of canned fruits and vegetables will be affected by the metal material. Once the paint on the inner wall of the tank does not qualify, the metal ions etched from

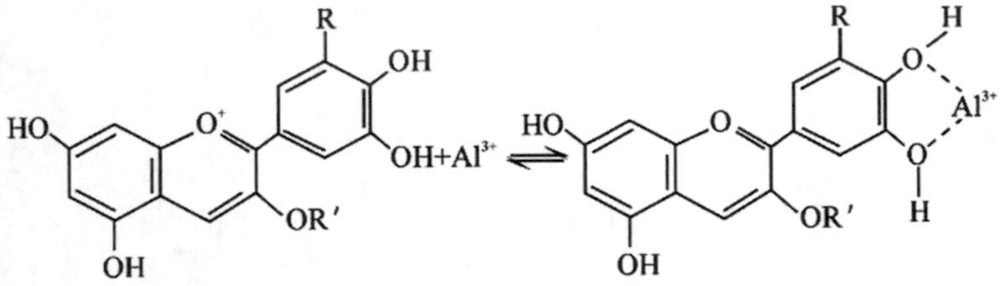

Fig. 9.24 Reactions between anthocyanins and metal ions

the inner wall of the tank often form complexes with anthocyanins, where in most cases this is undesirable, and in a few cases, the food color is beautified. It is advised to use stainless steel to process fruits and vegetables and apply citric acid to complex metal ions in food processing.

In the processing of peach, pear, lychee, cranberry and red cabbage, the problem of discoloration caused by the complexation of anthocyanin metal ions often occurs. The stability of this complex is higher than that of anthocyanin. Once it is formed, it is not easy to reverse, but citric acid can complex metal ions, which can reduce the formation of anthocyanin-metal complexes and can partially reverse them to anthocyanins.

8. Condensation of anthocyanins

Anthocyanins can undergo condensation reactions with themselves or other organic compounds to form weaker complexes such as proteins, tannins, other flavonoids and polysaccharides, by which the color of anthocyanins is modified by red shifting and the maximum absorption wavelength is increased. The formed pigments are also relatively stable during storage and processing. For example, the stable color in wine is partially due to the self-condensation of anthocyanins, and the polymer is insensitive to pH and somehow resistant to the bleaching of sulfur dioxide.

When 2-phenylbenzopyran cations and/or quinoid bases are adsorbed on a suitable substrate, such as glue or starch, the anthocyanins can be kept stable. When anthocyanins combine with certain nucleophilic compounds such as amino acids, phloroglucinol, catechol and ascorbic acid, they condense to produce a colorless substance, as shown in Fig. 9.25.

9. Hydrolysis of anthocyanins

The hydrolysis of anthocyanins includes acid hydrolysis and enzymatic hydrolysis. Generally, in a 1 mol/L HCl solution at 100 °C, anthocyanins are completely hydrolyzed to form corresponding anthocyanidins and sugars within 1 h. The higher the acidity, the faster the hydrolysis. Glucosidase and polyphenol oxidase can cause anthocyanin degradation. The former hydrolyzes the glyosidic bonds on the anthocyanin molecule to produce anthocyanidins and sugars; the latter catalyzes the oxidation of small molecular phenols to form orthoquinones, which can convert anthocyanins into oxidized anthocyanins and their degradation products by chemical oxidation.

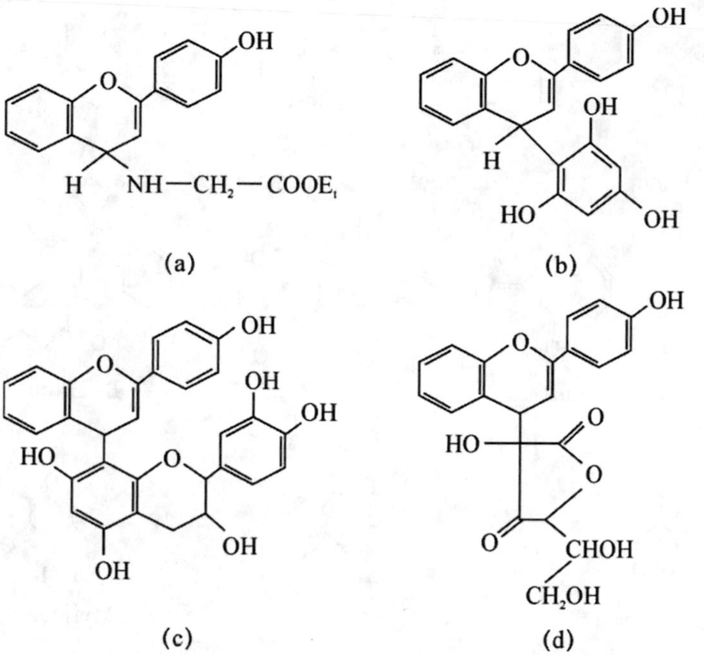

Fig. 9.25 The colorless condensate formed by 2-phenylbenzopyran cation with ethyl glycinate (**a**), phloroglucinol (**b**), catechin (**c**) and ascorbic acid (**d**)

Before processing, storage and packaging, preliminary steam bleaching can destroy and inhibit the anthocyanidase in fruits and vegetables. Glucose, gluconate and glucose delta-lactone are competitive inhibitors of glycosidase, and the activity of polyphenol oxidase can also be effectively inhibited by sulfur dioxide and sulfite. Therefore, in the process of fruit and vegetable processing, proper steam heating and the addition of enzyme inhibitors can effectively inhibit enzyme activity and protect the color of fruits and vegetables.

9.4.2 Flavonoid Pigments

9.4.2.1 Structure and Physical Properties

As shown in Fig. 9.8, flavonoids have the basic structure of C_6–C_3–C_6 framework, including flavonoid glycosides and free flavonoid aglycones in plant tissues. In flowers, leaves and fruits, most of them exist in the form of glycosides, while in xylem tissues, they are mostly in the form of free aglycones. The flavonoid compounds are further divided into the following subclasses as shown in Fig. 9.26 with their respective structures.

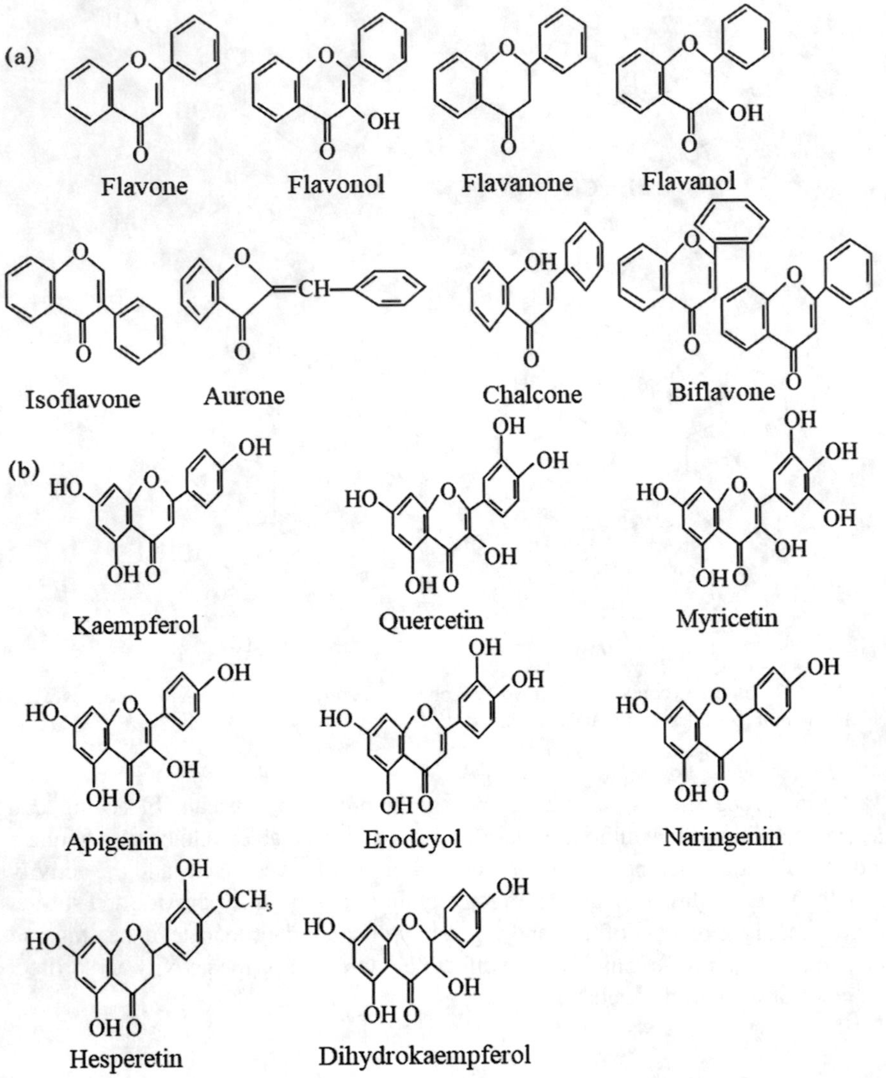

Fig. 9.26 Flavonoid compounds and their related structures. **a** The names and structures of some subtypes of flavonoids; **b** The names and structures of some common flavonoid compounds

It can be seen from Fig. 9.26 that flavonoids contain a wide range of natural pigments. Among these subclasses, anthocyanin has formed an independent category and has been introduced earlier in this chapter, and also, catechin belonging to flavanol will be discussed later. Flavonoids are also involved in the formation of another important phytochemical in vascular and non-vascular tissues, tannin. All these typical and important flavonoid-related pigments in the food industry will be discussed in detail in the following part.

	C position	Group	Color
Table 9.4 The effect of groups at different carbon positions on the color of flavonoids	3	– OH	Only gray-yellow
	3' or 4'	– OH or –OCH$_3$	Mostly dark yellow

Flavonoids have many special biological functions such as anti-oxidation, anti-tumor, anti-mutation and cardiovascular protection. Besides anthocyanin with blue, red or purple color, other colored flavonoids are generally found in flavonoids. Flavonols, isoflavones, chalcone and flavonoid glycosides are mostly yellow. And in these subclasses, the number and binding position of phenolic hydroxyl groups in the structure of flavonoids have a great influence on their coloration. If only hydroxyl groups are present at the C_3 positions, the flavonoids are only gray-yellow; if there are hydroxyl groups at the $C_{3'}$ or $C_{4'}$ position, the flavonoids of the methoxy group are mostly dark yellow, and the hydroxyl group at the C_3 carbon position can strengthen the color of the compound having a hydroxyl group at the $C_{3'}$ or $C_{4'}$ carbon position (Table 9.4).

Natural flavonoids are mostly in the form of glycosides; flavonoid glycosides are easily soluble in water, methanol and ethanol solutions, and are insoluble in organic solvents. The glycosyl group of flavonoid glycosides is often glucose, galactose, xylose, rutinose, neohesperidose and gluconic acid. There are changes in the position of glyosidic bonds, but the most common glycosides occur at the C_7, C_5 and C_3 positions of the parent nucleus structure. There are also acyl substitutions in flavonoid compounds. There are more than 1,670 known flavonoids including their glycosides, and more than 400 are colored substances, most of which are pale yellow, and a few are orange-yellow.

Some flavonoids have a certain contribution to the color of food, but due to their light color, their contribution is small when the concentration is low. The light-yellow color of cauliflower, onion and potato is mainly produced by flavonoids. Like anthocyanins, flavonoids can also form condensates. After condensation, the color changes. The condensed flavonoids in cauliflower, onion and potato are considered as an important substance for the visible color of these vegetables.

9.4.2.2 Flavonoid Changes in Food Processing and Storage

Flavonoids form complexes with a variety of metal ions, and these complexes have a stronger coloration effect. For example, the complexation of flavonoids with Al^{3+} enhances yellow color. The maximum absorption wavelength of erodcyol and Al^{3+} is 390 nm, which is very attractive. The flavonoids can be blue, purple, brown and black after complexation with iron ions; and the rutin (3-rutinosyl-quercetin) in asparagus produces an unsightly dark color when it encounters iron ions, causing dark spots in asparagus. On the contrary, when rutin is complexed with tin ions, it produces an ideal yellow color.

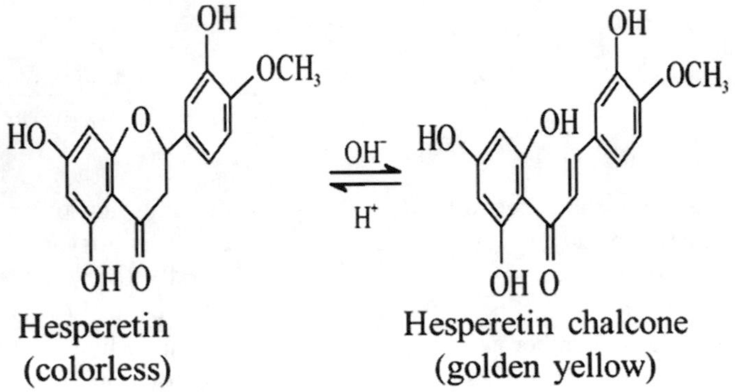

Hesperetin (colorless) ⇌ Hesperetin chalcone (golden yellow)

Fig. 9.27 Colorless flavanones are heated with alkali to transform into colored chalcone

In food processing, sometimes the pH is increased due to the use of sodium carbonate and sodium bicarbonate, or the high hardness of the water. Under such conditions, the originally colorless flavanone or flavanol can be converted into colored chalcone (Fig. 9.27). For example, potatoes, wheat flour, alfalfa, onion, cauliflower and kale will turn from white to yellow when processed (cooked) in alkaline water. This change is a reversible change that can be controlled and reversed with an organic acid.

The ethanol solution of flavonoids rapidly appeared red or purple under the reducing action of magnesium powder and concentrated hydrochloric acid. For example, flavonoids turn orange-red, flavonols turn red, and flavanones and flavanols turn purple-red. This is due to various anthocyanins that are formed after the reduction of flavonoids.

Flavonoids are also polyphenols. The intermediate products of enzymatic browning such as o-quinone or other oxidants can oxidize flavonoids to produce brown precipitates. The black color of mature olives is formed by the oxidation of erodcyol-7-glucoside during product fermentation and later storage; it is also one of the reasons for the long-term browning of the juice and the precipitation.

9.4.3 Catechins

Catechin, also called tea polyphenol, with the name derived from catechu of *Acacia catechu* L. extract, is 3,3′,4′,5,7-pentahydroxyflavan with two steric forms of (+)-catechin. In addition, catechin also contains its derivatives showing a similar molecular arrangement. There are six common catechins in tea, namely L-epigallocatechin, L-gallocatechin, L-epicatechin, L-Catechin, L-Epicatechin gallate and L-Epigallocatechin gallate (Fig. 9.28).

Catechins are widely distributed in food materials and herbs including tea, persimmons, apples, cacaos, berries and grapes. Catechins are high in tea. Catechin itself is

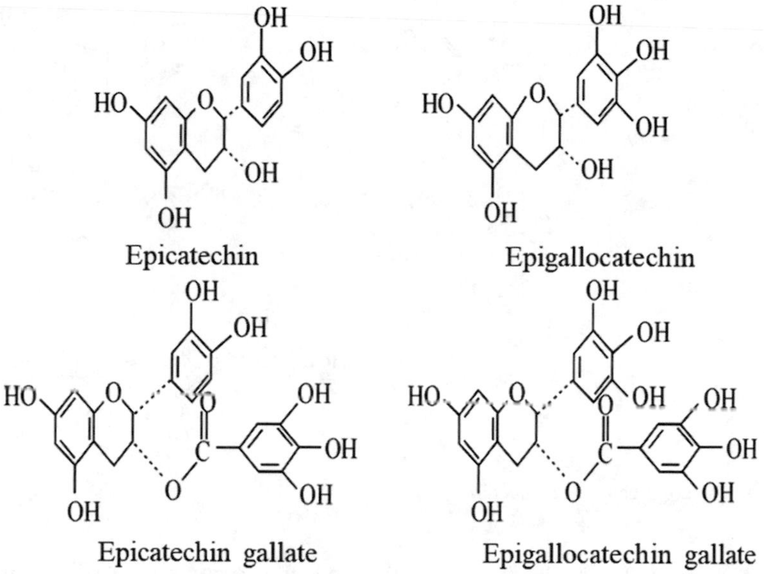

Fig. 9.28 The structure of several common catechins

colorless and has a slight astringency, while the combination of catechin and metal ions produces a white or colored precipitate. For example, a catechin solution reacts with ferric chloride to form a black-green precipitate, and a yellow precipitate is formed in the presence of lead acetate.

As a polyphenol, catechins are very easily oxidized to form a brown substance. Many catechin-containing plant tissues also contain polyphenol oxidase and/or peroxidase. When the tissue is damaged, catechins are oxidized by these enzymes to form a brown substance. The intermediate product of enzymatic browning, ortho-quinone, is an important substance that causes further oxidation of catechins or oxidative polymerization (Fig. 9.29). In the processing of black tea, the oxidation products of catechins are oxidized to form theaflavins and thearubigins, which help to form the color of black tea. Catechins can also be automatically oxidized when exposed to oxygen under high temperature and humid conditions.

9.4.4 Tannin

Tannin is an important kind of polyphenolic component with over 160 000 tons potentially biosynthesized each year in the world. It can be found not only in all vascular plant tissues but also in some non-vascular plants, i.e., marine algae. In food materials, tannins are extremely high in gallnuts and persimmons. Tradition-ally, tannins are divided into two types: *hydrolysable* and *condensed tannins* (or

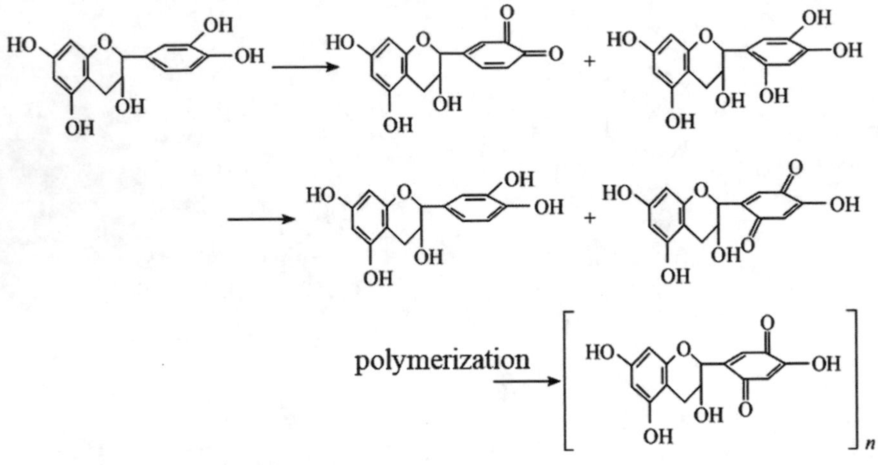

Fig. 9.29 The color change of catechin

anthocyanogen). The hydrolysable tannin molecule is composed of phenols and sugar (i.e., glucose) esters. According to the phenol type, hydrolysable tannins are further classified into two families, *gallotannins* that yield gallic acids and related derivatives after hydrolysis, and *ellagitannins* that produce ellagic acids and their derivatives.

The basic structural unit of condensed tannins is a flavan-3-ol or flavan-3,4-diol, and 2–8 of flavonoid repetitions are needed to form condensed tannins, which are generally in a complex with proteins and were first found in cocoa beans and later found to be ubiquitous in fruit juices, and now condensed tannins occupy more than 90% of worldwide tannin production. Besides hydrolysable and condensed tannins, another complex tannin should be also considered. *Complex tannins* are always formed by ellagitannin unit and flavan-3-ol unit with a typical representative as acutissimin A shown in Fig. 9.30, where the flavagallonyl group connects to a polyol derived from D-glucose by a glucosidic connection in C_1 and three other ester bonds.

Condensed tannins are converted to anthocyanins and catechins under acidic heating conditions, such as pelargonidin, petunidin and delphinidin. For example, the dimeric anthocyanogen in apples, pears and other juices can be converted into anthocyanidins and other polyphenols when heated under acidic conditions. The mechanism of this reaction is shown in Fig. 9.31.

Condensed tannins also produce oxidation products during processing and storage. For example, when the juice is exposed to the air or under light, they turn into a stable reddish-brown substance, which is responsible for the discoloration of apple juice. It is generally believed that the intermediates of enzymatic browning can also oxidize condensed tannins.

The color of tannin is yellow or slightly brown, which has a strong astringent taste, and they can precipitate with proteins and combine with various alkaloids or polyvalent metal ions to form colored insoluble precipitates. In food storage,

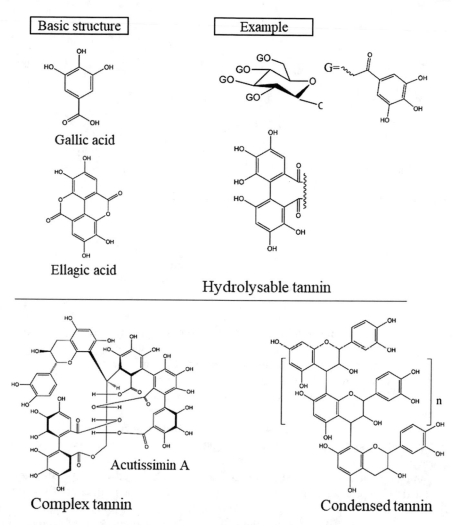

Fig. 9.30 Tannin classifications and their respective examples

tannins condense under certain conditions (such as heating and oxidation), thereby eliminating astringency. As polyphenols, tannins are also susceptible to oxidation, both enzymatic browning and non-enzymatic browning, and the former is dominant.

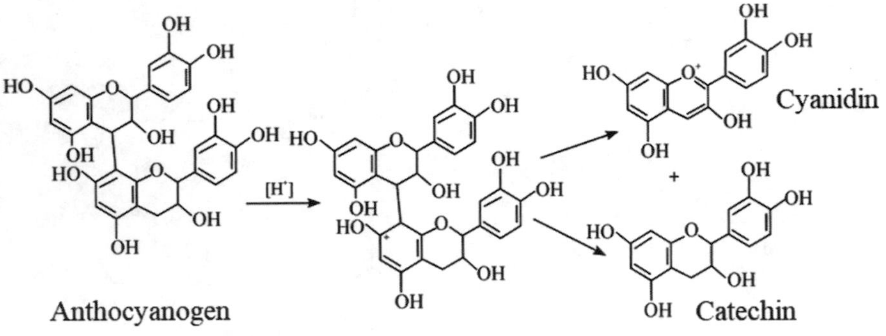

Fig. 9.31 The mechanism of acid hydrolysis of anthocyanogen

9.5 Food Colorant

9.5.1 Caramel Color

Caramel color (caramel) is a complex red-brown or dark-brown mixture formed by dehydration and condensation of carbohydrate raw materials, such as cerealose, sucrose, molasses, invert sugar, lactose, maltose syrup and starch hydrolysate, during the heating process. And it is a semi-natural food coloring agent that is widely used. According to the different catalysts used in the production process of caramel coloring, the Codex Alimentarius Commission (CAC) divides it into four categories (Table 9.5).

Caramel pigment is a dark brown gel or lump, with a special sweet aroma and pleasant bitterness, but it is rarely displayed under normal usage. It is easily soluble in water and has good stability to light and heat.

Caramel pigments have colloidal properties and are charged. The type of charge is related to the production process of caramel and the pH environment of the food. Therefore, when choosing a caramel colorant, it is necessary to consider that the charge of the caramel should be the same as that of the food, otherwise flocculation or precipitation will occur. For example, the caramel pigment added to beverages should have a strong negative charge, and the isoelectric point should be less than 1.5, and its pH range is mostly between 2.5 and 3.5, while the caramel pigment added to soy sauce and beer should usually be positively charged, and the pH range should be from 3.8 to 5.

The caramel pigment produced by the ammonia method currently occupies the largest market for caramel pigment. This type of caramel pigment may contain 4-methylimidazole, which is a convulsant. The results of chronic toxicity tests have confirmed that it will decrease the number of leukocytes, with slow growth. Therefore, in the caramel pigment produced by the ammonium salt method, the content of 4-methylimidazole must be strictly controlled.

Table 9.5 Types and characteristics of caramel colorant

Characteristics	Caramel type			
	Ordinary caramel (I)	Sulfite caramel (II)	Ammonia caramel (III)	Ammonium sulfite caramel (IV)
International code	ISN 150a	ISN 150b	ISN 150c	ISN 150d
	EEC No. E150a	EEC No. E150b	EEC No. E150c	EEC No. E150d
Typical use	Distilled alcohol, sweets, etc.	Alcohol	Baked goods, beer and soy sauce	Soft drinks, soup, etc.
Electric charge	Negative	Negative	Positive	Negative
Whether it contains ammonia compounds	No	No	Yes	Yes
Whether it contains sulfur compounds	No	Yes	No	Yes

Note ISN is the international numbering system for food additives adopted by the International Codex Alimentarius Commission (CAC) in 1989 (revised in 2001) and EEC is the European Community

According to the provisions of FAO/WHO, caramel pigment can be used in orange peel jelly, broth, cold drinks and other foods, and the dosage can be determined according to the normal production requirements.

9.5.2 Monascin Pigment

Monascin is derived from microorganisms. It is produced from a group of *Monascus* sp., *Monascus purpureus*, *Monascus anka* and *Monascus barkeri*, and it belongs to ketone pigments. There are six kinds of pigments, namely rubropunctamine, monascorubramine, rubropunctatin, monascorubrin, monascine and ankaflavine. There are red, yellow or purple pigments. Some of their structures are shown in Fig. 9.32.

The composition of monascin pigment obtained from different strains is different. For example, what is obtained from *Monascus purpureus* is ankaflavine, and what is obtained from *Monascus purpureus* is monascine. The physical and chemical properties of the above six monascin pigments are different from each other. The main ones with practical application value are rubropunctamine and monascorubramine.

Monascin pigment is a red or dark red powder, or liquid paste. The melting point is about 60 °C, and they are soluble in ethanol aqueous solution, ethanol, ether and glacial acetic acid. The color does not change with the pH value, and the

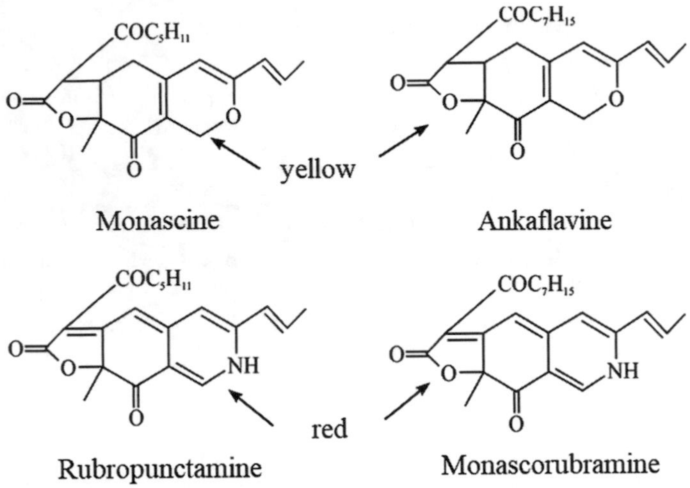

Fig. 9.32 The structure of monascin pigment

thermal stability is high. It is hardly affected by metal ions (such as Ca^{2+}, Mg^{2+}, Fe^{2+} and Cu^{2+}), and is also hardly affected by oxidants and reducing agents (except hypochlorous acid). However, under direct sunlight, the chromaticity is reduced. It has a good dye effect on protein. Once the protein is dyed, it will not fade after washing.

Monascin pigment also has an antiseptic effect, and has a strong inhibitory effect on *Bacillus cereus*, *Bacillus subtilis* and *Staphylococcus aureus*; secondly, it also has a certain inhibitory effect on *Pseudomonas aeruginosa*, *Pullorum* and *Escherichia coli proteus*; it has no inhibitory effect on *Sarcina*, *Saccharomyces cerevisiae* and *Penicillium chrysogenum*. The combination of monascin pigment, nisin and potassium sorbate can inhibit the growth of *Clostridium botulinum*. In addition, monascin pigment also has health effects such as reducing triglycerides, cholesterol and preventing arteriosclerosis. Monascin pigment can be used for the coloring of meat products. For example, the use of a coloring agent for fermented sausages can partially replace the amount of sodium nitrite. The color of fermented sausages made with 1 600 mg/kg monascin pigment as the coloring agent is close to that with 150 mg/kg sodium nitrite, and it has a certain inhibitory effect on *Clostridium botulinum*. Monascin pigment is used for the coloring of red fermented bean curd food, which not only provides the desired color of red preserved bean curd, a variety of aromas and fragrance components, but also has the functions of lowering blood pressure and cholesterol. Monascin pigment can also be used in the preparation of wine, fruit vinegar, beverages, various condiments, brewed foods, vegetable protein foods, etc.

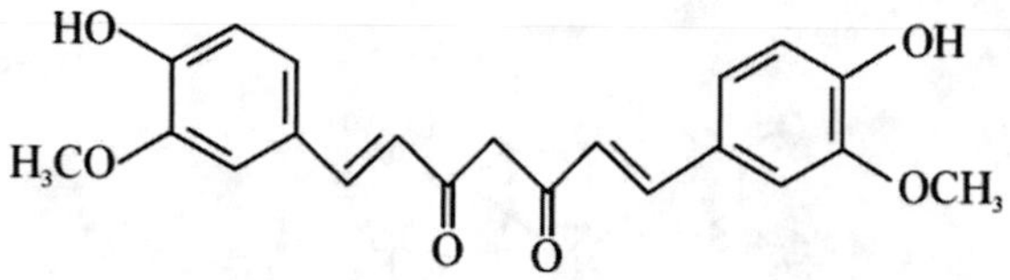

Fig. 9.33 The structure of curcumin

9.5.3 Curcumin

Curcumin, or turmeric yellow, is a yellow pigment extracted from the underground rhizomes of *Curcuma longa* in the ginger family. It is a group of ketone pigments. The main components are curcumin, demethoxycurcumin and bisdemethoxycurcumin. Its core structure is shown in Fig. 9.33.

Curcumin is an orange-yellow crystalline powder, almost insoluble in water, but soluble in ethanol, propylene glycol, glacial acetic acid and alkali solutions or ethers. It has a special fragrance, is slightly bitter, and yellow in neutral and acidic solutions, and brownish red in alkaline solutions. Curcumin is unstable to light, heat, oxidation and iron ions, but has good resistance to reduction. It has good coloring power to protein and is often used for coloring curry powder. Curcumin is also used in various oils to restore the color lost during processing.

9.5.4 Betalain

Betalain is a group of water-soluble pigments extracted from the tubers of the red beet from *Liliaceae*, and it is also widely present in flowers and fruits. It is present in these plants in the form of betacyanin and betaxanthin and their glycosides in the vacuole and the structures are shown in Fig. 9.34.

Betacyanin solution is purple-red in the pH range of 4–7. When the pH value is lower than 4 or higher than 7, the color changes to purple. When the pH value is above 10.0, betacyanin is hydrolyzed to betaxanthin. The solution immediately turned yellow. The heat resistance of betalain is not high, and it is relatively stable at pH 4.0–5.0. It will be converted into betalamic acid (BA) and cyclodopa-5-O-glucoside (CDG), and the reaction is reversible when the pH drops between 4 and 5.

Betanin can cause isomerization under the action of heating and acid, and two epimers can be formed in the chiral center of C_{15}. As the temperature rises, the proportion of isobetanin increases (Fig. 9.35), causing serious fading.

Betalain is also not resistant to oxidation, and the mechanism of oxidation is not clear, but it is not a free radical mechanism. For example, bleaching powder or sodium hypochlorite can make it fade, and the oxygen in the headspace of canned beets will speed up the fading of betalain. Light will accelerate oxidation, and ascorbic acid can slow down oxidation. The purple-red color fades after betalain is oxidized, and

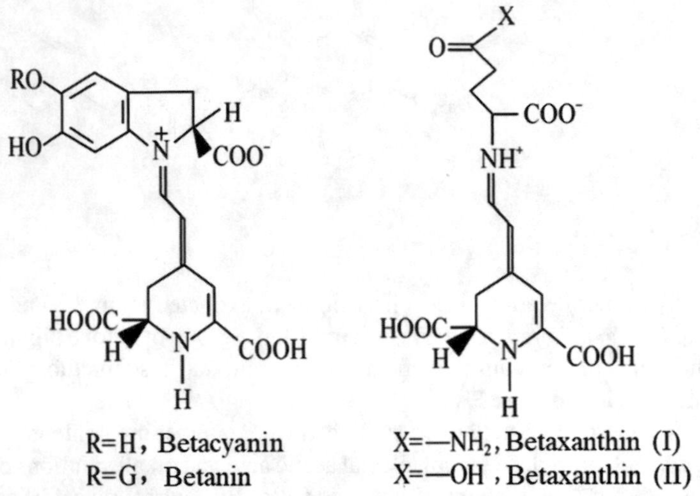

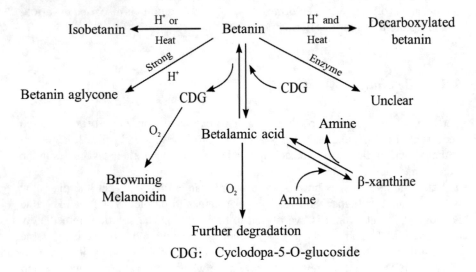

Fig. 9.34 The structure of betacyanin and betaxanthin

Fig. 9.35 Acid and/or thermal degradation of betanin

browness is often produced at the same time. If there is no oxidation condition, the stability of betalain to light is still good.

Certain metal ions also have a certain effect on the stability of betalain, such as Fe^{2+}, Cu^{2+} and Mn^{2+}. The mechanism is that these metal ions can catalyze the oxidation of ascorbic acid, thereby reducing the protective effect of ascorbic acid on betalain. The presence of metal chelating agents can greatly improve the effect of ascorbic acid as a betalain protector.

Betalain has good food coloring properties. The color is stable when used in foods with a pH from 3.0 to 7.0. In foods with low water activity, the color can be maintained for a long time.

9.5.5 Other Natural Colorants

Countries around the world also allow the use of a variety of other natural colorants, such as safflower yellow, shellac red, bilberry red, chili red, red-rice red, black-currant red, mulberry red, natural amaranth red, vine spinach red, black bean red, sorghum red, radish red, gardenia yellow, chrysanthemum yellow, corn yellow, *Hippophae rhamnoides* yellow, cocoa-shell pigment, tanoak brown, *Rosa laevigata michx* brown and *Quercus* brown.

9.6 The Principle and Practical Application of Food Toning

9.6.1 Preparation of Colorant Solution

The colorant powder is inconvenient to use directly, it is unevenly distributed in the food and may form pigment spots, and often needs to be formulated into a solution for use. The synthetic colorant solution generally uses a concentration of 1–10%, and it is difficult to adjust the hue if the concentration is too large.

In preparation, the weighing of the colorant must be accurate. In addition, it should be prepared according to the amount of each time, as the prepared solution is easy to precipitate after a long time. Due to the influence of temperature on the solubility of the colorant, the concentrated solution of the colorant that is prepared in summer will precipitate when stored in the refrigerator or during the winter. The aqueous solution of carmine turns black after long-term storage.

The water used in the preparation of the colorant aqueous solution should usually be boiled and cooled before use, or distilled water or the water treated with ion exchange resin should be used.

When preparing the solution, it should be avoided using metal utensils as the colorant may interact with metals and show different visual color; and in the storage, it should be avoided direct sunlight, preferably in a cool and dark place.

9.6.2 Principles of Hue Selection for Food Coloring

Hue is an attribute of visual perception that a surface presents one or two colors like red, yellow, green and blue. Most foods have rich colors, and their hue is closely related to the inner quality and outer aesthetic characteristics of the food. Therefore, in the production of food, what hue is used in the food is very important. The choice of food color is based on the psychological or customary requirements for food color, and the relationship between color, flavor and nutrition. The hue selection should be like the original color of the food or consistent with the name of the food, and the color matching principle should be used to formulate the corresponding characteristic color of the specific food. For example, canned cherries and bayberry jam should choose corresponding cherry red and bayberry red. Red wine should choose purple, and brandy, yellow–brown. Another example is that the color of candy can be selected according to its flavor characteristics. For instance, mint candies tend to be green; orange candies, red or orange; chocolate candies, brown, etc.

9.6.3 Hue Tuning

With red, yellow and blue as the basic colors, two or three of them can be selected to form a variety of different color spectrums according to different needs. The basic method is to combine the basic colors into secondary colors, or combine them into tertiary colors. The simple tuning principle is as follows.

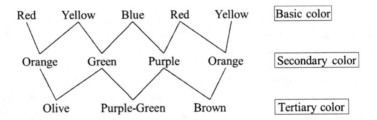

A variety of food synthetic colorants dissolved in different solvents can produce different hues and color intensities, especially when two or more food synthetic colorants are used for color matching, the situation effect is more obvious. For example, a certain proportion of a mixture of red, yellow and blue is yellower in aqueous solution, but redder in 50% ethanol. Due to the different alcohol content in food and aqueous solution, the hue after the colorant is dissolved is also different, so it is necessary to match colors according to the alcohol content and the intensity of the colorant. In addition, the food is moist when it is colored. When the moisture evaporates and gradually dries, the colorant will also concentrate on the surface layer, causing the so-called "concentration effect", especially when the affinity between the food and the coloring agent is low. When matching colors, pay attention to the different stability of various pigments, which will lead to changes in the hue of the synthetic color. For example, indigo fades faster, while lemon yellow is not easy to fade, so the initial matched green will gradually change to yellow-green. Synthetic pigments use the above principles to match the color. Natural pigments are not easy to match colors due to their low adherence, easy discoloration and strong sensitivity to the environment.

9.7   Summary

Food pigments refer to substances in food that absorb and reflect visible light waves and then present foods in various colors. There are various methods for classifying food pigments. According to their sources, food colorants are classified into two categories: natural food colorants and synthetic food colorants; natural food pigments are further classified into plant pigments, animal pigments and microbial pigments according to their origins. According to their solubility, food pigments are classified into water-soluble and fat-soluble ones. Based on the chemical nature of food natural pigments, they are mainly divided into tetrapyrrole, isoprene and polyphenol categories, and their representatives were discussed in this chapter. Changes in natural pigments during food processing such as chlorophyll, heme, carotenoids and flavonoid pigments are the basis for food color changes. Understanding their reaction mechanisms is important to control their changes and ensure food color.

Questions

1.   What are food pigments? What is their role in the food industry?
2.   Under what conditions are the main chlorophyll derivatives produced? How to control the conditions in food storage and processing to keep food green?
3.   What chemical changes can occur when meat is cured? Does excessive colorant cause any hazardous results?
4.   Carotenoid pigments are multi-functional natural pigments, what about their nutritional effects?
5.   What substances are included in polyphenol pigments? What are the factors that affect the color change of polyphenol pigments?
6.   What are the main similarities and differences between natural colorants and synthetic colorants? How to choose food colorants for color matching?
7.   What kind of color for food will be liked by people? Will the color of these foods change during processing and storage?
8.   Why do shrimp turn from cyan to red after high-temperature processing?
9.   Why does the surface color of meat change from purple-red to bright red, and again to brown, if it is placed at room temperature for a long time?
10.   Why does the color of purely fermented wine change from purple-red to blue after adding baking soda?
11.   What is the difference between synthetic pigments and natural pigments? Why are synthetic pigments replaced by natural pigments widely used in foods?
12.   Explain the following terms: chlorophyll, heme, carotenoids, anthocyanins, metmyoglobin, oxymyoglobin and flavonoids.

Bibliography

1. Adadi, P., Barakova, N.V., Krivoshapkina, E.F.: Selected methods of extracting carotenoids, characterization, and health concerns: a review. J. Agric. Food Chem. **66**, 5925–5947 (2018)
2. Ahmed, M., Eun, J.B.: Flavonoids in fruits and vegetables after thermal and nonthermal processing: a review. Crit. Rev. Food Sci. Nutr. **58**, 3159–3188 (2018)
3. Arbenz, A., Avérous, L.: Chemical modification of tannins to elaborate aromatic biobased macromolecular architectures. Green Chem. **17**(5), 2626–2646 (2015)
4. Bakoyiannis, I., Daskalopoulou, A., Pergialiotis, V., Perrea, D.: Phytochemicals and cognitive health: are flavonoids doing the trick? Biomed. Pharmacother. **109**, 1488–1497 (2019)
5. Buzala, M., Slomka, A., Janicki, B.: Heme iron in meat as the main source of iron in the human diet. Journal of Elementology **21**, 303–314 (2016)
6. Bylsma, L.C., Alexander, D.D.: A review and meta-analysis of prospective studies of red and processed meat, meat cooking methods, heme iron, heterocyclic amines and prostate cancer. Nutr. J. **14**, 125–143 (2015)
7. Chen, K., Ríos, J. J., Pérez-Gálvez, A., Roca, M.: Development of an accurate and high-throughput methodology for structural comprehension of chlorophylls derivatives. (I) Phytylated derivatives. Journal of Chromatography A **1406**, 99–108 (2015)
8. Chen, K., Roca, M.: Cooking effects on chlorophyll profile of the main edible seaweeds. Food Chem. **266**, 368–374 (2018)
9. Eghbaliferiz, S., Iranshahi, M.: Prooxidant activity of polyphenols, flavonoids, anthocyanins and carotenoids: updated review of mechanisms and catalyzing metals. Phytother. Res. **30**, 1379–1391 (2016)
10. Giuffrida, D., Donato, P., Dugo, P., Mondello, L.: Recent analytical techniques advances in the carotenoids and their derivatives determination in various matrixes. J. Agric. Food Chem. **66**, 3302–3307 (2018)
11. Gladwin, M.T., Grubina, R., Doyle, M.P.: The new chemical biology of nitrite reactions with hemoglobin: r-state catalysis, oxidative denitrosylation, and nitrite reductase/anhydrase. Acc. Chem. Res. **42**, 157–167 (2009)
12. Gorniak, I., Bartoszewski, R., Kroliczewski, J.: Comprehensive review of antimicrobial activities of plant flavonoids. Phytochem. Rev. **18**, 241–272 (2019)
13. Iwashina, T.: Contribution to flower colors of flavonoids including anthocyanins: a review. Nat. Prod. Commun. **10**, 529–544 (2015)
14. Holst, B., Williamson, G.: Nutrients and phytochemicals: from bioavailability to bioefficacy beyond antioxidants. Curr. Opin. Biotechnol. **19**(2), 73–82 (2008)
15. Kiokias, S., Proestos, C., Varzakas, T.: A review of the structure, biosynthesis, absorption of carotenoids-analysis and properties of their common natural extracts. Curr. Res. in Nutr. Food Sci. **4**, 25–37 (2016)
16. Khoo, H.E., Azlan, A., Tang, S.T., Lim, S.M.: Anthocyanidins and anthocyanins: colored pigments as food, pharmaceutical ingredients, and the potential health benefits. Food Nutr. Res. **61**, 1–21 (2017)
17. Krga, I., Milenkovic, D.: Anthocyanins: from sources and bioavailability to cardiovascular-health benefits and molecular mechanisms of action. J. Agric. Food Chem. **67**, 1771–1783 (2019)
18. Liu, Y., Tikunov, Y., Schouten, R.E., Marcelis, L.F.M., Visser, R.G.F., Bovy, A.: Anthocyanin biosynthesis and degradation mechanisms in solanaceous vegetables: a review. Front. Chem. **6**, 52 (2018)
19. Maiani, G., Periago-Castón, M.J., Catasta, G., Toti, E., Cambródon, I.G., Bysted, A., Granado-Lorencio, F., Olmedilla-Alonso, B., Knuthsen, P., Valoti, M., Böhm, V., Mayer-Miebach, E., Behsnilian, D., Schlemmer, U.: Carotenoids: actual knowledge on food sources, intakes, stability and bioavailability and their protective role in humans. Mol. Nutr. Food Res. **53**(S2), S194–S218 (2009)
20. Maqsood, S., Benjakul, S., Kamal-Eldin, A.: Haemoglobin-mediated lipid oxidation in the fish muscle: a review. Trends Food Sci. Technol. **28**, 33–43 (2012)

21. Moller, J.K.S., Skibsted, L.H.: Myoglobins - The link between discoloration and lipid oxidation in muscle and meat. Quim. Nova **29**, 1270–1278 (2006)
22. Nagula, R.L., Wairkar, S.: Recent advances in topical delivery of flavonoids: a review. J. Control. Release **296**, 190–201 (2019)
23. Odorissi, X., Augusta, A., Perez-Galvez, A.: Carotenoids as a source of antioxidants in the diet. In: Stange, C. (ed.) Carotenoids in Nature: Biosynthesis, Regulation and Function, pp. 359–375. Springer, New York (2016)
24. Papuc, C., Goran, G.V., Predescu, C.N., Nicorescu, V.: Mechanisms of oxidative processes in meat and toxicity induced by postprandial degradation products: a review. Compr. Rev. Food Sci. Food Saf. **16**, 96–123 (2017)
25. Reig, M., Aristoy, M.C., Toldra, F.: Variability in the contents of pork meat nutrients and how it may affect food composition databases. Food Chem. **140**, 478–482 (2013)
26. Sharma, K., Mahato, N., Lee, Y.R.: Extraction, characterization and biological activity of citrus flavonoids. Rev. Chem. Eng. **35**, 265–284 (2019)
27. Suman, S.P., Joseph, P.: Myoglobin chemistry and meat color. Annu. Rev. Food Sci. Technol. **4**(4), 79–99 (2013)
28. Yong, H.I., Han, M., Kim, H.J., Suh, J.Y., Jo, C.: Mechanism underlying green discolouration of myoglobin induced by atmospheric pressure plasma. Sci. Rep. **8**(1), 9790–9799 (2018)
29. Zhao, Y., Wu, Y.: Pigments. In: Chemistry, F. (ed.) Kan, J, pp. 257–290. Beijing, China Agricultural University Press (2016)

Dr. Kewei Chen associate professor and master supervisor in College of Food Science, Southwest University, China. He obtained his doctorate from Universidad de Sevilla, Spain in 2016, and from 2012 to 2016, he also did his research work in Instituto de la Grasa (CSIC, Spain). He has participated in several international book chapters about food industry and technology. His major research interests and areas are related with food chemistry & nutrition, including micronutrient bioavailability, function evaluation and food safety, and he has presided over four provisional and administerial projects related with food nutritional changes and food safety evaluation. Dr. Kewei Chen has published more than 10 research papers on Food Chemistry, Journal of Functional Foods, Molecular Nutrition & Food Research, etc.

Chapter 10
Food Flavor Substances

Liyan Ma and Jingming Li

Abstract Food flavor is the combination of all sensory attributes experienced while ingesting food, with gustatory and olfactory perceptions being the most important attributes. This chapter mainly focuses on common types of flavor substances present in food, the interactions between these different flavor substances, and the influencing factors and flavor perception mechanisms for flavor substances. This chapter also introduces aromas present in different foods, formation of aromas, and the influence of processing on aroma composition and stability.

Keywords Food flavor · Flavor substance · Flavor perception mechanisms · Aromas

10.1 Overview

10.1.1 An Overview of Flavor

In addition to providing necessities to human beings, food also provides people with psychological enjoyment and sensory pleasure. Food flavor is a comprehensive yet broad concept; it is the overall impression kept in the brain by all sense organs while consuming food, mainly through gustatory and olfactory perceptions, but also including pain, touch, temperature sensations, and those sensations triggered by the trigeminal nerves (Fig. 10.1). Flavor substances in food are mostly categorized into taste substances received by gustatory perceptions and aroma substances received by olfactory perceptions.

Gustatory perception is the feeling from food in the mouth stimulating the gustatory organs. This stimulus sometimes comes from only one aspect of taste perception, but in most cases, it comes from a complex taste perception, including psychological taste perception (shape, color, brightness, etc.), physical taste perception (degree

L. Ma (✉) · J. Li
College of Food Science & Nutritional Engineering, China Agricultural University, Beijing 100083, China
e-mail: lyma1203@cau.edu.cn

© The Author(s), under exclusive license to Springer Nature Singapore Pte Ltd. 2021
J. Kan and K. Chen (eds.), *Essentials of Food Chemistry*,
https://doi.org/10.1007/978-981-16-0610-6_10

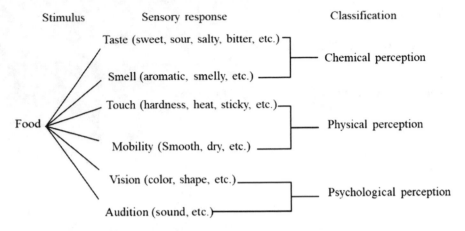

Fig. 10.1 Sensory response and classification of food

of hardness, sliminess, temperature, chewiness, texture, etc.), and chemical taste perception (sourness, sweetness, bitterness, saltiness, and other tastes).

Taste classifications are not uniform around the world. For example, it is divided into 5 tastes namely salty, sour, sweet, bitter, spicy in Japan; in Europe and the United States, it is divided into 6 flavors namely sweet, sour, salty, bitter, spicy, metal taste; it is divided into 8 flavors: sweet, sour, salty, bitter, spicy, light, astringent, and abnormal in India; while in China, in addition to the five flavors of sour, sweet, bitter, spicy and salty, there are also flavors of the fresh and astringent.

But physiologically, the four basic tastes are sweet, sour, salty, and bitter. In recent years, the concept of umami has become more accepted by the public and it has been named as the fifth basic taste. Spicy taste is caused by food stimulating the mucous membranes of the mouth and causing pain in both the nasal mucosa and the skin. Astringency taste refers to the astringent effect of the mucous membrane on the tongue.

However, in terms of food seasoning, spicy and astringent taste should be regarded as two separate tastes. As for umami and other flavors, they can make the whole flavor of food has a more delicious special effect, therefore, in Europe and the United States, umami substances can be used as flavor intensifier or synergistic agent, not as an independent taste of umami. It can be said that umami should also be regarded as an independent flavor in the flavoring of food.

Looking at human reactions to the basic tastes, saltiness is tasted first, while bitterness comes last. When looking at the sensitivity of tastes, people are most sensitive to bitterness; bitterness can be detected easily. Sensitivity for taste substances is often measured with threshold concentration as the standard. Threshold refers to the lowest concentration that could be detected. Because humans have different distribution of taste buds and different sensitivity to gustatory substances, the threshold and sensitivity for taste compounds will be different as well. For basic tastes, such as sour, sweet, bitter, and salty, the threshold for each representing substance is generally

Table 10.1 The range of taste perception thresholds in different parts of the tongue mol/L

Taste	Taste compounds	Tongue tip	Margin of tongue	Root of tongue
Salty	Salt	0.25	0.24	0.28
Sour	Hydrochloric acid	0.01	0.006 ~ 0.007	0.016
Sweet	Saccharose	0.49	0.72 ~ 0.76	0.79
Bitter	Quinine Sulfate	0.000 29	0.000 2	0.000 5

considered to be 0.3% for sucrose, 0.02% for citric acid, 16 mg/kg for quinine, and 0.2% for sodium chloride. (Table 10.1).

Olfactory perception is an important sense to the human body; it is produced by volatile substances stimulating the olfactory nerve cells in the nasal cavity and triggering a sensation by the central nervous system. Aromas in food are created by a combination of multiple aroma substances; it is rarely created by a single aroma compound. When these substances are combined in the perfect proportions, they can emit an attractive aroma. If not properly combined, the smell of food can be unbalanced, even causing off-flavors. Similarly, the relative concentration of fragrant substances in food can only reflect the strength of food aroma, but cannot completely and truly reflect the degree of pros and cons of food aroma. Therefore, the value of determining the role of an aromatic substance in a food aroma is called the aroma value (fragrance value). The aroma value is the ratio of the concentration of the aromatic substance to its threshold, that is

$$\text{Aroma value} = \frac{\text{The concentration of a fragrant substance}}{\text{Threshold value}}$$

In general, when the aroma value is less than 1, people's smelling organs will not cause sensation to this fragrant substance.

Flavor is one of the important aspects in assessing food quality and plays a decisive role in the selection, acceptance, and intake of food. Since flavor is a sensory attribute, the understanding and evaluation of flavor are often strongly biased depending on the individual, region, and ethnicity. Although modern analytical techniques provide a great convenience for in-depth study of flavor chemistry, it is difficult to accurately determine and describe the flavor of food by either qualitative or quantitative methods, because flavor is the physiological result of certain compounds acting on the human sensory organs. Therefore, sensory evaluation is still an important measure to flavor research.

10.1.2 Characteristics of Flavoring Substances

The flavor compounds reflecting the flavor of food generally are called flavor substances. There are generally many kinds of flavor substances interacting with each other in food, some of which play a leading role, others as a supporting role. If one or more compounds in food represent its food flavor, these compounds are called characteristic compounds. For example, the characteristic compound of sweet taste in a banana is isoamyl acetate, the characteristic compounds of cucumber are 2, 6-nonadienal, etc. The number of characteristic compounds in food is limited, existing at very low concentrations and sometimes not very stable. However, their existence provides an important basis for us to study the chemical basis of food flavor. The flavor substances reflecting the flavor of food generally have the following characteristics:

(1)   There are many kinds of flavor substances that can interact obviously each other. For example, flavor substances reached more than 500 in the cooked coffee. In addition, the antagonistic or synergistic effects among flavor substances make it difficult to reproduce the original flavor with recombined monomer components.

(2)   Small amount but significant effect. The content of flavoring substances in food varies greatly and the proportion is very low, but the flavor produced is obvious. For example, the aroma characteristic compounds of banana will give water a banana flavor at a concentration of only 5×10.6 mg per kilogram.

(3)   Many flavoring substances are easily decomposed by oxidation, heating, and so on, with poor stability. For example, the flavor of tea will become worse due to the automatic oxidation of its flavor substances.

(4)   The molecular structure of flavoring substances lacks general regularity. The molecular structure of flavoring substances is highly specific, a slight change in structure can make a big difference in flavor, even it is difficult to find regularity among the molecular structure of compounds with the same or similar flavor.

(5)   Flavor substance is also affected by its concentration, medium, and other external conditions.

10.2   Food Taste

10.2.1   Physiological Basis of Taste

Gustatory perception is produced by food's soluble substances dissolved in saliva or food liquid stimulating taste receptors inside the oral cavity and then transmitted to the gustatory cortex in the brain through the gustatory sensory system. Lastly, through analysis by the brain's central nervous system, gustation or taste is produced.

Taste bud is the part where the taste receptors and taste substances interact. Human's taste buds are mainly located on the papillae, found on the surface of the tongue; only a small portion of taste buds are distributed in the soft palate, throat, pharynx, and other places. Filiform papillae, located on the surface of the tongue, is responsible for the sensation of touch but does not contain any taste buds. Near the filiform papillae, especially at the tip and sides of the tongue, fungiform papillae can be found. Along the two sides of the tongue, from the tip of the tongue to two-thirds after the end of the tongue, are foliate papillae, and circumvallate papillae are located at the end of the tongue (Fig. 10.2). Taste buds are elliptical, with supporting cells on the outer layer and multiple spindle-shaped gustatory cells on the inside. Gustatory hair is on top of gustatory cells, and nerve endings are distributed throughout the bottom of a taste bud (Fig. 10.3). Gustatory cell surface consists of protein, fat, and small amount of sugars, nucleic acids, and inorganic ions that can combine with different taste substances.

Different taste substances interact with different components on the receptors of taste cells. For example, the receptor for sweet substances is a protein, while the receptor for bitter and salty substances is a lipid. Some people think that the receptors of bitter substances are also may be related to protein. The experiments

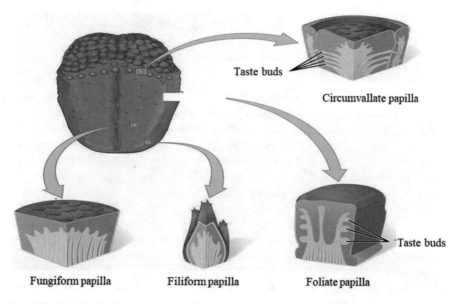

Taste buds

Circumvallate papilla

Taste buds

Fungiform papilla   Filiform papilla   Foliate papilla

Fig. 10.2 The distribution of the papilla in the tongue

Fig. 10.3 The structure of
taste buds

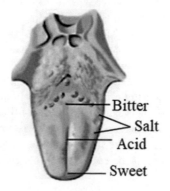

Bitter
Salt
Acid
Sweet

Fig. 10.4 Taste sensitivity of different parts of tongue

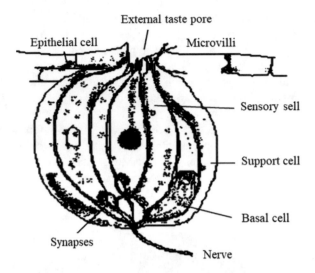

also showed that different taste substances have different binding sites on taste buds, especially sweet, bitter, and delicious substances. Their molecular structure has strict spatial specificity requirements, which reflects that different parts of the tongue have different sensitivity. There are different sensitivities. At the same time, the papillae on the surface of the tongue can be divided into fungiform papillae, filiform papillae, and foliate papillae according to their shapes. They exist in different parts of the tongue, respectively. Due to the uneven distribution of nipples, the perceptibility and sensitivity of different parts of the tongue to taste are also different (Fig. 10.4).

Only when the taste substances are dissolved in water, they can enter the orifice of taste buds and stimulate taste cells. When a piece of very dry sugar is placed on the surface of the tongue dried with filter paper, the sweetness of the sugar will not be felt. The saliva secreted by parotid gland, submandibular gland, sublingual gland, and numerous small salivary glands is the natural solvent of food. The activity of the secretory glands and the composition of saliva is also adapted to the type of food to a large extent. The drier the food is, the more saliva is secreted per unit time. When you eat egg yolk, the saliva secreted is thick and rich in proteases, while when you eat a sour plum, it secretes thin saliva with less enzymes. Saliva can also wash the mouth, so that the taste buds can distinguish taste more accurately. Therefore, saliva also has a great relationship with taste.

Experiments have shown that it takes only 1.5 to 4.0 ms for human taste from stimulating taste buds to perceiving taste, which is much faster than vision (13–15 ms), hearing (1.27–21.5 ms) or touch (2.4–8.9 ms). This is because taste is transmitted by nerves, almost reaching the limit speed of nerve transmission, while vision and hearing are transmitted by sound waves or a series of secondary chemical reactions, so they are slow. The bitter taste is the slowest, so in general, the bitter taste is always felt at the end. But people are often more sensitive to bitter substances than sweet ones.

The physiological mechanism of taste production has been basically confirmed, as shown in Fig. 10.5. For sweet compounds, the results showed that the taste receptors were combined with G-proteins (also for umami and bitterness). Once the sweet compounds are combined with the proteins of receptors on the surface of taste cells, the configuration of receptor proteins would change and then interact with G-proteins, activates adenyl cyclase to synthesize 3 ', 5'- cyclic AMP (cAMP) from ATP. After that, cAMP stimulates the cAMP-dependent kinase, which leads to the phosphorylation of K^+ channel protein, and the K^+ channel is finally closed. As a result, the reduction of K^+ delivered to the cell leads to depolarization of the cell membrane, which activates the potential-dependent calcium channel and Ca^{2+} flows into the cell, releasing neurotransmitters (norepinephrine, norepinephrine) at the synapse. Therefore, an action potential is generated in the nerve cells, thereby generating corresponding conduction, and finally forming a corresponding sensation in the central nervous system.

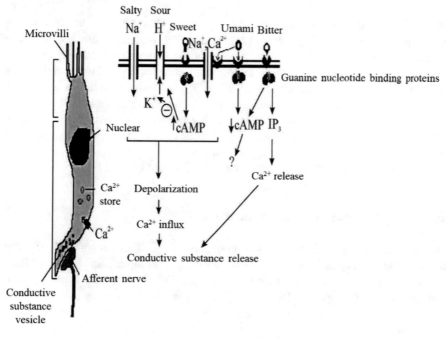

Fig. 10.5 The physiological mechanism of gustation generation

10.2.2   The Main Factors that Affect Taste

10.2.2.1   Taste Substance Structure

The structure of taste substance is an internal factor affecting gustatory perception. Sometimes, a minute change in molecular structure can cause magnificent changes in taste. In general, sugars, such as glucose and sucrose, mostly provide sweet taste; carboxylic acids, such as acetic acid and citric acid, mostly provide sour taste. Salts, such as sodium chloride and potassium chloride, give off a salty taste, while alkaloids and heavy metals give off mostly bitter taste. However, there are many exceptions: saccharin, lead acetate, and other non-sugar organic salts are sweet, oxalic acid is not sour but astringent, potassium iodide is bitter but not salty, and so on.

10.2.2.2   Temperature

In general, gustatory perception is sensitive between 10 to 40 °C, with 30°C having the highest sensitivity. All gustatory perceptions will weaken with higher or lower temperature. At 50°C, all gustatory perceptions become slower. Sweet and sour tastes have an optimum perception temperature between 35 to 50 °C. Salty taste have an optimum temperature between 18 to 35 °C; however, bitter taste have an optimum temperature at 10°C. Every gustatory perception threshold will change as temperature changes and temperature will have varying effects on different gustatory perceptions, but the change is regular over a range of temperatures. The degree that different feeling of flavors is affected by temperature varies, generally, saccharin had the greatest effect on sweetness, while hydrochloric acid had the least effect.

10.2.2.3   Taste Substances' Concentration

Odorant substances in an appropriate concentration will usually give a pleasant feeling, and inappropriate concentrations will generate unpleasant feelings. Taste substances' concentration has varying effects on different gustatory perceptions. Usually, sweetness gives a pleasant feeling at perceived concentration, and pure bitterness will almost always be unpleasant. Sourness and saltiness, at low concentration, will bring happiness, but at high concentration, they will make people feel unpleasant.

10.2.2.4   Taste Substances' Solubility

Flavoring substances can only stimulate the taste buds after being dissolved. Therefore, solubility and dissolution rate will affect gustatory sensation's perceived rate and duration. For example, sucrose can be easily dissolved; therefore, sweetness is perceived quickly but also disappears quickly. On the other hand, saccharin is more difficult to dissolved; therefore, gustation appears slower while lasting longer. Because flavoring substances can only diffuse to taste receptors when they are dissolved to create a taste sensation, taste is also influenced by the medium in which the flavoring substance is located. The viscosity of solvent can influence taste substances' contact with taste receptors. Different solvents could lower taste substances' solubility or prevent the release of taste substances.

10.2.2.5 Age, Gender, and Physiological Conditions

Age influences taste sensitivity, which is mostly seen in people over 60. Generally, there was no significant change in taste sensitivity under the age of 60. Because taste buds, located on the tongue's papillae, will decrease as age increases. This decrease in taste buds will reduce gustatory perception sensitivity. Under normal circumstances, when above 60 years of age, gustatory sensation sensitivity for salty, sour, sweet, bitter, and other tastes will be significantly reduced.

There are two different views on the effect of gender on taste. Some researchers believe gender does not have any influence on basic gustatory perceptions, while others believe gender does not influence bitterness, but women are more sensitive to saltiness and sweetness compare to men. Men are more sensitive to sour taste.

To some extent, gustatory perception sensitivity is dependent on body conditions; when the human body suffers from certain diseases or abnormalities, it can lead to lost or dullness of taste and changes in gustatory perception. For example, in the case of jaundice, the perception of bitterness is significantly reduced or even lost; when suffering from diabetes, the sensitivity of the tongue to sweet stimuli is significantly reduced; if ascorbic acid is lacking for a long time, the sensitivity to citric acid is significantly increased; after the increase of blood sugar content, the sensitivity to sweet feeling will be reduced. These facts also prove that, in a sense, taste sensitivity depends on the body's needs. Changes in gustatory perception due to illness can be temporary or permanent,taste can return to normal after the disease is cured, but some are permanent changes.

People's taste sensitivity increases when they are hungry, but has little effect on the preference of certain taste. Some experiments showed that the sensitivity of the four basic flavors reached the highest at 11:30 a.m. The degree of decrease was related to the caloric value of the food. People have high taste sensitivity before eating, which proves that taste sensitivity is closely related to the physiological needs of the body. One reason for the decrease in taste sensitivity after eating is that the intake of food meets the physiological needs; the other reason is that the eating leads to fatigue of taste receptors, thereby reducing taste sensitivity.

10.2.3 Interactions Between Taste Substances

The formation of taste, in addition to physiological phenomena, is also related to the chemical structure and physical properties of flavoring substances. Just as a substance may not taste the same because of its optical properties, while different substances can present the same taste.

In terms of the speed of basic taste perception, salty taste is the fastest while bitter taste is the slowest. But in terms of sensitivity, bitterness is the most sensitive and more perceptible. Now we use the threshold, which is the minimum concentration at which the substance can be perceptible (mol/m^3, %或mg/kg). Due to the differences among animal species, people, race, habits, etc., the threshold value of various literature will be different to some extent.

The composition of food varies widely, and the ingredients can interact with each other. The taste of each food component cannot be simply combined; various factors must be considered.

10.2.3.1 Synergistic Taste Effect

The taste of one substance can be significantly enhanced by the presence of another substance, and this phenomenon is called the synergistic taste effect. For example, sodium glutamate (MSG) and 5'-inosinic acid (5'-IMP) can work synergistically to enhance umami taste. Maltol can synergize almost any other flavor; adding maltol to beverages and juices can enhance sweetness.

10.2.3.2 Antagonistic Taste Effect

One substance weakening or inhibiting the effect of another substance' taste is known as the antagonistic taste effect. For example, when mixing any two of the following, sucrose, citric acid, sodium chloride, and quinine, in adequate concentration, will weaken each individual taste perception.

10.2.3.3 Taste Contrast Effect

The existence of two taste substances at the same time can influence feelings and psychology, and this is called the contrasting taste effect. For example, when salt is present in MSG, umami flavor is enhanced; when a small amount of salt is sprinkled onto watermelon, its sweetness can be enhanced. Coarse granulated sugar sometimes tastes sweeter than pure granulated sugar due to the presence of impurities.

10.2.3.4 Taste Alteration Effect

It has been found that the leaves of the tropical plant Gymnema sylvestre contain gymnemic acid. Sweet and bitter food can no longer be detected after chewing Gymnema sylvestre leaves, it can suppress sweet and bitter tastes for hours. However, these leaves have no alteration effect on sourness and saltiness. Furthermore, the interaction between two substances can sometimes change gustatory perception, such as the "miracle fruit" from Africa. This fruit contains an alkaline protein that will make sour substances taste sweet after consumption. Sometimes a sour orange gives you a sweet taste in your mouth, this phenomenon is known as the taste alternation effect or inhibitory effect. Alteration effect is a change in taste substance, while contrast effect is a change in the intensity of taste substance.

10.2.3.5 Taste Fatigue

After being stimulated by certain gustation substances for a long period of time, when consuming the same gustation substance afterwards, the intensity of the taste will often reduce; this phenomenon is known as taste fatigue. The phenomenon of taste fatigue involves psychological factors, for example, the second piece of candy will not taste as sweet as the first piece. Some people have a habit of eating MSG; even with more MSG added to a dish, the perception of umami can decrease.

When all kinds of sweeteners are used together, they can improve each other's sweetness. For example, although D.E.42 starch syrup is much less sweet than sucrose at the same concentration, the sweetness of the mixture of 26.7% sucrose solution and 13.3% D.E.42 starch syrup was equal to that of 40% sucrose in the liquid phase.

A small amount of polysaccharide thickener was added to the sugar solution, for example, the sweetness and viscosity can be slightly improved when 2% starch or a small amount of gum is added to 1 ~ 10% sucrose solution. Sweeteners at appropriate concentrations (especially below the threshold) often have the effect of improving flavor when they are used with salty, sour, and bitter substances. However, when the concentration is high, the effect of other flavor sensitive substances on sweetness is not regular. For example, if 0.5% salt is added to 5 ~ 7% sucrose, the sweetness will increase, while 1% salt will give the opposite effect.

In addition, there is also an interaction between gustation substances and olfaction substances. Physiologically, although gustatory and olfactory perceptions are completely different, the complex feelings created by the mixture of tastes and aromas during food chewing and the transformation by taste substances and aroma compounds make the two perceptions promote each other.

In a word, the various flavoring substances and their taste sensation interacts with each other, the psychological effects they cause are all very subtle, but much remains unclear and needs further study.

10.3   Food Taste and Taste Substances

10.3.1   Sweet Taste and Sweet Taste Substances

Sweet taste is one of the most popular basic tastes, and sugars are the most common natural sweetening substances. Besides sugar and its derivatives, there are many non-sugar natural compounds, derivatives of these natural compounds, and synthetic compounds that can also possess a sweet taste.

10.3.1.1   Sweetening Mechanism (Shay's Theory)

There are several hypotheses explaining the mechanisms of sweet taste perception. Earlier hypothesis believed sweet taste perception is related to the presence of multiple hydroxyl groups on sugar molecules, but this hypothesis was quickly rejected because the sweetness of different hydroxyl-containing compounds varied greatly. Many amino acids, certain metal salts, and non-hydroxyl-containing compounds, such as chloroform and saccharin, also have a sweet taste.

Currently, the AH/B theory proposed by Shallenberger et al. has the greatest influence; it explains the relationship between sweetness and its molecular structure (Fig. 10.6). This theory believes there is a hydrogen bond forming group (-AH), such as -OH, $-NH_2$, $= HN$, in the molecule of a sweet compound called a proton donor. There is also an electronegative atom (-B), such as oxygen and nitrogen, called a proton acceptor; this atom is 0.25–0.4 nm from the -AH group. These two groups in sweet taste compounds must meet stereochemistry requirements to bind with receptors. Inside the sweet taste receptor, there are also AH/B structural units, and the distance between -AH and -B is 0.3 nm. When sweet taste compounds' AH/B structure hydrogen bonds the AH/B structure on the receptor, gustatory nerves become stimulated and produce sweetness. The AH-B structure of other compounds, such as chloroform, saccharin, and glucose, can be represented by Fig. 10.7.

Shellenberger's theory explains from a molecular level whether a substance can have a sweet taste but cannot explain the intrinsic factors causing compounds with AH/B structure having vastly different sweetness intensities. Kier, later, modified and developed the AH/B theory. He believed there is a lipophilic region for specific steric structures in sweet taste compounds besides just having the- AH and -B groups. There is a hydrophobic group (-X), such as $-CH_2CH_3$ and $-C_6H_5$, existing 0.35 nm

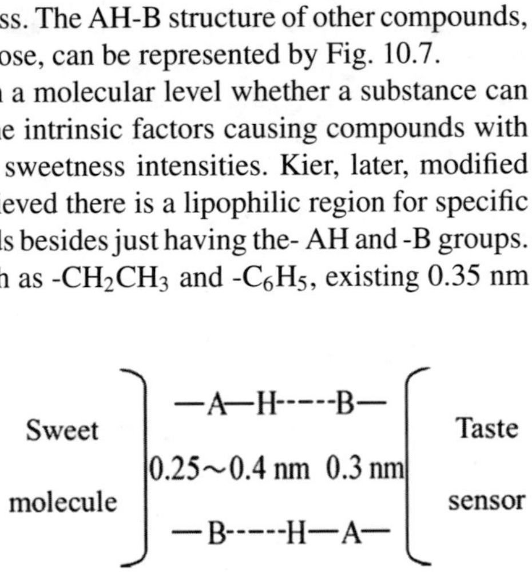

Fig. 10.6 AH/B theoretical diagram

$$\left.\begin{array}{c} \text{Sweet} \\[1em] \text{molecule} \end{array}\right\} \begin{array}{c} -A-H\text{-----}B- \\ 0.25\sim0.4\text{ nm}\quad 0.3\text{ nm} \\ -B\text{-----}H-A- \end{array}\left\{\begin{array}{c} \text{Taste} \\[1em] \text{sensor} \end{array}\right.$$

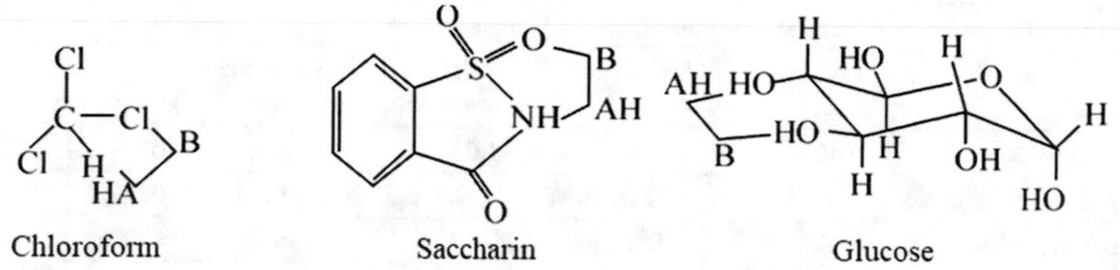

Fig. 10.7 Relationship diagram of several compounds's AH/B

Fig. 10.8 The relationship between AH/B and X in the sweetness unit of β-D-fructopyranose

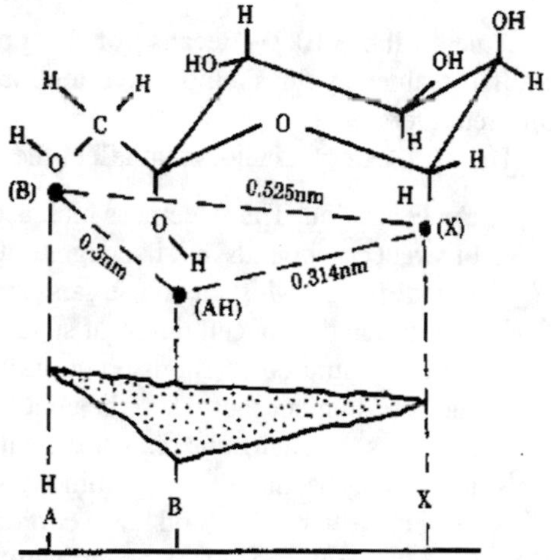

from -AH group and 0.55 nm from -B group; this group can form hydrophobic interactions with sweet taste receptor's lipophilic parts to produce a third contact point forming a triangular contact surface (Fig. 10.8). The -X group seems to promote interactions between certain molecules with sweet taste receptors, thus affecting perceived sweetness intensity. Therefore, the site of X is an extremely important property of a strongly sweet compound, it may be an important explanation for the difference in the quality of sweetness among sweet compounds. After this addition, the theory now becomes the AH-B-X theory.

10.3.1.2 Sweetness Intensity and Its Influencing Factors

The intensity of sweetness can be expressed by "sweetness", but the sweetness cannot be measured quantitatively by physical or chemical methods at present, and can only be judged by human taste. It is usually based on the non-reducing natural sucrose that is more stable in water (for example, the sweetness of 15% or 10% sucrose aqueous solution at 20 °C is 1.0 or 100) when compared to other sweeteners at the same temperature and concentration. This kind of sweetness is called relative

Table 10.2 Relative sweetness of some sugars and sugar alcohols

Sweetener	Relative sweetness	Sweetener	Relative sweetness	Sweetener	Relative sweetness
α-D-glucose	0.40 ~ 0.79	sucrose	1.0	Xylitol	0.9 ~ 1.4
β-D-Frutofuranose	1.0 ~ 1.75	β-D-maltose	0.46 ~ 0.52	Sorbitol	0.5 ~ 0.7
α-D-galactose	0.27	β-D-lactose	0.48	Mannitol	0.68
α-D-Mannose	0.59	Raffinose	0.23	Maltitol	0.75 ~ 0.95
α-D-Xylose	0.40 ~ 0.70	Invert syrup	0.8 ~ 1.3	Galactitol	0.58

sweetness (Table 10.2). Because of the great influence of subjective factors, the results obtained by this method are often inconsistent and sometimes vary greatly in different literature.

The main external factors that affect the sweetness of sweet compounds:

(1) Concentration. The sweetness increases with the increase in the concentration of sweet compounds, but the degree of sweetness of various sweet compounds was different. Most of the sugars and their sweetness were higher with the increase in the concentration of sugar, especially glucose. For example, when the concentration of sucrose and glucose is less than 40%, the sweetness of sucrose is higher; but when the concentration of both is greater than 40%, the sweetness of sucrose is almost the same. However, the bitterness of synthetic sweeteners becomes very prominent when the concentration is too high, so the use of sweeteners in food has a certain range of dosages.

(2) Temperature. The effect of temperature on the sweetness of sweeteners is shown in two aspects. One is the effect on taste organs, the other is the effect on the structure of compounds. Generally, the sensitivity of sensory organs is the highest at 30 °C, so the evaluation of taste is more appropriate at 10 ~ 40 °C. The taste perception becomes dull under high and low temperature, which cannot truly reflect the actual situation. For example, ice cream has a high sugar content, but because we eat it at a low temperature, it doesn't feel very sweet. In the lower temperature range, the temperature has little effect on sucrose and glucose, but the sweetness of fructose is significantly affected by temperature. This is because in the equilibrium system of fructose, with the increase of temperature, the percentage of high sweetness β - d-fructopyranose decreases, while the content of unsweetened β - d-fructofurane increases (Fig. 10.9).

(3) Dissolution. Sweet compounds, like other flavor compounds, can interact with receptors on taste cells only when they are dissolved, thus producing corresponding signals and being recognized. Therefore, the solubility of sweet compounds will affect the production speed and maintenance time of sweetness. Sucrose produced sweetness quickly but maintained for a short time, while saccharin produced sweetness slowly but maintained for a longer time.

(4) The interaction of sweet substances also affects the sweetness.

Fig. 10.9 The relationship between sweetness and temperature of 4 kinds of sugar

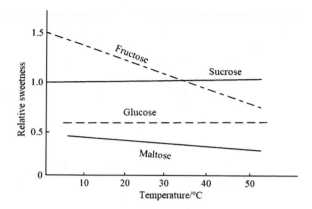

10.3.1.3 Common Sweeteners and Their Applications

There are numerous kinds of sweeteners. Common sweeteners include monosaccharides and disaccharides (such as glucose, fructose, xylose, sucrose, maltose, and lactose), sugar alcohols (including xylitol, sorbitol, maltitol, lactitol, D-mannitol, isomalt, erythrose), corn syrup, stevioside, neotame, sodium cyclamate, aspartame, etc. Sweeteners can be divided into two categories according to their sources: one is natural sweeteners, such as sucrose, starch syrup, fructose, glucose, maltose, glycyrrhizin, and stevioside; the other is synthetic sweeteners, such as sugar Alcohol, saccharin, cyclamate, palatinose, etc. Synthetic sweeteners have low calorific value and no fermentability, which are beneficial to diabetics and cardiovascular patients. According to their physiological and metabolic characteristics, sweeteners can also be divided into nutritional sweeteners and non-nutritional sweeteners.

1.   Monosaccharide and disaccharide in monosaccharide,

Among monosaccharides, glucose has a cool feeling, and its sweetness is 65 ~ 75% of that of sucrose. It is suitable for direct consumption and intravenous injection. Fructose (fructose) exists in fruits and honey together with glucose. It is sweeter than other sugars, and can be directly metabolized in the human body without pressure, so it is suitable for children and diabetic patients. Xylose is produced by the hydrolysis of xylan. It is soluble in water and has a sweet taste like fructose. Its sweetness is about 65% of that of sucrose. It has high solubility and permeability but low hygroscopicity. It is easy to cause a browning reaction and cannot be fermented by microorganisms. It is a sweetener that does not produce heat in the human body and can be consumed by patients with diabetes and high blood pressure.

Among disaccharides, sucrose has pure sweetness and high sweetness. It is rich in cane sugar and beet sugar. In industry, sucrose is often used as raw material to produce sucrose, which is the most used natural sweetener. Maltose has the highest nutritional value among the sugars. It is sweet, refreshing, and mild. Unlike sucrose, maltose can stimulate the gastric mucosa. Its sweetness is about 1/3 of that of sucrose. Lactose is a special sugar in milk. Its sweetness is 1/5 of that of sucrose. It is a kind of sugar with low sweetness and poor water solubility. After eating, lactose is decomposed into galactose and glucose in the small intestine and absorbed by the human body, which is conducive to the absorption of calcium. It has strong adsorption to gas and colored substances and can be used as a protective agent for meat flavor and color; It is easy to react with protein by Maillard reaction and form attractive golden yellow when added to baked food.

2.   Starch syrup

Starch syrup (starch syrup), also known as conversion syrup, is made up of glucose, maltose, oligosaccharide, and dextrin. Glucose value (D.E.) is commonly used to express the degree of starch conversion in the industry. D.E. refers to the percentage of dry matter of invert sugar (in terms of glucose) contained in starch conversion solution. D. If E. is less than 20%, it is called low conversion syrup; if D.E. = 38 ~ 42%, it is called medium conversion syrup; when D.E. is more than 60%, it is called high conversion syrup. Medium conversion syrup, also known as ordinary syrup or standard syrup, is the main product of starch syrup. D. Syrups with different E. values are different in sweetness, viscosity, thickening, hygroscopicity, permeability, and storability, and can be selected according to the use. Isomeric syrup is a part of glucose isomerization to fructose under the action of isomerase, also known as fructose syrup. At present, the conversion rate of fructose of isomeric syrup is generally over 42%, even more than 90% (called high fructose syrup). The isomeric syrup has pure sweetness, good crystallinity, fermentability, permeability, moisture retention, and storability, and has developed rapidly in recent years.

3.   Glycyrrhizin

Glycyrrhizin (glycyrrhizin) is formed by the condensation of glycyrrhizic acid and two molecules of glucuronic acid, with a relative sweetness of 100 to 300, and its disodium or trisodium salt is commonly used. It has a good fragrance-enhancing effect, can alleviate the salty taste of table salt, is not fermented by microorganisms, and has detoxification and liver protection effects. However, it is rarely used alone because of its slow sweetness and long retention time. When it is used together with sucrose, it is helpful to develop sweetness and save about 20% sucrose. When it is combined with saccharin, the ratio of glycyrrhizin/saccharin (3 ~ 4): 1, plus sucrose and sodium citrate, the sweetness is better. It can be used for flavoring dairy products, cocoa products, egg products, beverages, soy sauce, pickles, etc.

4.   Stevioside

Stevioside exists in the stem and leaf of stevioside, which is the dried powder of water extract of Stevia rebaudiana leaves. The sugar base is sophorose and glucose, and the aglucone is diterpene steviol. The specific sweetness is 200–300, which is one of the sweetest natural sweeteners. The sweetness of stevioside is close to that of sucrose. It is stable to heat, acid, and alkali, has good solubility, no bitterness and foaming, and has curative effects in lowering blood pressure, promoting metabolism, and treating hyperacidity. It is suitable for sweeteners and low energy food for diabetic patients.

The structure of stevioside

5. Sugar alcohol

At present, there are four kinds of sugar alcohol sweeteners (alditols) put into practical use, mainly including D-Xylitol, D-sorbitol, D-Mannitol, and maltitol. Their absorption and metabolism in the human body are not affected by insulin, nor do they hinder the synthesis of glycogen. They are a kind of sweeteners that do not increase blood sugar. They are ideal sweetener for patients with diabetes, heart disease, and liver disease. They all have moisture retention and can maintain certain moisture content in food and prevent drying. In addition, sorbitol can prevent the crystallization of sucrose and salt from the food, keep the balance of sweet, sour, and bitter taste, maintain food flavor and prevent starch aging. Xylitol and mannitol have cool taste and aroma, and can also improve food flavor; they are not easy to be used and fermented by microorganisms, so they are good anticaries sweeteners.

Sugar alcohol sweeteners also have a common feature, that is, excessive intake can cause diarrhea, so it has the effect of defecation with moderate intake.

6. Saccharin

Saccharin is currently the most used synthetic sweetener. Its molecules have a bitter taste, but the anions dissociated in water have a sweet taste, with a relative sweetness of 300 to 500, and a slightly bitter aftertaste. When the concentration is greater than 0.5%, the bitter taste of the molecule is easy to appear. After people consume saccharin, it will be excreted in feces and urine as it is, so it has no nutritional value.

7. Sodium cyclamate

Sodium cyclamate is a kind of non-nutritive sweetener. Its chemical name is Sodium cyclamate. It is a safe food additive with less toxicity. Its sweetness is 30 ~ 50 times of that of sucrose, slightly bitter. It is soluble in water, stable to heat, light, and air, and slightly bitter after heating. It is widely used in the production of beverage, ice cream, candied fruit, candy, and medicine.

$$\text{⬡—NHSO}_3\text{Na}$$

8.   Aspartame

Aspartame (AMP) is also known as protein sugar and aspartame. The chemical name of the active ingredient is aspartyl phenylalanine methyl ester. Its sweetness is 100 to 200 times that of sucrose, and its sweetness is refreshing. Pure, soluble in water, white crystals. But the stability is not high, it is easy to decompose and lose its sweetness. Aspartame is safe and has a certain degree of nutrition. It is widely used in the beverage industry. It can be added according to normal production needs in China.

$$\text{HO}-\overset{\overset{\displaystyle O}{\|}}{C}-CH_2-\underset{\underset{\displaystyle NH_2}{|}}{CH}-\overset{\overset{\displaystyle O}{\|}}{C}-NH-\underset{\underset{\displaystyle CH_2}{|}}{CH}-\overset{\overset{\displaystyle O}{\|}}{C}-O-CH_3$$

L-ASP   L-PHE   MET—OH

9.   Parachine

Palaginose, also known as isomaltulose, is a white crystal with a sweet taste and no peculiar smell. Its biggest characteristic is anticaries, slow absorption by the human body, and slow rise of blood glucose, which is beneficial for the prevention and treatment of diabetes patients and the prevention of excessive accumulation of fat. As anti-caries and functional sweetener, palaginose is widely used in chewing gum, high-grade candy, sports drinks, and other food.

10.   Others

Honey is the nectar collected from honeybees' nectaries. It is a light yellow to red yellow strong viscous transparent paste and crystallizes at low temperature. The total sugar content is about 80%, in which glucose is 36.2%, fructose is 37.1%, sucrose is 2.6%, dextrin is about 3.0%. Honey has its own special flavor because of the different kinds of flowers which contains more fructose. Due to its feature that hard to crystallize and easy to absorb moisture in the air, it can prevent food from drying and is mostly used in the processing of cakes and pills.

In addition to the above sweeteners, there are also some natural derivatives sweeteners, such as some amino acid and dipeptide derivatives, dihydrochalcone derivatives, Perillaldehyde derivatives, sucralose, etc.

10.3.2   *Sour Taste and Sour Taste Substances*

Sour taste is a chemical gustation caused by taste receptors stimulated by hydrogen ions. Any compounds that can release H + in solution will have a sour taste. Vinegar is regarded as one of the representatives and reference materials to distinguish the taste of food. Since human beings have adapted to acidic foods, proper sourness can give people a refreshing feeling and promote appetite. Sour taste intensity can be evaluated by some evaluation methods, such as tasting method or measuring salivary flow rate. The subjective equivalent value (P.S.E) is often used in the tasting method, which refers to the concentration of acid when the same sour taste is felt; the flow rate of saliva secretion is expressed by measuring the number of milliliters of saliva flowing out of each parotid gland within 10 min.

Different acids have different tastes, and there is not a simple relationship between acid concentration and sourness. The sour taste is related to the characteristics of acidic groups, pH, titer acidity, buffering effect, and other compounds, especially the presence or absence of sugar. The main factors affecting sourness:

(1)   Hydrogen ion concentration. All the acid taste agents can dissociate hydrogen ion, which shows that the acid taste is related to the concentration of hydrogen ion. When the concentration of hydrogen ion in the solution is too low (pH > 5.0 ~ 6.5), it is difficult to feel a sour taste; when the concentration of hydrogen ion in the solution is too high (pH < 3.0), the intensity of sour taste is too strong to be tolerated; however, there is no functional relationship between the concentration of hydrogen ions and the acid taste.

(2)   Total acidity and buffering effect. Usually, when the pH value is the same, the sour agent with greater total acidity and buffering effect has stronger sourness. For example, succinic acid is more acidic than malonic acid because the total acidity of succinic acid is stronger than that of malonic acid at the same pH value.

(3)   The properties of anions in acidizing agents. The anions of acidizing agents have a great influence on the strength and quality of acid taste. When the pH value is the same, the acidity of organic acids is stronger than that of inorganic acids. Adding hydrophobic unsaturated bonds in the structure of anions, the acidity is stronger than that of carboxylic acids with the same carbon number; if hydrophilic hydroxyl groups are added to the structure of anions, the acidity is weaker than the corresponding carboxylic acids.

(4)   Other factors. When sugar, salt, and ethanol are added to the acidizing agent solution, the acidity will be reduced. The proper mixing of sour and sweet taste is an important factor to form the flavor of fruits and drinks; the appropriate salty and sour taste is the flavor characteristic of vinegar; if a proper amount of bitter substances is added into the acid, the special flavor of food can also be formed.

Different acids have different gustatory perceptions, and there is a complex relationship between the concentration of acid and the taste of sourness. The taste of sourness is related to the concentration of hydronium ion, total acidity, action of buffer, nature of anions, and other factors. When a solution's hydronium ion concentration is too low (pH > 5.0–6.5), sourness cannot be detected; when hydronium ion concentration in solution is too high (pH < 3.0), sourness becomes unbearable. Usually, when pH values are the same, the acidifier with higher total acidity and higher buffering capacity will taste sourer. When pH values are the same, organic acids will have stronger sourness compare to inorganic acids. If hydrophobic unsaturated double bonds are added to an anion, then sourness taste will be stronger than carboxylic acid with the same carbon number. If hydrophilic hydroxyl groups are added to an anion, then sour taste will be weaker compare to the corresponding carboxylic acid. Furthermore, when sugar, salt, or ethanol is added to acidic solutions, the taste of sourness will be reduced.

10.3.2.1   Mechanism of Acid Formation

It is currently believed that H^+, in an acidifying agent HA, is the taste determining group and A^- is the taste assisting group. Sour taste receptors are the phospholipids on the taste buds. Sour sensation occurs when the cation interacts and exchanges with the phospholipid heads in the receptor. When pH is the same, organic acids will have stronger sourness compared to inorganic acids, It is due to the strong adsorption of the taste assisting group A- of organic acids on the surface of phospholipid receptors, which can reduce the positive charge density on the membrane surface, that is to say, the repulsion to H + is reduced. For diprotic acids, the longer the carbon chain, the longer the sourness sensation will last. This long-lasting sourness sensation is caused by anion A^- forming intramolecular hydrogen-bonded cyclic chelate or metal chelate that can adhere to the lipid membrane, thus reducing positive charges on the membrane surface. If a carboxyl or hydroxyl group is added to A^-, it will weaken the lipophilicity of A^- and reduce sourness. On the contrary, if a hydrophobic group is added to A^-, A^- can adhere better onto lipid membranes. Anions have an influence on sourness sensation as well. Organic acid anions usually provide a refreshing acidity, Of course, there are certain exceptions.

The order of acid intensity obtained by taste method and saliva flow rate method is not consistent, so some people think that the two reactions come from different parts of stimulation. It has also been proved that most of the protons bound to the acid receptor membrane are ineffective and can not cause local conformation changes on the membrane. Since the unsaturated hydrocarbon chain in the membrane structure is easy to combine with water, the proton in the acid also has a tunneling effect. So, some people also believe sour taste reception might not be located on the phospholipid head, but on the double bonds of the phospholipid chain. This is because conformational changes in sections of lipid membrane require strong electrostatic repulsion created by π complex formed after protonation of the phospholipid double bonds.

Although the previous mechanism explained some sour taste sensations, it is not enough to justify whether H^+, A^-, or HA has the most influence on sourness perception. There are many properties of acidulating molecules, such as molecular weight, molecular structure, and polarity, and their influence on sourness perception is still unknown. Sour taste perception mechanism has yet to be clarified.

10.3.2.2 Important Acidifiers and Their Application

Common acidifiers include vinegar, citric acid, malic acid, tartaric acid, lactic acid, ascorbic acid, gluconic acid, and many more. Vinegar is the most common acidifiers used in China; its main component is acetic acid. Citric acid and malic acid are mainly present in fruits and vegetables, while tartaric acid content is higher in fruits, such as grapes. Lactic acid in fermented kimchi and sauerkraut is not only for flavoring, but also to prevent the growth of undesired microorganisms. Ascorbic acid can be used as an acidifier in foods and can also prevent oxidation and browning. Gluconic acid is a sugar acid formed by the oxidation of aldehyde group in glucose; it is easily dehydrated in dry environments forming gluconolactone, and this reaction is reversible. Gluconolactone could be used as a coagulator in making lactone tofu and as a swelling agent for baking cookies.

1. Vinegar. Vinegar is the most commonly used sour flavor material in China. In addition to 3 ~ 5% acetic acid, it also contains a small amount of other organic acids, amino acids, sugars, alcohols, esters, etc. Its sour taste is mild, and in addition to being used as a seasoning agent in cooking, it also has the functions of preventing corruption and removing fishy smell. Acetic acid is highly volatile and has a strong sour taste. The industrially produced acetic acid is a colorless irritating liquid, which can be mixed with water as well, and can be used to prepare synthetic vinegar, but it lacks the flavor of vinegar. Acetic acid with a concentration above 98% can freeze into an ice-like solid, so it is called glacial acetic acid.

2. Citric acid. Citric acid is one of the most widely distributed organic acids in fruits and vegetables. It can be completely dissolved in water and ethanol at 20 °C, and is more soluble in cold water than in hot water. Citric acid can form three forms of acid salts, but most of them are insoluble or hardly soluble in water except alkali metal salts. The sour taste of citric acid is round, nourishing, refreshing, and delicious. It reaches the highest acidity immediately after entering the mouth, and the aftertaste lasts for a short time. It is widely used in the preparation of cool drinks, canned fruits, candies, and so on, with a normal dosage of 0.1 ~ 1.0%. It can also be used to prepare fruit juice powder as a synergist of antioxidants. Citric acid has good anti-corrosion performance, anti-oxidation, and synergistic effect with high safety.

3. Malic acid. Malic acid usually coexists with citric acid. It is a colorless or white crystal, and easily soluble in water and ethanol, and can be soluble in 55.5% at 20 °C. Its sour taste is 1.2 times stronger than citric acid, refreshing, slightly irritating, slightly bitter, and astringent, and has a long taste time. When combined with citric acid, it has the effect of strengthening sour taste. Malic acid

is highly safe and is often used in beverages, especially jelly. Sodium malate has a salty taste and can be used as a salting agent for kidney patients.

4. Tartaric acid. Tartaric acid is widely present in many fruits, and it dissolves 120% in water at 20 °C. Tartaric acid has a stronger sour taste, about 1.3 times that of citric acid, but it has a slightly astringent feeling. Tartaric acid is highly safe, and its use is the same as citric acid, and it is mostly used in combination with other acids, but it is not suitable for preparing foaming beverages or as a food expander.

5. Lactic acid. Lactic acid is rarely found in fruits and vegetables. It is mostly artificial synthetic products. It is soluble in water and ethanol and has an antiseptic effect. The acid taste is slightly stronger than citric acid. It can be used as a pH regulator. Used in refreshing drinks, synthetic wine, synthetic vinegar, spicy soy sauce, etc. Use it to make kimchi or sauerkraut, not only for seasoning, but also to prevent the reproduction of bacteria.

6. Ascorbic acid. Ascorbic acid (ascorbic acid) is white crystals, easily soluble in water, has a refreshing sour taste, but is easily oxidized. It can be used as a sour agent and vitamin C additive in food, and it also has the effect of preventing oxidation and browning and can be used as an auxiliary sour agent.

7. Gluconic acid. Gluconic acid is a colorless or light yellow liquid, easily soluble in water, slightly soluble in ethanol, because it is not easy to crystallize, its products are mostly 50% liquid. It is easy to be dehydrated to produce γ- or δ-gluconolactone when it is dried, and this reaction is reversible. Using this characteristic, it can be used in some foods that cannot be acidic at first but need acidity after being heated in water. For example, adding gluconolactone to soy milk will generate gluconic acid when heated to coagulate soy protein to obtain lactone tofu. In addition, gluconolactone is added to biscuits, which becomes a bulking agent during baking. Gluconic acid can also be used directly in the preparation of refreshing drinks, vinegar, etc., can be used as a preservative flavoring agent for instant noodles, or as a substitute for lactic acid in nutritious foods.

10.3.3   Bitterness and Bitter Taste Substances

Bitter taste has a very low threshold concentration but is a ubiquitous taste in food. Bitter taste lasts longer than sweetness, saltiness, and sourness. Although pure bitterness is unpleasant, when combined in proper proportions with sweetness, sourness, and other tastes, it can create a special flavor. For example, Momordica charantia, ginkgo, tea, coffee, and so on all have a certain bitter taste, but are regarded as delicious food. Most bitter substances have pharmacological effects and can regulate physiological functions. For example, some people with digestive disorders and taste weakening or declining often need strong stimulation of receptors to return to normal. Because of the minimum threshold of bitterness, it is easy to achieve this purpose.

10.3.3.1 The Mechanism of Bitterness

The mechanisms of bitter taste perception can mainly be explained by steric effect theory, intramolecular hydrogen bond theory, and the three-point contact theory.

1. Steric effect theory. Shallenberger et al. believe that, bitter taste and sweet taste, are both depended on the stereochemistry of molecules; these two taste perceptions can be stimulated by similar molecules. Some molecules can produce both sweetness and bitterness.

2. The intramolecular hydrogen bond theory. Kubota et al. in the investigation of enmein molecular, believe any hydrogen bond separated by 0.15 nm will have a bitter taste. Intramolecular hydrogen bonds can increase the hydrophobicity of molecules and form a chelate with transition metal ions easily which have similar structures with bitter-tasting molecules.

3. The three-point contact theory. Lehmann et al. find that there is a linear relationship between the sweetness intensity of several D-amino acids and the bitterness intensity of their L-isomers. So, it is believed that, like the sweet taste, bitter taste molecules and bitter taste receptors produce bitterness by three-point contact; however, the third contact point of bitter taste substances is in the opposite direction of that for sweet taste substances.

Although the above-mentioned bitterness theories can explain the production of bitterness to a certain extent, most of them break away from the structure of the taste cell membrane and only focus on the molecular structure of the stimulus, and do not consider the existence of some bitter inorganic salts.

4. Theory of Induced Adaptation Guangzhi Zeng proposed the bitter taste molecular recognition theory based on his taste cell membrane induction adaptation model. The main points are as follows:

    (1) Bitter taste receptors are "water holes" formed by polyene phospholipids on the membrane surface, which provide a nest for the coupling between bitter substances and proteins. At the same time, inositol phospholipids (PI) can generate PI-4-PO_4 and PI-4, 5-$(PO_4)_2$ through phosphorylation, and then combine with Cu^{2+}, Zn^{2+}, Ni^{2+}, etc., to form the "lid" of the acupoint. The bitter molecules must first push the lid off before they can enter the hole and interact with the receptor. In this way, the inorganic ions bound to the lid in the form of salt bonds, which become the monitoring indicator of molecular recognition. Once it is replaced by some transition metal ions, the lid on the taste receptor no longer receives the stimulation of the bitter substance, resulting in an inhibitory effect.

    (2) The receptor acupoints composed of coiled polyene phospholipids that can form various multipolar structures and interact with different bitter substances. The results showed that the taste of quinine sulfate did not affect the bitter taste of urea or magnesium sulfate, and vice versa. If quinine and urea are tasted together, the synergistic effect will be produced and the bitterness will be enhanced. It is proved that quinine and urea

have different action sites or water holes on taste receptors. However, if you drink coffee after tasting quinine, the bitterness of coffee will be weakened, which indicates that the two have the same action site or water hole on the receptor, and they will produce competitive inhibition.

(3) The receptor acupoints composed of polyene phospholipids have a side that adheres to the surface protein and has a wider contact with the lipid block. Compared with the specificity requirements of sweet substances, the requirements of polar base position distribution and three-dimensional direction order of bitter substances are not very strict. Any stimulant that can enter any part of the bitter receptor will cause "hole closure", which changes the conformation of phospholipid and produces bitter information through the following ways.

① Salt bridge conversion. Cs^+, Rb^+, K^+, Ag^+, Hg^{2+}, R_3S^+, R_4N^+, $RNH-NH^{3+}$, $Sb(CH_3)_4$, etc., belong to structure destroying ions. They can destroy the ice crystal structure around the hydrocarbon chain, increase the water solubility of organic matter, and freely enter and exit the biofilm. When they open the salt bridge into the bitter receptor, they can induce conformational changes. Although Ca^{2+}, Mg^{2+}, etc., have the same structure as Li^+ and Na^+ to produce ions, they have a salting-out effect on organic matter, but Ca^{2+} and Mg^{2+} can make phospholipids agglomerate with some anions. Facilitate structural destruction of ions to enter the receptor, and produce a bitter taste.

② The destruction of hydrogen bonds. $(NH_2)_2C = X$, $RC(NH_2) = X$, $RC = NOH$, $RNHCN$, etc., where X or O, NH or S group, can be used as hydrogen bond donors.

And the above structures can be used as hydrogen bond acceptors. Since the bitter receptor is a curly polyene phospholipid hole, there is no obvious spatial selectivity, so that the above stimulus with a multipolar structure can also open the lid salt bridge to enter the receptor (larger bitter peptides can only have a part of the side chain to enter) Then destroys the hydrogen bonds and lipid-protein interactions, which gives a great impetus to the change of receptor conformation.

③ The formation of hydrophobic bonds. Hydrophobic bond stimulants are mainly esters, especially lactones, thiols, amides, nitriles and isonitriles, nitrogen heterocycles, alkaloids, antibiotics, terpenes, amines, etc. Hydrophobes without polar groups cannot enter the receptor, because the ligands of the salt bridge and the phospholipid head have chirality, which makes the receptor surface have a certain degree of selectivity for the hydrophobe. However, once these hydrophobes penetrate deep into the pore lipid layer, they do not have any spatial specificity requirements, and can cause the conformation of the receptor to change through the action of hydrophobic bonds.

The theory of induced adaptation has further developed the bitterness theory and made a great contribution to the explanation of the complex phenomenon of bitterness.

For examples:

(1)   It broadly summarizes various types of bitter substances, which provides convenience for further research on the relationship between structure and taste.
(2)   The view that there are transition metal ions on the receptor provides an explanation for the fact that thiols, penicillamine, acidic amino acids, oligopeptides, etc., can inhibit bitterness and certain metal ions can affect bitterness.
(3)   A possible explanation is made for the phenomenon that the sweet blind cannot feel any sweeteners, while the bitter blind cannot perceive the few bitter substances with conjugated structure. Bitterness blindness is inherited congenitally. When Cu^{2+}, $Zn^{2+,}$ and Ni^{2+} form a strong complex with the protein on the patient's receptor, when the receptor surface is used as a monitoring ion, some bitter substances are difficult to open the lid to enter the acupuncture point.
(4)   The view that the bitter taste receptor is mainly composed of phospholipid membrane also provides an explanation for the intensity of bitter taste. Because bitter substances have a cohesive effect on the lipid membrane and increase the surface tension of the lipid membrane, there is a corresponding relationship between the two; the greater the surface tension produced by the bitter substance, the greater the intensity of its bitterness.
(5)   Explain the phenomenon that the intensity of bitterness increases with the decrease of temperature, which is just the opposite of the effect of temperature on sweetness and spiciness. Because the process of bitter substances condensing the lipid film is an exothermic effect, which is opposite to the endothermic effect of sweet and spicy substances that make the film expand.
(6)   It also explains why the effect of anesthetics on various taste receptors is the fastest disappearance of bitterness and the slowest recovery. This is because polyene phospholipids have greater solubility for anesthetics, and the receptor loses the law of changing conformation after swelling, and can no longer trigger bitter information, and so on.

10.3.3.2   Common Bitter Substances and Their Applications

There are 4 major bitter taste substances derived from plants that are present in food and medicine, including alkaloids, anthraquinones, glycosides, and bitter peptides. The animal sources include picric acid, formanilide, formamide, phenylurea, and urea.

Almost all alkaloids are bitter, and bitterness increases as alkalinity increases. Quinine is the most commonly used bitter reference substance. There are as many as tens of thousands of terpenoids, generally containing lactones, internal acetals, internal hydrogen bonds, glycoside hydroxyl groups, and other structures that can form chelates and have a bitter taste. For example, the bitter component of hops is terpenoids. Most of the glycosides have a bitter taste, such as amygdalin and glucosinolate. Flavonoids and flavanones are widely found in citrus peels and Chinese herbal medicines, most of which are bitter molecules. When the amino acid side chain group has more than 3 carbon atoms and has a base, it is a bitter molecule. When the side chain group is not hydrophobic, its bitterness is not strong.

Many of the salts have a bitter taste, which may be related to the sum of their anion and cation radius. As the sum of ionic radii increases, the saltiness decreases and the bitterness increases. For example, NaCl and KCl have a pure salty taste, the sum of their radii is less than 0.658 nm, while KBr is salty and bitter, the sum of radii is 0.658 nm, and the bitterness of CsCl and KI is large, and the sum of radii is greater than 0.658 nm.

1. Caffeine and Theobromine. Both caffeine and theobromine are purine derivatives and the main alkaloid bitter substances in food. When the concentration of caffeine in water is 150–200 mg/kg, it has a moderate bitter taste, and it is present in coffee, tea, and kola nuts. Theobromine (3,7-dimethylxanthine) is like caffeine and has the highest content in cocoa, which is the cause of the bitter taste of cocoa.

2. Amygdalin. Amygdalin is a glycoside formed by cyanobenzyl alcohol and gentiobiose. It is found in the cores, seeds, and leaves of many Rosaceae plants such as peaches, plums, and apricots. Especially bitter almonds are the most. The seed kernel also contains enzymes that decompose it. Amygdalin itself is nontoxic and has antitussive effect. Too much raw almonds and peach kernels can cause poisoning, because the ingested amygdalin is decomposed into glucose, benzaldehyde, and hydrocyanic acid under the action of emulsions in the body at the same time.

3. Naringin and nehoesperidin. Naringin and nehoesperidin are the main bitter substances in citrus peels. The bitterness of naringin pure product is bitter than quinine, and the detection threshold can be as low as 0.002%. The type of glycoside in the flavonoid glycoside molecule has a decisive relationship with whether it has a bitter taste. After hydrolysis, the bitterness disappears. Using this principle, enzyme preparations can be used to remove the bitterness of orange juice (Fig. 10.10). Rutose and hesperetose are normally in rhamnose glucoside form, rutose is presented as rhamnose (1 → 6) glucose while hesperetose is presented as rhamnose (1 → 2) glucose.

4. Bile. Bile is a liquid secreted by the liver of animals and stored in the gallbladder, with a very bitter taste. The initially secreted bile is a clear and slightly viscous golden yellow liquid with a pH between 7.8 and 8.5. Due to dehydration and oxidation in the gallbladder, the color turns green and the pH drops to 5.50. The main components in bile are cholic acid, chenodecholic acid, and deoxycholic acid.

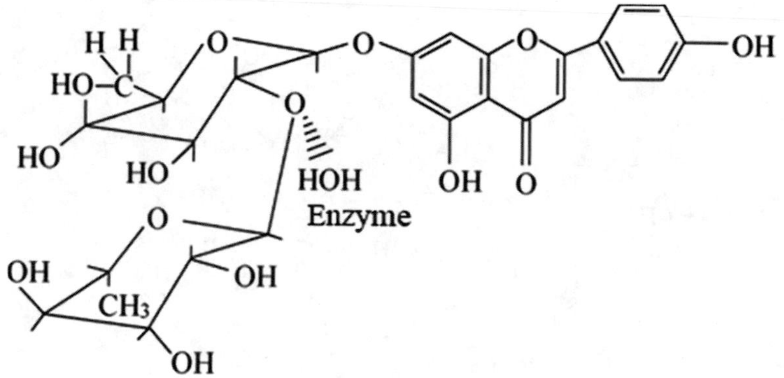

Fig. 10.10   The structure of enzymatic hydrolysis of naringin to produce non-bitter derivatives

Fig. 10.11   Quinine
hydrochloride

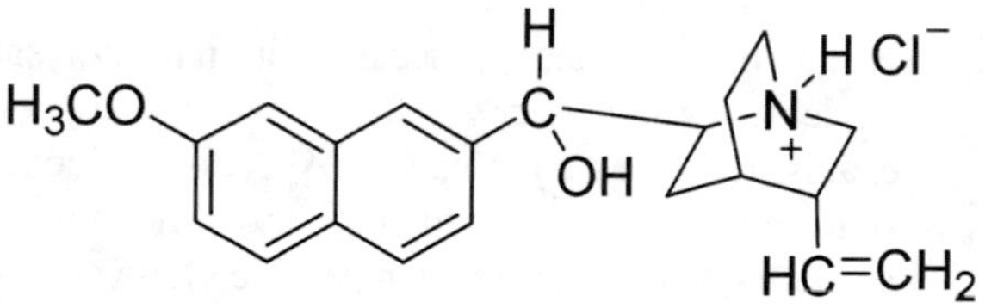

5.   Quinine. Quinine is a widely used standard substance for bitterness. The bitterness threshold of quinine hydrochloride is about 10 mg/kg. Bitter substances have lower taste thresholds than other taste substances and are less soluble in water than other taste active substances. In soft drinks with the characteristics of sweet and sour taste, the bitterness can be reconciled with other tastes, so that this type of drink has a cooling and exciting effect.

6.   Bitter hops. Hops are widely used in the beer industry to give the beer a characteristic flavor. The bitter substance of hops is a derivative of humulone or lupulone. Humulone is the most abundant in beer. When the wort is boiled, it is converted into isohumulone through an isomerization reaction. Isohumulone is the precursor of the skunk odor and sun-flavored compounds produced by beer under light irradiation. When there is hydrogen sulfide produced by yeast fermentation, the ortho-carbon atom of the ketone group on the isohexene chain occurs. The photocatalytic reaction produces a 3-methyl-2-butene-1-thiol (isoprenethiol) compound with a skunk smell. In the pre-isomerized hop extract, the selective reduction of ketones can prevent this reaction from happening, and the use of clean brown glass bottles to package beer will not produce skunk smell or sun smell. Whether the volatile hop aroma compounds remain in the malt boiling process is a question that has been debated for many years. It has now been fully proved that the compounds that affect the flavor of beer do remain in the process of full boiling of the wort. Together with other compounds formed by the bitter hop substances, the beer has a flavor.

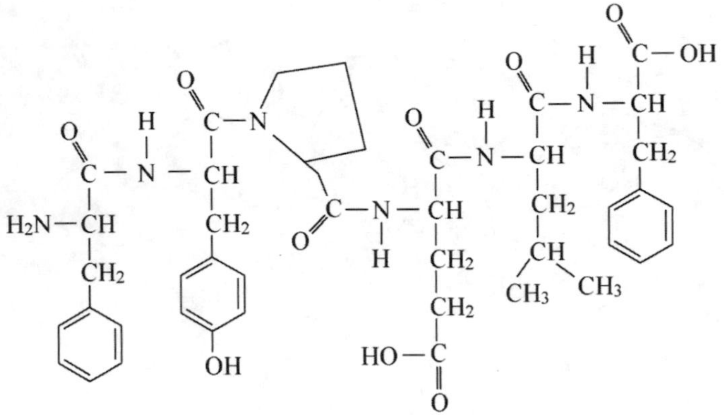

Fig. 10.12 Bitter peptides of strong polar αs1 casein derivatives

7. Protein hydrolysate and cheese. Protein hydrolysate and cheese have an obvious unpleasant bitter taste, which is caused by the total hydrophobicity of the side chains of peptide amino acids. All peptides contain a considerable number of AH-type polar groups, which can meet the requirements of the position of polar receptors, but the size of each peptide chain and the nature of their hydrophobic groups are very different. Therefore, the ability of these hydrophobic groups to interact with the main hydrophobic positions of the bitter taste sensor is also different. It has been proved that the bitterness of peptides can be predicted by calculating their hydrophobic value. Figures 10.12 characterize the bitter peptides of αs1 casein derivatives, and their hydrophobic amino acids trigger strong hydrophobicity.

8. Hydroxylated fatty acids. Hydroxylated fatty acids often have a bitter taste. The bitter taste of these substances can be expressed by the ratio of the number of carbon atoms in the molecule to the number of hydroxyl groups or the R value. The R value of sweet compound is 1.00 ~ 1.99, the value of bitter compound is 2.00 ~ 6.99, and there is no bitterness when it is greater than 7.00.

9. The bitter taste of salts. The bitter taste of salts is related to the sum of the ion diameters of the salt anions and cations. The salt whose ion diameter is less than 0.65 nm shows a pure salty taste (LiCl = 0.498 nm, NaCl = 0.556 nm, KCl = 0.628 nm). Therefore, KCl has a slightly bitter taste. As the sum of ion diameter increases (CsCl = 0.696 nm, CsI = 0.774 nm), the bitter taste of its salt gradually increases, so magnesium chloride is (0.850 nm) quite bitter salt.

Quinine is an alkaloid and is also recognized as a standard for bitter taste perception. The bitterness threshold of quinine hydrochloride is about 10 mg/kg (Fig. 10.10). Terpenoid compounds are abundant in plants, and the ones containing lactone, aldolactol, intramolecular hydrogen bond, or glycosidic hydroxyl group, and other groups capable of forming chelates are bitter. For example, bitterness produced by some isoprenoid derivatives in hops is an important feature of its flavor profile,these bitter compounds are mainly derivatives of humulone or lupulone. Most ligands in glycosides have a bitter taste, such as amygdalin and glucosinolate. Amygdalin is composed of cyanbenzyl alcohol and gentiobiose; it is non-toxic, but can be toxic when excessive intake because it can be broken down by amygdalase in the body into glucose, benzaldehyde, and hydrocyanic acid. Naringin and nehoesperidin are flavanone glycosides; they are the main bitter taste substances present in citrus fruit peels. Bitter taste disappears when nehoesperidin glycosides are hydrolyzed (Fig. 10.11). Some L-amino acids have bitter tastes, such as leucine, phenylalanine. When the side chain is not strongly hydrophobic, bitterness is not strong either. Protein hydrolysates and fermented hard cheese have a distinct bitter taste; this is related to the peptide side chain's total hydrophobicity and relative molecular mass.

10.3.4 Salty Taste and Salty Taste Substances

Salt taste is very important in food seasoning. Salty taste is the taste displayed by neutral salt. Only sodium chloride produces a pure salty taste. It is not easy to simulate this salty taste with other substances. Such as potassium bromide, ammonia iodide, etc., in addition to salty taste, but also has a bitter taste, which is not a simple salty taste, the taste is present in coarse salt. The taste characteristics of various salt solutions with a concentration of 0.1 mol/L are shown in Table 10.3.

10.3.4.1 Patterns of Saltiness

The saltiness is jointly determined by the dissociated anions and cations. Although the production of salty taste is related to the interdependence of cations and anions, the cations are easily adsorbed by the carboxyl or phosphate groups of the protein of taste receptors and present a salty taste. Therefore, salty taste is more closely related to the cations dissociated from salt, while anions affect the strength and side taste of salty taste, cations are the positioning groups of salts, and anions are the taste assisting groups. The strength of salty taste is related to the relative size of the taste nerve's induction of various anions. From the comparison of several salty substances, it is found that the salt with a small anion and cation radius has a salty taste, the salt with large radius has a bitter taste, and the salt in the middle has a salty bitter taste.

Taste	Types of salt
Salty	$NaCl$, KCl, NH_4Cl, $NaBr$, NaI, $NaNO_3$, KNO_3
Salty and bitter	KBr, NH_4I
Bitterness	$MgCl_2$, $MgSO_4$, KI, $CsBr$
Unpleasant and bitter	$CaCl_2$, $Ca(NO_3)_2$

Table 10.3 The taste characteristics of salt

The larger the atomic weight of the cation and anion of the salt, the more it tends to increase the bitterness.

10.3.4.2   Common Salty Substances

The salt used for food seasoning should be pure salt. Table salt is often mixed with other salts such as potassium chloride, magnesium chloride, and magnesium sulfate, and their content increases, which in addition to salty taste, but also brings bitterness; but if they are present in trace amounts, they are beneficial to the appearance when processed or eaten directly. Taste effect. Therefore, table salt needs to be refined to reduce the content of these bitter salts.

Although many neutral salts show salty taste, their taste is not as pure as sodium chloride, and most of them have bitter or other tastes.

Due to the adverse effects of excessive salt intake on the body, people are interested in salt substitutes. In recent years, there have been many varieties of table salt substitutes. Several organic sodium salts such as sodium gluconate and sodium malate also have the same salty taste as table salt. They can be used as salt free soy sauce and salt food for patients with kidney disease to limit the intake of salt.

In addition, the salt of amino acid also has a salty taste, such as adding 15% of 5'-nucleotide sodium with 86% $H_2NCOCH_2N^+H_3Cl^-$, its salty taste is no different from table salt, which may become a food salty agent in the future. Potassium chloride is also a relatively pure salty substance, which can partially replace NaCl in athlete drinks and low-sodium foods to provide salty taste and supplement potassium in the body. However, there is still a big difference between the taste of food using salt substitutes and the taste of foods using NaCl, which will limit the use of salt substitutes.

Neutral salts provide salty taste. Sodium chloride is the most common salty taste substance. Although saltiness is present in other neutral salts, their salty flavor is not as pure as sodium chloride; many exhibit bitter flavor and other flavors. Bitterness in salt is related to the sum of ion diameters in between anions and cations. When the sum of ion diameter is less than 0.65 nm, salt will have a pure salty taste. As the sum of ion diameter increases, bitterness in salt gradually intensifies as well.

Salty taste perception mechanism is mainly the interaction of hydrated anion–cation complex with AH/B receptor. The cations and anions in salt can both bind to the receptor, and the cation is more likely to bind to the taste perception receptor producing a salty taste. Anions have an influence on salty taste intensity and food flavors.

10.3.5   Umami Taste and Umami Taste Substances

Umami taste is a complex gustatory perception; umami mainly refers to the taste of specific chemicals such as glutamic acid, inosinic acid and aspartic acid, sodium

L-glutamate, commonly known as MSG. Umami was recognized as the fifth basic taste of food in recent years. When an umami taste substance is used in concentration higher than the threshold, food's umami flavors can be enhanced. When used in concentration below the threshold, it can only enhance flavors that are already present. In Europe and the United States, umami taste substances are referred to as flavor enhancers. Common umami taste substances can be distinguished by their chemical structures, mainly including amino acids, nucleotides, peptides, and organic acids.

10.3.5.1 Mechanism of Umami

The backbone structure of umami taste substances is $^-O\text{-}(C)_n\text{-}O^-$, n = 3-9. When n = 4-6, umami taste becomes noticeable; when n = 5, umami taste is the strongest. The aliphatic chain is not limited to a straight chain, but can also be a part of an alicyclic ring; the C can be substituted by O, N, S, P, etc. Maintaining the negative charge at both ends of the molecule is very important for the umami taste. If the carboxyl group is esterified, amidated, or heated to form lactones and lactams, the umami taste will be reduced. However, the negative charge at one end can also be replaced by a negative dipole, such as tricholic acid and ibotenic acid, whose umami taste is 5-30 times stronger than monosodium glutamate. This general formula can summarize all the peptides and nucleotides with umami taste. At present, for reasons of economic efficiency, side effects, and safety, the main commercial umami flavor agents are glutamic acid type and nucleotide type.

10.3.5.2 Common Umami Agents

Umami agents can be divided into amino acids, peptides, nucleotides, and organic acids if they are distinguished from their chemical structure characteristics.

1. **Amino acids and peptides**

Among the natural amino acids, the sodium salt of L-glutamic acid and L-aspartic acid and their amides have umami taste. Sodium L-glutamate is commonly known as monosodium glutamate, which has a strong meat flavor. Glutamic acid-type umami flavors (MSG) are aliphatic compounds, which have spatial specificity requirements in structure. If they exceed the specificity range, they will change or lose their taste. Their taste determining groups are the negatively charged functional groups at both ends, such as –C=O, –COOH, –SO_3H, and –SH. Taste assisting groups have some hydrophobic properties, such as α–L–HN_2 and –OH. The electrostatic attractions between NH_3^+ and COO^- are what give MSG its savory flavor. When pH reaches its isoelectric point of 3.2, umami taste will be at its minimum. When pH is at 6, the compound almost completely dissociates, and umami taste intensity will be at its maximum. When pH is above 7, umami taste disappears due to the formation of disodium salt. Table salt is an adjuvant of MSG, MSG also has the effect of alleviating saltiness, sourness, and bitterness, so that food has a natural flavor. The sodium salt and amide of L-aspartic acid also have umami taste and are the main umami taste substances in plant foods such as bamboo shoots.

The dipeptides and tripeptides with hydrophilic amino acids connected to the carboxyl end of glutamic acid also have umami taste, such as L-α-aminoadipate, disodium succinate, glutathione-glycerine, and glutathione Silk tripeptide, tricholic acid, etc. If it connects to a hydrophobic amino group, it will produce a bitter taste. In addition to the two amino acids mentioned above, L-alanine, glycine, theanine, and other amino acids all have unique savory flavors.

Umami peptides are umami taste producing micromolecular peptides; they can be extracted from food or synthesized by amino acids. The taste characteristics of umami peptides are related to their amino acid composition and generally contain one or two acidic groups from glutamic acid and aspartic acid. Umami peptides' flavor profile is not only related to the type of amino acid, but also its primary structure and spatial arrangement. Studies have found that umami peptides have good thermal stability. In addition to having umami taste, it can also mask or weaken bitterness and improve food flavor.

2. Nucleotides

Nucleotide based umami taste substances are aromatic heterocyclic compounds that are structurally spatially specific. Its food determining group is the hydrophilic ribose phosphate, and the flavor assisting group is the hydrophobic substituent on the aromatic heterocyclic ring. Common nucleotide substances exhibiting savory tastes are 5'-IMP, 5'-GMP, and 5'-xanthic acid; the first two compounds have the most intensive umami taste (Fig. 10.13), which, respectively, represent the umami taste of fish and mushrooms. Furthermore, 5'-deoxyinosinic acid and 5'-deoxyguanosic acid also have umami taste. Inosinic acid-type umami flavor (IMP) is an aromatic heterocyclic compound, and its structure also requires space specificity. Its positioning group is hydrophilic ribose phosphate, and the taste assisting group is a

Fig. 10.13 Nucleotide based umami substances

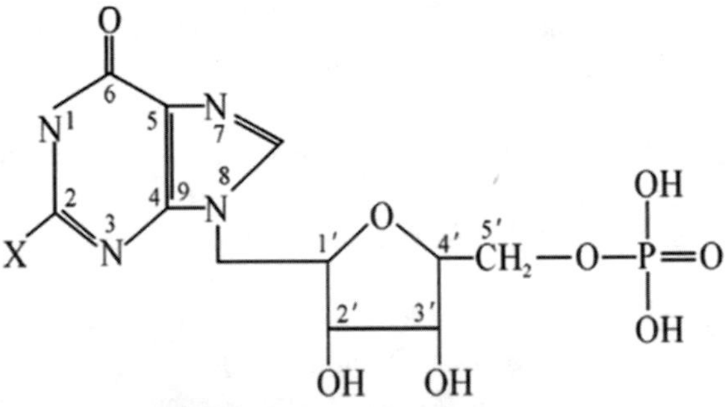

X=-H (5'-IMP, 5'-Inosinic acid)
X=-NH₂ (5'-GMP, 5'-Guanylic acid)
X=-OH (5'-AMP, 5'-Adenine nucleotides)

Table 10.4   Synergistic effect of MSG and IMP

MSG dosage/g	IMP dosage/g	Mixture dosage/g	Equivalent to MSG amount/g	Multiplying effect/time
99	1	100	290	2.9
98	2	100	350	3.5
97	3	100	430	4.3
96	4	100	520	5.2
95	5	100	600	6.0

hydrophobic substituent on the aromatic heterocyclic ring. When these 5'-nucleotides are combined with sodium glutamate, they can significantly increase the umami taste of sodium glutamate (Table 10.4). For example, the umami taste of a mixture of 1% IMP +1% GMP + 98% MSG is simply MSG 4 times. These 5'-nucloetides have synergistic effects with MSG; when used together, savory taste intensifies as concentration increases.

3.   **Organic acids**

The main organic acid having umami taste is succinic acid and its sodium salts; it is the main umami taste substance present in shellfish. There is also a small amount of food fermented by microorganisms such as soy sauce, sauce, rice wine, and so on. They can be used as seasonings, and used for the flavoring of alcoholic refreshing drinks and candy, and their sodium salts can be used for brewing and meat food processing. If used in combination with other umami agents, it will help the freshness effect.

Maltol and ethyl maltol are commercially used as flavor enhancers in fruits and sweets. Commercial maltol is a white or colorless crystalline powder, and its solubility in water at room temperature is 1.5%. When heated, its solubility in water and oil increases. The appearance and chemical properties of ethyl geritol are like maltol. Both two substances have the structure of o-hydroxy ketene, and a small number of isomers formed by o-diketone exist in equilibrium with it. Because the structure is like phenols, it has some chemical properties like phenols. For example, maltol can react with terpene salts to become purple-red, and can form salts with alkalis. The high concentration of maltol has a pleasant caramel aroma, while the dilute solution has a sweet taste. 50 mg/kg of maltol can make the juice have a round and soft taste. Both maltol and ethyl maltol can match the AH/B part of the sweetness receptor, but as a sweetness enhancer, ethyl maltol is much more effective than maltol. Maltol can reduce the detection threshold concentration of sucrose by half. The actual flavor enhancement mechanism of these compounds is still unclear.

In addition, the umami taste of the compound can change with the change of structure. For example, although sodium glutamate has umami taste, glutamic acid and disodium glutamate have no umami taste.

466 L. Ma and J. Li

10.3.6 Spicy Taste and Spicy Taste Substances

Spicy taste is the combination of a special burning sensation and a sharp tingling sensation caused by the consumption of certain compounds. Spicy taste substances can not only stimulate the tactile nerves of the tongue and the oral cavity, but also stimulate the nasal cavity and sometimes causing burning sensations on the skin. Appropriate level of spiciness can increase appetite and promote the secretion of digestive fluid.

10.3.6.1 Mechanism of Spicy Taste

Common spicy taste substances are generally amphiphilic molecules with both polar and non-polar groups. The polar head is the flavor determining group, while the non-polar tail is the flavor assisting group. Studies have shown that increasing the chain length of the non-polar tail will increase the molecule's overall perceived spiciness. When the chain length (C_n) in the non-polar tail is $n = 9$, then spicy taste will reach its maximum intensity and dramatically drops afterwards (Figs. 10.14 and 10.15), known as the C9 law. The spiciness of several substances such as capsaicin, piperine, xanthophylline, gingerin, cloves, allicin, and mustard oil conforms to the C9 hottest law.

For spicy taste substances containing double bonds, the more cis double bonds present, the spicier the substance will be; trans-double bonds have little effect on spiciness intensity. Double bond has the greatest influence when on C_9 position. When a spicy taste molecule's end chain contains no cis double bonds or branched chains, the spicy taste will be completely lost as the number of carbons in the molecule becomes greater than 12. Similarly, the hydrocarbon chain lengths of aliphatic aldehydes, alcohols, ketones, and carboxylic acids can have changes in spicy taste. If the chain length exceeds C_{12} but there is a cis double bond near ω-position, then there can still be spicy taste present. Some of the less polar molecules, such as BrCH =

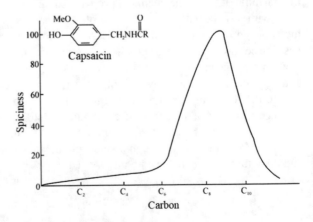

Fig. 10.14 Relationship between capsaicin and tail chain Cn

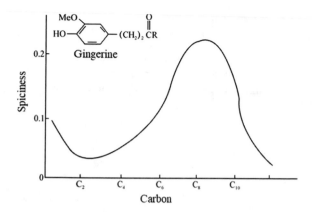

Fig. 10.15 Relationship between gingerine and tail Cn

CHCH₂Br, and CH₂ = CHCH₂X(X being NCS, OCOR, NO₂, and ONO),and (CH₂ = CHCH₂)₂Sn(*n* = 1,2,3), Ph(CH₂)ₙNCS, etc., can also have spicy taste.

The polarity and position of the polar groups of the spicy substances have a great relationship with the taste. When the polarity of the polar head is large, it is a surfactant; when the polarity is small, it is an anesthetic. Symmetrical molecules with a central polarity such as

$$RCON \begin{array}{c} \diagup \\ \diagdown \end{array} NCOR \quad \text{or} \quad RCOO - \langle \ \rangle - NHCOR$$

Their spiciness is only equivalent to half a molecule, and its spiciness is greatly reduced because of its reduced water solubility. Symmetrical molecules with polar groups at both ends such as:

In this case, their taste becomes weaker. Increase or decrease the hydrophilicity of the polar head, such as the transform from to , where the spiciness is reduced. Even the position change of the hydroxyl group may lose the spiciness and produce sweetness or bitterness.

10.3.6.2 Common Spicy Substances

Common spices and vegetables containing spicy taste substances include chili pepper, black pepper, ginger, nutmeg, clove, garlic, onion, chives, wasabi, and radish.

1. Hot and spicy substance Hot and spicy substance is a non-aromatic spiciness that can cause a burning sensation in the mouth. There are:

(1)   Capsicum. The main spicy taste substance in capsicum is capsaicin, a class of unsaturated monocarboxylic vanillyl-amides with different carbon chain lengths (C_8-C_{11}); a small amount of dihydrocapsaicin, a linear saturated carboxylic acid, is

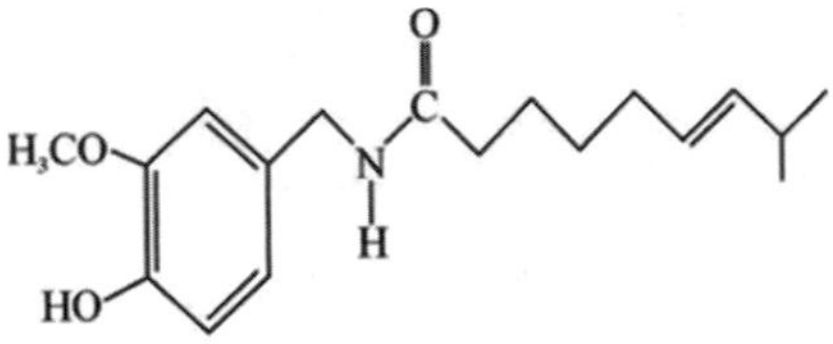

Capsaicin

also present. Capsaicinoids have different spiciness intensities, with the spiciest having a side chain of C_9-C_{10}. The double bond is not necessary for spiciness. The capsaicin content of different peppers varies greatly. Sweet peppers are usually very low in content. Generally, red peppers contain 0.06%, horn red peppers contain 0.2%, Indian sam peppers contain 0.3%, and Uganda peppers can be as high as 0.85%.

(2)   Pepper. There are two common types of black pepper and white pepper, both of which are processed from fruits. Among them, black pepper can be made from unripe green fruits; white pepper can be made from mature fruits harvested when the color changes from green to yellow but not red. In addition to a small amount of capsaicinoids, the spicy components of pepper are mainly piperine. Piperine is an amide compound, and its unsaturated hydrocarbon group has *cis-trans* isomers. Among them, the more *cis* double bonds it has, the hotter it is; the all-*trans* structure is also called isopiperine. The spicy taste of pepper will decrease after exposure to light or storage, which is caused by the isomerization of *cis*-piperine to *trans* structure. Synthetic piperine has been used in food (Fig. 10.16).

Fig. 10.16 The main spicy compounds in pepper and their intensity

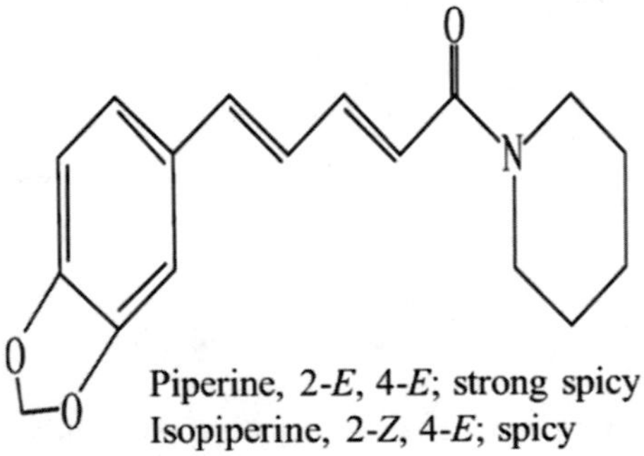

Piperine, 2-*E*, 4-*E*; strong spicy
Isopiperine, 2-*Z*, 4-*E*; spicy

(3)   Xanthoxylum. The main spicy ingredient of Xanthoxylum is sanshool, which is an amide compound. The amides found in Xanthoxylum are shown in Table 10.5. In addition, there is a small amount of alkyl propyl isothiocyanate and so on. Like pepper and chili, it also contains some volatile fragrance components in addition to spicy components.

2. Aromatic spicy substances Aromatic spicy substances are a class of substances that are accompanied by strong volatile aromatic substances in addition to the spicy taste, and are ingredients that have the dual effects of taste and smell.

(1)   Ginger. The pungent ingredient of fresh ginger is a type of o-methoxyphenol alkyl ketone, the most representative of which is 6-gingerol. The length of the carbon chain outside the hydroxyl group on the side chain of the ring in the molecule is different ($C_5 \sim C_9$). Fresh ginger is dried and stored, gingerol will be dehydrated to produce gingerol compounds, which are more pungent than gingerol. When ginger is heated, the side chain on the gingerol ring breaks to produce gingerone, and the pungent taste is milder. Among gingerol and gingerenol, the spicy taste is strongest when n = 4 (Fig. 10.17).

The spiciness in fresh ginger is produced by a class of o-methoxyl-phenol alkyl ketones; the most commonly known compound is 6-gingerol. Dehydrated gingerol can form shogaol phenolic compounds, which are spicier compare to gingerol. When ginger is heated, the side chain of the shogaol ring breaks to form zingerone; zingerone has a moderate spicy taste. Spicy taste intensity is strongest when n = 4 in gingerol and shogaol (10.11). The main spicy taste substances in nutmeg and clove are eugenol and isoeugenol; these compounds also contain o-methoxyl-phenol groups.

(2)   Nutmeg and clove. The pungent components of nutmeg and cloves are mainly eugenol and isoeugenol, and these compounds also contain o-methoxyphenol groups.
(3)   Mustard glycosides. There are two types of glucosin, sinigrin and sinalbin, which produce glucose and mustard oil during hydrolysis. Myrosin is present in the seeds of mustard (brassica juncea), black mustard (sinapic niqra) and horseradish (horse radish), and other vegetables. Glucosinolates are found in white mustard seeds (sinapis alba).

S-methyl-cysteine-S-oxide is also contained in cruciferous vegetables such as glycoside blue, radish, and cauliflower.

3. Stimulating spicy substances stimulating spicy substances is a class of substances that can stimulate the nasal cavity and eyes in addition to stimulating the tongue and oral mucosa, with taste, smell, and tearing properties. Mainly includes:

(1)   Garlic, green onion, and leeks. The main spicy components of garlic are allicin, diallyl disulfide, and propyl allyl disulfide. Among them, allicin has the largest physiological activity. The main spicy components of green onions and onions are dipropyl disulfide and methyl propyl disulfide. Leek also contains a small amount of the above disulfide compounds. These disulfides will decompose to

Table 10.5 The amides in Xanthoxylum

Serial number	Name	Types	Substituents and double bond types (Z/E)
1	α- sanshool	I	R = H,2E,6Z,8E,10E
2	Hydroxy-α-Sanshool	I	R = OH,2E,6Z,8E,10E
3	Hydroxy-β-Sanshool	I	R = OH,2E,6E,8E,10E
4	β-Sanshool	I	R = H,2E,6E,8E,10E
5	γ-sanshool	II	R = H,2E,4E,8Z,10E,12E
6	Hydroxy-γ-Sanshool	II	R = OH,2E,4E,8Z,10E,12E
7	2'-Hydroxy-N-isobutyl-2,4,8,10,12-tetradecylpentaenamide	II	R = OH,2E,4E,8E,10E,12E
8	N-isobutyl-2,4,8,10,12-tetradecylpentaenamide	II	R = H,2E,4E,8E,10E,12E
9	2'-Hydroxy-N-isobutyl-2,4,8,11-tetradecyltetraenamide	III	R = OH,2E,4E,8Z,11Z
10	2'-Hydroxy-N-isobutyl-2,4-tetradecanedienamide	IV	R = OH,2E,4E
11	N-isobutyl-2,4-tetradecanedienamide	IV	R = H,2E,4E
12	2'-Hydroxy-N-isobutyl-2,4,8-tetradecyltrienamide	V	R = OH,2E,4E,8Z

(continued)

Table 10.5 (continued)

Serial number	Name	Types	Substituents and double bond types (Z/E)
13	N-isobutylene-2,4,8,10,12-tetradecylpentaenamide	VI	2E,4E,8E,10E,12E

Note

$$CH_3\text{-}\underset{10}{CH}=CH\text{-}\underset{8}{CH}=CH\text{-}\underset{6}{CH}\text{-}CH_2\text{-}CH_2\text{-}\underset{2}{CH}\text{-}CONH\text{-}CH_2\text{-}C(Me)_2\text{-}R(I)$$

$$CH_3\text{-}\underset{12}{CH}=CH\text{-}\underset{10}{CH}=CH\text{-}\underset{8}{CH}=CH\text{-}CH_2\text{-}CH_2\text{-}\underset{4}{CH}=CH\text{-}\underset{2}{CONH\text{-}CH_2\text{-}(Me)_2\text{-}R(II)}$$

$$CH_3\text{-}CH_2\text{-}\underset{11}{CH}=CH\text{-}CH_2\text{-}CH_2\text{-}\underset{8}{CH}=CH\text{-}CH=CH\text{-}CH=CH\text{-}\underset{4}{CONH\text{-}CH_2}\text{-}\underset{2}{C(Me)_2}\text{-}R(III)$$

$$CH_3\text{-}(CH_2)_8\text{-}\underset{8}{CH}=CH\text{-}CH=CH\text{-}\underset{4}{CONH\text{-}CH_2\text{-}CH=CH\text{-}\underset{2}{C(Me)_2}\text{-}R(IV)$$

$$CH_3\text{-}(CH_2)_4\text{-}\underset{12}{CH}=CH\text{-}CH_2\text{-}CH_2\text{-}\underset{8}{CH}=CH\text{-}CH=CH\text{-}\underset{4}{CONH\text{-}CH_2}\text{-}\underset{2}{C(Me)_2}\text{-}R(V)$$

$$CH_3\text{-}\underset{12}{CH}=CH\text{-}CH=CH\text{-}\underset{10}{CH}=CH\text{-}CH_2\text{-}CH_2\text{-}\underset{8}{CH}=CH\text{-}CH=CH\text{-}\underset{4}{CONH\text{-}CH_2}\text{-}\underset{2}{CH\text{-}C(Me)}=CH_2(VI)$$

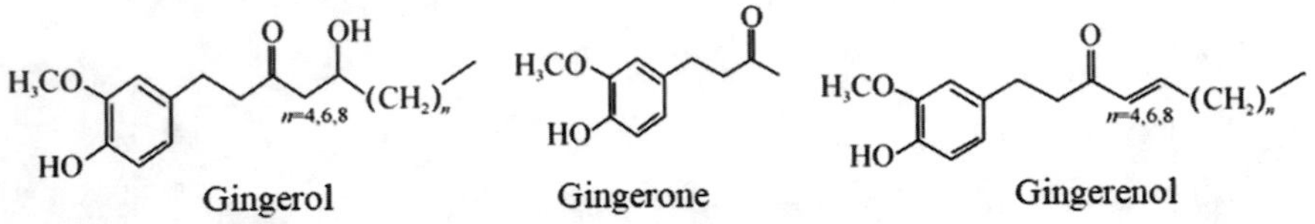

Fig. 10.17 Spicy ingredients in ginger

produce mercaptan when heated, so garlic and shallots will not only weaken the spicy taste, but also produce sweetness after being cooked.

(2)   Mustard and radish. The main spicy component is isothiocyanate compounds, among which propyl isothiocyanate is also called allyl mustard oil, which has a strong pungent spicy taste. They are hydrolyzed to isothiocyanate when heated, and the spicy taste is weakened.

Looking at garlic, onions, and chives, the main spicy taste substances in garlic are allicin, diallyl disulfide, and propyallyl disulfide, and of the three, allicin has the highest physiological activity. Leeks and onions' main spicy taste compounds are dipropyl disulfide and methyl propyl disulfide. Chives also have a small amount of disulfide compounds. Disulfides decompose to form the corresponding mercaptan when heated; therefore, garlic and onions' spiciness intensity weakens after cooking and sweetness is produced. The main spicy taste substances in wasabi and radish are isothiocyanate compounds, and one of the compounds, propyl isothiocyanate, is also called allyl mustard oil; it has a pungent spiciness. When heated, these compounds hydrolyze to isothiocyanic acid and reduce the spiciness.

10.3.7   Other Gustatory Perceptions

10.3.7.1   Refreshing Taste

Refreshing taste is produced by the stimulation of special taste receptors in the nasal cavity and oral cavity by some compounds. The typical refreshing taste is mint flavor, including spearmint and wintergreen oil flavor. With menthol and D-camphor as representatives (Fig. 10.18), they have both a refreshing smell and a refreshing taste. Among them, menthol is a commonly used cooling flavor in food processing, and it is widely used in candies and cooling drinks. The mechanism by which such flavor products produce a cooling sensation is unclear. Menthol can be obtained by steam distillation from the stems and leaves of peppermint. It has 8 optically active bodies, and L-menthol exists in nature.

Some sugars also have a cooling sensation when they are crystallized into the mouth, but this is because they absorb a lot of heat when they dissolve in saliva. For example, the heat of dissolution of sucrose, glucose, xylitol, and sorbitol crystals

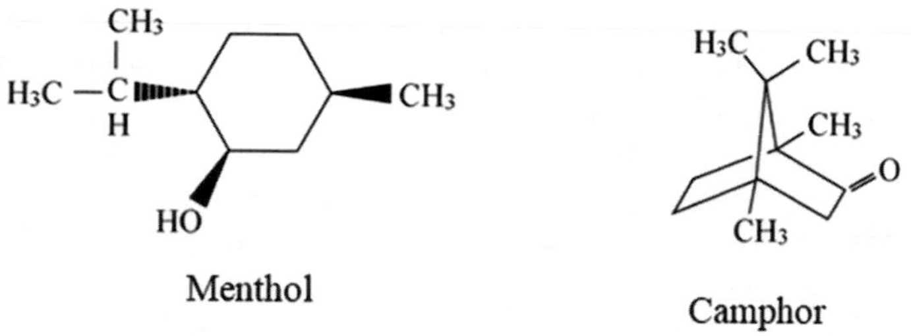

Fig. 10.18   Example of the structure of mint-like cooling sensation substances

are 18.1, 94.4, 153.0, and 110.0 (J/g), respectively, and the latter three sweeteners obviously have this cool flavor.

10.3.7.2   Astringency

When the oral mucosa protein is coagulated, it will cause convergence, and the taste is astringency. Therefore, astringency is not produced by the action of taste buds, but by the stimulation of tactile nerve endings, which is characterized by astringency and dryness in the mouth.

The main chemical components that cause the astringency of food are polyphenols, followed by iron metals, alum, aldehydes, phenols, and other substances. Some fruits and vegetables also cause astringency due to the presence of oxalic acid, coumarin, and quinic acid. The astringent effect of polyphenols is directly related to the property of being hydrophobically bound to proteins. For example, tannin molecules have a large cross-section and are easy to have hydrophobic interactions with protein molecules. Phenol groups can also cross-link with proteins. Generally, tannins with a moderate degree of condensation have this effect, but when the degree of condensation is too large, the solubility will no longer appear astringent.

The astringency of immature persimmon is a typical astringency, and its astringent components are glycosides with leucoanthocyanins as the basic structure, which are polyphenolic compounds and are easily soluble in water. When the cell membranes of astringent persimmon and immature persimmon rupture, the polyphenol compounds gradually dissolve in water and present an astringent taste. In the process of persimmon ripening, intermolecular respiration or oxidation causes the polyphenols to oxidize and polymerize to form water-insoluble substances, and the astringency disappears.

Tea also contains more polyphenols. Due to the different processing methods, the polyphenols contained in the various teas made are different, so their astringency is also different. In general, green tea contains a lot of polyphenols, while black tea is oxidized after fermentation, and its content is reduced, and the astringency is not as strong as green tea.

Astringency is the desired flavor in some foods, such as tea and red wine. In some foods, it has an impact on the quality of the food. For example, when there is protein, there will be precipitation between the two.

10.3.7.3 Metallic

Since there may be an ion exchange relationship between the metals in contact with food and the food, there is often an unpleasant metal taste in canned foods that have been stored for a long time, and some foods will also have peculiar smells due to the introduction of metals into the raw materials.

Refreshing taste is produced by certain compounds being in contact with oral tissue or nerves stimulating special receptors. The most common refreshing taste is mint flavor, including menthol and camphor; they are considered as both refreshing aroma and refreshing taste. The mechanism by which these flavor products are produced is not known. Some sugar crystals can produce refreshing feels after entering the mouth, but that is due to heat absorption during dissolution of crystals.

Astringent taste is not produced by taste buds, but by the stimulation of tactile nerve endings causing the oral cavity to feel astringent and dry. The main chemical composition of astringent taste in food is polyphenolic compounds, and others include ferrous metals, alums, aldehydes, and phenols. Some fruits and vegetables contain oxalic acid, coumarin, and quinic acid, which can cause astringent taste.

Due to possible ion exchange occurring between metal in contact with food and the food, canned foods with long storage times often have an unpleasant metallic taste. Some foods may also have off-flavors due to raw materials contaminated with metals.

10.4   Olfaction

10.4.1   Olfactory Organs

Olfaction is an important sensation to the human body. It is a sensation caused by the central nervous system as volatile substances stimulate the olfactory nerve cells in the nasal cavity. Among them, the pleasant sense of smell is called fragrance, and the unpleasant sense of smell is called stink. Smell is a more complex and sensitive sensory phenomenon than taste.

According to the anatomical structure, the olfactory system can be divided into 3 parts, the olfactory epithelium, the olfactory bulb, and the olfactory cortex. The olfactory epithelium is composed of olfactory cells, supporting cells, basal cells, and mucous glands. There are specific receptors on the olfactory epithelial ciliary membrane called the olfactory receptor. The first step of aroma perception begins with olfactory receptors interacting with taste substances and the olfactory receptor molecules will determine the specificity of olfactory perception signals. When the olfactory receptor is activated, electrical signal will be generated; this signal will reach the olfactory center through the olfactory conduction pathway and cause a sense of smell. Current research suggests there are at least four different systems involved in the sensation and conduction of olfactory signals, including the main olfactory system, the accessory olfactory system, terminal nerve system, and the trigeminal nervous system. The main olfactory system mostly senses aroma-containing and volatile substances and is also the major focus in flavor chemistry research. The accessory olfactory system is mainly composed of vomeronasal organs, and it is an independent olfactory system that senses odorless substances and substances that are difficult to volatilize, such as pheromones. Terminal nerve system is an independent chemosensory system present in all vertebrates and is accompanied by the main olfactory system and vomeronasal organs. In addition to feeling pain, coldness, temperature, and touch, the trigeminal nervous system is also involved in olfaction. Some taste substances can stimulate the trigeminal nerve to produce a sensation, which is often considered as part of the olfactory perception.

10.4.2 Theories of Olfaction

Several olfaction theories have been proposed based on aroma substances' molecular characteristics and the relationship between their odors. The theories with the most influences are sterochemical theory and vibrational theory of olfaction.

Olfactory stereochemical theory was proposed by Amoore in 1952. It is the first time that the olfaction produced by substances is related to its molecular shape, and the concept of primary odors is proposed for the first time in olfactory research. Therefore, this theory is also called the main aroma theory, which is like the visual perception of color Amoore's theory holds that the odor of different substances is a different combination of a limited number of dominant odors, and each dominant odor can be perceived by a different primary odor receptor in the nasal cavity. According to the frequency of various odors in the literature, Amoore proposed seven dominant odors, including ethereal, camphoraceous, musty, floral, minty, pungent, and putrid. To prove that dominant odors do exist and how to distinguish them, Amoore also conducted a "specific anosmia" experiment. After that, guillott's analysis of amoore's experimental results suggested that the lack of a certain dominant odor receptor was the cause of the specific olfactory anosmia with the lack of a certain dominant odor receptor. So, Olfaction stereochemical theory believes aroma characteristics are determined by taste substances' molecular weights and structure. The reaction of different molecularly sized, shaped, and charged taste substances with corresponding receptors of the olfactory system is a lock and key mechanism. When a gas molecule can be properly embedded into a receptor like a key in a lock, this gas molecule's unique smell can then be detected. The theory of olfactory stereochemistry explains to some extent that substances with similar molecular shapes have different odors because they have different functional groups.

The vibrational theory of olfaction was first proposed by Dyson in 1937 and further developed by Wright in the 1950s and 1960s. The theory holds that olfactory receptor molecules can resonate with aroma molecules. This theory is based primarily on the comparative study of optical isomers and isotopic substitution. In general, enantiomers have the same far-infrared spectrum, but their aromas can vary widely. The replacement of aroma molecules with hydrazine can change the vibrational frequency of the molecules, but this has little effect on the aroma of the substance.

10.4.3 Characteristics and Classification of Olfaction

10.4.3.1 Characteristics of Olfaction

1. Acuity. People have a very acute sense of smell. Some odor compounds can be detected even at very low concentrations. It is said that individual trained experts can distinguish 4000 different odors. Some animals have an acuter sense of smell, and sometimes even modern instruments can't catch up. The olfactory sensitivity of dogs is well known, while the olfactory ability of eels is almost equal to that of dogs, and they are about 1 million times more sensitive than that of humans.

2. Fatigue and adaptation. when the olfactory central nervous system falls into negative feedback status due to the long-term stimulation of some odors, the sensation is inhibited and adaptation is produced. Perfume, though fragrant, is not known for a long time, but it can endure for a long time even though it is smelly. This indicates that olfactory cells are prone to fatigue and are not sensitive to specific odors. In addition, when people's attention is distracted, they will not feel the smell, and they will form a habit of the smell when they are stimulated by a certain smell for a long time. Fatigue, adaptation, and habit work together and are difficult to distinguish.

3. Great individual differences. Different people have different senses of smell, even those who have a keen sense of smell will also vary according to the smell. The extreme case of being insensitive to smell forms olfactory blindness, which is also genetic. Some people think that women's sense of smell is sharper than men's, but there are different opinions.

4. The threshold will change with people's physical condition. When people are tired or malnourished, their olfactory function will be reduced; when they are sick, they will feel that the food is not fragrant; women may have hyposmia or hypersensitivity during menses, gestation or menopause, etc. All these indicate that the physiological condition of human beings also has an obvious influence on olfaction.

10.4.3.2 Classification of Olfaction

In fact, olfactory classification is to divide the odor like substances into a group and make semantic description of their characteristic odors. At present, there is no authoritative olfactory classification method. However, Amoore analyzed the odors of 600 compounds and their chemical structures and proposed that there were at least seven basic odors: light smell, camphor smell, moldy smell, flower smell, mint smell, pungent smell, and rotten smell. Many other odors may be caused by the combination of these basic odors. However, in the study of structure–odor relationship, some people often divide the odor into ambergris, bitter almond, musk, and sandalwood. Boelens studied 300 kinds of aroma compounds and found that the odorants could be classified into 14 basic odors, while Abe classified 1 573 odorants into 19 categories by cluster analysis. In the classification of smell, the most important thing is how to measure the similarity between two kinds of smell, that is, the standard of classification, which is also an important reason for the different classifications of odor.

10.4.3.3 Olfaction Substances

There are many unique types of olfaction substances in foods, mainly from flavor substances present in food raw materials. These flavor substances can also come from the processing and storage of food and be produced by microorganisms. Although most flavor substances only require a low usage concentration, this low concentration can still have an important influence on the food product's flavor and a person's appetite.

Olfaction substances usually have a low boiling point and high volatility. Some are water-soluble and others are fat-soluble. They can pass through the mucous membrane of the olfactory receptor and the lipid membrane of receptor cells. The molecular weight of these substances is relatively small, normally not greater than 300. Most aroma substances are organic; only a small number of inorganic substances have aromas, such as H_2S and NO_2. The smell of food is usually the result of a combination of different volatile substances, and different types of flavor substances will produce different aromas.

Odor activity value (OAV) is the detection of an olfaction substance having an effect in food aroma. OAV is the ratio of aroma substance concentration and its threshold value. The aroma threshold value refers to the concentration of aroma compounds in air or water causing an odor detection. The smaller the threshold concentration, the stronger the odor detection. It is commonly believed that when OAV is lower than 1, human's olfactory organs will not be able to detect this substance. The larger the OAV, the more contributions it will bring to food aroma.

There are numerous kinds of olfaction substances, and the perceptions created by them are very different. It is difficult to accurately classify these substances. Currently, there are physicochemical classification, psychological classification, and olfactory blindness classification. Looking at the properties of aroma substances, they can be classified into alcohol, ester, acid, ketone, terpene, heterocyclic (such as pyrazine, pyrrole, pyrroline, and imidazole), sulfur containing matter, and aromatic hydrocarbon. The relationship between the structure of aroma substances and their aroma is extremely complicated and is still inconclusive.

10.5 Aroma Components in Food

10.5.1 Aroma Components in Fruits and Vegetables

The flavor of fruits and vegetables is one of the main factors contributing to their quality. The types and quantities of flavor substances present in different fruits and vegetables vary. The aroma components of fruits are produced during the metabolic processes of plants and increase in concentration as fruits ripen. Aroma components are mainly organic acid esters, aldehydes, anthraquinones, and volatile phenols, followed by alcohols, ketones, and volatile acids.

Small esters are the main components of many fruit aromas, such as apples, strawberries, pears, melons, bananas, and cherries. 78 to 92% of the volatile substances in apple are esters formed by reactions of acetic acid, butyric acid, and carpoic acid with ethanol, butanol, and hexanol, respectively. Among the volatile components in pineapple, esters account for 44.9%; ethyl acetate accounts for more than 50% of muskmelon's volatile components. Some of the small esters in fruit aromas are methyl or methylthio-branched esters; for example, apple volatile substances contain 3-methylbutyl acetate, tertiary 3-methylbutyrate, and butyl 3-methylbutyrate, and they create the typical apple aroma with low threshold concentration. In these compounds, ethyl 3-methylbutyrate's threshold concentration is only at 1×10^{-7} mg/kg and is recognized as one of the most important components of apple aroma. The six important thioester aroma components in melons are methyl methylthioacetate, ethyl methylthioacetate, 2-methylthioethyl acetate, methyl 3-methylthiopropionate, ethyl 3-methylthiopropionate, and 3-methylthioacetate. Methyl 3-methylthiopropionate and ethyl 3-methylthiopropionate have the most influence on pineapple aroma.

Certain varieties of strawberry and citrus fruit also contain thioesters in their volatile substances. Alcohol in apples account for 6–12% of total volatile substances, and the main alcohols are butanol and hexanol. Ripe bananas contain large amount of syringol, syringol methyl ester, and its derivatives. Volatile substances in grapes contain benzyl alcohol, pheylethyl alcohol, vanillin, vanilone, and its derivatives. Derivatized esters of cinnamic acid are also found in ripe strawberries mainly including methyl esters and ethyl esters. Terpenoid compounds are an important part of grape aroma. Thirty-six monoterpenoid compounds are identified from grape volatiles, and it is believed that linalool and geraniol are the main aroma components.

Compared to fruits, the aroma of vegetables is not strong but some vegetables have unique aromas, such as garlic and onions. These vegetables contain sulfur compounds; when tissue cells are damaged, enzymes will combine with aroma precursor substrates in the cytoplasm and catalyze the production of volatile aroma substances. These enzymes are usually multi-enzyme complexes or multi-enzyme systems with differences in types and varieties. If dried cabbages are treated with enzymes from leaf mustard, these cabbages will have leaf mustard's aroma. The volatile substances in tomatoes are mainly alcohols, ketones, and aldehydes, including cis-3-hexenal, hexenal, hexenol, cis-3-hexenol, 1-hepten-3-one, 3-methylbutanol, 3-methylbutanal, acetone, and 2-heptenal.

10.5.2 Meat Aroma and Aroma Components

The aroma of meat is mainly produced after processing, especially after heat treatment. Raw meat exhibits the rawness aroma present in animals and gamey odor from blood. Meat aroma is mainly formed by flavor precursors in meat through lipid thermal degradation, thiamine pyrolysis, Maillard reaction, and other pathways. The juice in meat contains many kinds of amino acids, peptides, nucleotides, acids, and sugars. Among these components, inosinic acid content is quite high and can be mixed with other compounds to create the aroma of meat. When fat is removed from different types of meat and heated afterwards, the aroma components produced are very similar. Sulfur-containing compounds, such as 2-methyl-3-furanthiol, furfuryl mercaptan, 3-mercapto-2-pentanone, and thiomethanine, are essential flavor substances in all meats. The aroma variation in different types of meat is mainly caused by fat composition. The aroma components produced by heating meat fat are mainly carbonyl compounds, lipids, and lactones. The main compounds in beef aroma are mercaptothiophene and mercaptofuran. The unique flavor in lamb is related to medium-chain fatty acids, especially those with methyl branches. 4-methyloctanoic acid is one of the most important fatty acids in generating lamb flavor. Pork contains a high content of γ-C_5, C_9, and C_{12} lactones, and they produce a sweet aroma. When the meat is heated in different ways, the resulting aroma can have either similar components or components with unique characteristics. The main characteristic components of boiled meat aroma are sulfides, furan compounds, and benzene ring compounds. The main characteristic components of barbecue aroma are pyrazine compounds; this aroma also contains ketone compounds and carbonyl compounds, such as isovaleraldehyde.

10.5.3 Dairy Products Aroma and Aroma Components

The aroma components in fresh milk are mainly short-chain saturated fatty acids and carbonyl compounds, such as 2-hexanone, 2-pentanone, methyl ethyl ketone, acetone, acetaldehyde, and formaldehyde. There is also trace amount of ether, ethanol, chloroform, acetonitrile, vinyl chloride, and methyl sulfide in milk. Although there is only small amount of dimethyl sulfide in cow's milk, it is the main aroma component. The aroma threshold value for dimethyl sulfide in distilled water is approximately 1.2×10^{-4} mg/L; if present in a concentration higher than the threshold, there will be off-flavors and malt flavors present in cow milk.

The fat and lactose in milk have a strong ability to absorb external odors, especially cow's milk. When the temperature is about 35 °C, its absorption capacity is the strongest, and the temperature of the milk that has just been expressed is exactly in this range. Therefore, it is necessary to prevent contact with materials with peculiar smell.

Milk contains lipase, which can hydrolyze dairy fat into short-chain saturated fatty acids; among these fatty acids, butyric acid has a strong rancid odor. When dairy cows are fed with green fodder, it can inhibit the hydrolysis-type rancid odor of milk. This may be related to the carotene content in the fodder, because carotene can inhibit hydrolysis. On the contrary, when fed with dry feed, the milk is prone to hydrolytic rancidity. In addition to feeding factors, the temperature fluctuations, lack of timely cooling, and long-term stirring that cause the hydrolysis-type rancid odor of cow's milk promote the hydrolysis of milk fat, which makes the milk produce a rancid odor.

Long-term exposure of cow's milk and other dairy products to air can also produce rancid odors, also known as oxidative odor. Rancidity is caused by the auto-oxidation of unsaturated fatty acids in milk fat that produces α-, β-unsaturated aldehydes (RCH = CHCHO) and unsaturated aldehydes with two double bonds. Among them, octadienal with 8 carbon atoms and nonadienal with 9 carbon atoms are the most prominent. Even if the two are below 1 mg/kg, the oxidative smell of dairy products can be smelled. Trace metals, ascorbic acid, and light all promote the oxidative odor of dairy products, especially the strongest catalytic effect of divalent copper ion. When the copper content of dairy products is one part per million, it can form a strong catalytic effect. Ferric ion also has a catalytic effect, but it is weaker than copper.

When cow's milk is exposed to sunlight, it will produce a sun-burning smell. This is because methionine in cow's milk undergoes oxidative decomposition under the action of vitamin B_2 (i.e., riboflavin) to produce β-methylthiopropionaldehyde which has a cabbage odor. if it is highly diluted, it will have a sun odor. Even if β-methylthiopropionaldehyde in milk is diluted to 0.05 mg/kg, the smell can be felt. Cow's milk must have the following 4 factors to produce the smell of sunlight: light energy, free amino acids or peptides, oxygen, and vitamin B_2. β-Methylthiopropionaldehyde can be decomposed to produce methanethiol and dimethyldisulphide and other pungent odor compounds (Fig. 10.19).

In addition, the effect of bacteria can decompose leucine in milk into 3-methylbutanal, making milk produce a malty odor, and the reaction process is shown in Fig. 10.20.

The main aroma components in fresh butter are volatile acids (such as n-butyric acid, n-pentanoic acid, isovaleric acid, n-octanoic acid), alcohols (such as ethanol and isobutanol), isovaleraldehyde (1.0 ~ 10 mg/kg), diacetyl (0.001 46 mg/kg), and acetoin (0.004 47 mg/kg). Among these compounds, aldehydes are derived from amino acid degradation and ketones are derived from the oxidative decomposition of fatty acids such as oleic acid and linoleic acid. Diacetyl and acetoin are the main aroma

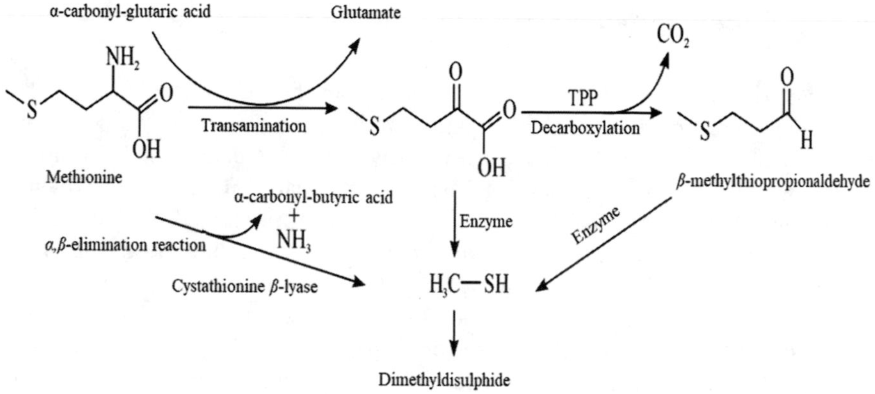

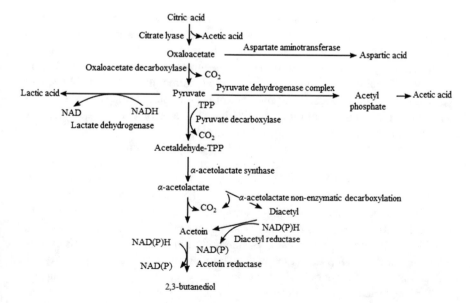

Fig. 10.19 The formation mechanism of sun odor in cow milk

$$H_3C-\underset{CH_3}{\overset{H}{C}}-CH_2-\underset{NH_2}{\overset{H}{C}}-COOH \xrightarrow[\text{Enzyme}]{[O]} H_3C-\underset{CH_3}{CH}-CH_2-CHO+NH_3+CO_2$$

Fig. 10.20 The formation mechanism of malty odor in cow's milk

components of fermented dairy products; they are formed by microbial fermentation of citric acid. The reaction process is shown in Fig. 10.21.

Fig. 10.21 The formation path of diacetyl in milk and wine in fermented products

10.5.4 Aquatic Products Aroma and Aroma Components

The range of aquatic product aroma is broader than that of livestock and poultry meat. There are copious types of aquatic products, including not only animal species, such as finfish, shellfish, and crustaceans, but also aquatic plants. The aroma changes with freshness and processing. The smell of aquatic products can be roughly divided into the aroma of fresh products and aroma of processed products. Fresh aquatic products have plant-like scent and melon-like aroma like the aroma of C_6, C_8, and C_9 compounds produced by lipoxygenase in plants; these compounds are mainly aldehydes, ketones, and alcohols. Fish will begin to have fishy off-flavor as it becomes less fresh, and this off-flavor is mainly caused by amines, including ammonia, dimethylamine, and trimethylamine as representing compounds. Compare to fresh raw fish, cooked fish has a higher content of volatile acids, nitrogenous compounds, and carbonyl compounds. The main aroma components in crab meat include pentanol, valeraldehyde, aldehyde, furfural, and trimethylamine; Aroma components in different kinds of crab will vary greatly. Shrimp contains hydrocarbons, alcohols, ketones, lipids, naphthalenes, and other compounds, with more than 65% of hydrocarbons giving shrimps a sweet aroma. 1-penten-3-ol, (cis, trans) 3,5-Octadien-2-one, (trans, trans) 3,3-octadien-2-one, and ester compounds impart a nice flavor to shrimps.

10.5.5 Tea Aroma and Aroma Components

Tea aroma is one of the important indicators of tea quality. The characteristic aroma of tea is related to various factors such as tea tree variety, harvesting season, climate conditions, and processing. Current research shows that aroma components in fresh tea are relatively small, and most of the aroma substances in tea are formed during processing. The aroma components in processed tea mainly include alcohols, aldehydes, ketones, esters, acids, nitrogen-containing compounds, and sulfur-containing compounds. Due to different processing techniques used on each unique variety of tea, the aroma components are also widely different.

The aroma components in each variety of green tea differ greatly due to variation in processing techniques. In roasted green tea, caramel aroma substances are high in concentration, such as benzyl alcohol, geraniol, pyrazine, and parole, and they usually have a chestnut aroma or fresh odor. There is a high concentration of linalool and its oxides in steamed green tea providing a strong grassy aroma. Black tea generally has floral and fruity aromas, and the main aroma substances are geraniol, linalool, and its oxides including benzyl alcohol, 2-phenylethyl alcohol, and methyl salicylate. Oolong tea is a semi-fermented tea with a floral aroma as its unique property. The main aroma substances in oolong tea include methyl jasmonate, hydrazine, linalool, and its oxides such as benzyl alcohol, phenylethyl alcohol, jasmone, jasmolactone, nerolidol, and geraniol.

10.5.6 Baked Goods Aroma and Aroma Components

The aroma of baked goods is mainly the products of carbonylation (Maillard reaction), carbohydrate pyrolysis, oil decomposition, and sulfur compounds (thiamine and sulfur-containing amino acids) decomposition during heat treatment.

The Maillard reaction not only produces brown-black pigments, but also forms a variety of aroma substances. Baked goods aroma is mostly produced by pyrazine compounds. Sugar is an important precursor to the formation of aroma. When the temperature is above 300 °C, sugar can be pyrolyzed to form a variety of aroma substances; the most important ones are furan derivatives, ketones, aldehydes, and diacetyl. Carbonylation reactions not only produce brown-black colored pigments, but also forms a variety of aroma substances. The product of carbonylation varies with the reactants, such as leucine, valine, lysine, and proline that can react during moderate heating with glucose to produce attractive odors. However, cystine and tryptophan can produce off-flavor aromas.

In addition to alcohols and esters formed during fermentation, the aroma substances of bread and other flour products also contain many carbonyl compounds produced during the baking process. Addition of leucine, valine, and lysine into fermented dough can enhance the aroma of bread. Heating dihydroxyacetone and proline together can produce cookie aromas.

Peanuts and sesame have a strong unique aroma after roasting. The flavor of roasted peanut is closely related to pyrazines, of which 2,5-diethylpyrazine is the most relevant substance to roasted peanut flavor. The main characteristic component of sesame aroma is sulfur-containing compounds.

10.5.7 Fermented Foods Aroma and Aroma Components

The effect of fermentation on food flavor is mainly shown in two aspects; first is flavor substances produced by microorganisms acting on raw material's protein, sugar, fat, and other substances, and examples of these flavor substances include the acidic taste in vinegar and pleasant aroma in soy sauce. On the other hand, fermentation by microorganisms can convert non-flavor substances into flavor substances during the ripening and storage of products such as aroma compounds in Chinese spirits. The aroma components in fermented foods and seasonings are mainly alcohols, aldehydes, ketones, acids, and esters. Due to the large number of different metabolites produced by microorganisms in different proportions, the aroma of each fermented food is unique.

10.5.7.1 The Aroma of Alcohol

The aromatic components of various wines are very complex, and their components vary with the varieties. For example, the main aroma components of Maotai liquor are ethyl acetate and ethyl lactate; the main aroma producing substances of Luzhou Daqu are ethyl caproate and ethyl lactate; the contents of acetaldehyde and isoamyl alcohol are relatively high in the two kinds of liquor; in addition, there are dozens of other microscales and trace volatile components identified in the liquor.

10.5.7.2 The Aroma of Sauce and Soy Sauce

The sources of flavor substances in sauce and soy sauce include: aroma components produced by raw material components; aroma components produced by microbial metabolism; and chemical reactions produced during the fermentation of sauce mash. The main volatile flavor components in sauces include esters, alcohols, aldehydes and ketones, phenols, organic acids, sulfur-containing compounds, furans, and nitrogen-containing heterocyclic compounds. The aroma substances of soy sauce include organic acids, alcohols, esters, phenols, aldehydes, furans, pyrazines, pyridines, and other heterocyclic substances, and their content in soy sauce is inconsistent, which together constitute the special aroma of soy sauce.

10.6 Aroma Formations Pathways in Food

There are numerous varieties of aroma substances in foods and they vary widely; formation of these aroma substances is also extremely complicated. Many reaction mechanisms and pathways are still unknown. However, the basic pathways are divided into two types; one of them is biosynthesis under direct or indirect action of enzymes, such as natural flavor substances produced during the growth, ripening, and storage of raw materials. The other type of pathway is the non-enzymatic chemical reaction pathway, such as aroma substances formed during food processing due to physical and chemical changes. For example, the aroma components of peanuts, sesame, coffee, and bread are produced during roasting and baking; meat and fish are braised and cooked; aldehydes, ketones, acids, and other aroma components formed when fat is oxidized by air.

In general, the pathways or sources of aroma substances in food are roughly as follows: biosynthesis, enzyme action, fermentation, pyrolysis, and food flavoring. This section will focus on the biosynthetic pathway of aromas.

10.6.1 Biosynthesis

The main source of aroma substances in foods is the biosynthesis of raw materials during growth, ripening, and storage; for example, the formation of aroma substances in fruits, such as apples and pears, and the production of aroma substances in certain vegetables, such as onions, garlic, and cabbage, all follow this formation pathway. Different biosynthesis precursors and pathways will produce completely different aroma substances. The aroma components in foods are mainly formed by further biosynthesis using amino acids, fatty acids, hydroxyl acids, monosaccharides, glycosides, and pigments as precursors.

10.6.1.1 Biosynthesis with Amino Acids as Precursors

Amino acid metabolism can form aroma components, such as fruity aroma, ester aroma, and spicy aroma, in fruits and vegetables. The amino acids involved in the synthesis of aroma compounds are mainly branched-chain amino acids (such as leucine), aromatic amino acids (such as phenylalanine and tyrosine), and sulfur-containing amino acids. During the metabolism of these amino acids, aldehydes are formed by transamination and decarboxylation under the action of enzymes; these aldehydes further react and form branched, aromatic, or aliphatic alcohols, carbonyl compounds, acids, and esters. During amino acid metabolism, enzyme activity and substrate specificity determine the concentration and type of branched alcohols and esters that can form.

1. **Branched-chain amino acids**

Fruits and vegetables contain many short-chain alcohols, aldehydes, acids, esters, and other aroma components. Most of the biosynthetic precursors for these substances are derived from branched-chain amino acids. The aroma components of many fruits, such as bananas and apples, are rapidly formed as they mature during the ripening process, especially at the climacteric stage. For example, one of the characteristic aroma compounds in apple, 3-methylbutyrate, is formed in the post-ripening stage. Banana peel changes from green to yellow as the fruit ripens, and its characteristic aroma substance, isoprene acetate, also rapidly increases. These characteristic aroma compounds in apples and bananas are produced by biosynthesis using branched-chain amino acid, L-leucine, as the precursor (Fig. 10.22).

During the ripening of tomato, the content of isoamyl alcohol, isoamyl acetate, and isobutyric acid or isoamyl butyrate increased, and the key substance was isovaleraldehyde. The 14C-labeled leucine was added to fresh tomato extracts, and 14C-containing isovaleraldehyde was obtained; but when the leucine was added to the boiled tomato crude extracts, there was no such phenomenon. This shows that the process of producing isovaleraldehyde from leucine in tomato has the nature of enzymatic reaction. Pyrazine compounds give characteristic flavors to certain vegetables.

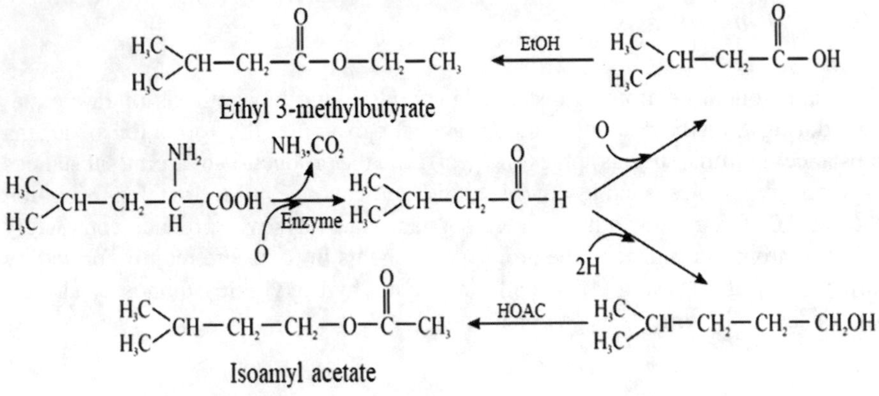

Fig. 10.22 Aroma substances of apple and banana with leucine as a precursor

For example, 2-methoxy-3-isobutylpyrazine in sweet peppers and peas, 2-methoxy-3-sec-butyl-pyrazine in sugar beets, and 2-methoxy-3-isopropylpyrazine in potatoes are all pyrazine compounds. These compounds are also biosynthesized in plants using branched-chain amino acids are precursors (Fig. 10.23).

Some microorganisms, including yeasts and certain malt aroma producing *Streptococcus* strains, can also convert amino acids with the method described previously. Plants can also convert amino acids, except leucine, into similar derivatives using the previously discussed biosynthetic pathway to produce aroma substances. For example, rose-scented 2-phenylethanol is synthesized by the previously discussed pathway using phenylalanine.

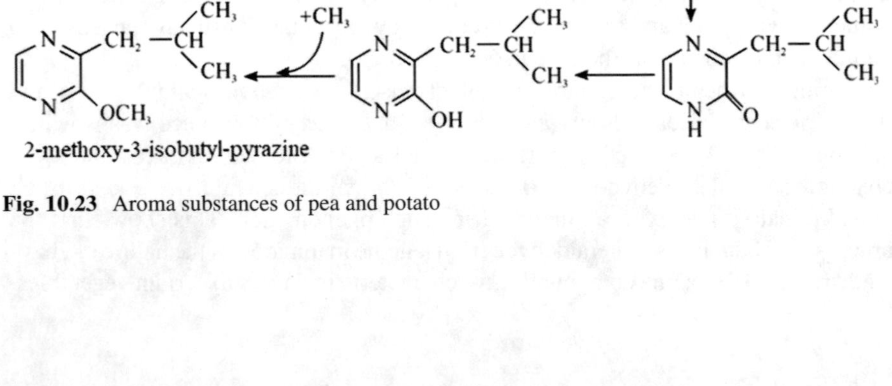

Fig. 10.23 Aroma substances of pea and potato

2.   **Aromatic amino acid**

Aromatic amino acids are synthetic precursors for volatile phenols, ethers, and certain aromatic aroma substances, such as elemicin and 5-methyl eugenol in bananas, cinnamic acid esters in strawberries and grapes, and vanillin in some fruits and vegetables. These are all aromatic amino acids with phenylalanine and tyrosine as precursors (Fig. 10.24). Since these aromatic amino acids are produced from shikimic acid in higher plants and microorganisms, this synthetic pathway is also known as the shikimate pathway. Using this pathway, aroma components associated with essential oils can also be produced.

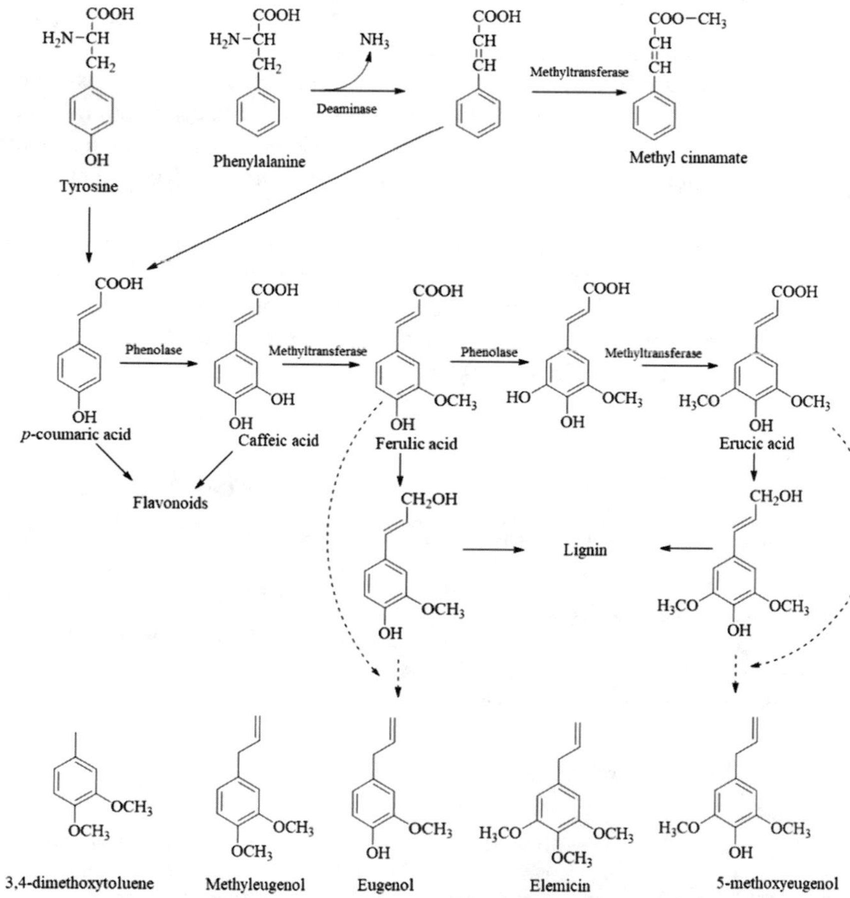

Fig. 10.24   Biosynthesis pathway of phenolic ether compounds in fruits and vegetables (shikimic acid pathway)

3. Sulfur-containing amino acids

The characteristic flavor substances in onion, scallion, garlic, chives, and other allium plants have a strong and penetrating aroma, and the main components are sulfur-containing flavor compounds. These flavor compounds are formed after plant tissue ruptures and the separation of enzyme from flavor precursor is destroyed; flavor precursors can then be converted into volatile substances. The flavor precursor for onion is S-(1-allyl)-L-cysteine sulfoxide; this compound is also present in chives. Alliinase can hydrolyze S-(1-allyl)-L-cysteine sulfoxide to form unstable sulfenic acid intermediates, ammonia, and pyruvic acid. Sulfenic acid is further rearranged into Syn-propanethial-S-oxide; this compound triggers tearing and is one of the characteristic components in fresh onion. Sulfonic acid can also form thiols, disulfide compounds, trisulfide compounds, and thiophene compounds; these compounds and their derivatives are flavor substances of cooked onions.

The flavor precursor of garlic is S-(2-allyl)-L-cysteine sulfoxide, and its flavor formation process is like an onion. The formation of diallyl thiosulfinate (allicin) under the action of allinase exhibits a freshly cut garlic odor. Under the action of enzyme or heat, allicin produces trisulfide, disulfide compounds, and thioethers, and these aroma also produced after garlic has been cooked or stored.

Lentinetic acid is the main flavor substance in shiitake mushrooms, and its synthetic precursor is a peptide formed by thio-L-cysteine sulfoxide and γ-glutamyl group. Lentinetic acid produces the active flavor substance lentionione in raw shiitake mushroom under the action of S-alkyl-L-cysteine sulfoxide lyase. These pathways are shown ins Figs. 10.25, 10.26, and 10.27.

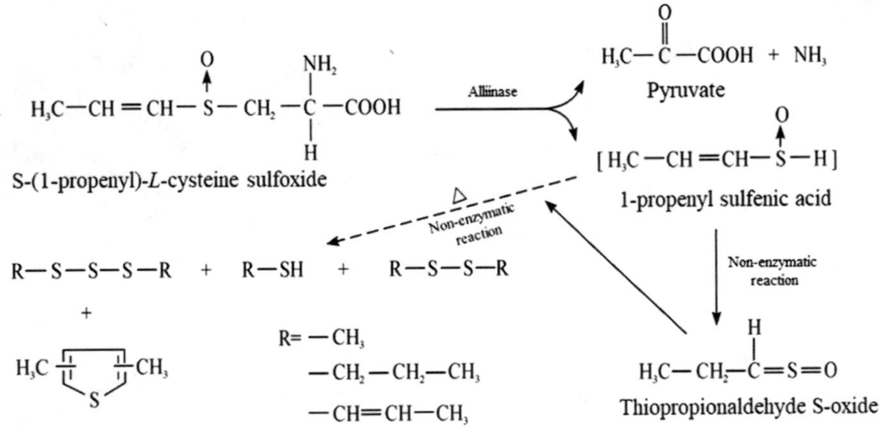

Fig. 10.25 The characteristic pathway of formation of aroma substances in onion

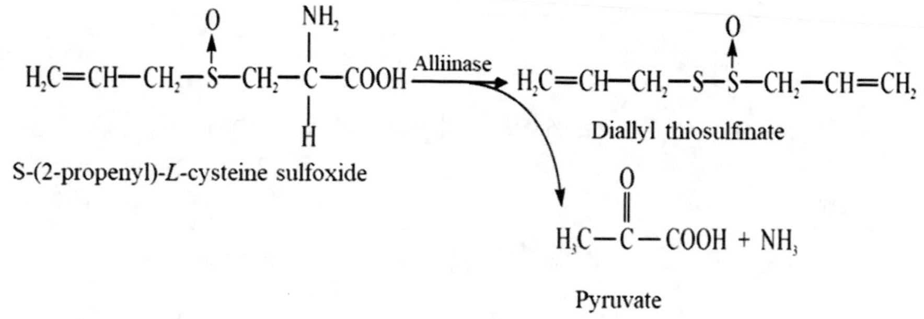

$$H_2C=CH-CH_2-\overset{O}{\underset{}{\overset{\uparrow}{S}}}-CH_2-\overset{NH_2}{\underset{H}{\overset{|}{C}}}-COOH \xrightarrow{\text{Alliinase}} H_2C=CH-CH_2-S-\overset{O}{\overset{\uparrow}{S}}-CH_2-CH=CH_2$$

S-(2-propenyl)-L-cysteine sulfoxide

Diallyl thiosulfinate

$$H_3C-\overset{O}{\overset{\parallel}{C}}-COOH + NH_3$$

Pyruvate

Fig. 10.26 The characteristic pathway of formation of aroma substances in garlic

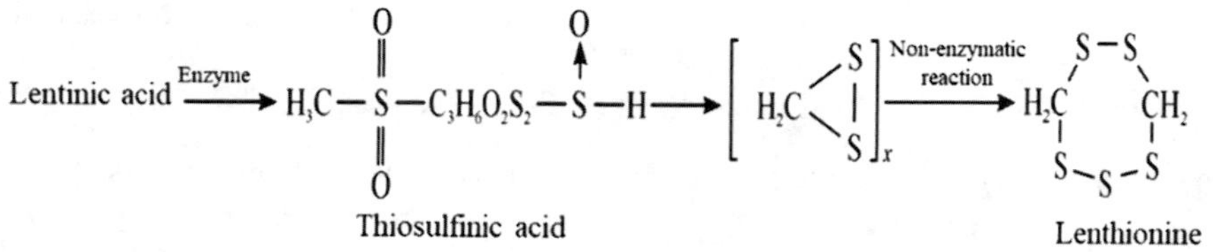

$$\text{Lentinic acid} \xrightarrow{\text{Enzyme}} H_3C-\overset{O}{\underset{O}{\overset{\parallel}{\underset{\parallel}{S}}}}-C_3H_6O_2S_2-\overset{\uparrow}{S}-H \longrightarrow \left[H_2C\underset{S}{\overset{S}{\diagdown}} \middle| \right]_x \xrightarrow[\text{reaction}]{\text{Non-enzymatic}} H_2C\underset{S_{\diagdown}S}{\overset{S-S}{\diagup}}CH_2$$

Thiosulfinic acid

Lenthionine

Fig. 10.27 The characteristic pathway of aroma substances formation of lentinus edodes

10.6.1.2   Biosynthesis with Fatty Acids as Precursors

Aroma components in fruits and some squash vegetables often contain C_6 and C_9 alcohols, aldehydes, and esters formed by C_6 and C_9 fatty acids; these compounds are mostly formed by biosynthesis with fatty acids as the precursors. The catabolism of fatty acids is mainly through two pathways, lipoxygenase oxidation and β-oxidation.

1.   Aroma components produced by lipoxygenase

In plant tissues, oxidative cleavage of unsaturated fatty acids by enzymes is very common. The aroma substances synthesized by enzymatic reactions have unique aromas compared with compounds produced by auto-oxidation of lipid. The fatty acids used as precursors are mainly linoleic acid and linolenic acid, such as hexanal in apple, banana, grapes, pineapple, and peach, characteristic aroma substances trans-2-nonenal and cis-3-nonenol in melon and watermelon, cis-3-hexene and cis-2-hexenol in tomato, and trans-2-cis-6-nonadienal in cucumber. These characteristic aroma compounds are all produced by lipoxygenase, lyase, isomerase, and oxidase using fatty acids, linoleic acid, and linolenic acid, as precursors (Fig. 10.28). In general, C_6 compounds exhibit grass aroma, C_9 compounds exhibit melon-like and cucumber-like aroma, and C_8 compounds have the aroma of mushroom, hoary stock, or geranium leaf.

The characteristic aroma substances in shiitake mushroom, 1-octen-3-ol, 1-octen-3-one, and 2-octenol, are also formed by enzymatic cleavage of linoleic acid. However, some studies have found that the degradation product of linoleic acid

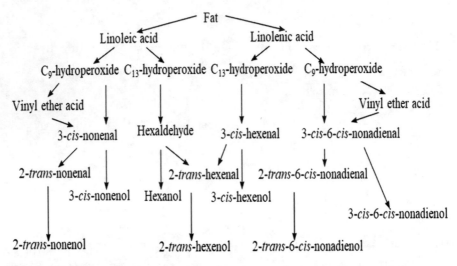

Fig. 10.28 The characteristic pathway of aroma substances formation of lentinus edodes

C_{13}-hydroperoxide only found the corresponding hydroxy acid in the mushroom homogenate, and no 1-octen-3-ol was detected, and the linoleic acid C_9 hydrogen Peroxide is also not a precursor of octenol.

The aroma substances of vegetables such as cucumbers and tomatoes include C_6 and C_9 saturated and unsaturated aldehydes and alcohols. In addition to the synthesis of linoleic acid as the precursor through the above pathways, linolenic acid can also be used as the precursor for biosynthesis. The products (3Z)-vinyl alcohol and (2Z)-hexenal are characteristic aroma substances of tomato, and (2E, 6Z)-Nadienal (alcohol) is the characteristic aroma component of cucumber.

The main component causing green-bean-like flavor in soybean products is hexanal; this compound is also formed by the action of lipoxygenase with unsaturated fatty acids (linoleic acid and linolenic acid) as precursors (Fig. 10.29).

2. Aroma components produced by β-oxidation of fatty acids

Many fruits, such as pears and peaches, produce a pleasant fruity aroma after ripening and many of these aroma components are medium-chain (C_6-C_{12}) compounds formed by β-oxidation of long-chain fatty acids. For example, ethyl (2E,4Z)-decadienoate is formed by linoleic acid through the β-oxidation pathway and it is a characteristic aroma component of pears (Fig. 10.30). In this pathway, $C_8 \sim C_{12}$ hydroxy acids are also generated at the same time. These hydroxy acids can also be cyclized under enzyme catalysis to generate γ-lactone or δ-lactone. $C_8 \sim C_{12}$ lactones have obvious characteristics of coconut and peach aroma. Generally, naturally mature fruits are more fragrant than artificially ripened fruits. For example, the content of lactones (especially γ-lactone) in naturally mature peaches increases rapidly, and the content of esters and benzaldehyde is 3 to 5 times more than that of artificially ripened peaches, which is related to the activity of related enzymes.

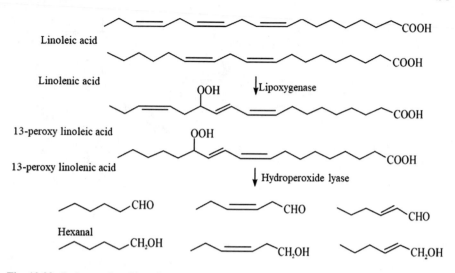

Fig. 10.29   Pathway of soybean flavor formation

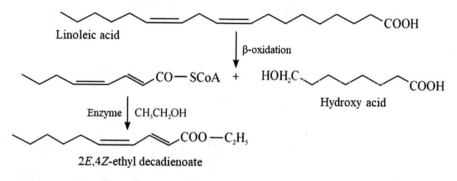

Fig. 10.30   Flavor formation pathway of fatty acid β-oxidation

10.6.1.3   Biosynthesis with Hydroxy Acids as Precursors

Terpene compounds are important flavoring substances in foods and exhibit a special aroma in fruits, vegetables, spices, and essential oils. Terpene compounds are mostly synthesized via the isoprenoid pathway; monoterpenoids consist of 10 carbon atoms, while sesquiterpene consists of 15 carbon atoms. The precursor is mevalonate (also known as mevalonic acid) and it is enzymatically catalyzed into isoamyl pyrophosphate, then synthesized in two different pathways (Fig. 10.31). Products of these reactions mostly exhibit natural aromas, such as nerol in lemon, β-sinensal in sweet orange, and nootkatone in grapefruit. These compounds are all characteristic aroma components in different fruits.

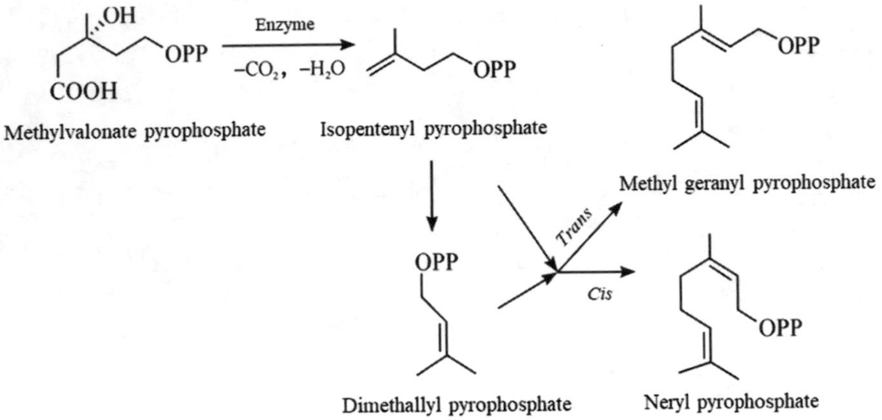

Fig. 10.31 Pathway of hydroxy acids forming terpene aromatic substances

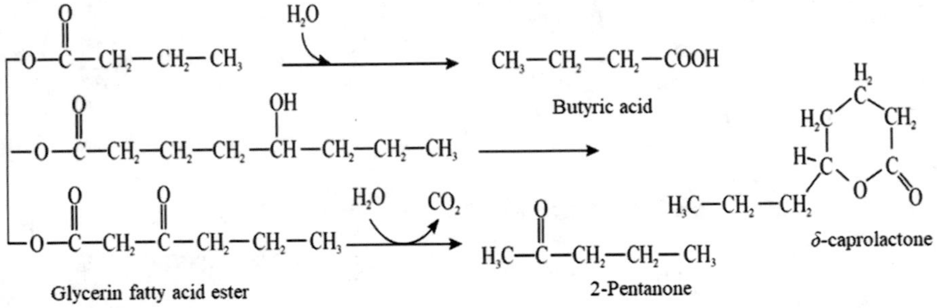

Fig. 10.32 Pathways of cyclization of hydroxy acids to form aroma substances

During β-oxidation of fatty acids, C_8-C_{12} hydroxy acids are also formed. These hydroxyl acids can undergo cyclization reaction under enzyme catalysis to form γ-lactone or δ-lactone. C_8-C_{12} lactone has characteristic aromas of coconut and peach, and δ-octanolactone is an important aroma component in dairy products (Fig. 10.32).

10.6.1.4 Biosynthesis with Monosaccharides and Glycosides as Precursors

Monosaccharides are not just sweeteners in fruits and vegetables, but also the main synthetic precursors of many olfactory components. After monosaccharide is glycolytically converted into pyruvic acid, it is then oxidatively decarboxylated by dehydrogenase to form acetyl-CoA. Under the action of an acylase, capable of converting alcohol to esters, and a reductase, acid acetate, and acid ethyl ester are formed (Fig. 10.33).

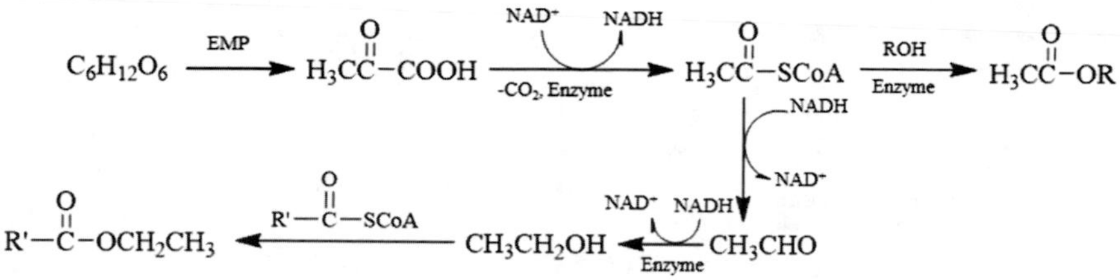

$$C_6H_{12}O_6 \xrightarrow{EMP} H_3C\overset{O}{\overset{\|}{C}}-COOH \xrightarrow[-CO_2, \text{ Enzyme}]{NAD^- \quad NADH} H_3C\overset{O}{\overset{\|}{C}}-SCoA \xrightarrow[\text{Enzyme}]{ROH} H_3C\overset{O}{\overset{\|}{C}}-OR$$

$$R'-\overset{O}{\overset{\|}{C}}-OCH_2CH_3 \xleftarrow{R'-\overset{O}{\overset{\|}{C}}-SCoA} CH_3CH_2OH \xleftarrow[\text{Enzyme}]{NAD^- \quad NADH} CH_3CHO$$

Fig. 10.33 Ester biosynthesis pathway using monosaccharide as precursor

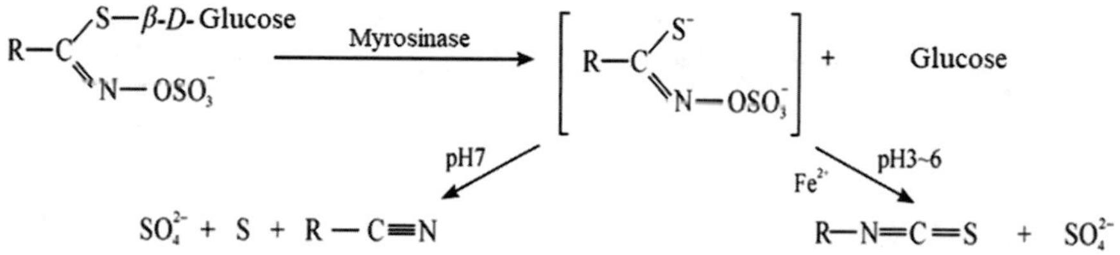

$$R-C\overset{S-\beta\text{-}D\text{-Glucose}}{\underset{N-OSO_3^-}{}} \xrightarrow{\text{Myrosinase}} \left[R-C\overset{S^-}{\underset{N-OSO_3^-}{}} \right] + \quad \text{Glucose}$$

pH7 → $SO_4^{2-} + S + R-C\equiv N$

pH3~6, Fe^{2+} → $R-N=C=S + SO_4^{2-}$

Fig. 10.34 The formation of characteristic aroma substances in cruciferous plants

Characteristic aroma substances are widely produced in cruciferous plants, such as cabbage, red cabbage, daikon, mustard leaf, and horseradish, are isothiocynate, thiocyanate, and nitrile compounds. Normally these spicy substances are not directly present in plants, but when plant cells are destroyed, glucosinolate, the precursor for spicy substances, is produced by myrosinase. Myrosin degrades glucosinolate into one molecule of glucose, one molecule of HSO_4^-· and one molecule of an unstable intermediate, non-sugar ligand. Depending on reaction conditions, non-sugar ligand can form isothiocyanate and thiocyanate. When pH is less than 4, nitrile compounds and elemental sulfur can be easily formed (Fig. 10.34). Cabbage and red cabbage contain alyl isothiocyanate and butyronitrile; their concentrations vary as growth conditions and processing conditions change. The spicy taste in daikon is produced by 4-methylthio-3-trans-butenyl isothiocyanate; the main flavor substance in mustard leaves and horseradish is α-phenylethyl isothiocyanate and allyl Isothiocyanate.

10.6.1.5   Biosynthesis with Pigments as Precursors

Some aroma substances in foods are formed with pigments as precursors, including lycopene and carotene. For example, under the catalysis of enzymes, lycopene forms 6-methyl-2-heptene-2-oxo and farnesylacetone (Fig. 10.35). In black tea, oxidation of carotenoids produces β-ionone and β-damasone.

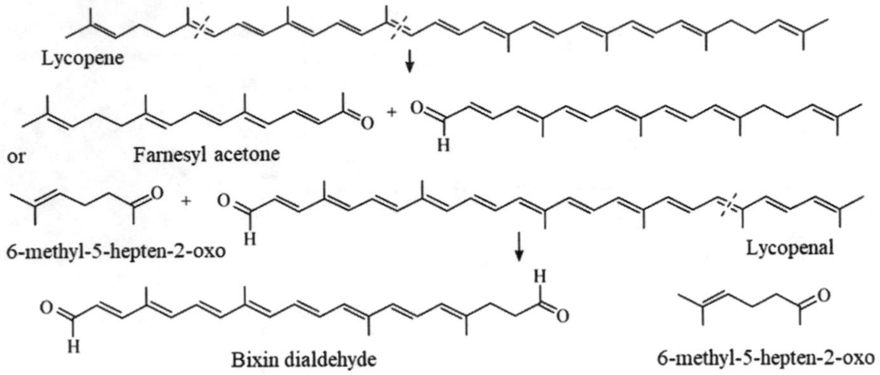

Fig. 10.35 Pathways of lycopene degradation to form aroma substances

10.6.2 The Role of Enzymes

The effect of enzyme on food aroma mainly refers to the process of forming aroma substances under the catalysis of a series of enzymes during the processing or storage of food materials after harvest, including the direct action of enzymes and the indirect action of enzymes. The so-called direct action of enzymes refers to the action of enzymes catalyzing a certain aroma substance precursor to directly form aroma substances, while the indirect action of enzymes mainly refers to the role of oxidation products catalyzed by oxidase to oxidize the aroma substance precursors to form aroma substances. The aroma formation of onion, garlic, cabbage, and mustard greens belongs to the direct action of enzymes, while the aroma formation of black tea is a typical example of indirect enzyme action.

10.6.3 Fermentation

The aroma components of fermented food and its flavoring are mainly produced by microorganisms acting on proteins, sugars, fats, and other substances in the fermentation substrate, mainly including alcohols, aldehydes, ketones, acids, esters, and other substances. Due to the wide variety of products metabolized by microorganisms and the different proportions of various components, the aroma of fermented foods also has their own characteristics. The influence of fermentation on food aroma is mainly reflected in two aspects: on the one hand, certain substances in the raw materials are fermented by microorganisms to form aroma substances, such as the sourness of vinegar and the aroma of soy sauce; on the other hand, some non-aroma formed by microbial fermentation Substances are further transformed during the maturation and storage of the product to form aroma substances, such as the aroma components of liquor. A typical example of microbial fermentation to form aroma

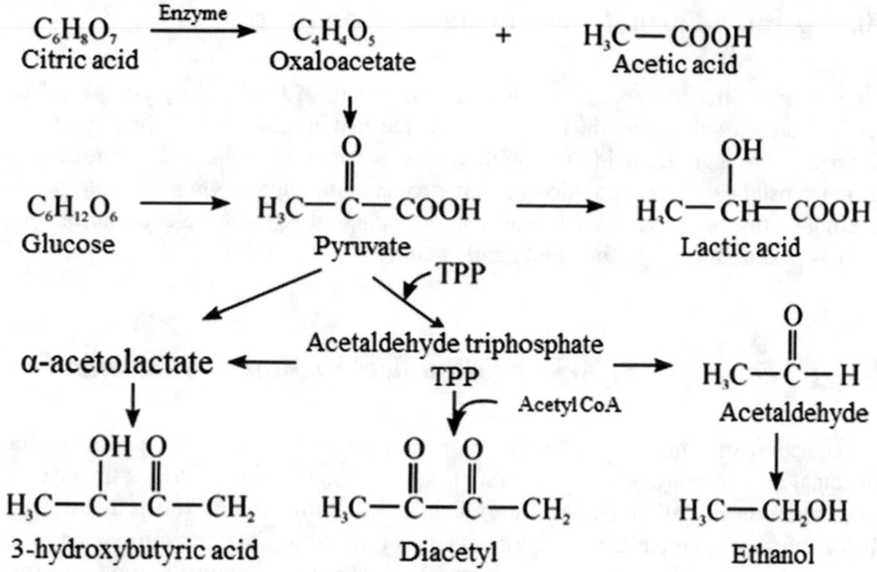

$$C_6H_8O_7 \xrightarrow{\text{Enzyme}} C_4H_4O_5 \quad + \quad H_3C-COOH$$

Citric acid   Oxaloacetate   Acetic acid

$$C_6H_{12}O_6 \longrightarrow H_3C-\overset{\overset{\displaystyle O}{\|}}{C}-COOH \longrightarrow H_3C-\overset{\overset{\displaystyle OH}{|}}{CH}-COOH$$

Glucose   Pyruvate   Lactic acid

TPP

α-acetolactate ⟵ Acetaldehyde triphosphate
                              TPP ⟶ $H_3C-\overset{\overset{\displaystyle O}{\|}}{C}-H$

AcetylCoA   Acetaldehyde

$$H_3C-\overset{\overset{\displaystyle OH}{|}}{C}-\overset{\overset{\displaystyle O}{\|}}{C}-CH_2 \qquad H_3C-\overset{\overset{\displaystyle O}{\|}}{C}-\overset{\overset{\displaystyle O}{\|}}{C}-CH_2 \qquad H_3C-CH_2OH$$

3-hydroxybutyric acid   Diacetyl   Ethanol

Fig. 10.36 The main aroma substances produced by lactic acid fermentation

substances is lactic acid fermentation (Fig. 10.36). Lactic acid, diacetyl, and acetalde-hyde together constitute most of the aroma of hetero-lactic fermented butter and cheese, while lactic acid, ethanol, and acetaldehyde constitute the aroma of homo-lactic fermented yogurt, of which acetaldehyde is the most important. Diacetyl is a characteristic aroma substance of draft beer and most foods with multi-strain lactic acid fermentation.

10.6.4   *Food Flavoring*

The flavor of food is mainly by some aroma enhancers or odor masking agents to significantly increase the aroma intensity of the original food or to mask the unpleasant smell of the original food. There are many types of aroma enhancers, but the main ones that are widely used are sodium L-glutamate, 5'-inosinic acid, 5'-guanylic acid, maltol, and ethyl maltol. Aroma enhancers themselves can also be used as odor masking agents. In addition, there are many odor masking agents used. For example, when cooking fish, adding a proper amount of vinegar can significantly reduce the fishy smell.

10.7   Aroma Formations During Heating Process

Changes in aroma components during heat treatment of food is very complex. The original aroma substances in food are lost due to volatilization by heat, and raw materials can degrade or interact with each other to form a large amount of new aroma substances. The formation of new aroma components is related not only to internal factors, such as raw material composition, but also due to external factors, such as heat treatment method and heating time.

10.7.1   Formation of Aroma Substances by Maillard Reaction

The reaction of amino compounds (such as amines, amino acids, peptides, and proteins) and carbonyl compounds (such as reducing sugars) in foods under suitable conditions is called the Maillard reaction. Maillard reaction forms a variety of flavor substances and at the same time, causes browning. This browning reaction is the main source of food color and flavor. Maillard reaction plays an important role in food flavor, such as the aroma from freshly baked bread, grilled steak, and freshly brewed coffee; these foods are appealing and are loved by consumers. However, before processing or heating, these characteristic flavors are not present in the food.

Products from the Maillard reaction are complicated, with carbonyl compounds, nitrogen-containing heterocyclic compounds, oxygen-containing heterocyclic compounds, sulfur-containing heterocyclic compounds, and oxygen-containing compounds. Products from the reaction are related to the carbonyl compounds and amino compounds involved in the reaction, and it is also related to heating temperature, time, food pH, and moisture content (Fig. 10.37). In general, when heating time is short and heating temperature is low, the main products of the reaction, in addition to Strecker aldehydes, will also include characteristic aroma substances, such as lactones and furan compounds. When the temperature is high and heat treatment

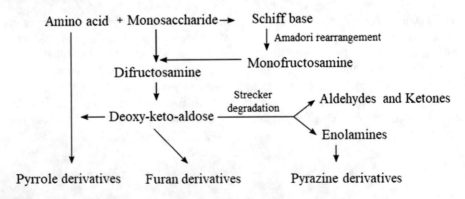

Fig. 10.37 Aroma formation pathway in Maillard reactions

time is long, aroma content will increase; baked goods aroma substances pyrazine, pyrrole, and pyridine can also be formed.

Different types of sugars and amino acids will produce different aroma substances. Maltose reacts with phenylalanine to produce a pleasant caramel aroma. Fructose reacts with phenylalanine to produce an unpleasant caramel odor, but when dihydroxyacetone is present, aroma of hoary stock can be produced. Dihydroxyacetone and methionine form a roasted potato-like aroma, while glucose and methionine react to give a burnt potato aroma. When glucose is present, proline, valine, and isoleucine can produce a pleasing baked bread aroma, but if a reduced disaccharide, such as maltose, is present, burnt cabbage odor can be produced; if a non-reducing disaccharide, such as sucrose is present, an unpleasant coke odor can be detected. When ribose is heated with other amino acids, large amount of aroma changes can occur; however, when sulfur-containing amino acids are heated under the same conditions without ribose, sulfur odor will be produced without changes to other aromas.

Different amino acids will have varied Millard reaction rates. In general, the order of degradation rate for amino acids in descending order is hydroxyl amino acids, sulfur-containing amino acids, acidic amino acid, basic amino acid, aromatic amino acid, and aliphatic amino acid. Figures 10.38, 10.39, 10.40, 10.41 and 10.42 show the formation pathways of main aroma substances imidazole, pyrroline, pyrrole, pyrazine, oxazole, and thiazole during the Maillard reaction.

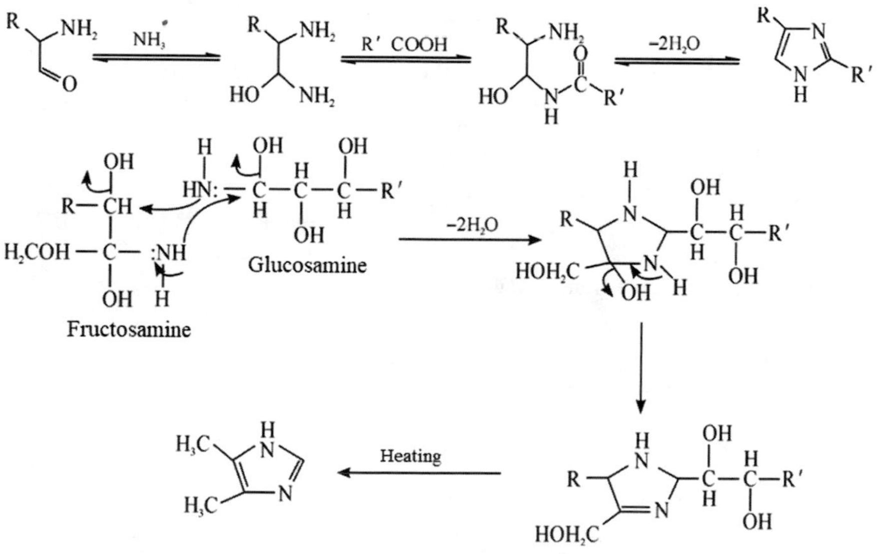

Fig. 10.38 The two pathways of imidazole formation in Maillard reactions

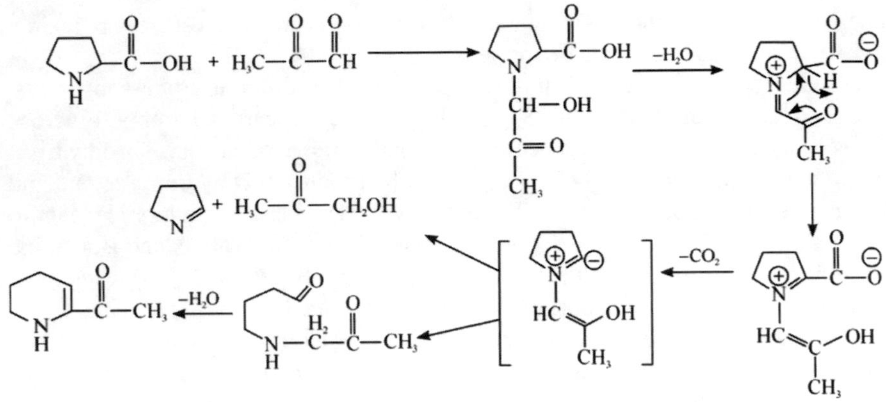

Fig. 10.39 The transformation from proline to pyrroline according to Strecker degradation in Maillard reactions

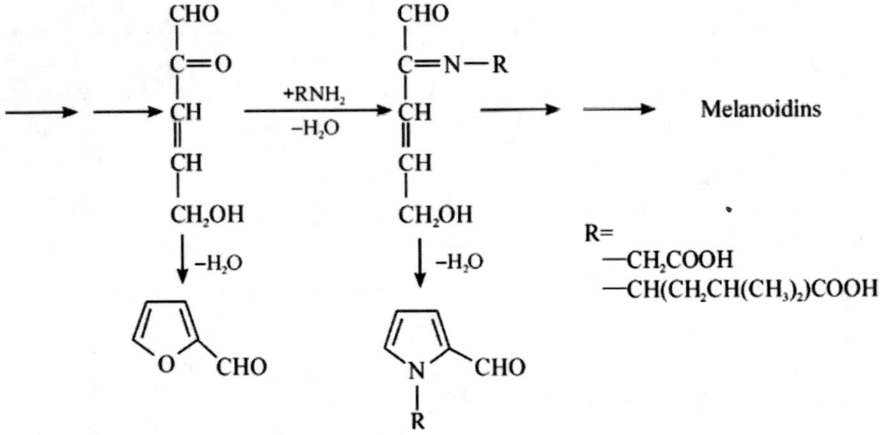

Fig. 10.40 The formation of pyrrole in Maillard reactions

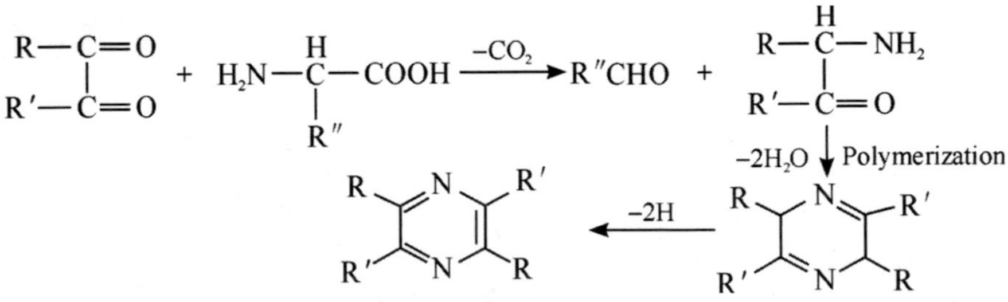

Fig. 10.41 The formation of pyrazine in Maillard reactions

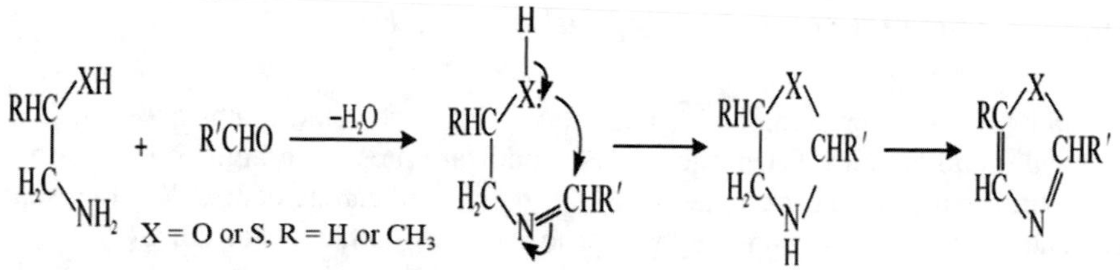

Fig. 10.42 The formation of oxazole and thiazole in Maillard reactions

10.7.2   Thermal Degradation of Sugar

When carbohydrates, especially sucrose and reducing sugars, are heated in the absence of nitrogen oxides, a series of degradation reactions can occur. Different aroma substances are formed when different heating conditions are applied, such as temperature and time; the main aroma substances are furan compounds, but lactones and cyclic diketones are also formed. Monosaccharides and disaccharides usually must be melted to go through thermal decomposition. When the heating temperature is too low or heating time is too short, a milk candy-like aroma will be produced; if continued to be heated, the carbon chain of monosaccharide will be cleaved to form low molecular substances, such as pyruvic aldehyde, glyceraldehyde, and glyoxal. If heated at a high temperature or too long, a burnt caramel smell will appear.

10.7.3   Thermal Degradation of Amino Acids

When amino acids are heated at high temperature, decarboxylation, deamination, and decarbonylation reactions generally occur first; however, the amine products formed at this step often have an unpleasant odor. If continuously heated, other products can interact with each other to form compounds with a pleasant aroma. The amino acids with the most impact on food aroma are sulfur-containing amino acids and heterocyclic amino acids. Besides hydrogen sulfide, ammonia, and acetaldehyde, the thermal decomposition products of sulfur-containing amino acids also contain thiazoles, thiophenes, and other sulfur-containing compounds; most of these compounds are highly volatile aroma substances and are important parts of cooked meat aroma. Proline and hydroxyproline in heterocyclic amino acids further react with methylglyoxal, formed from food during heating, to produce pyrrole and pyridine compounds with aromas of bread, cookie, grilled corn, and grain crop. In addition, thermal decomposition products of threonine and serine are mainly pyrazine compounds with baked goods aroma; the thermal decomposition products of lysine are mainly pyridine, pyrrole, and lactam compounds with baked goods and cooked meat aroma.

10.7.4 Thermal-Oxidative Degradation of Fat

Oxidation or decomposition of fats and oils can produce volatile substances such as aldehydes and ketones. When these compounds are present in high concentration, they will produce off-aromas like paint, fat, metal, and candle odors. When present in an adequate concentration, they will produce pleasing flavor substances.

Lipids will decompose into free fatty acids during heating; unsaturated fatty acids such as oleic acid, linoleic acid, and arachidonic acid contain double bonds and can be easily oxidized to form hydroperoxides. Hydroperoxides can easily decompose at a temperature above 150°C to produce volatile aroma substances such as carbonyl compounds, ketones, aldehydes, and acids (Fig. 10.43). Lipid thermal degradation products mainly include hydrocarbons, β-keto acids, methyl ketones, lactones, and esters. Most of the lactones are produced from linoleic acid; therefore, lipids containing linoleic acid will provide better aroma when deep fried. Thermal degradation products can continue to undergo non-enzymatic browning reactions with the small amount of proteins and amino acids present in oil forming hetero-cyclic compounds with a characteristic aroma. When lipids from saturated fatty acid (such as glyceryl stearate) are heated with air present, its pyrolysis products are mainly C_3-C_{17} methyl ketones, C_4-C_{14} lactones, C_2-C_{12} fatty acids, and others. Low concentration of γ-lactone has peach and dairy aroma, while at high concentration, there is a deep fried food aroma. Aroma substances formed by lipid degradation in cooked meat products include aliphatic hydrocarbons, aldehydes, ketones, alcohols, carboxylic acids, and esters (Fig. 10.44).

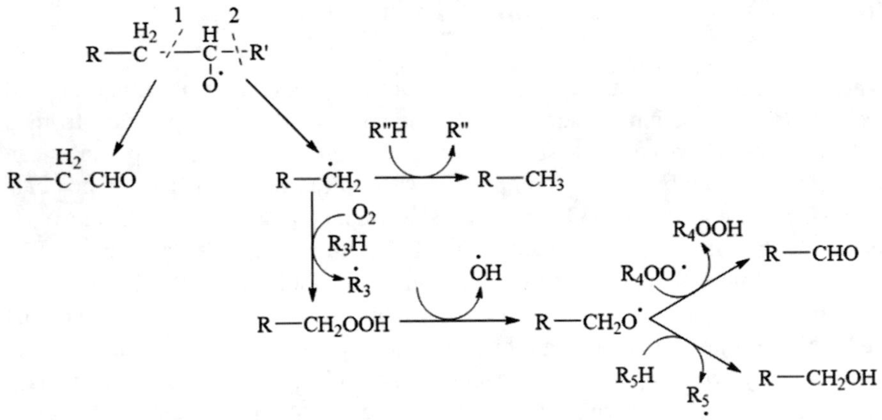

Fig. 10.43 Thermal degradation of fat hydroperoxide to produce volatile compounds

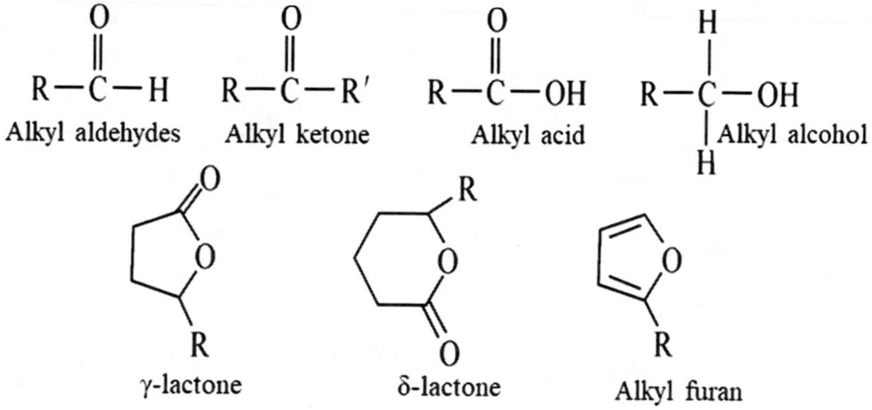

Fig. 10.44 Aroma compounds formed from fat oxidation in heat degradation

10.7.5 Aroma Substances Formed by Degradation of Other Food Components

In addition to the three major nutrients mentioned above, other components in food will also form aroma substances during thermal degradation. The following are some of the degradation pathways of several components that have been most extensively studied and have the greatest impact on food aroma.

10.7.5.1 Thermal Degradation of Thiamine

Thiamine itself does not have any aroma, but when affected by positively charged nitrogen atoms, a typical degradation reaction occurs with a nucleophilic substitution on the methylene carbon connecting the two rings. Thermal degradation products of thiamine are very complex, mainly including furans, pyridines, thiophenes, and aliphatic sulfur-containing compounds (Fig. 10.45). Some of these compounds are flavor substances for meaty aroma.

10.7.5.2 Thermal Degradation of Ascorbic Acid

Ascorbic acid is extremely unstable and can be easily degraded under heat, oxygen, and light to form furfural and small molecule aldehydes. Furfural compounds are one of the most important components of roasted tea leaves, peanuts, and cooked beef aroma. When ascorbic acid is heated in aerobic conditions, dehydration and decarboxylation reactions occur forming furfural, glyoxal, glyceraldehyde, and other compounds. Under anaerobic conditions, the degradation products are mainly

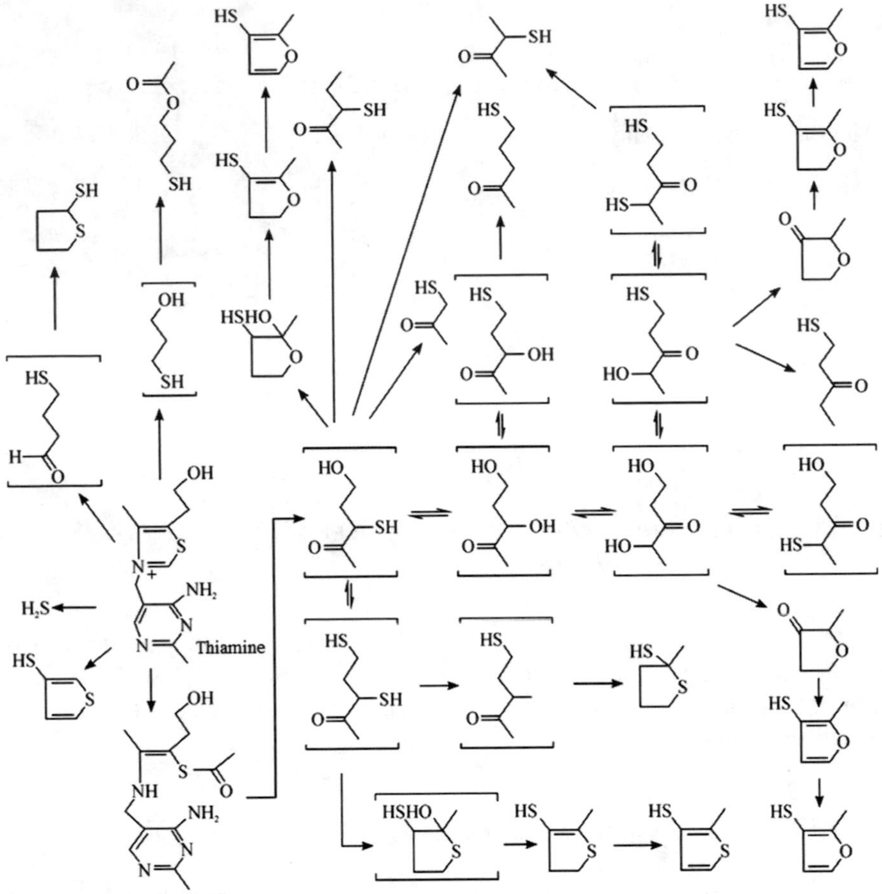

Fig. 10.45 Thermal degradation of thiamine

furfural. Low-molecular-weight aldehydes are aroma components on their own; they can also react with other compounds to form new aroma substances.

10.7.5.3 Oxidative Degradation of Carotenoids

Carotenoids are very unstable and are susceptible to heat or oxidative degradation during storage and processing. Products, such as cis-spirulina and β-ionone, are derived from the oxidative decomposition of β-carotene or lutein; these products can impart a rich sweet and floral aroma to tea leaves (Fig. 10.46). Even though these compounds are present in low concentrations, they are widely distributed, allowing many foods to have complete and harmonious aroma profiles.

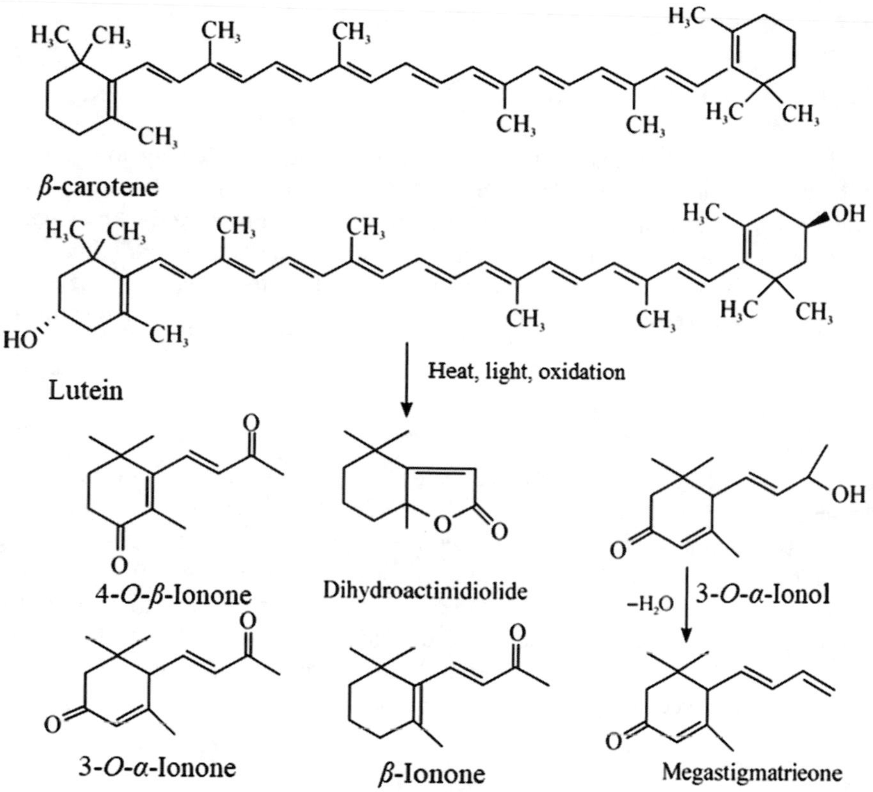

Fig. 10.46 Degradation products of carotene and lutein

10.8   Aroma Control in Food Processing

Food processing is an extremely complicated procedure; series of changes occur during processing in food's shape, structure, texture, nutrition, and flavor. Some processes can greatly enhance food aromas, such as roasting peanuts, baking bread, cooking beef, and deep-frying foods, while some processes can cause food aroma loss or off-flavor production, such as cooked-flavor from juice pasteurization, aroma deterioration of green tea stored in room temperature, over-cooked flavor of steamed beef, and burnt smell of dehydrated products. Therefore, aroma control during food processing is particularly important.

10.8.1   Control of Food Aroma

10.8.1.1   Raw Materials Selection

Food raw materials are one of the most important factors affecting food aroma. Different raw material type, origin, ripening stage, harvest time, and condition will produce distinct aromas. Different varieties of the same raw materials may cause great variances in aromas as well. Fruits harvested during respiration climax will have a much better aroma compare to fruits harvested before. Therefore, selecting the correct raw materials is one of the important factors to ensure great food aromas.

10.8.1.2   Processing Techniques

The impact of food processing techniques on food aroma is also significant. When the same raw materials are processed differently, aromas produced will vary as well, especially under heat treatment. In roasted green tea, the product made with rolling techniques often has a fresh aroma, while tea made without rolling step often exhibit a floral aroma. Fixation and drying are two key processes in the formation of roasted green tea aroma. Moderate spreading of tea leaves can increase the content of the main free-form aroma compounds in tea. Different drying methods will impose a significant difference on tea aromas.

10.8.1.3   Storage Conditions

Storage conditions also have a significant impact on the aroma. Tea will oxidize during storage, resulting in deteriorated and decreased quality. The aroma of apples stored in a controlled atmosphere is worse compare to refrigerated apples. If apples stored in a controlled atmosphere are later stored in refrigerated conditions for about 15 days, its aroma will not be significantly different from the apples that were refrigerated the entire time. An ultra-low oxygen environment is beneficial for maintaining the crunchiness of fruits, but it has an adverse effect on fruit aroma formation. Under different storage conditions, the composition of aroma components in fruits will also be different. This is mainly due to different storage conditions selectively inhibit or accelerate the pathways for the formation of certain aroma substances.

10.8.1.4   Packaging Methods

The influence of packaging methods on food aroma mainly affects two areas. Firstly, packaging changes the environmental conditions of food and this change will lead to material transformation or metabolism inside the food, eventually lead to changes in the aroma. Secondly, different packaging materials for packaged food will cause selective absorption of aroma substances. Packaging methods can selectively affect certain metabolic processes of food. For example, there is no significant difference in aldehyde, ketone, and alcohol content in apples with different bagging types; however, ester content in apples with double-layered bag is lower. Oxygen-free, vacuum, and nitrogen-filled packaging can effectively slow down quality deterioration in tea. For foods with high fat content, hermetic seal, vacuum, and nitrogen-filled packaging can significantly inhibit aroma deterioration.

10.8.1.5 Food Additives

Food ingredients or additives can interact with aroma components. There is strong binding between protein and aroma substances. Fresh milk should avoid contact with odorous substances. β-cyclodextrin has a special molecular structure and stable chemical properties; it cannot be easily decomposed by enzymes, acids, alkalis, light, and heat. It can embed aroma substances, thus reducing its volatile loss and make aroma longer lasting.

10.8.2 Food Aroma Strengthening

10.8.2.1 Aroma Recovery and Re-Addition

Aroma recovery process refers to the extraction of aroma substances initially and then adding the recovered aroma substances back into the product to maintain the original aroma profile. Main methods of aroma substance extraction include steam distillation, solvent extraction, molecular distillation, and supercritical fluid extraction. Because supercritical fluid extraction has many advantages such as high extraction rate, fast mass transfer, non-toxic, non-harmful, no residual, and no pollution, it has a broad application potential in aroma recovery.

10.8.2.2 Addition of Natural Flavors

Addition of essences is a common method of food aroma enhancement; it is also known as flavoring. Artificial essences are inexpensive, but their uses are becoming more limited due to safety reasons. The essences obtained from natural plants, microorganisms, or animals has a characteristic natural aroma and is very safe. Natural essences are becoming more popular.

10.8.2.3 Addition of Aroma Enhancer

Although aroma enhancers themselves hardly show aroma, they can significantly enhance or improve the aromatic effect in food. The mechanism of aroma enhancement is not to increase the content of aroma substances, but to improve the sensitivity of olfactory receptors to aroma substances by acting on olfactory receptors and reducing the sensory threshold of aroma substances. At present, the main applications in practice are sodium L-glutamate, 5'-inosinic acid, 5'-guanylic acid, maltol, and ethyl maltol. The most used for aroma enhancement are maltol and ethyl maltol. Maltol has a good flavor enhancement and flavor adjustment effect under acidic conditions; under alkaline conditions, its flavor adjustment effect is reduced due to the formation of salt; when it encounters iron salts, it is purple-red, so the amount of product should be appropriate to avoid affecting the color of food. Currently, the most used aroma enhancers are maltol and ethyl maltol. Ethyl maltol's chemical properties are like maltol, but its aroma enhancement property is six times that of maltol.

10.8.2.4 Addition of Aroma Precursors

Addition of substances, such as carotene and ascorbic acid, to fresh tea leaves after fixing can enhance the aroma of black tea. The biggest difference between addition of aroma precursor and direct addition of food essence is that the aromas formed by precursors are more natural and harmonious. Currently, research in this area is also an important part of food flavor chemistry.

10.8.2.5 Enzyme Technology

Flavor enzymes are those enzymes that can be added to food to significantly enhance the flavor of food. The basic principle of using flavor enzymes to enhance food aroma is mainly in two aspects. On the one hand, the aroma substances in food may be free or bonded, and only free aroma substances can cause olfactory stimulation. Bonded aroma substances affect food aroma. The presentation is not contributing. Therefore, under certain conditions, the aroma substances existing in the form of bonding state in food are released to form free aroma substances, which will undoubtedly greatly improve the aroma quality of food; on the other hand, there are some precursors of aroma substances that can be transformed by enzymes in food. Under the action of specific enzymes, these precursors will be transformed into aroma substances and enhance the aroma of food. This research is also a hot spot in flavor chemistry.

The bonded aroma substances in food mainly exist in the form of glycosides. For example, many fruits and vegetables such as grapes, apples, tea, pineapple, mangoes, passionfruit, etc., have a certain amount of bonded aroma substances. Adding a certain amount of glycosidase to the wine can significantly improve the aroma, and adding a certain amount of glucosidase to the dried cabbage can make the aroma of the product stronger. In addition, some bonded aroma substances in food may also exist in the form of being embedded, adsorbed or wrapped on some macromolecular substances. For the release of this kind of bonded aroma substances, the corresponding macromolecular substances are generally hydrolyzed by hydrolytic enzymes. For example, adding pectinase in the processing of green tea beverages can release a large amount of linalool and geraniol.

There are many enzymes that catalyze the conversion of aroma precursors in foods, but more research has focused on polyphenol oxidase and peroxidase. Studies have shown that polyphenol oxidase and peroxidase can be used to improve the aroma of black tea, and the effect is obvious. Catalase and glucose oxidase can be used for terpene aroma substances in tea beverages and have a fixed aroma effect on tea beverages.

Any enzyme that can affect the flavor of food can be called a flavor enzyme. Under specific conditions, flavor enzymes can release bound aroma substances in food to form free aroma substances and can produce aroma components by acting on aroma precursors. Ripe fruits and vegetables contain free-form flavor substances; they also contain flavor precursors that can combine with sugars to form glycosides and other substances. Addition of flavor enzymes to foods can release bound aromas thus enhance and improve food flavor. By adding an amount of glycosidase to red wine can significantly increase wine aroma, while adding glucosidase to dried cabbage can intensify the product's aroma.

10.9 Summary

Flavor is an important indicator of food quality. It does not only affect appetite, but also has the potential to impact people psychologically and physiologically. Food flavor is the combination of all sensory aspects in the food ingested, with the most important being gustatory and olfactory senses. Gustatory perception is the sense produced when water-soluble compounds in foods stimulate chemoreceptors in the mucous membranes of the tongue; while olfactory perception is mainly caused by certain volatile compounds in foods stimulating the olfactory neurons in the nasal cavity. In most cases, the taste or smell produced by food is the result of a combination of numerous flavor substances and aroma substances. Different types of substances have varied flavor mechanisms, and different gustatory senses can interact with each other. Compare to gustation, olfaction is much more complex. This is not only reflected in the complexity of olfaction production mechanisms, but also the difficulties in quantifying the compounds contributing to food aroma. The aroma substances in foods are mainly formed by biosynthesis, pyrolysis, and addition of food flavoring. Food processing has a major impact on the formation of food aromas; therefore, measures should be taken to enhance and maintain food aroma.

Questions

1. What do the threshold value and aroma value of food refer to, and how does the interaction of taste substances affect the flavor?
2. Briefly describe the mechanism of sweet, sour, bitter, and fresh taste substances.
3. Briefly describe the formation and control methods of food aroma substances.
4. Taste characteristics of common sweeteners, sour agents, and umami agents in food.
5. What factors are related to the flavor of food?
6. Why do people always feel sweet and spicy first, then sourness and bitterness last?
7. Why does the dough emit an attractive fragrance after baking?
8. Why are artificially ripened fruits not as strong as naturally ripened fruits?
9. Why does the saying "If you want to be sweet, add salt first"?
10. Explanation of terms: flavor, threshold, aroma value, relative sweetness, contrast effect of taste, alternation effect of taste, elimination effect of taste, multiplication effect of taste, adaptation phenomenon of taste, spiciness, astringency, Umami.

Bibliography

1. Belitz, H.D., Grosch, W., Schieberle, P.: Food Chemistry. Springer-Verlag Berlin, Heidelberg (2009)
2. Bi, S., Xu, X., Luo, D., Lao, F., Pang, X., Shen, Q., Hu, X., Wu, J., Characterization of key aroma compounds in raw and roasted peas (*Pisum sativum* L.) by application of instrumental and sensory techniques. J. Agricul. Food Chem. **68**(9), 2718–2727 (2020)
3. Damodaran, S., Parkin, K.L., Fennema, O.R.: Fennema's Food Chemistry. CRC Press/Taylor & Francis, Pieter Walstra (2008)
4. Ding, N.K.: Food Flavor Chemistry. China Light Industry Press, Beijing (2006)
5. Furusawa, R., Goto, C., Satoh, M., Nomi, Y., Murata, M.: Formation and distribution of 2,4-dihydroxy-2,5-dimethyl-3(2H)-thiophenone, a pigment, an aroma and a biologically active compound formed by the Maillard reaction, in foods and beverages. Food Funct. **4**(7), 1076–1081 (2013)
6. Gonzalez-Barreiro, C., Rial-Otero, R., Cancho-Grande, B., Simal-Gandara, J.: Wine aroma compounds in grapes: A critical review. Crit. Rev. Food Sci. Nutr. **55**(2), 202–218 (2015)
7. Granvogl, M., Christlbauer, M., Schieberle, P.: Quantitation of the intense aroma compound 3-mercapto-2-methylpentan-1-ol in raw and processed onions (*Allium cepa*) of different origins and in other *Allium* varieties using a stable isotope dilution assay. J. Agric. Food Chem. **52**(10), 2797–2802 (2004)
8. Hu, K., Jin, G.J., Mei, W.C., Li, T., Tao, Y.S.: Increase of medium-chain fatty acid ethyl ester content in mixed H. *uvarum*/S. *cerevisiae* fermentation leads to wine fruity aroma enhancement. Food Chem. **239**, 495–501 (2018)
9. Kan, J.: Food Chemistry. China Agricultural University Press, Beijing (2016)
10. Li, M., Yang, R., Zhang, H., Wang, S., Chen, D., Lin, S.: Development of a flavor fingerprint by HS-GC-IMS with PCA for volatile compounds of tricholoma matsutake singer. Food Chem. **290**, 32–39 (2019)

11. Mahmoud, M.A.A., Buettner, A.: Characterisation of aroma-active and off-odour compounds in German rainbow trout (*Oncorhynchus mykiss*). Part II: Case of fish meat and skin from earthen-ponds farming. Food Chem. **232**, 841–849 (2017)
12. Nishimura, T., Egusa, A.S., Naga, A., Odahara, T., Sugise, T., Mizoguchi, N., Nosho, Y.: Phytosterols in onion contribute to a sensation of lingering of aroma, a koku attribute. Food Chem. **192**, 724–728 (2016)
13. Renault, P., Coulon, J., de Revel, G., Barbe, J.C., Bely, M.: Increase of fruity aroma during mixed T-delbrueckii/S-cerevisiae wine fermentation is linked to specific esters enhancement. Int. J. Food Microbiol. **207**, 40–48 (2015)
14. Smid, E.J., Kleerebezem, M.: Production of aroma compounds in lactic fermentations. In: Doyle, M.P., Klaenhammer, T.R. (eds.) Annual Review of Food Science and Technology, vol. 5, pp. 313–326 (2014)
15. Spaggiari, G., Di Pizio, A., Cozzini, P.: Sweet, umami and bitter taste receptors: State of the art of in silico molecular modeling approaches. Trends Food Sci. Technol. **96**, 21–29 (2020)
16. Wang, Z., Xu, S.Y., Tang, J.: *Food Chemistry*. China Light Industry Press, Beijing (2007)
17. Xie, B.Y.: *Food Chemistry*. China Science Press, Beijing (2011)
18. Yu, A.N., Tan, Z.W., Wang, F.S.: Mechanism of formation of sulphur aroma compounds from L-ascorbic acid and L-cysteine during the Maillard reaction. Food Chem. **132**(3), 1316–1323 (2012)
19. Zeng, L., Watanabe, N., Yang, Z.: Understanding the biosyntheses and stress response mechanisms of aroma compounds in tea (*Camellia sinensis*) to safely and effectively improve tea aroma. Crit. Rev. Food Sci. Nutr. **59**(14), 2321–2334 (2019)
20. Zhang, Q., Zhang, F., Gong, C., Tan, X., Ren, Y., Yao, K., Zhang, Q., Chi, Y.: Physicochemical, microbial, and aroma characteristics of Chinese pickled red peppers (*Capsicum annuum*) with and without biofilm. Rsc Advances **10**(11), 6609–6617 (2020)
21. Zhao, M.M.: Food Chemistry. China Agricultural Press, Beijing (2012)

Dr. Liyan Ma is a professor-level senior technician in the College of Food Science and Nutritional Engineering, China Agricultural University. Her research areas are mainly focused on food nutrition and safety. Dr. Liyan Ma has been engaged in the research of mycotoxins in food, the risk assessment of endogenous toxins, and the research of plant growth regulators in fruits and vegetables. She has published more than 70 research papers.

Dr. Jingming Li professor and doctoral supervisor in China Agricultural University, is an expert of Fruit Wine Technical Committee, branch of China Wine Industry Association, and a member of Wine and Fruit Wine Expert Committee in China Food Association. He is the national first-class wine taster. His research field is food flavor and functionality. He has carried out the research about the flavor of fruit wine (wine) and the functionality of fruits and vegetables, with the aid of new flavor research technologies such as electronic nose and electronic tongue. It has been established wine origin and variety identification, brandy aging year identification, and food safety evaluation, according to the methodology of volatile component analysis. Dr. Jingming Li and his team has investigated the essential oils, polyphenols and saponins in the by-products of fruit and vegetable processing, about their efficient extraction, purification and industrial applications. The relevant results have been awarded by the Ministry of Education, and Chinese Institute of Food Science and Technology as the first prize for technology invention. In total, Dr. Jingming Li has published more than 140 scientific papers.

Chapter 11
Harmful Food Constituents

Kewei Chen and Jianquan Kan

Abstract In addition to nutrients and non-nutritive ingredients that can impart color and flavor to food, food often contains some harmful constituents, which come from either the food materials themselves or food processing process, or microbial and environmental pollution. When the harmful ingredients in food exceed a certain amount, they can cause a hazard to human health. This chapter will describe the concept and source of harmful constituents from food according to their classification, and the relationship between the structure of harmful substances and their toxicity will also be discussed. After that, readers could be familiar with the characteristics of harmful ingredients in food and their absorption, distribution and excretion in the human body. Finally, the safety assessment methods of harmful ingredients in food will also be introduced.

Keywords Natural harmful substances · Lethal dose · Microbial toxins · Chemical toxins · Safety evaluation

11.1 Overview

11.1.1 The Concept of Harmful Constituents

Harmful constituents, or harmful ingredients or harmful substances, in food refer to substances that have been proven to cause considerable harm to humans and animals when ingested in enough amounts, which are also known as undesirable constituents or toxic substances or toxicants.

A substance that causes cell or tissue damage through mechanisms other than physical damage is called a *toxic substance*. The clinical state of toxic substances under certain conditions is called *intoxication* or *poisoning*. The ability of toxic substances to cause damage to cells and/or tissues is called *toxicity*. Substances with higher toxicity can cause damage with a smaller dose, while substances with

K. Chen · J. Kan (✉)
College of Food Science, Southwest University, Chongqing 400715, China
e-mail: ganjq1965@163.com

© The Author(s), under exclusive license to Springer Nature Singapore Pte Ltd. 2021 511
J. Kan and K. Chen (eds.), *Essentials of Food Chemistry*,
https://doi.org/10.1007/978-981-16-0610-6_11

lower toxicity require larger doses to show toxic effects. Therefore, when discussing the toxicity of a substance, other factors should also be taken into consideration, such as the amount (dose), the entrance mode (oral, respiratory or transdermal) and time distribution (administered once or repeatedly). The most basic factor is dose. Therefore, the following concepts should also be clarified:

(1) Lethal dose (**LD**): it literally refers to the dose that can cause animal death. But in fact, different LD levels reflect different results in the animal investigations, therefore, the following concepts should be further clarified for a lethal dose. Absolute lethal dose (**LD_{100}**): the lowest dose that can cause all deaths in a group of animals.
Median lethal dose (**LD_{50}**): the lowest dose that can cause 50% of the death in a group of animals.
Minimum lethal dose (**MLD**): the highest dose that can cause only individual deaths in a group of animals.
Maximum tolerable dose (**LD_0**): the highest dose that can cause a group of animals to survive severe poisoning, but none of them die.

(2) **Maximum no-effect level**: it refers to the highest dose at which a certain substance can no longer cause biological changes to the body. Based on the maximum inactive amount, the acceptable daily intake (ADI) and the maximum allowable content or maximum residue limit in a certain food can be formulated.

(3) **Minimal effect level**: it refers to the lowest dose that can cause the body to begin to have a toxic reaction, that is, the minimum dose necessary to cause an observation index of the body to change beyond the normal range.
Among the abovementioned various dose-related concepts, LD_{50}, maximum non-effect level and minimum effect level are the three most important dose parameters.

(4) **Non-adverse effect**: it refers to not causing changes in the aspect of morphology, growth, development and lifespan to the body, and to not causing the reduction of functional capacity to the body or damage to the compensatory ability of additional stress state. The biological changes caused are generally reversible. After cessation of exposure to related chemicals, damage to the body's ability to maintain homeostasis cannot be detected, nor can the body susceptibility to adverse effects of other environmental factors be increased.

(5) **Adverse effect**: it is contrary to the non-adverse effect.

(6) **Effect**: it refers to the biological changes caused by exposure to a certain dose of chemical substances in the body. For example, exposure to certain organophos-phorus pesticides can cause cholinesterase activity to decrease, which is the effect caused by organophosphorus pesticides.

(7) **Response**: it refers to the proportion of individuals in a group that exhibits a certain degree of effect after exposure to a certain dose of chemical substances.

Therefore, the effect only involves an individual, that is, a person or an animal, and its intensity can be expressed in a certain measurement unit; while the response involves a population, such as a group of people or a group of animals, only a percentage (%) or ratio indicates the intensity of the reaction.

11.1.2 Source and Classification

Harmful constituents in food can be divided into four categories according to their sources: **natural toxicants**, **derived toxicants**, **contaminated toxicants** and **added toxicants**. Derivatives are produced during food storage, processing and cooking procedure. Contaminated toxicants and added toxicant pollutants and additives are extraneous sources.

According to the structure of harmful constituents, they can be divided into two categories: **organic toxicants** and **inorganic toxicants**.

According to toxicity, harmful constituents can be divided into extremely toxic, highly toxic, toxic, low toxic, practically non-toxic, non-toxic, etc.

11.1.3 The Impact of Harmful Food Constituents on Food Safety

In China, it has been stipulated that food safety refers to the actual assurance where it will not produce adverse reactions to the eater after long-term consumption under the prescribed usage and dosage conditions. Adverse reactions include not only general toxicity and specific toxicity, but also acute toxicity caused by accidental ingestion and chronic toxicity caused by long-term trace intake.

Current food safety issues involve **acute foodborne illness**, and **chronic food-borne hazard** with long-term effects. Acute foodborne illness includes food poisoning, intestinal infectious diseases, zoonotic diseases, enteric viral infections and parasitic diseases caused by intestinal infections. Chronic foodborne hazard includes the potential damage to health caused by toxic and harmful ingredients in food, such as interference with metabolism and physiological functions, carcinogenesis, teratogenesis and mutagenesis.

Therefore, there are many factors affecting food safety, including microorganisms, parasites, biotoxins, pesticide residues, heavy metal ions, food additives, releases from packaging materials and radionuclides and compounds produced during food processing and storage. In addition, deficiency or insufficient amounts of nutrients in food can also easily cause metabolic diseases such as malnutrition and growth retardation, which are also unsafe factors in food.

11.1.4 Research Methods of Harmful Constituents in Food

The study of harmful constituents in food includes three parts:

(1) The composition, structure, content, physical and chemical properties of harmful constituents, the form of existence in food or the external environment and the changes that occur during food processing and storage.

(2)   The process of distribution, metabolic transformation and excretion of harmful constituents in the body after being absorbed into the body along with the food.

(3)   The biological changes in the body caused by harmful components and their metabolites that enter the body with food, that is, the possible toxic damage to the body and its mechanism.

Therefore, the research methods of harmful ingredients in food should include two aspects: ***experimental research*** and ***population survey***. In terms of experimental research, the use of physical and chemical methods is to study the abovementioned harmful ingredients in the (1) part of the content. The use of biological methods, combined with physical, chemical, physiological and biochemical or molecular biology methods, such as animal experiments, is to investigate the (2) part, and the use of animal, microbe, insect or animal cell strain to carry out toxicity test is to study the (3) part. Population surveys are direct observations of the human body. It is mainly through the treatment of poisoning accidents that relevant information is directly obtained to understand the general health status, morbidity, possible related special diseases or other abnormal phenomena. The results of the population survey can be mutually confirmed with the results of animal toxicity tests.

Finally, based on the results of the above experiments to clarify the composition, structure, physical and chemical properties of the toxic ingredients in the food and their changes during food processing and storage, and to determine their safety or toxicity to the human body, preventive measures or relevant safety standards will be formulated scientifically.

11.2   The Relationship Between the Structure of Hazardous Substances and Their Toxicity

Different chemical structures result in different physical and chemical properties, which makes the toxicity of various harmful ingredients in food vary in the body. If the law of the relationship between chemical structure and toxicity can be found, it is possible to estimate or predict the toxic effects of chemical substances based on their structures, providing a basis for the design of animal toxicity experiments.

11.2.1   Functional Groups and Their Toxicity in the Structure of Organic Compounds

11.2.1.1   Hydrocarbons

The higher the degree of hydrocarbon unsaturation, the more active the chemical properties, and therefore the stronger the toxicity. When the carbon chain length is the same, unsaturated olefins are more toxic than alkanes, and alkynes are the most toxic.

Cycloalkanes are generally less toxic than corresponding alkanes. Hydrocarbons with side chains are generally less toxic than straight-chain hydrocarbons with the same number of carbon atoms.

Aromatic hydrocarbons are more toxic, but they mainly show inhaled toxicity. For example, benzene shows strong neurological and blood toxicity after inhalation. Those with alkyl side chains on the benzene ring are generally less toxic in the body, mainly exhibiting chronic toxicity, as the side chains are easy to oxidize, and finally form benzoic acid, which combines with glycine to form hippuric acid that is finally excreted in urine.

Polycyclic aromatic hydrocarbons with less than three rings, such as biphenyl, naphthalene and anthracene, are not carcinogenic. They are less water-soluble and difficult to absorb, so oral toxicity is not great, and they all have strong irritating odors, so they are not easy to cause acute poisoning.

Terpene olefins are cyclopentane compounds with unsaturated bonds, which have a strong tendency to generate peroxides and are therefore more toxic.

Contaminated hydrocarbons in food, mainly come from the residues of light gasoline (hexane and heptane) used as the extraction solvent of edible vegetable oil, and from the release of polyethylene, polypropylene and polystyrene from food packaging films, but they all show low oral toxicity.

11.2.1.2 Halogenated Hydrocarbons

Halogen has a strong electron-withdrawing effect, which can increase the polarity of halogenated hydrocarbon molecules and can be easily involved in enzyme systems in the body. Therefore, halogen is a strong toxic group. Therefore, halogenated hydrocarbons are generally more toxic than their parent hydrocarbons.

The toxicity of halogenated hydrocarbon compounds can be different due to different halogen elements, which are generally enhanced in the order of fluorine, chlorine, bromine and iodine; the more the halogen atoms, the higher the toxicity.

The toxic effects of various halogenated hydrocarbon compounds are not the same, but generally, they show the irritating and corrosive effect on the skin, mucous membranes and respiratory system, and most of them have anesthetic and invasive effects on the nervous system, and damage other organs such as liver and kidney.

There are many important substances related to contaminated toxicants in the form of halogenated hydrocarbons, such as organochlorine pesticides (BHC, DDT, etc.), herbicides containing various halogens (trifluralin, chlornidine, etc.), fumigants (methyl bromide, etc.), food packaging plastics (polyvinyl chloride, polytetrafluoroethane, etc.) and certain mycotoxins and related substances in industrial wastes (polychlorinated biphenyl, etc.).

11.2.1.3   Nitro and Nitroso Compounds

Nitro compounds are very toxic and are poisons that act on the central nervous system of the liver and kidneys and in the blood. They are mainly poisoned by inhalation and absorption through the skin.

When halogen, amino and hydroxyl are introduced into the molecule of an organic nitro compound, its toxicity can be increased, and when alkyl, carboxyl and sulfonic acid groups are introduced, the toxicity is weakened; generally, the more the nitro groups, the stronger the toxicity.

Nitroso compounds are like nitro compounds but are more toxic. Nitrosamines are carcinogens.

These nitro compounds have little chance of being present in food by direct contamination, but they still need special attention as they could be the precursors of pesticides and other food contaminants. For example, both nitrochloroform used as a food fumigant and the herbicide trifluralin have the precursors of dinitrotoluidine derivatives.

11.2.1.4   Amino Compounds and Azo Compounds

The toxicity of amino compounds is different. Aliphatic amines and aromatic amines without substituents are toxic, especially aromatic amines, such as aniline, toluidine, benzidine, β-naphthylamine, β-naphthol, chloroaniline, diphenylamine and 2-Phenylbenzylamine and so on. The introduction of carboxyl or hydroxyl groups can reduce toxicity.

Azobenzene, aminoazobenzene and diaminoazobenzene have similar toxic effects with aniline.

Aliphatic amines mostly appear in the process of food spoilage, such as methylamine, dimethylamine, trimethylamine and various amines with longer carbon chains. Synthetic pigments such as basic malachite green, methyl violet, basic fuchsin and basic bright green have the precursors of aniline; butter yellow (ρ-dimethylaminoazobenzene) is a typical oil-soluble azo pigment.

11.2.1.5   Nitrile and Urea

Aliphatic nitriles, aromatic nitriles and dinitriles all have obvious toxicity, and cyanohydrin is very toxic. One of the confirmed toxic substances in *Lathyrus sativas* L. is β-N-(γ-L-glutamyl)-aminopropionitrile.

Urea itself is not very toxic, but urea herbicides, barbiturates, pyrimidines and xanthines have urea precursors, which are generally not highly toxic in the mouth.

11.2.1.6 Alcohol and Phenol

Among the aliphatic monohydric alcohols, butanol and pentanol are the most toxic, and other monohydric alcohols with carbon atoms have lower toxicity, except for methanol, which is more toxic.

Among alcohols with the same carbon number, isomeric alcohols are less toxic than n-alkanols. The toxicity of cycloalkanol compounds such as cyclopentanol and cyclohexanol is like that of cycloalkanes. Polyols are generally very toxic. Aromatic monohydric alcohols, such as benzyl alcohol and phenethyl alcohol, have a certain degree of oral toxicity. Halohydrins are very toxic.

The toxicity of phenols is stronger than that of the corresponding aromatic hydrocarbons and similar cycloalkanols, and gradually decreases with the increase of the number of carbon atoms in the side chain. The toxicity of polyphenols is mostly less than that of phenol. The toxicity of naphthol is like that of phenol, but lower.

The toxicity of halogenated phenol compounds is higher than that of the parent phenol, and it increases with the increase of the number of halogen atoms.

11.2.1.7 Ethers

Aliphatic lower ethers have anesthetic and stimulating effects, and their anesthetic effects are stronger than the corresponding alcohols. If there are double bonds and halogens in the molecule, the anesthetic effect will be weakened and the irritation will increase.

Aromatic ethers can be used as raw materials for food flavor, with varying toxicity. Cyclic ethers have certain toxicity, such as double bonds and halogens in the molecule can enhance their irritation. Ether, mainly with a narcotic effect, has little oral toxicity and may cause less food contamination.

11.2.1.8 Aldehydes and Ketones

The toxicity of aldehydes gradually decreases with the lengthening of the molecular carbon chain; when there are double bonds or halogens in the molecule, the toxicity increases. The toxicity of ketones is like that of aldehydes. The increase in molecular weight, the presence of unsaturated bonds and the substitution of halogens can all increase the toxicity. Generally, aliphatic ketones are more toxic than aromatic ketones. The toxicity of aliphatic lower ketones and their halogen substitutions increases in the following order: acetone, monochloroacetone, monobromoacetone and monoiodoacetone.

11.2.1.9   Carboxylic Acids and Esters

If carboxyl groups are introduced into organic compounds, their toxicity is reduced or disappears. The oxalic acid and citric acid in the polybasic acid can combine with the calcium in the blood and tissues, so they have special toxicity, but they are also normal metabolites in the body, so the poising effect mainly depends on the dosage.

Aromatic monobasic acids are generally not very toxic. Phenylene dibasic acids, such as the meta and para isomers of phthalic acid, have low oral toxicity. The toxicity of aromatic acids above ternary is unclear. The oral toxicity of hydroxycarboxylic acid is lower than that of carboxylic acid.

The toxicity of esters is generally more closely related to acids than alcohols, and generally stronger than acids. Methyl esters are more toxic than higher fatty acid esters, and most of the ethyl, propyl, butyl and pentyl esters of other lower fatty acids (below capric acid) have little oral toxicity. Salicylate has chronic toxicity, and the toxicity of oxalate is like that of oxalic acid. Lactone is generally toxic, and some have carcinogenic or pro-carcinogenic effects.

Phosphate pesticides, or organophosphorus pesticides, show the general formula as follows:

$$\begin{array}{cc}
\underset{R_2O}{\overset{R_1O}{>}}\overset{\overset{\displaystyle O\ or\ (S)}{\|}}{P}\!-\!X
&
\underset{R_2O}{\overset{R_1O}{>}}\overset{\overset{\displaystyle O\ or\ (S)}{\|}}{P}\!-\!O\!-\!\overset{\overset{\displaystyle O\ or\ (S)}{\|}}{P}\underset{OR_4}{\overset{OR_3}{<}}
\end{array}$$

Phosphate esters   Pyrophosphate ester

The toxicity of phosphate pesticides is mainly related to its R group and non-alkyl X group. In the R group, as the number of carbon atoms increases, the toxicity increases accordingly. In terms of non-alkyl X groups, for example, when X is a benzene ring, the toxicity of the substituents on the benzene ring is different; in general, the approximate order of toxicity of various substituents is $-NO_2$, $-CN$, $-Cl$, $-H$, $-CH$, tertiary-C_4H_9, $-CH_2O$ and $-NH_2$ in decreasing manner, and the relationship among the position of the $-NO_2$ group on the benzene ring and the toxicity is para position > ortho position > meta position; in addition, $-P=O$ is more toxic than $-P=S$.

11.2.1.10   Mercaptans, Thioethers and Thioureas

Mercaptans are mainly malodorous and not toxic. Although they have the paralysis effect on the central nervous system, they show mainly inhalation toxicity, which is similar to the toxicity of aromatic thiols.

Both sulfide and disulfide are narcotic but only have inhalation toxicity. The typical representative of halogenated sulfide gas is mustard gas (dichlorodiethyl sulfide), which is a highly toxic gas that can corrode skin and mucous membranes, so it is called an erosive gas.

Thiourea is highly toxic and can cause cancer, and various derivatives of thiourea vary in toxicity.

11.2.1.11 Sulfonic and Sulfinic Acids, Sulfones and Sulfoxides

After the toxic compound is introduced with the sulfonic acid group, the toxicity will be reduced, and in some cases, the carcinogen can also lose its carcinogenicity. Hydrocarbons, esters and compounds containing nitro and amino groups that are generally toxic to blood and nerves can reduce their toxicity or even completely lose their toxicity after being sulfonated. Sulfuric acid is like sulfonic acid, and oral toxicity is generally not significant.

Sulfones and sulfoxides are not toxic by themselves, and their toxicity is determined by other substances combined with them. Diphenyl sulfone or diphenyl sulfoxide is not very toxic, but when the sulfone or sulfoxide is combined with halogen or reduced to sulfide, the irritation is much more obvious.

11.2.2 Toxicity of Inorganic Compounds

The toxicity of inorganic compounds has no certain rules, but the toxicity of various inorganic compounds can be roughly predicted based on the following points.

11.2.2.1 Metal Poison

Firstly, the toxicity of inorganic compounds is related to their solubility. Generally, metals themselves are less soluble in water than their salts, so they are less toxic.

Secondly, some metal-organic compounds are easier to absorb than inorganic compounds, so they are more toxic. For example, the absorption rate of inorganic mercury is only 2%; mercury acetate, about 50%; phenyl mercury, 50–80%; and methyl mercury, even 100%. Therefore, its toxicity increases in the above order, and the difference is more obvious.

Thirdly, the same metal often has different valences. For example, arsenic has trivalent and pentavalent. In general, the lower valence is more toxic, except for chromium. The hexavalent chromium is more toxic than the trivalent chromium.

11.2.2.2   Redox Agents, Acids and Bases

Compounds with strong oxidizing ability are generally more toxic; the toxicity of acids or bases depends on the degree of dissociation in water. The greater the dissociation degree of strong acid and strong base, the greater the harm to the body. So, weak acids and weak bases shows less toxicity compared with strong ones.

11.2.3   The Relationship Between the Physical and Chemical Properties and Toxicity of Harmful Substances in Food

11.2.3.1   Oil–Water Partition Coefficient

The distribution rate of a substance in oil and water is often a constant ratio, which is the oil–water partition coefficient of the substance. A large oil–water partition coefficient indicates that it is easily soluble in oil; otherwise, it is easily soluble in water.

Since lipophilic substances are easier to penetrate the lipid bilayer of biological membranes and enter tissue cells, their toxicity is relatively stronger than that of hydrophilic substances. Therefore, substances with a larger oil–water partition coefficient are more toxic than those with a small oil–water partition coefficient.

As for the hydrophilic substances themselves, those with higher solubility in water are relatively easier to be absorbed than those with lower solubility in water, and their toxicity is also higher. On the other hand, substances with higher water solubility are easily excreted from the body, so their toxicity can be reduced. If a chemical substance contains polar groups, its toxicity can be reduced.

Substances that are not easily soluble in oil and water have low toxicity, such as many metal elements and higher alkanes such as paraffin wax.

11.2.3.2   Optical Isomerism

If there is optical isomerism of harmful substances in food, body tissues or enzymes usually can only interact with one optical isomer, always with L-isomer, while D-isomer is not very active in vivo, showing low or even no biological activity at all. But there are exceptions. For example, the L-isomer and D-isomer of nicotine are equally toxic in the body.

11.2.3.3 Electronegativity of Groups

If harmful substances in food are combined with negatively charged groups, they will form a "positive center of electricity" in the molecule due to the influence of electron attraction. Here, the electron cloud density is significantly reduced, and the negative charge of the receptor is attracted to each other and firmly binds, that is, toxicity occurs, and the stability and toxicity of the binding substance to the receptor can be predicted.

11.3 Overview of Various Harmful Constituents in Food

As mentioned earlier, harmful substances in food come from the raw food materials themselves, or are produced during food processing, or microbial contamination or environmental pollution, etc. This section will briefly introduce them.

11.3.1 Natural Harmful Constituents in Food

Natural harmful substances in food mainly refer to some toxic natural components contained in some animals and plants, such as tetrodotoxin. Some animal and plant foods may also form certain harmful substances due to improper storage. For example, improper storage of potatoes can produce solanine after germination. In addition, due to some special reasons, food can also be poisonous. For example, honey is non-toxic, but when the plant in the nectar source contains toxins, it will cause poisonous honey, and it will also cause poisoning after accidental ingestion.

11.3.1.1 Natural Harmful Substances in Plant Foods

1. Toxic plant protein and amino acids

(1) Lectin

Beans and some bean seeds (such as castor) contain a protein that can agglutinate red blood cells, called hemagglutinins, or lectins for short.

Lectin can agglutinate blood cells through highly specific binding with the blood cell membrane and can stimulate the division of cultured cells. After oral administration of black bean agglutinin to rats, the absorption of all nutrients was significantly reduced. In the isolated bowel test, it was observed that the glucose absorption rate through the intestinal wall was 50% lower than that of the control group. Therefore, it is speculated that the role of lectin is to combine with intestinal wall cells, thereby affecting the absorption of nutrients by the intestinal wall.

Eating raw beans can cause symptoms such as nausea and vomiting, and in severe cases can be fatal. All lectins are destroyed during wet heat treatment, but not easily destroyed during dry heat treatment. Therefore, heat treatment, hot water extraction and other measures can be taken to detoxify.

(a)   **Soybean lectin**. Soybean lectin is a glycoprotein with a molecular weight of 110,000. The sugar part occupies 5%, mainly with mannose and N-acetylglucosamine. Animals that eat raw soybeans need more vitamins, minerals and other nutrients than animals that eat cooked soybeans. Steam treatment under normal pressure for 1 h or high-pressure steam (9.8×10^4 Pa) for 15 min can inactivate it.

(b)   **Lectin from *Phaseolus***. Legumes of the genus *Phaseolus*, such as *Phaseolus vulgaris*, mung bean, kidney bean and runner bean, all have lectins, which have a significant effect on inhibiting the growth of feeding animals and can be fatal at high doses. The treatment of high-pressure steam for 15 min can completely deactivate lectins. Other legumes such as lentils and broad beans have similar toxicity.

(c)   **Ricin**. Although castor beans are not edible seeds, there are cases where castor oil is heated and eaten in folk cultures. Ingestion of castor bean or castor oil can cause poisoning, vomiting, diarrhea, and even death for animals and human beings. The harmful ingredient in castor is ricin, which is the first plant lectin to be discovered. It is extremely toxic and is 1000 times more toxic than legume lectin to mice. Heat treatment with steam can be used for detoxification.

(2)   **Protease inhibitor**

Substances that can inhibit the activity of certain proteases are widely found in plants, called proteinase inhibitors, which belong to the category of anti-nutrient substances and have a relatively important impact on the nutritional value of food.

The most important protease inhibitors are trypsin inhibitor, chymotrypsin inhibitor and α-amylase inhibitor. Trypsin inhibitors are mainly found in foods such as soybeans and other legumes and potato tubers. When these foods are eaten raw, trypsin is inhibited, which causes pancreatic enlargement. Alpha-amylase inhibitors are mainly found in foods such as wheat, kidney beans, taro, immature bananas and mangoes, and affect the digestion and absorption of sugars.

High-pressure steam treatment or normal pressure cooking or microbial fermentation after soaking can effectively eliminate the effect of protease inhibitors.

(3)   **Allergens**

Allergy refers to the immunological reaction caused by the exposure (ingestion) of a foreign substance. This foreign substance is called an allergen.

The immune response caused by food ingredients is mainly an immediate allergic reaction mediated by immunoglobin E (IgE). The process firstly is that B lymphocyte secretes allergen-specific IgE antibodies. The sensitized IgE antibodies and allergens

Table 11.1 Allergens in food

Food	Allergen	Food	Allergen
Milk	β-Lactoglobulin, α-whey protein	Peanut	α-Conarachin
Egg	Ovomucin, Ovalbumin	Soybean	Kunitz inhibitor, β-conglycinin
Wheat	Albumin, globulin	Kidney bean	albumin (molecular weight of 18000)
Rice	Gluten component, albumin (molecular weight of 15,000)	Potato	Undetermined protein (molecular weight of 16,000–30,000)
Buckwheat	Trypsin inhibitor		

are cross-linked on the surface of mast cells and basophils so that the mast cells release sensitive mediators such as histamine, resulting in an allergic reaction.

The main symptoms of allergies are skin eczema, neurogenic edema, asthma, abdominal pain, vomiting, diarrhea, dizziness, headache, etc. In severe cases, joint swelling and bladder inflammation may occur, leading to death. The specific allergic reaction is related to the physical fitness and the physiological period of the individual; and food allergies in general are more widely found in children than in adults.

In theory, any kind of protein in food is suitable for the immune system of a special population to produce IgE antibodies, thereby causing allergic reactions. But in fact, only a few types of food allergens are found, including animal foods such as milk, eggs, shrimp and marine fish, and plant foods such as peanuts, soybeans, kidney beans and potatoes (Table 11.1).

(4) **Toxic peptide**

The most typical toxic peptides are amatoxins (Fig. 11.1) and phalloidins (Fig. 11.2) that are present in toadstools.

Amatoxins are cycloheptapeptides, with six homologs. Phalloidin toxins are cyclooctides, with five homologs. Their toxic mechanisms are basically the same. Amatoxin acts on the nucleus of liver cells, and phalloidin acts on the microsomes of liver cells. Amatoxin is more toxic than phalloidin, but its action speed is slower and its incubation period is longer.

The clinical course of toxic peptide poisoning is generally divided into six stages: incubation period, gastroenteritis period, false healing period, visceral damage period, mental symptom period and recovery period.

The length of the incubation period varies with the proportion and the content of the two types of toxic peptides in toadstools, generally 10 ~ 24 h. Nausea, vomiting, diarrhea, abdominal pain, etc. appear at the beginning, which is called the gastroenteritis period. After the symptoms of gastroenteritis disappear, the patient has no obvious symptoms, or only has fatigue, and has no appetite, but the toxic peptide gradually invades the substantial organs, which is called the false healing period. At this time, patients with mild poisoning are not seriously damaged and can enter the recovery period. Severe patients enter the stage of visceral damage, with the liver,

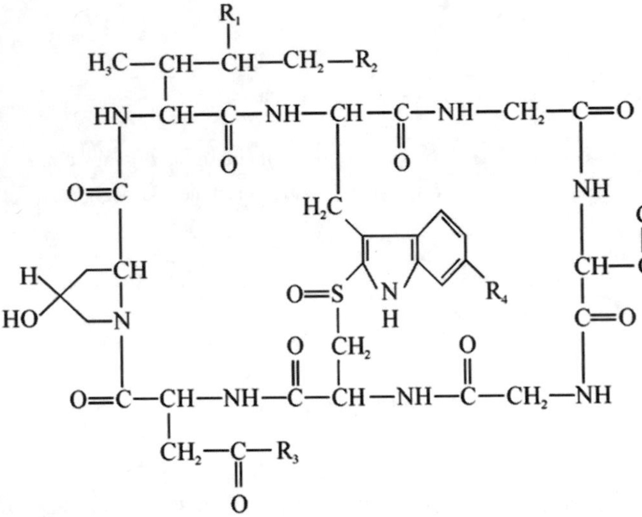

	R₁	R₂	R₃	R₄
α-amanitin	OH	OH	NH₂	OH
β-amanitin	OH	OH	OH	OH
γ-amanitin	OH	H	NH₂	OH
ε-amanitin	OH	H	OH	OH
amanin	OH	OH	OH	H
amaninamide	H	H	NH₂	OH

Fig. 11.1 Amatoxins

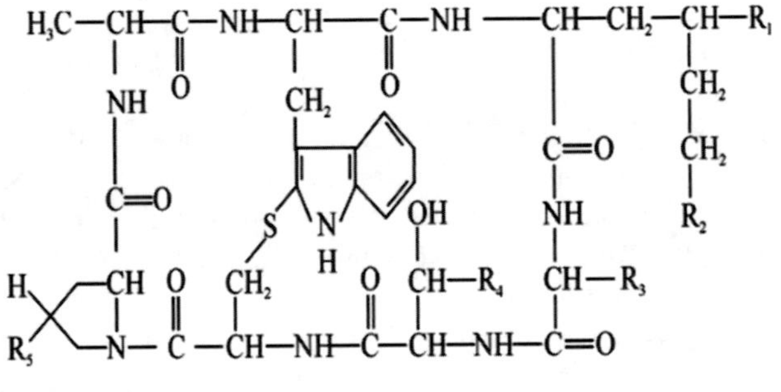

	R₁	R₂	R₃	R₄	R₅
Phalloidin	OH	H	CH₃	CH₃	OH
Phalloidin derivative I	H	H	CH₃	CH₃	OH
Phalloidin derivative II	OH	OH	CH₃	CH₃	OH
Phalloidin derivative III	OH	H	CH(CH₃)₂	COOH	OH
Phalloidin derivative IV	H	H	CH₂C₆H₅	CH₃	H

Fig. 11.2 Phalloidins

kidneys and other organs damaging, causing liver enlargement, and even acute liver necrosis, with a mortality rate as high as 90%. After active treatment, the patients generally enter the recovery period after 2–3 weeks, with the symptoms and signs gradually disappearing, and finally heal.

(5)   **Toxic amino acids and their derivatives**

(a)   *Lathyrus* **toxins**. Lathyrism is a food poisoning phenomenon caused by eating beans of the genus *Lathyrus*, such as vetch, chickpeas and garbanzos. There are two manifestations of lathyrism, osteopathic *Lathyrus* poisoning or osteo-lathyrism, and neuropathic *Lathyrus* poisoning, or neurolathyrism. The typical symptoms of poisoning are muscle weakness, irreversible paralysis of the legs and feet and even death.

There are two main types of toxins that cause lathyrism. one is the toxin that causes nerve paralysis, with three types of amino acids, namely α, γ-diaminobutyric acid, γ-N-oxalyl-α, γ-diaminobutyric acid and β-N-oxalyl-α, β-diaminopropionic acid; the others are toxins that cause bone deformities, such as β-N-(γ-glutamyl)-aminopropionitrile, γ-methyl-L-glutamic acid, γ-hydroxyvalerine and *Lathyrus* alanine.

$$H_2N \quad N \quad CH_2-CH-COOH$$
$$\underset{N}{\qquad} \qquad \underset{NH_2}{\qquad}$$

Lathyrus alanine

(b)   **β-cyanoalanine**. β-cyanoalanine is a neurotoxin found in broad beans and can cause the same symptoms as *Lathyrus* toxins.

$$CH_2-CH-COOH$$
$$\underset{CN}{\qquad} \underset{NH_2}{\qquad}$$

β-cyanoalanine

(c)   **Canavanine**. Canavanine is a homologue of arginine that exists in the genus *Canavalia*. For this reason, some species of canavalin are poisonous when eaten raw. Roasting or boiling for 15–45 min can destroy most of the canavanine.

$$HN=C-NH-O-CH_2-CH_2-CH-COOH$$
$$\underset{NH_2}{\qquad} \qquad\qquad \underset{NH_2}{\qquad}$$

Canavanine

(d)   **L-3,4-Dihydroxyphenylalanine (L-DOPA)**. Many plants contain a small amount of L-DOPA, but broad bean pods contain up to 0.25% of free or β-glycoside L-DOPA, which is an important cause of fava bean disease (favism).

The symptom is acute hemolytic anemia, occurring after 5–24 h of ingestion, and the acute attack period can be 24–48 h, and then patients can heal on their own. L-DOPA is also a medicine that can treat tremor paralysis (Parkinson's disease).

2. Toxic glycosides

Toxic glycosides in plant foods mainly include cyanogenic glycosides, glucosides and saponin.

(1) Cyanogenic glycosides

Many plant foods such as the nucleus of apricots, peaches, plums, loquats, cassava roots and flax seeds contain cyanogentic glycosides, such as amygdalin contained in bitter almonds, and linamarin in cassava and flax seeds.

The basic structure of cyanogenic glycosides is a glycoside containing α-hydroxy nitrile, and its carbohydrate components are often glucose, gentiobiose or viburnum. Due to the unstable chemical properties of α-hydroxy nitrile, it is decomposed into aldehydes or ketones and hydrocyanic acid by the enzymes and acids in the gastrointestinal digestion. After hydrocyanic acid is absorbed by the body, its cyanide ion combines with the iron of cytochrome oxidase, thereby destroying the role of cytochrome oxidase in transferring oxygen and affecting the normal respiration of tissue, and finally cause the body to suffocate and die. The clinical symptoms after poisoning are confusion, muscle paralysis, difficulty in breathing, convulsions and coma.

Cyanogentic glycosides can also be hydrolyzed to produce hydrocyanic acid under the action of acid, but the acidity in the stomach of ordinary people is not enough to hydrolyze them totally. Heating can inactivate the enzyme that converts cyanogenic glycosides into hydrocyanic acid to achieve the purpose of detoxification. As cyanogenic glycosides have good water solubility, cyanogenic glycosides can also be removed by rinsing.

Cyanogenic glycosides in common foods are shown in Table 11.2.

(2) Glucosinolate

Glucosinolates are found in *Cruciferae* such as cabbage, radish, mustard, green onions, garlic and other plants. They are the main components of the pungent taste of these vegetables and all contain β-D-thioglucose substitute. Various natural glucosinolates exist simultaneously with one or more corresponding glycosidases, but in intact tissues, these glycosidases do not contact the substrate, and only when the tissue is destroyed, such as homogenization, crushing or slicing process, the glycosidases contact with glucosinolates in broken and unheated tissue and quickly hydrolyze them into aglycone, glucose and sulfate.

Table 11.2 Cyanogenic glycosides in common foods

Glycosides	Plant sources	Hydrolysates
Amygdalin	*Rosaceae* plants, including almonds, apples, pears, peaches, apricots, cherries, plums, etc.	Gentiobiose + HCN + benzaldehyde
Prudomenin	*Rosaceae* plants, including laurocerasus, etc.	Glucose + HCN + benzaldehyde
Vicianin	*Vicia* plants	Podbiose + HCN + benzaldehyde
Dhurrin	*Sorghum moench* plants	D-glucose + HCN + p-hydroxybenzaldehyde
Linamarin	Tapioca, white clover, etc.	D-glucose + HCN + acetone

$$R-C=N-O-SO_3H \xrightarrow[\text{Glycosidase}]{H_2O} H_2SO_4 + \text{Glucose} + \left[\begin{array}{c} R-C=NH \\ | \\ SH \end{array} \right] \begin{array}{l} R-S-C\equiv N \quad \text{Thiocyanate} \\ R-N=C=S \quad \text{Isothiocyanate} \\ R-C\equiv N+S \quad \text{Nitrile} \end{array}$$

$S\text{-}\beta\text{-}D\text{-}$Glucose

Glucosinolates   Aglycone   Rearrangement

The aglycone undergoes molecular rearrangement to produce thiocyanate and nitrile. Thiocyanate inhibits iodine absorption and has an anti-thyroid effect. Nitrile decomposition products are toxic, and isothiocyanate can become goitrin (5-vinyloxazolidine-2-thione) through cyclization, preventing the absorption of iodine by the thyroid gland when blood iodine is low, thereby inhibiting the synthesis of thyroxine, and in this case, the thyroid gland also undergoes metabolic enlargement.

The edible parts of rapeseed, mustard, radish and other plants contain very little goitrin, but the content in their seeds is higher, reaching more than 20 times that of stems and leaves. In the comprehensive utilization of rapeseed meal, the development of rapeseed protein resources, or the use of rapeseed meal as feed, anti-thyroid substances must be removed.

(3) **Saponins**

This type of substance can be dissolved in water to form a colloidal solution. When stirred, it will produce foam like soap, so it is called saponin. Saponins can damage red blood cells and cause hemolysis, which is extremely toxic to cold-blooded animals. Saponins are widely found in the plant kingdom, but most of the saponins in food are not toxic to humans and animals when taken orally (such as soy saponins), and a few are highly toxic (such as solanin).

There are five known saponins in soybeans. The glycoside sugars include xylose, arabinose, galactose, glucose, rhamnose, glucuronic acid, etc. Aglucone is soyasaponin alcohol, including five kinds of homologues.

Solanin is a cholinesterase inhibitor. Excessive intake by humans and animals can cause poisoning. At first, the tongue and pharynx become numb, and heartburn, vomiting and diarrhea will occur. Then it will induce dilated pupils, tinnitus and excitement; in extreme cases, it will lead to severe convulsions and even death. Solanin is stable to heat and will not be damaged by cooking.

The content of solanin in potatoes is generally 30–100 mg/100 g, but the content of solanin in the area around the sprouts of germinated potatoes where the area turns green is extremely high, reaching 5 g/kg. It is generally considered safe to consume within the content of 200 mg/kg for potatoes.

β-D-glucose

3

D-galactose — O —

2

β-L-rhamnose

Solatrise

Solanidine

Solanin

3.   Alkaloids

Alkaloids are nitrogen-containing alkaline compounds present in plants, most of which are toxic.

There are not many varieties of alkaloids in food. The more important ones are solanine in potatoes and toxic alkaloids in certain mushrooms.

Solanine is higher in green and sprouted potatoes. Patients who accidentally ingest germinated potatoes have vomiting, diarrhea and dyspnea. In severe cases, they can die due to cardiopulmonary failure.

Xanthine derivatives such as caffeine, theophylline and theobromine are the most widely distributed excitatory alkaloids in food, and relatively speaking, these alkaloids are harmless.

Muscarine and bufotenine, which are found in *Amanita muscaria* and other poisonous *Agaric* mushrooms, are poisons with the symptoms of profuse sweating. In severe cases, nausea, vomiting and abdominal pain occur, and these alkaloids have hallucinogenic effects.

The psilocybin and dephosphorylated psilocybin, which are found in Mexican *psilocybe, Panaeolus retirugis* and other mushrooms, may cause confusion after being ingested. *Panaeolus retirugis* is distributed widely and is born on dung piles, so it is also called fecal fungus, laughing fungus or dancing fungus.

Toxic alkaloids mainly include pyrrolidine alkaloids, colchicine and saddle rhzomorph. Colchicine itself is non-toxic to the human body, but it is highly toxic after being oxidized into dicolchicine oxide in the body, with a lethal dose of 3–20 mg/kg body weight. Colchicine is present in fresh daylily, and the onset is a few minutes to ten hours after eating plenty of fresh fried daylily. The symptoms are mainly nausea, vomiting, abdominal pain, diarrhea, dizziness, etc. Fresh daylily is non-toxic after drying.

4. Gossypol

Cotton seed contains 0.15–2.8% of free gossypol, and most of it is transferred to cotton seed oil when the seed is freshly pressed. Cotton seed oil contains gossypol up to 1–1.5%.

Gossypol can cause swelling and bleeding of human tissues, neurological disorders, loss of appetite, weight loss and affect fertility.

The removal of gossypol can be done by chemical methods such as $FeSO_4$ treatment, alkali treatment, urea treatment, ammonia treatment, heat steaming, frying treatment and microbial fermentation.

Gossypol structure

11.3.1.2 Natural Harmful Constituents in Animal Food

Almost all poisonous animal foods are aquatic products. The toxins of aquatic animals can be divided into two categories: fish toxins and shellfish toxins.

1. Shellfish Toxins

Although the poisoning of marine shellfish toxin is caused by ingestion of shellfish, such toxins are not shellfish metabolites. This is due to that, the shellfish have ingested toxic dinoflagellate and effectively concentrated the toxins contained therein. This shellfish toxin is called saxitoxin, and its molecular formula is $C_{10}H_{17}N_7O_4 \cdot 2HCl$. Several derivatives of saxitoxin (STX) are also found in some other shellfishes (Fig. 11.3).

Saxitoxin is stable to heat and will not be destroyed during cooking. It is a neurotoxin, and nearly the most toxic among low-molecular toxins. A few minutes to a few hours after ingestion, the onset will begin with numbness of the lips, tongue and fingertips, followed by numbness of the legs, arms and neck, and then general movement disorders, accompanied by headache, dizziness, nausea and vomiting. In severe cases, death can occur within 2–24 h due to breathing difficulties.

Poisoned shellfish can be detoxified by stocking them in clean water for 1–3 weeks.

STX	R₁	R₂	R₃
Saxitoxin	H	H	H
Gonyautoxin-II	H	H	OSO_3^-
Gonyautoxin-III	H	OSO_3^-	H
Neosaxitoxin	OH	H	H
Gonyautoxin-I	OH	H	OSO_3^-
Gonyautoxin-IV	OH	OSO_3^-	H

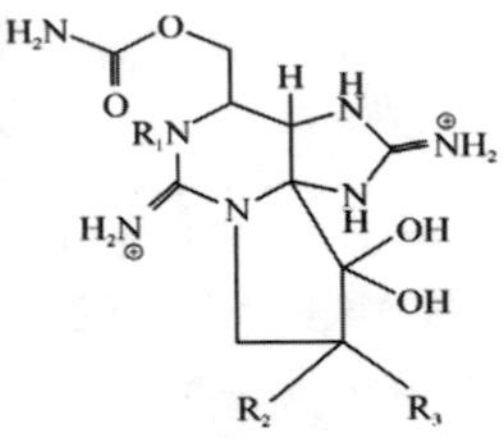

Saxitoxin structure

Fig. 11.3 Saxitoxins and their derivatives

2.　Fish toxins

It is known that about 500 species of marine fish can cause human body poisoning, and their toxin sources are either endogenous or exogenous.

Tetrodotoxin is one of the most studied fish toxins. It mainly exists in the ovaries, liver, intestines, skin and eggs. Most of the pufferfishes, either from freshwater or seawater, are poisonous. The muscles of pufferfish are generally non-toxic, but some are also toxic.

Tetrodotoxin is amino perhydroquinazoline ($C_{11}H_{17}N_3O_8$), with molecular weight of 319. The pure product is colorless crystal, slightly soluble in water, soluble in dilute acetic acid and insoluble in absolute ethanol and other solvents; it is unstable when the pH is above 7 and below 3, and decomposes into tetrodonic acid, but the toxicity does not disappear. Tetrodotoxin shows extremely high temperature resistance, and it can be destroyed by heating at either 100 °C for 4 h, or 115 °C for 3 h, or 120 °C for 20–60 min, or above 200 °C for 10 min, so the sterilization conditions of canned food cannot make it completely inactivated.

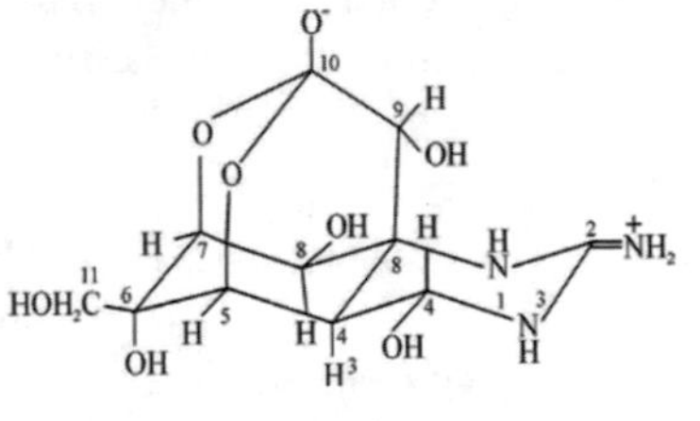

Tetrodotoxin structure

The toxicity of tetrodotoxin is mainly manifested in paralysis of the nerve center and nerve endings, and finally death due to paralysis of the respiratory center and vascular motor center.

Fish histamine toxins are produced by free histidine in fish tissues under the action of histidine decarboxylase in bacteria such as *Streptococcus* and *Salmonella*, as shown in Fig. 11.4.

The formation of fish histamine toxins is related to fish species and microorganisms. Mackerel, tuna, sardines and other fish can produce 1.6–3.2 mg/g histamine

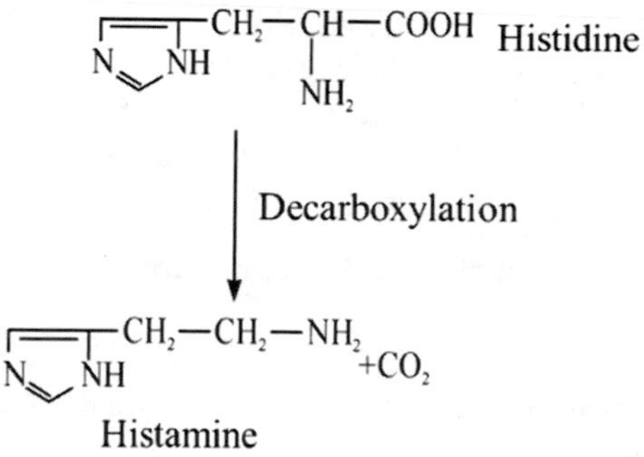

Fig. 11.4 Formation of fish histamine toxins

$$\underset{N\diagdown NH}{\boxed{}}\!-CH_2-\underset{\underset{NH_2}{|}}{CH}-COOH \quad \text{Histidine}$$

Decarboxylation

$$\underset{N\diagdown NH}{\boxed{}}\!-CH_2-CH_2-NH_2 \quad +CO_2$$

Histamine

when placed at 37 °C for 96 h. Under the same circumstances, sea bass only produces 0.2 mg/g histamine, and other freshwater fishes such as carp, crucian carp and eel also produce less histamine, 1.2–1.6 mg/kg. When the fish is not fresh or spoiled, the histamine content is higher.

Histamine toxin poisoning is caused by the expansion of capillaries and bronchoconstriction caused by histamine, which is mainly manifested as the flushing of the face, chest, whole-body skin and conjunctival hyperemia. The patient can recover within 1–2 days.

11.3.2 Microbial Toxins

Many food-contaminated microorganisms can produce toxins harmful to humans and animals, some of which are carcinogens and highly toxic.

11.3.2.1 Bacterial Toxins

The most important bacterial toxins are *salmonella* toxin, staphylococcal enterotoxin and *botulinum* toxin.

1. **Salmonella toxin**

Among bacterial food poisoning, the most common is food poisoning caused by *Salmonella*. The most common ones are *Salmonella typhimurium*, *Salmonella enteritidis*, *Salmonella choleraesuis* and *Salmonella paratyphi C*.

Salmonella itself does not secrete exotoxins, but it produces highly toxic endotoxins. Food poisoning caused by *Salmonella* is usually caused by swallowing many bacteria at one time. After the bacteria are destroyed in the intestine, they release enterotoxin and cause symptoms. The incubation period is generally 8–24 h. The general symptoms are acute gastroenteritis symptoms such as sudden onset, nausea, vomiting, diarrhea and fever. The course of the disease is short, and usually recovers within 2–4 days. In severe cases, it can be fatal.

Salmonella poisoning is mostly caused by animal food. Although these bacteria breed in meat, milk, eggs and other foods, they do not decompose protein to produce indole-like odorous substances. Therefore, if cooked meat and other foods are contaminated by *Salmonella* and have even grown to a considerable degree, they usually have no sensory abnormality, so the change is difficult to detect.

2. **Staphylococcal enterotoxin**

The most famous *Staphylococcus* is *Staphylococcus aureus*, some of which produce enterotoxins, known as A, B, C, D, E and other types, and the most common are A and D.

Enterotoxin is a protein with a molecular weight of $3–3.5 \times 10^4$. It is relatively heat-resistant and will not be destroyed under normal cooking conditions. It needs to be cooked at 100 °C for 2 h or heated at 218–248 °C for 30 min to inactivate.

Symptoms of enterotoxin poisoning include salivation, nausea, vomiting, cramps and diarrhea. Most patients return to normal state after 1–2 days, and rarely die.

3. *Botulinum* **toxin**

There are seven serotypes of toxin-producing *botulinum* known in nature: A, B, C, D, E, F and G, of which A, B and E are often related to human botulism.

Clostridium botulinum can produce highly toxic protein-like exotoxins, and its spores are extremely heat-resistant. Insufficient sterilization of canned meat often causes canned food to deteriorate, which can cause poisoning if eaten. *Botulinum* toxin mainly acts on the synapses of the peripheral nervous system, blocking the release of acetylcholine from nerve endings, causing muscle paralysis, and most patients die from suffocation due to paralysis of the diaphragm and other respirators.

11.3.2.2 Mycotoxins

Mycotoxins refer to toxic substances produced by molds in the metabolic process. There are more than 150 types of mycotoxins that have been discovered, including liver toxins such as aflatoxin, sterigmatocystin, luteoskyrin and oxygenated peptides; some are kidney toxins such as citrinin; some are neurotoxins such as citreoviridin; some are hematopoietic tissue toxins, such as photoallergic dermatitis toxins; and some of these toxins are carcinogenic.

1. *Aspergillus* **toxin**

(1) **Aflatoxin**

Aflatoxin is a hepatotoxic metabolite produced by a few strains of *Aspergillus flavus* and *Aspergillus parasiticus*. According to the different colors of fluorescence emitted by aflatoxin under ultraviolet light, it is divided into two major groups: B group and G group. At present, 17 kinds of aflatoxins B_1, B_2, G_1, M_1, etc. have been determined, among which B_1 is the most toxic and carcinogenic, followed by G_1 and M_1, and the others are relatively weak.

The molecular weight of various aflatoxins is from 312 to 346, and the melting point is 200–300 °C. They are hardly soluble in water, hexane petroleum ether, and soluble in solvents such as methanol, ethanol, chloroform, acetone and dimethylformamide. It has strong heat resistance, and starts to crack when heated to the melting point temperature, and it is rarely destroyed at normal cooking temperature. Sodium hydroxide can open the six-membered ring of the lactone of aflatoxin to form the corresponding sodium salt, which is soluble in water and can be washed away with water. Therefore, vegetable oil can be detoxified by alkali refining, and its sodium salt can be acidified with hydrochloric acid and then can be lactonized again.

Aflatoxin B_1 Aflatoxin B_1 salt

Humans are more sensitive to aflatoxin B_1. Acute poisoning can occur with a daily intake of 2–6 mg, which is mainly manifested as hepatitis symptoms such as vomiting, anorexia, fever, jaundice and ascites, and severe cases can lead to death. In addition, aflatoxin is the most carcinogenic substance currently known, which can induce tumors in a variety of animals and induce multiple tumors at the same time.

Aflatoxins mainly pollute grains, oils and their products, such as peanuts, peanut oil, corn, rice and cotton seeds. Beans are generally not susceptible to contamination.

(2)   Sterigmatocystin

Sterigmatocystin is mainly the final metabolite of *Aspergillus versicolor* and *Aspergillus nidulans*, and it is also the intermediate for the synthesis of aflatoxin by *Aspergillus flavus* and *Aspergillus parasiticus Speare* in the late stage. There are 14 homologues, and they are also liver cancer-causing toxins, which can also damage the kidneys, but their toxicity is relatively low and sterigmatocystin mainly appears on corn and other grains.

(3)   Ochratoxin

This is a class of toxins produced by *Aspergillus ochraceus* and *Penicillium viridicatum*. There are seven compounds with similar structures. Among them, Ochratoxin A is the most toxic. Animal experiments have shown that it can cause liver and kidney damage and enteritis. Ochratoxin mainly exists in corn, wheat, peanuts, soybeans, rice and other grains.

2. *Penicillium* Toxin

After harvesting and storage, rice is easily contaminated by *Penicillium* due to excessive moisture and mildew. The rice becomes yellow, which is called yellow rice. Various *Penicillium* toxic metabolites can often be separated from moldy yellow rice, among which are the toxins produced by *Penicillium islandicum*, *Penicillium citrinum* and *Penicillium chrysogenum*.

(1) **Silanditoxin**

A variety of toxins can be isolated from *Penicillium islandicum*, among which the important ones are luteoskyrin, cyclochlorotin and silanditoxin. Other toxins such as erythroskyrin, are also important.

(a) **Luteoskyrin**. It is also known as yellow-rice toxin, a bis-polyhydroxydihydroanthraquinone derivative with a molecular formula of $C_{30}H_{22}O_{12}$ and a melting point of 287 °C. It is soluble in fatty solvents. The oral LD_{50} of mice is 221 mg/kg, and LD_{50} of the intraperitoneal injection is 40.3 mg/kg. When in poisoning, it mainly causes liver disease.

(b) **Cyclochlorotin**. It is a highly toxic chlorinated peptide compound. The pure product is white needle-like crystals, soluble in water, and has a melting point of 251 °C. Cyclochlorotin is a fast-acting liver toxin that can interfere with glycogen metabolism. The oral LD_{50} of mice is 5.6 mg/kg body weight.

(c) **Silanditoxin**. It is also a chlorinated cyclic peptide, with similar physical and chemical properties as cyclochlorotin, and it is also a hepatic toxin that acts quickly.

(2) **Citrinin**

Citrinin has an inhibitory effect on the central nerve and spinal cord motor cells. When the poisoning starts, the limbs are paralyzed, and then the patient may die of breathing difficulties. After oral administration of the toxin at 5 mg/kg per day, half of the mice died.

The pure product of citrinin is lemon yellow needle-like crystals, with a melting point of 172 °C and a molecular weight of 250. It is soluble in absolute ethanol, chloroform and ether, but hardly soluble in water.

3. Fumonisin

Fusarium is one of the common molds that contaminate food and feed. According to the chemical structure and toxicity of fumonisin, it can be roughly divided into four categories: *Trichothecene* mycotoxins, zearalenone, butanolide and moniliformin.

(1) *Trichothecene* **mycotoxins**

Trichothecene mycotoxins are known to have about 40 homologues, all of which are colorless crystals, slightly soluble in water. They are stable in nature, not easily destroyed by common cooking methods, and show strong cytotoxicity, which makes the nucleus collapse in the bone marrow cells, thymus and the intestinal epithelial cells. Symptoms of poisoning are mainly dermatitis, vomiting, diarrhea, refusal to eat, etc.

(2)   **Zearalenone**

Zearalenone, also known as F-2 mycin, is a metabolite produced by *Fusarium*, such as *F. graminearum* and *F. tricinctum*, which is one of the most common contaminated toxins in corn, barley and other grains. The molecular formula is $C_{18}H_{22}O_5$, with a molecular weight of 318, and the melting point is 164–165 °C. It is insoluble water, carbon disulfide and carbon tetrachloride, and soluble in alkaline aqueous solution, ether, benzene, chloroform, dichloromethane, ethyl acetate, acetonitrile and ethanol.

Zearalenone has an estrogenic effect, and can make the uterus hypertrophy, inhibiting the normal function of the ovaries, thus causing miscarriage and infertility. Eating various pasta made from flour containing zearalenone can cause symptoms of central nervous system poisoning, such as nausea, chills, headache, mental depression and ataxia.

(3)   **Butenolide**

Butenolide is produced by a variety of *Fusarium* species, and it is a blood toxin that can cause inflammation and necrosis of animal skin.

(4)   **Moniliformin**

Fusarium moniliforme is one of the pathogens that parasitize plants, and it can produce moniliformin. Feeding horses with corn contaminated with *Fusarium moniliforme* may cause subcutaneous hemorrhage, jaundice, heart bleeding, liver damage, etc.

4.   Sweet potato poisoning

If moldy sweet potato is ingested by the body, this may lead to sweet potato poisoning. But this is not due to the toxin produced by the mold, but due to the secondary metabolite produced by the physiological response of sweet potato to the parasitic fungus after the sweet potato is contaminated with black spot pathogen and *Fusarium solani*. There is mainly ipomeamarone, ipomeanol and ipomeanine. The first two are liver toxins, and the latter one is a pulmonary edema factor.

Ipomeamarone, ipomeanol and ipomeanine are all oily liquids. The pure products are odorless and stable when the pH is neutral. They are easily destroyed by acid and alkali.

11.3.3   Chemical Toxins

11.3.3.1   Pesticide Residues in Food

Pesticides are divided into insecticides, fungicides, herbicides, acaricides, molluscicides, rodenticides, etc. with preparations of organic chlorine, organophosphate, organic mercury and inorganic arsenic have the strongest residual toxicity. Grain is the most polluted food by pesticides, followed by fruits and vegetables.

1.   **Organophosphate pesticides**

Organophosphate pesticides are the earliest pesticides synthesized by humans and are still widely used. Most of the early developments were high-efficiency and high-toxic varieties, such as parathion, methamidophos, chlorpyrifos and phorate, and then they gradually developed many high-efficiency, low-toxic and low-residue varieties, such as dimethoate, trichlorfon, malathion, diazinon and cyanophos.

Organophosphate pesticides have good solubility, are easily hydrolyzed, have a short residence time in the environment, can be quickly degraded, have small accumulation in animals, have the characteristics of rapid degradation and low residues, and have become the mainstay in many countries. Some of organophosphate pesticides have replaced organochlorines in the produce of insecticides.

Organophosphate pesticides have less pollution to food. Generally speaking, except for organophosphorus pesticides with strong systemic properties, the residues in food are reduced to varying degrees after washing and cooking.

Organophosphate pesticides are neurotoxins, which mainly competitively inhibit the activity of acetylcholinesterase, accumulating acetylcholine, a neurotransmitter in the nerve synapse and the central nervous system, and cause the central nervous system to become over-excited and cause symptoms of poisoning.

2.   **Organochlorine pesticides**

Organochlorine pesticides mainly include BHC, chlordane, aldrin, dieldrin, Lindane, heptachlor and toxaphene. Organochlorine pesticides have stable properties, strong fat solubility and long residual period. The long-term and large-scale use of organochlorine pesticides has caused long-term pollution of the environment, food and human body. Therefore, the use of organochlorine pesticides has been stopped since 1984 in China.

The overall situation of organochlorine pesticide residues in food is as follows. Animal food is higher than plant food. Food with more fat is higher than food with less fat. Pork is higher than beef and lamb. Freshwater products in aquatic products are higher than marine products. Pond products are higher than river and lake products. The degree of pollution in plant foods decreases in the order of vegetable oil, grain, vegetables and fruits.

Organochlorine pesticide poisoning mainly causes neurological diseases. In addition, it can also cause liver fatty disease, liver and kidney organ enlargement.

3.   Carbamate pesticides

Carbarmate is a new type of insecticide developed in response to the shortcomings of organochlorine and organophosphorus pesticides. It has the characteristics of strong selectivity, high efficiency, broad spectrum, low toxicity to humans and animals, easy decomposition and low residual toxicity. It has been widely used in agriculture, forestry and animal husbandry. The main varieties are metolcarb, carbaryl, aldicarb, carbofuran, isoprocarb. pirimicarb, etc.

Carbarmate pesticides are relatively stable under acidic conditions. They are easily decomposed when exposed to alkali, air and sunlight. The half-life in the soil is several days to several weeks.

The toxicity mechanism of carbamate is the same as that of organophosphate, which is a blocker of mammalian acetylcholinesterase and has mutagenic, teratogenic and carcinogenic effects. The toxic symptoms include characteristic cholinergic tears, salivation, miosis, convulsions and death.

4.   Pyrethroid pesticides

Currently, there are nearly 20 pyrethroid insecticides in use, and the main ones are cypermethrin, fenvaletate, deltamethrin, cyhalotrin, etc.

Pyrethroid pesticides are easily converted into polar compounds under the action of light and soil microorganisms, and are not easy to cause pollution. The residual period in crops is 7–30 days. It basically does not produce an accumulation effect in the organism, and it is not very toxic to mammals, and it shows mainly central nervous system toxicity.

5.   Herbicide

(1)   Chlorophenate

Chlorophenoxy acid esters are currently widely used herbicides, mainly 2,4-D (2,4-dichlorophenoxyacetic acid) and 2,4,5-T (2,4,5-trichlorophenoxyacetic acid). They are easily hydrolyzed into acids and are directly excreted in the urine. They have poor accumulation in the human body, so chronic poisoning is not common. Ingestion of lower doses of these substances can cause uncharacteristic muscle weakness; and large doses of these substances can cause progressive stiffness, ataxia, paralysis and coma.

(2)   Tetrachlorodiphenyl-P-dioxin

Tetrachlorodibenzo-P-dioxin (TCDD) is an important class of herbicides, with 22 different isomers. The chemical properties of TCDD are quite stable, and chemical decomposition occurs only at temperatures exceeding 700 °C. It is lipophilic and is closely combined with solids and other substances in the soil, and is easy to diffuse in the environment.

The acute toxicity of TCDD is not as sensitive for human as that for animals, but long-term intake of low-dose TCDD may cause harm to humans, including carcinogenic effects. Studies have shown that phenoxy herbicides containing TCDD can increase the occurrence of sarcoma in human muscle, nerve and adipose tissue.

11.3.3.2 Polychlorinated Biphenyl Compounds and Polybrominated Biphenyl Compounds

Polychlorinated biphenyl (PCB) and polybrominated biphenyl (PBB) are stable and inert molecules with good insulation and flame resistance, and are widely used in the industry. For example, they can be added to paint as a flame retardant and antioxidant; as a softener, they can be added to plastics, rubber, ink, paper and packaging materials. PCB and PBB are not easily decomposed by biological and chemical means, and they easily become pollutants by industrial waste. Due to their high stability and lipophilicity, it can accumulate in the food chain through various means, especially in aquatic organisms. Fish is the main source of PCB and PBB for human consumption. Poultry, milk and eggs also contain these substances. After entering the human body, PCB and PBB are mainly accumulated in adipose tissue and various organs. The poisoning is manifested as rash, pigmentation, edema, weakness and vomiting. The content of PCB and PBB in the patient's fat tissue is 13.1–75.5 mg/kg. The United States stipulates that the PCB residue in poultry should be less than 5 mg/kg body weight.

11.3.3.3 Heavy Metals

Metals with specific gravity above 4.0 are collectively referred to as heavy metals. The general mechanism of heavy metal damage to the body is to combine with proteins and enzymes to form insoluble salts to denature proteins. When the functional proteins of the human body, such as enzymes and immune proteins are denatured and inactivated, the damage to the human body is great, and severe cases can often cause death.

Heavy metals enter the environment mainly due to industrial pollution, and enter the food chain through multiple channels. Humans and animals absorb and accumulate a large amount of heavy metals through food. In severe cases, symptoms of poisoning can occur. Among them, mercury, cadmium and lead are most important.

1. **Mercury**

Mercury mainly comes from the natural release of the environment and industrial pollution. Fish growing in industrially polluted waters can accumulate organic mercury. Mercury was once used as a fungicide to treat seeds, which also pollutes food crops. Mercury in the environment, through the methylation of the microbial community, forms more toxic alkyl mercury compounds, such as dimethyl mercury.

2.   Cadmium

Cadmium is one of the most important heavy metal polluting foods, which is mainly caused either through polluted water bodies and aquatic organisms, or by the absorption of cadmium-containing wastewater, waste residue, and waste gas by crops and pastures.

The content of cadmium in the human body increases with age. Cadmium enters the body with water and food, and mainly accumulates in organs such as the kidney and liver. The half-life of cadmium in the human body is 16 to 31 years. Long-term intake of food containing trace amounts of cadmium can accumulate in the body and cause chronic poisoning, which mainly damages the epithelial cells of the renal proximal tubules, and the symptom will manifest as proteinuria, diabetes and amino aciduria. As cadmium has a certain affinity for phosphorus, it can precipitate calcium in bones and cause osteoporosis, resulting in severe back pain, joint pain and tingling all over the body. It is now known that cadmium is harmful to individual development, so it can be teratogenic, affects zinc-related enzymes, interferes with metabolic functions, changes blood pressure, has carcinogenic effects and causes anemia. The human tolerance to cadmium is 0.5 mg/week.

3.   Lead

Lead in the human body mainly comes from food. According to measurement, the amount of lead absorbed by each person through contact per day is 300 μg, 90% of which comes from food. There are three main sources of lead in food. One is the residues of lead-containing pesticides on food and fruits; another is the environmental pollution caused by lead oxide emitted from the combustion of leaded gasoline and leaded paints and coatings. The other is the lead contained in the equipment and utensils used in food processing, storage and transportation, or in the ingredients used in food processing. For instance, lead alloys, enamels, ceramics, tin plates and preserved egg wraps, all contain lead.

The half-life of lead in organisms is about 4 years, and the half-life of lead deposited in bones is 10 years, so lead is easier to accumulate in the body, and it can become toxic when it reaches a certain amount. Lead mainly damages the nervous system, hematopoietic organs and kidneys. At the same time, there will be oral metallic taste, gingival lead line, gastrointestinal disease, neurasthenia, muscle aches and anemia. In severe cases, shock and death may occur. The human tolerance to lead is 3.5 mg/week.

4.   Arsenic

The main reasons for arsenic-contaminated food are the use of arsenic-containing pesticides in the field and the use of arsenic-containing auxiliary agents that do not meet sanitary standards during food processing. After arsenic enters the human body, it is mainly concentrated in the hair and nails rich in colloids, followed by bones and skin. The average arsenic content in adults is 18–21 mg. Arsenic is excreted slowly in the body and can cause chronic poisoning due to accumulation. The mechanism of poisoning is as follows. Trivalent arsenic combines with sulfhydryl-containing enzymes in the body to form a stable complex, which makes the enzyme inactive, hinders cell respiration, causes cell death and becomes toxic; it can also cause nerve cell metabolism to become obstructed, causing neurological diseases, such as polyneuritis. Pentavalent arsenic can also be toxic after being reduced to trivalent arsenic in the body. The human tolerance to arsenic is 0.35 mg/week.

11.3.4 Harmful Substances Produced in Food Processing

Due to the need for food processing, some toxic chemical components are intentionally or unintentionally introduced during food processing. For example, nitrites are often added to meat products, causing them to potentially form strong carcinogens, nitrosamines. Smoking introduces carcinogen benzopyrene and other fused ring aromatic hydrocarbons. In the food processing process, raw food materials will also change and produce some toxic substances, such as oxidized products of oils and fats.

11.3.4.1 Toxicity Caused by Food Additives

In the process of food production and processing, a certain dose of food additives is used within a certain range, which is harmless to the human body. But if abused, it may also cause various forms of toxic results, such as chronic poisoning, teratogenic, carcinogenic, and mutagenic effects. The main reasons for the toxic effects of food additives are as follows.

1. **Toxic damage caused by the metabolism and transformation of food additives**

After food additives enter the human body with food, some metabolites and chemical conversion products are toxic, and they can generally be divided into the following categories:

(1) Impurities generated in the manufacturing process, such as o-toluene sulfonamide in saccharin and 4-methylimidazole in the production of caramel pigment by ammonia method.

(2) The conversion of additives during food processing and storage, such as the conversion of asparagine sweet peptide to dicarboxypiperazine and the conversion of erythrosin into fluorescein.

(3) Food additives react with food ingredients to produce toxic products, such as the formation from diethyl pyrocarbonate to a strong carcinogen, ethyl carbamate, and the conversion from nitrite to yield nitroso compounds.

(4) Metabolic conversion products. For instance, saccharin is metabolized into cyclohexylamine in the body. Azo dyes metabolize to form free aromatic amines, etc.

2. Toxic damage caused by contaminated impurities in food additives

Contamination of harmful impurities in harmless additives can often cause serious poisoning events. For example, in 1955, in Japan, blended milk powder of "Morinaga" brand was added with disodium hydrogen phosphate containing 3–9% of arsenic as a stabilizer, resulting in a severe poisoning incident. The number of poisoned infants nationwide reached 12,131, where 131 infants died. Therefore, the specifications and grades of chemical substances used as food additives must not be ignored, and they must not be substituted casually.

3. Toxic effects of excessive nutritional additives

Some nutrients are often added as fortifiers in food processing, such as vitamins. There have been several reports of poisoning caused by excessive intake of vitamins. For example, excessive intake of vitamin A can cause chronic poisoning. The main symptoms are loss of appetite, headache, blurred vision, insomnia, hair loss, rash on the shoulders and back, dry skin, dandruff, cleft lip bleeding, nose bleeding, red gums and anemia.

Excessive intake of vitamin D can cause increased serum calcium, increased total cholesterol and excessive bone marrow calcium deposition. It can also cause lack of appetite, vomiting, irritability, constipation, weight loss and growth arrest.

4. Additives make some people allergic

A large part of allergic reactions with unknown causes may be caused by food additives. It is known that beverages with a synthetic pigment of lemon yellow can cause allergic reactions such as bronchial asthma, urticaria and angioedema. Saccharin can cause skin itching, sun allergy dermatitis and other symptoms. Many spices can also cause allergic reactions, such as rhinitis, cough, throat edema, bronchial asthma, constipation and edema arthritis.

Allergic reactions can occur in most tissues and organs, but the skin, respiratory organs and gastrointestinal system are most likely to occur.

11.3.4.2 Harmful Substances Produced in Food Processing

1. The formation of nitrates and nitrosamines

The sources of nitrates and nitrites in food contain two parts. One is the use of coloring agents in cured meat products; the other is excessive fertilization transferred from soil to vegetables. Under the action of nitrate reductase, nitrate is reduced to nitrite.

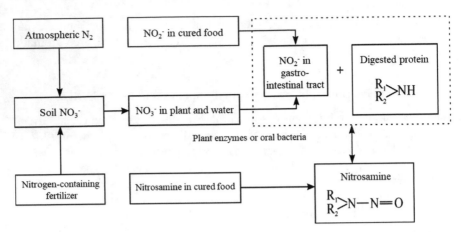

Fig. 11.5 Conversion of nitroso compounds

Under suitable conditions, nitrite can either react with amino acids in meat or cause secondary and quaternary amine reactions with the digested products of protein in the human gastrointestinal tract to form nitroso compounds (NOC). Especially, it produces carcinogens such as nitrosamine, so nitrite is also called an endogenous carcinogen. Figure 11.5 shows the conversion between nitrate, nitrite and nitrosamine.

The acute toxic effect of nitrite is to cause methemoglobinemia, that is, nitrite causes the ferrous ion of hemoglobin to be oxidized to ferric ion, and blood oxygen transport is severely blocked. This symptom is particularly likely to occur in infants, as they have low intestinal acidity and lack of diaphorase that reduces the ferric ions of heme to ferrous ions.

The chronic toxicity of nitrates and nitrites has three aspects. Firstly, it causes goiter, which interferes with normal iodine metabolism when the concentration of nitrate is high, resulting in a compensatory increase in the thyroid gland. Secondly, it induces insufficient intake of vitamin A, as long-term excessive intake of nitrate causes the oxidative destruction of vitamin A and hinders the conversion of carotene into vitamin A. Thirdly, it combines with secondary or tertiary amines to form nitroso compounds, many of which have strong carcinogenic effects.

2.   **Polycyclic aromatic hydrocarbons**

Polycyclic aromatic hydrocarbons (PAH) are volatile hydrocarbons produced when organic matter such as coal, petroleum, wood, tobacco and organic polymer compounds are incompletely burned. They are important environmental and food pollutants. Fats, sterols and other ingredients in food undergo high-temperature pyrolysis or thermal polymerization during cooking and processing to form fused ring aromatic hydrocarbons, which are the main source of fused ring aromatic hydrocarbons in food. There are many kinds of PAH, and more than 10 kinds have strong carcinogenic effects, among which the most detailed research is on benzo[a]pyrene.

Benzo[a]pyrene is a polycyclic aromatic hydrocarbon composed of five benzene rings.

Structure of benzo[a]pyrene

Benzo[a]pyrene is light yellow needle-like crystals at room temperature, with very stable properties. Its boiling point is 310–320 °C (1.3×10^3 Pa); melting point, 179–180 °C; and solubility in water, 0.004–0.012 mg/L. Benzo[a]pyrene is slightly soluble in ethanol and methanol and easily soluble in organic solvents such as cyclohexane, hexane, benzene, toluene, xylene and acetone. Benzo[a]pyrene does not interact with concentrated sulfuric acid at room temperature, but it can be dissolved in concentrated sulfuric acid and can react with nitric acid, perchloric acid and chlorosulfonic acid. This property can be used to eliminate benzo[a]pyrene. Benzo[a]pyrene is relatively stable under alkaline conditions.

PAH is a strong carcinogenic and mutagenic chemical substance, which can be dispersed to all parts of the body through the skin and fat with the blood circulation. PAH in the environment mainly causes skin cancer and lung cancer, while PAH in food mainly causes stomach cancer.

To prevent and reduce the pollution of PAH in food, attention should be paid to improving food cooking and processing methods to reduce pyrolysis and thermal aggregation of food ingredients. For example, direct flames should be not used to grill food. The management should be strengthened to monitor the water area and environment in the breeding industry. Active detoxification measures should be taken. For instance, grease can be detoxified with activated carbon, and grains can be detoxified by milling, sunlight or ultraviolet radiation, to reduce the content of PAH in food.

3. Oil oxidation and oil heating products

Many products from auto-oxidation and thermal changes of oils are extremely harmful to the human body.

(1) Auto-oxidation products of oil and fat and their toxicity

In the presence of oxygen, oils are prone to free radical reactions to produce various types of hydroperoxides and peroxides, which are then further decomposed to produce low molecular weight aldehydes and ketones. While the peroxide decomposes, it may also polymerize to form dimers and polymers. The auto-oxidation of fat not only reduces the nutritional value of the fat and deteriorates the smell and taste, but also produces toxic substances. Fat auto-oxidation mainly produces peroxide and 4-hydroperoxide alkenal. Peroxide has strong toxicity, which can destroy some enzymes of the body, such as succinate dehydrogenase and cytochrome oxidase.

Peroxide can also make vitamin A, vitamin D and vitamin E in the oil lose their activity, and induce illness due to a lack of essential fatty acids. It is generally believed that oils with a peroxide value below 100 will not cause adverse symptoms to animals, but considering factors such as individual differences and food intake, the peroxide value should not exceed 30. 4-Hydroperoxide alkenal is a secondary oxidation product produced by the oxidation of fats, and its toxicity is stronger than that of hydroperoxide. Mice were fed with grease containing 4-hydroperoxide alkenal, and died within 2 h. The reason is due to its small molecular weight, and it is easy to be absorbed by the intestines, making an indispensable enzyme system inactivated.

(2)   **Heating products of lipid and their toxicity**

If the lipid is heated for a long time at a high temperature above 200 °C, it is easy to cause various reactions such as thermal oxidation, thermal polymerization, thermal decomposition and hydrolysis, which will cause the lipid to foam, smoke, color and reduce the storage stability. The degraded fats have reduced nutritional value and may produce poisons as follows.

(a)   **Glyceride polymer**. This substance will be hydrolyzed into diglyceride or fatty acid polymer components in the digestive tract. Fatty acid polymers are difficult to decompose and are directly absorbed, and then transferred to tissues related to lipid metabolism, forming copolymers with various enzymes, thus hindering the action of enzymes.

(b)   **Cyclic compound**. The cyclic monomer is extremely toxic, and the thermal polymer above the dimer is less toxic as it is not easily absorbed.

4.   Maillard reaction products

In the baking process of foods such as bread, pastries and coffee, the Maillard reaction can produce attractive browning and unique flavor. The Maillard reaction is also the main cause of food browning during heating or long-term storage.

In addition to the formation of brown pigments, flavor substances and polymers, the Maillard reaction can also form many heterocyclic compounds. The mixture obtained from the Maillard reaction exhibits many different chemical and biological properties, including pro-oxidant and antioxidant properties, mutagenic and carcinogenic properties, and anti-mutagenic and anti-carcinogenic properties. In fact, the Maillard reaction induces the reaction of amino groups and carbonyl groups in biological tissues and leads to tissue damage, which was later proved to be one of the causes of damage to the biological system. During food processing, some products formed by the Maillard reaction have strong mutagenicity, suggesting that they may form carcinogens.

5.   Solvent extraction and toxin formation

Some compounds themselves are not toxic within a certain range of concentration but can produce toxic products after chemical reactions with food components. For example, in some countries, trichloroethylene has been used to extract oil from the seeds of various oil crops and to extract coffee from coffee beans. During the extraction process, the solvent interacts with the cysteine in the raw material to produce toxic S-(dichloroethylene)-L-cysteine, which induces aplastic anemia in animals when fed with the extracted residue (cake).

11.4 Safety Assessment Methods of Harmful Food Constituents

The safety evaluation method of hazardous substances in food is the same as the safety toxicology evaluation method of food, and can be carried out in accordance with the standard method in respective countries. In China, for instance, it is GB15193.1. Basically, it is stipulated in the standard that the toxicology evaluation procedure for food safety includes four stages, namely **acute toxicity test**, **genotoxicity test**, **subchronic toxicity test** and **chronic toxicity test**. In addition, there are the following principles and regulations.

(1) All newly created substances, especially those whose chemical structure suggests chronic toxicity, genotoxicity, or carcinogenicity, or those with large output, wide range of use and many intake opportunities, must undergo all four stages of toxicology evaluation.

(2) For derivatives or analogues that show basically the same chemical structure as known substances that present known results of safety evaluations and have been allowed for use, the first three stages of tests should be carried out, and the test results are used to determine whether the fourth stage test should be continued or not.

(3) For known chemical substances, where WHO has announced ADI for them, there have been data to prove that the quality specifications of domestic products are consistent with international products, and then the first and second stage tests can be carried out first. If the product quality or test results are consistent with international data, generally no further toxicity test is required, otherwise, the third stage test should be carried out.

To establish an effective method for food safety and toxicological analysis, that is, to rely on the effective system to make correct judgments in food toxicity assessment, and at the same time to reduce the number of experimental animals required for testing, saving expenses and time, the proposal called the "decision tree" is widely accepted by countries all over the world. The general procedure of the decision tree is shown in Fig. 11.6.

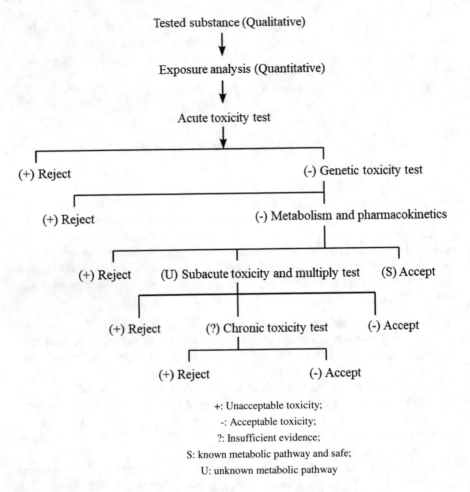

Fig. 11.6 Decision tree for food safety assessment

11.4.1 Qualitative Analysis of Harmful Substances in Food

The safety evaluation of hazardous substances in food firstly depends on the determination of hazardous substances in food, that is, the separation of hazardous substances from food and correct quantitative determination. Hazardous substance analysis methods include toxicity detection methods and hazardous substance separation methods.

First, it is necessary to accurately identify the substance to be tested. It is easier when the sample purity is high, as the chemical identification and standard procedures for pure substances are relatively complete. For a complex mixture, it is more complicated. The most important thing is to figure out the composition of the mixture and determine which component is toxic. Toxicity testing is usually to observe the effects of poisoning. Since humans are rarely used for testing, an "animal model" (usually a rat or mouse) must be selected for verification experiments.

Food should first be divided into different components, and the toxicity of each component should be monitored. The toxic components are further separated and tested until the pure poison is completely separated. The chemical structure of the poison can be confirmed by various spectroscopic analysis such as UV, IR, NMR and GC-MS.

11.4.2 Quantitative Analysis of Hazardous Substances in Food

The quantitative analysis method of each type of poison has a series of government supervision or stipulated quality standards to ensure that the poison in food is below the legal level.

The FAO/WHO International Codex Committee on Pesticide Residues (CCPR) published 183 recommended pesticide residue inspection methods in 1993. In China, the National Standard GB/T5009 also provides guidelines for the analysis of most controlled substances in food. And it is required to observe the following guidelines when choosing analysis methods.

(1) It should be used as per the publicly published analytical methods that have been proven by the collaborative research of multiple laboratories to be effective, and standard methods, such as AOAC methods, are preferred.
(2) It should be chosen as the method that can determine more than one residue, that is, multiple residues, and is applicable to as many commodities as possible under the conditions of or below the prescribed maximum residue limit.
(3) It can be used in laboratories equipped with daily work analytical instruments, and GC or HPLC is preferred as the recommended method for determination.

11.4.2.1 Sampling

If a certain type of poison is widely present in any food, the purpose of chemical analysis is to determine the amount of the poison to determine whether its content exceeds the standard. To achieve this goal, sampling methods and related statistical methods are very important to the analysis, as they are directly related to the determination of the measurement results. A repeated sampling at different sampling points helps to find differences. Since all parts of the food sample need to be equally exposed to the extraction process, it is often necessary to mix or chop the sample to achieve homogenization.

Samples should be screened and collected for a specific compound, as different compounds require different sample processing methods or their sampling methods conflict with each other. For example, samples containing some chemicals require alkali treatment to prevent acid degradation, while other chemicals require acid treatment to prevent alkali degradation. In this case, a general screening test may be required to detect a large group of chemical substances, and then samples need to be collected again to perform specific tests and analysis on the chemical substances found in the screening.

11.4.2.2 Extraction and Purification

Once the appropriate number of representative samples has been selected, the next step is usually to separate the analyte from the food matrix. The detection ability of an analytical method often depends on the recovery rate of the analyte and the sensitivity of the detector to the analyte. Since the levels of toxicants in food are generally low or even in trace amounts, it is often necessary to enrich and concentrate the toxicants in food before analysis.

After the sample is extracted, any processing step before placing the sample in the final analysis instrument is called purification. The purpose of this step is to reduce the amount of foreign chemical substances entering the sensitive analysis system, and to avoid the contamination of the injection port and analytical chromatographic column by the sample as much as possible. Purification can also prepare a certain amount of analyte for further separation. Liquid–liquid extraction, solid-phase extraction (SPE), supercritical fluid extraction (SFE) and purge-trap method (P&T method) are currently commonly used as extraction and purification methods.

Among the abovementioned methods, SPE is currently designated by the US Environmental Protection Agency (EPA) to extract and purify pesticide residues. The general procedure is as follows. The liquid or dissolved solid sample is poured into the activated solid-phase extraction column, and then vacuum or pressure is used to make the sample enter the stationary phase. The solid-phase extraction step will retain the components of interest and similar components, and meanwhile minimize the retention of unwanted sample components. The weakly retained sample components can be washed away with a solvent, and then another solvent is used to elute the analyte of interest from the stationary phase. Alternatively, the component of interest (analyte) can be passed directly through the stationary phase without being retained, while most of the interfering substances are retained on the stationary phase for separation. In most cases, the analytes are retained on the stationary phase, which is much better for sample purification.

11.4.2.3 Chromatographic Analysis

Chromatography is currently one of the highest resolution chemical separation methods. Chromatography can be divided into many types according to its mobile phases, such as high-performance liquid chromatography (HPLC), capillary gas chromatography (CGC), thin-layer chromatography (TLC), supercritical chromatography (SFC) and capillary electrophoresis (CE). Chromatography can separate and purify almost all chemical substances. Due to the flexibility of the separation method, the high efficiency of the analysis and the wide range of separated substances, the chromatographic method has a profound impact on chemical toxicology analysis.

11.4.3 Safety Evaluation of Hazardous Substances in Food

11.4.3.1 Preliminary Work

Before conducting a food safety assessment, another problem that needs to be solved first is to determine the intake level of the population for a certain substance. One of the methods to do this measurement is to conduct a whole diet analysis, that is, interview respondents to obtain information about the types of food they consume, and conduct a comprehensive analysis of all the chemical components (including residues of toxic substances) contained in such foods. Another method is the so-called "vegetable basket analysis", that is, buying food from retailers, processing them with traditional or representative methods and then analyzing certain food ingredients in question. Through analysis, the annual amount per capita consumption or exposure of a specific food ingredient can be calculated.

Exposure assessment refers to the qualitative and quantitative assessment of the biological, chemical and physical factors that may enter through food or other related routes. For example, the dietary pesticide residue exposure assessment should be based on the pesticide residue level and dietary consumption structure. The level of pesticide residues is mainly obtained by monitoring and analyzing the specific residues in food (mg/kg or μg/kg). Dietary consumption can be obtained through the whole diet study, expressed in kg/(person days).

11.4.3.2 Acute Toxicity

The first toxicity test is usually an acute toxicity test. Generally, a single dose of the test substance is repeatedly fed to rats or mice of both sexes within 24 h, and the poisoning effect that occurs in 7 days is recorded. The main purpose of the acute toxicity test is to determine the dose level of the test substance that causes the death of the experimental animal, that is, to determine the LD_{50}. If the LD_{50} of the substance is less than ten times the human possible intake, the next step will not be tested. In rare cases, a substance has strong acute toxicity and cannot be used in food, and the substance should be discarded. The data obtained from the acute toxicity test is usually used to determine the dose and method of the subsequent long-term toxicity test. Generally, the test substance is usually followed by a genotoxicity test, and a metabolism and pharmacokinetic study. The results of the acute toxicity test can only be used as a reference for the next phase of the test, and cannot be used as a basis for the safety evaluation of a certain test substance.

11.4.3.3 Genotoxicity

The primary purpose of the genotoxicity test is to determine the possibility that the tested chemical substance induces mutations in the test organism. The mutagenicity test is used to qualitatively indicate whether the test substance has mutagenic effects or potential carcinogenic effects.

Mutagenicity tests include point mutation analysis of microorganisms and mammalian cells; chromosome aberration of cultured mammalian cells and animal models, and transformation studies of human and other mammalian cells (tumor transplantation). The mutagenicity test is generally first chosen with the *Salmonella typhimurium*/mammalian microsomal enzyme test, namely Ames test, and other tests can be selected if necessary, such as the measurement of mouse bone marrow micronucleus rate and bone marrow cell chromosome aberration analysis, and mouse sperm abnormality analysis and analysis of testicular chromosomal aberrations.

If genetic toxicology studies find that a substance is mutagenic and has possible carcinogenicity, then the hazard evaluation of the substance can be carried out. If a substance has shown mutagenicity in several tests and is related to human carcinogenicity, and the use of the substance can significantly increase the exposure of the population to the substance, even if no further chronic tests have been made, this substance should be removed from the list for further use. If a substance is tested with a low risk of mutagenicity, for example, the substance has only been observed to be mutagenic in one test, or only in large doses that have shown mutagenicity in several tests, then it is necessary to make further analysis for this substance.

11.4.3.4 Metabolism Test

Metabolic testing is part of subchronic testing and should be performed after mutagenicity testing. The purpose of this test is to obtain qualitative and quantitative data on the absorption, biotransformation, deposition (storage) and removal characteristics of a substance in the organism after a single dose or repeated doses are ingested, such as gastrointestinal absorption, blood concentration, distribution of major organs and excrement content. If the biological effects of the metabolites from the substance are known, a decision can be made to use or reject the substance. For example, if all metabolites of the substance can be analyzed and they are known to be non-toxic substances, then the substance to be tested can be considered safe. If some metabolites of the substance are toxic or most of the parent substance is deposited in some tissues, further tests are required. Before evaluating the potential hazards of a substance, it is necessary to determine whether the metabolism of the substance in the experimental animal is the same as that in the human body. Therefore, an understanding of the metabolism and pharmacokinetics of a substance is essential to infer from the results of animal experiments to the result of similar hazards in the human body.

It is required in many countries that for the final evaluation of newly created chemical substances, at least the following metabolic tests should be carried out:

the absorption of the substance in the gastrointestinal tract, the concentration and biological half-life in the blood, and the distribution and excretion of the substance in major organs and tissues (urine, feces and bile). Under some specific circumstances, further separation and identification of metabolites can be carried out. For chemical substances that have been approved for use in many countries and have relatively complete toxicity evaluation data, metabolic tests may not be required for the time being.

11.4.3.5 Subchronic Toxicity Test

The period of the subchronic toxicity test ranges from several months to one year. The purpose is to determine the possible cumulative effect of the intake of a substance on the tissue or metabolic system. To measure the safety of a certain food ingredient, the subchronic toxicity test generally conducts a 90-day diet study on two test subjects, one of which should be a rodent. The subchronic toxicity test includes daily inspection of the appearance and behavior changes of the test animals, and weekly recording of body weight, food consumption and excrement characteristics. In addition to blood and urine biochemical tests, hematology and eye examinations are also performed regularly. In some cases, liver, kidney and gastrointestinal function tests, as well as blood pressure and body temperature measurements are also performed. After the experiment, all animals were autopsied again to check the total pathological changes, including changes in the weight of major organs and glands.

The maximum non-effect dose (MNEL) of a substance can be determined through a subchronic toxicity test. If the MNEL of a substance is less than 100 times the human possible intake, it means that the toxicity is strong and should be abandoned; if it is greater than 100 times but less than 300 times, a chronic toxicity test can be carried out; if it is greater than or equal to 300 times, it is not necessary to carry out chronic testing, and toxicity evaluation can be performed directly.

11.4.3.6 Teratogenicity Test

Teratogenic testing is an important aspect of subchronic testing. Teratogenesis can be explained as abnormal development at any time between the formation of the fertilized egg and the mature birth of the fetus.

Teratogenicity tests include short-term (1–2 days) tests in pregnant female animals during the period of fetal organ formation and tests during the entire pregnancy. The teratogenic test of short-term administration avoids the adaptation system of the mother to the metabolism and excretion of toxins, and avoids the occurrence of stillbirth. The subsequent administration should cover the various key periods of embryonic organ development, and monitor the cumulative effect of the poison in the maternal and fetal systems, that is, simultaneously monitor the metabolic activity of the substance in the maternal liver during pregnancy and the fetal system, and changes in the concentration and composition of metabolites. It is also necessary to monitor the saturation level of the substance in the storage site of the pregnant mother, which is closely related to the increase in the concentration of the test substance in the fetal system.

If the test conditions are consistent with the actual exposure of the substance, the toxic effects of the substance can be evaluated through the above acute and subchronic tests. If the consumption of a certain substance is large, or the substance has a compound structure that may cause cancer, or the substance shows cumulative toxicity in a subchronic toxicity test, or shows a positive result of a genotoxicity test, it cannot be made the final decision as non-toxic to the human body.

11.4.3.7 Chronic Toxicity Test

The main purpose of the chronic toxicity test is to evaluate the toxic effects that cannot be verified in the subchronic test when exposed to a relatively low level of a substance for a long time. The chronic toxicity test helps to find out whether a substance has progressive and irreversible toxic effects and carcinogenic effects, and finally determine MNEL, which provides a basis for the final evaluation of whether the tested substance can be used in food. The chronic toxicity test is a critical period in the life cycle of experimental animals. Appropriate methods and dosages of tested substances are used to feed the animal to observe the cumulative toxicity effects. Sometimes it can include several generations of tests. The carcinogenicity test is a chronic toxicity test to determine whether the tested substance or its metabolites have carcinogenic or tumor-inducing effects.

For deciding whether to accept a substance for food, the result of the chronic toxicity test is the conclusion. If the substance is not found to be carcinogenic in the chronic toxicity test, then based on the abovementioned acute and chronic toxicity test data and the intake level of the substance, a general risk assessment of the substance in food will be made. If a substance is proved to have carcinogenic activity, and there is a relationship between dose and effect, in most cases the substance is not allowed to be used as a food additive. If it is found that the design of the toxicity test is wrong or there are unexpected discoveries in the future, further tests are required.

11.4.4 Prospects for Safety Evaluation Methods of Hazardous Substances in Food

What has been introduced above is the traditional evaluation method, and now some alternative test methods that reduce or partially replace the use of laboratory animals have been proposed. This alternative method is to use microorganisms, cells, tissues, genetic animals (including virtual databases), etc. to predict the toxicity of exogenous chemicals on humans.

At present, there are government organizations in Europe and North America that are working in this area, such as the European Centre for the Validation of Alternative Methods (ECVAM) and the Interagency Coordinating Committee on the Validation of Alternative Methods (ICCVAM). The innovation of toxicology test methods requires mechanism knowledge, especially at the molecular level and genetic level, which is the basis for the establishment of alternative methods.

It has been proposed by ICCVAM that the determination of exogenous chemicals and carcinogens should reduce the dependence on animal tests, and that more use of new molecular biology techniques and more understanding of how exogenous chemicals damage human cells and control the genetics of cell proliferation should be explored to measure the carcinogenic potential of exogenous chemicals.

It is necessary to obtain the possible cell damage caused by exposure to a small dose, instead of relying on the traditional large-dose animal carcinogenic test results to extrapolate the effect on the human body. This requires an in-depth understanding of the molecular biological changes and toxic mechanisms of pre-carcinogenicity. The toxicological evaluation about the safety of exogenous chemicals requires such qualitative and quantitative data related to the mechanism, and of course toxicokinetic data.

Therefore, the current development trend is to pay attention to studying the action mechanism of exogenous chemicals that occur under small doses in the earlier molecular/gene changes, and to improve the current toxicity test methods (such as mutagenicity test and carcinogenicity test) and conduct safety toxicological evaluation based on the toxic action mechanism, and gradually establish an internationally accepted "alternative test method" and evaluation standard.

The key points of alternative test methods are comparative toxicology. Comparative toxicology refers to the comparison of the concentration of endogenous toxins (such as naturally occurring and endogenous toxins) in a new food with the concentration of toxins in a traditional food corresponding to the new food. Compared with the traditional method, this has a different basic concept change. The traditional method emphasizes that there is no toxicity in a wide range of safety limits (subjectively set as 100 times), while the alternative method allows endogenous toxins in new foods, as long as their content does not exceed the amount of toxins contained in the traditional food, where the traditional food corresponds to the food that is still being eaten while will be replaced by the new food. The credibility of the alternative method depends on whether there is a reliable, comparable database of toxic substances naturally present in common food.

11.5 Summary

There are many types of harmful constituents in food. Some are inherent in raw food materials and some come from environmental pollution and microbial contamination, while the others are produced during food processing. These harmful substances have a significant impact on food safety.

The safety evaluation methods and procedures of hazardous substances in food are the same as those of food toxicology and have been developed to use molecular biology techniques for their evaluation.

Questions

1. What are the natural harmful substances in legumes? What is their main toxicity?
2. What kind of harmful substances are most likely produced during the storage period of peanuts, corn and their products? What are the types?
3. What are the main harmful substances brought into the food chain by fruits and vegetables, and how to prevent them?
4. What are the harmful substances produced by food processing?
5. How to evaluate the safety of hazardous substances in food?
6. How about the absorption, distribution and excretion of harmful substances in food?
7. Explain the following terms: absolute lethal dose (LD_{100}), median lethal dose (LD_{50}), minimum lethal dose (MLD), maximum tolerable dose (LD_0), maximum non-effect dose, minimum effective dose, non-adverse effect, effect, response, hemagglutinins, protease inhibitor, allergy and allergen.

Bibliography

1. Bane, V., Lehane, M., Dikshit, M., O'Riordan, A., Furey, A.: Tetrodotoxin: Chemistry, toxicity, source, distribution and detection. Toxins **6**(2), 693–755 (2014)
2. Belitz, H.D., Grosch, W., Schieberle, P.: Food Chemistry. Springer, Berlin, Heidelberg (2009)
3. Bernuau, J.R., Francoz, C., Durand, F.: Amatoxin poisoning: Immediate transfer to intensive care or liver unit of patients at early risk of severe acute liver injury. Liver Int. **39**(6), 1016–1018 (2019)
4. Damodaran, S., Parkin, K.L., Fennema, O.R.: Fennema's Food Chemistry. CRC Press/Taylor & Francis, Pieter Walstra (2008)
5. Flora, S.J.S., Pachauri, V.: Chelation in metal intoxication. Int. J. Environ. Res. Public Health **7**(7), 2745–2788 (2010)
6. He, X.-Y., Wu, L.-J., Wang, W.-X., Xie, P.-J., Chen, Y.-H., Wang, F.: Amygdalin—a pharmacological and toxicological review. J. Ethnopharmacol. **254**, 112717 (2020)
7. Idowu, O., Semple, K.T., Ramadass, K., O'Connor, W., Hansbro, P., Thavamani, P.: Beyond the obvious: Environmental health implications of polar polycyclic aromatic hydrocarbons. Environ. Int. **123**, 543–557 (2019)
8. Ismail, A., Goncalves, B.L., de Neeff, D.V., Ponzilacqua, B., Coppa, C.F.S.C., Hintzsche, H., Sajid, M., Cruz, A.G., Corassin, C.H., Oliveira, C.A.F.: Aflatoxin in foodstuffs: Occurrence and recent advances in decontamination. Food Res. Int. **113**, 74–85 (2018)
9. Junker, Y., Zeissig, S., Kim, S.-J., Barisani, D., Wieser, H., Leffler, D.A., Zevallos, V., Libermann, T.A., Dillon, S., Freitag, T.L., Kelly, C.P., Schuppan, D.: Wheat amylase trypsin inhibitors drive intestinal inflammation via activation of toll-like receptor 4. J. Exp. Med. **209**(13), 2395–2408 (2012)
10. Kan, J.: Food Chemistry. China Agricultural University Press, Beijing (2016)

11. Katsikantami, I., Colosio, C., Alegakis, A., Tzatzarakis, M.N., Vakonaki, E., Rizos, A.K., Sarigiannis, D.A., Tsatsakis, A.M.: Estimation of daily intake and risk assessment of organophosphorus pesticides based on biomonitoring data—the internal exposure approach. Food Chem. Toxicol. **123**, 57–71 (2019)

12. Khaneghah, A.M., Fakhri, Y., Sant'Ana, A.S.: Impact of unit operations during processing of cereal-based products on the levels of deoxynivalenol, total aflatoxin, ochratoxin A, and zearalenone: a systematic review and meta-analysis. Food Chem. **268**, 611–624 (2018)

13. Khosa, S., Mishra, S., Trikamji, B., Singh, S., Dwivedi, M., Moheb, N.: A clinical overview of lathyrism. J. Neurol. Sci. **381**, 564–565 (2017)

14. Pearson, L., Mihali, T., Moffitt, M., Kellmann, R., Neilan, B.: On the chemistry, toxicology and genetics of the cyanobacterial toxins, microcystin, nodularin, saxitoxin and cylindrospermopsin. Marine Drugs **8**(5), 1650–1680 (2010)

15. Perez de Albuquerque, N.C., Carrao, D.B., Habenschus, M.D., Moraes de Oliveira, A.R.: Metabolism studies of chiral pesticides: a critical review. J. Pharm. Biomed. Anal. **147**, 89–109 (2018)

16. Rushing, B.R., Selim, M.I.: Aflatoxin B$_1$: A review on metabolism, toxicity, occurrence in food, occupational exposure, and detoxification methods. Food Chem. Toxicol. **124**, 81–100 (2019)

17. Svircev, Z., Drobac, D., Tokodi, N., Mijovic, B., Codd, G.A., Meriluoto, J.: Toxicology of microcystins with reference to cases of human intoxications and epidemiological investigations of exposures to cyanobacteria and cyanotoxins. Arch. Toxicol. **91**(2), 621–650 (2017)

18. Wiese, M., D'Agostino, P.M., Mihali, T.K., Moffitt, M.C., Neilan, B.A.: Neurotoxic alkaloids: Saxitoxin and its analogs. Marine Drugs **8**(7), 2185–2211 (2010)

19. Zevallos, V.F., Raker, V., Tenzer, S., Jimenez-Calvente, C., Ashfaq-Khan, M., Ruessel, N., Pickert, G., Schild, H., Steinbrink, K., Schuppan, D.: Nutritional wheat amylase-trypsin inhibitors promote intestinal inflammation via activation of myeloid cells. Gastroenterology **152**(5), 1100–1113 (2017)

Dr. Kewei Chen associate professor and master supervisor in College of Food Science, Southwest University, China. He obtained his doctorate from Universidad de Sevilla, Spain in 2016, and from 2012 to 2016, he also did his research work in Instituto de la Grasa (CSIC, Spain). He has participated in several international book chapters about food industry and technology. His major research interests and areas are related with food chemistry & nutrition, including micronutrient bioavailability, function evaluation and food safety, and he has presided over four provisional and administerial projects related with food nutritional changes and food safety evaluation. *Dr.* Kewei Chen has published more than 10 research papers on Food Chemistry, Journal of Functional Foods, Molecular Nutrition & Food Research, etc.

Dr. Jianquan Kan professor and doctoral supervisor in College of Food Science, Southwest University, China. He is currently the director of the Quality and Safety Risk Assessment Laboratory for Agricultural Products Storage and Preservation of the Ministry of Agriculture and Rural Affairs (Chongqing), and the academic leader in the discipline of "Food Science and Engineering" belonging to Chongqing's key disciplines. *Dr.* Jianquan Kan also leads the "Excellent Teaching Team" in the field of Food Science and Safety, and He was awarded "Person of the Year (2015)" as Scientific Chinese. He is the special expert candidate for high-level talents in Sichuan Province of the "Thousand Talents Program" and currently is the chief scientist in the Innovation Team of Chongqing Modern Special High-efficient Agricultural Flavoring Industrial Technology System. Dr. Jianquan Kan has been the Editor-in-Chief in 6 textbooks related with food chemistry. For instance, Practical Chemistry of Oil and Fat (1997, Southwest Normal University Press: Chongqing), Food Chemistry (2002 1st edition, 2008 2nd edition, 2016 3rd edition, China Agricultural University Press: Beijing; 2006 1st edition, 2017, 3rd edition, New Wenjing Development and Publishing Co., Ltd.: Taiwan; 2009, 1st edition, China Metrology Press: Beijing; 2012, NOVA Science Publishers, Inc.: New York), Advanced Food Chemistry (2009, Chemical Industry Press: Beijing), etc.. He has published more than 150 SCI and EI papers, and obtained or applied in total 32 national invention patents, and presided over 56 provincial or ministerial research projects, and besides 71 research projects commissioned by enterprises. He won one first prize for technological invention from the Ministry of Education of the People's Republic of China, one first prize for technological invention from Chinese Institute of Food Science and Technology, two second prizes and four third prizes for Chongqing Science and Technology Progress.

Index